MW00846269

Ergebnisse der Mathematik und ihrer Grenzgebiete

3. Folge · Band 9

A Series of Modern Surveys in Mathematics

Mikhael Gromov

Partial Differential Relations

Springer-Verlag
Berlin Heidelberg New York
London Paris Tokyo

Mikhael Gromov

Institute des Hautes Etudes Scientifiques
35, route de Chartres
F-91440 Bures-sur-Yvette
France

Mathematics Subject Classification (1980): 53, 58

ISBN 3-540-12177-3 Springer-Verlag Berlin Heidelberg New York
ISBN 0-387-12177-3 Springer-Verlag New York Heidelberg Berlin

Library of Congress Cataloging in Publication Data
Gromov, Mikhael, 1943–
Partial differential relations.
(Ergebnisse der Mathematik und ihrer Grenzgebiete; 3. Folge, Bd. 9)
Includes index.
1. Geometry, Differential. 2. Differential equations, Partial. 3. Immersions (Mathematics)
I. Title. II. Series: Ergebnisse der Mathematik und ihrer Grenzgebiete; 3. Folge, Bd. 9.
QA641.G76 1986 515.3'53 86-13906
ISBN 0-387-12177-3 (U.S.)

Printed in Germany

Typesetting: ASCO, Hong Kong
Offsetprinting: Mercedes-Druck, Berlin · Bookbinding: Lüderitz & Bauer, Berlin
2141/3020-543210

Foreword

The classical theory of partial differential equations is rooted in physics, where equations (are assumed to) describe the laws of nature. Law abiding functions, which satisfy such an equation, are very rare in the space of all admissible functions (regardless of a particular topology in a function space).

Moreover, some additional (like initial or boundary) conditions often insure the uniqueness of solutions. The existence of these is usually established with some *apriori estimates* which locate a possible solution in a given function space.

We deal in this book with a completely different class of partial differential equations (and more general relations) which arise in differential geometry rather than in physics. Our equations are, for the most part, undetermined (or, at least, behave like those) and their solutions are rather dense in spaces of functions.

We solve and classify solutions of these equations by means of direct (and not so direct) geometric constructions.

Our exposition is elementary and the proofs of the basic results are selfcontained. However, there is a number of examples and exercises (of variable difficulty), where the treatment of a particular equation requires a certain knowledge of pertinent facts in the surrounding field.

The techniques we employ, though quite general, do not cover all geometrically interesting equations. The border of the unexplored territory is marked by a number of open questions throughout the book.

I am grateful to my friends and colleagues with whom I have discussed various aspects of the subject in the course of years. The book took final shape under unrelenting criticism by Nico Kuiper directed at earlier drafts. I thank Mme V. Houllet for typing the manuscript, Mari Anne Gazdick for rectifying my English and Mme J. Martin for the help with a multitude of last minute corrections.

Bures-sur-Yvette, May 1986 M. Gromov

Contents

Part 1. A Survey of Basic Problems and Results

1.1 Solvability and the Homotopy Principle

1.1.1 Jets, Relations, Holonomy

Consider a C^∞-smooth fibration $p: X \to V$ and let $X^{(r)}$ be the space of *r-jets* (of germs) of smooth sections $f: V \to X$. Denote by $p^r: X^{(r)} \to V$ and $p_r^s: X^{(s)} \to X^{(r)}$ for $s > r \geq 0$, where $X^{(0)} \overset{\text{def}}{=} X$, the natural projections. The rth-*order jet* (or *r*-jet) of a C^r-smooth section $f: V \to X$ is denoted by $J_f^r: V \to X^{(r)}$. A section $\varphi: V \to X^{(r)}$ is called *holonomic* if it is the *r*-jet of some C^r-section $f: V \to X$. This f, if it exists, is unique. Namely $f = p_0^r \circ \varphi$, since $p_0^r \circ J_f^r = f$ for all sections $f: V \to X$. Thus, sections $f: V \to X$ are identified with holonomic sections $V \to X^{(r)}$ by $f \mapsto J_f^r$.

Recall basic properties of jets which, in fact, uniquely define the fibration $X^{(r)} \to V$ and the jet operator $f \mapsto J_f^r$.

(a) The space $X^{(1)}$ consists of the linear maps $L: T_v(V) \to T_x(X)$ for all $x \in X$ and $v = p(x) \in V$, such that $D_p \circ L = \text{Id}: T_v(V) \circlearrowleft$, where $T_v(V)$ and $T_x(X)$ are the respective tangent spaces. Here $D_p: T(X) \to T(V)$ denotes the differential of the map p and Id stands for the identity map. Thus, the fibration $p_0^1: X^{(1)} \to X$ carries a natural structure of an *affine bundle* with fibers $X_x^{(1)} \approx \mathbb{R}^{nq}$, $x \in X$, where $n = \dim V$ and $q = \dim X - \dim V = \dim X_v$ for the fiber $X_v = p^{-1}(v) \subset X$, $v \in V$.

The 1-jet of a smooth section $f: V \to X$ sends each point $v \in V$ into the fiber $X_x^{(1)} \subset X^{(1)}$ for $x = f(v) \in X$, by $J_f^1(v) = D_f: T_v(V) \to T_x(X)$.

(b) The space $X^{(r+1)}$, for all $r = 0, 1, 2, \ldots$, is naturally embedded into $(X^{(r)})^{(1)}$ as it consists of the 1-jets of the smooth *holonomic* sections $V \to X^{(r)}$. The operator J^{r+1} on sections $V \to X$ is the composition of J^r with the 1-jet operator on sections $V \to X^{(r)}$.

(b′) The fibration $p_r^{r+1}: X^{(r+1)} \to X^{(r)}$ carries a natural structure of an affine bundle over $X^{(r)}$ which makes it a subbundle in $(X^{(r)})^{(1)} \to X^{(r)}$, whose fiber $X_x^{(r+1)} \subset X^{(r+1)}$, $x \in X^{(r)}$, has $\dim X_x^{(r+1)} = q\left(\dfrac{(n+r)!}{(n-1)!(r+1)!}\right)$, for $n = \dim V$ and $q = \dim X_v$.

(c) Take a trivial (split) subfibration $Y = U \times W_0 \subset X$ over an open subset $U \subset V$, for an open subset W_0 in the fiber $X_{v_0} \subset X$ over some $v_0 \in U$, and identify $Y^{(r)} \to Y$ with the *restriction* $X^{(r)} | Y = (p_0^r)^{-1}(Y) \to Y$. If $u_1, \ldots, u_n$ and $y_1, \ldots, y_q$ are local coordinates in U and in W_0 respectively, then sections $U \to Y^{(r)} \subset X^{(r)}$ are given by d_r-tuples of functions $U \to \mathbb{R}$ for $d_r = \dim X_{v_0}^{(r)} = q\left(1 + n + \frac{n(n+1)}{2} + \cdots + \frac{(n+r-1)!}{(n-1)!r!}\right)$, such that the *holonomic* sections $J_f^r\colon U \to Y^{(r)}$ are represented by strings of partial derivatives,

$$J_f^r = \left(f_\mu, \frac{\partial f_\mu}{\partial u_i}, \frac{\partial^2 f_\mu}{\partial u_{i_1} \partial u_{i_2}}, \ldots, \frac{\partial^r f_\mu}{\partial u_1, \ldots, \partial u_{i_r}}\right),$$

for $\mu = 1, 2, \ldots, q$ and $i = 1, 2, \ldots, n$, where the functions $f_\mu = f_\mu(u_1, \ldots, u_n)$ represent sections $f\colon U \to Y \subset X$.

Furthermore, the (trivial) fibration $Y^{(r)} \to U$ admits a (non-unique) *holonomic splitting*, $Y^{(r)} = U \times \mathbb{R}^{d_r}$, such that the submanifold $U \times z \subset Y^{(r)}$ is the image of a holonomic section $J_f^r\colon U \to Y^{(r)}$ for some C^∞-section $f = f_z\colon U \to Y \subset X$, and for all $z \in \mathbb{R}^{d_r}$. This may be achieved, for example, with the space $P_r = P_r(\mathbb{R}^n \to \mathbb{R}^q) \approx \mathbb{R}^{d_r}$ of maps $f\colon U \to W_0$ whose components f_μ, $\mu = 1, \ldots, q$, are polynomials in $u_1, \ldots, u_n$ of degree $\leq r$, as the (tautological) map $U \times P_r \to Y^{(r)}$ for $(u, f) \mapsto (u, J_f^r(u))$ is a diffeomorphism. In fact, each C^∞-section $f_0\colon U \to X$ admits a split neighborhood $Y = U \times W_0 \supset f_0(U)$ in X, such that $f_0(U) = U \times w_0$, for $w_0 = (0, 0, \ldots, 0) \in W_0$. Thus every holonomic section $J_{f_0}^r\colon U \to X^{(r)}$ admits a *holonomically* split neighborhood (for example $Y^{(r)} = U \times P_r$) in $X^{(r)}$.

Definition. A differential *relation* (or *condition*) imposed on sections $f\colon V \to X$ is a subset $\mathscr{R} \subset X^{(r)}$, where r is called the *order* of $\mathscr{R}$. A C^r-section f is said to *satisfy* (or to be a *solution* of) $\mathscr{R}$ if the jet $J_f^r\colon V \to X^{(r)}$ maps V into $\mathscr{R}$. Thus solutions of $\mathscr{R}$ are (naturally identified with) *holonomic* sections $V \to \mathscr{R} \subset X^{(r)}$.

Example. Let $\Psi\colon X^{(r)} \to \mathbb{R}$ be a continuous function. Then the relation imposed by the (zero) set $\mathscr{R} = \{x \in X^{(r)} \mid \Psi(x) = 0\}$ is expressed by the (partial differential) equation $\Psi(J_f^r) = 0$, that is

$$\Psi\left(u_1, \ldots, u_n, f_1, \ldots, f_q, \ldots, \frac{\partial^r f_\mu}{\partial u_{i_1}, \ldots, \partial u_{i_r}}\right) = 0$$

in local coordinates.

Solving a relation $\mathscr{R} \subset X^{(r)}$ can be done in two stages. We start with the *topological* problem of constructing a section $V \to \mathscr{R}$. The topology can be hard, but even so this first stage looks analytically easy. For example, when $\mathscr{R}$ is represented by the equation $\Psi\left(\frac{\partial f}{\partial u_1}, \ldots, \frac{\partial f}{\partial u_n}\right) = 0$, then sections $V \to \mathscr{R}$ correspond to solutions of $\Psi(\varphi_1, \ldots, \varphi_n) = 0$ with *arbitrary* functions φ_i in place of the partial derivatives $\partial f/\partial u_i$.

Our real problem appears at the second stage, when we try to pass from an arbitrary section $V \to \mathscr{R}$ to a holonomic one. The most optimistic expectation is expressed in the following

Homotopy Principle. We say that $\mathscr{R}$ *satisfies the h-principle* and (or) that the *h-principle holds* for (obtaining) solutions of $\mathscr{R}$ if every continuous section $V \to \mathscr{R}$ is homotopic to a holonomic section $V \to \mathscr{R}$ by a continuous homotopy of sections $V \to \mathscr{R}$.

If X is a *trivial* fibration, $X = V \times W \to V$, then sections $V \to X$ correspond to maps $V \to W$ and the *h-principle for maps $V \to W$ which satisfy $\mathscr{R}$*, by definition, refers to the h-principle for sections $V \to X$.

Remarks on C^s-Solutions for $s = r + k$. Let us *lift* (*or prolong*) a given relation $\mathscr{R} \subset X^{(r)}$ to a relation $\mathscr{R}^1 \subset X^{(r+1)}$ by mimicking the differentiation of an equation (like $\Psi(u_i, f(u_i), \partial f/\partial u_i, \ldots) = 0$ in the variables u_i) as follows:

Let $\mathscr{R}' \subset (X^{(r)})^{(1)}$ consist of the 1-jets of germs of C^1-sections $V \to \mathscr{R}$ and put $\mathscr{R}^1 = \mathscr{R}' \cap X^{(r+1)}$ for the canonical embedding $X^{(r+1)} \subset (X^{(r)})^{(1)}$. Then repeat this and define $\mathscr{R}^k = (\mathscr{R}^{k-1})^1 \subset X^{(r+k)}$ for all $k = 1, 2, \ldots$. Now, *the h-principle for C^s-solutions* of $\mathscr{R}$ for $s = r + k$, by definition, refers to the h-principle for (solutions of) $\mathscr{R}^k$.

Call $\mathscr{R}$ *stable* if $\mathscr{R}^{k+1} \subset X^{(r+k+1)}$ is an affine subbundle in the bundle $X^{(r+k+1)} \to X^{(r+k)}$ restricted to $\mathscr{R}^k \subset X^{(r+k)}$ for all $k = 1, 2, \ldots$. Observe that the intersection $\mathscr{R}_x^{k+1} = \mathscr{R}^{k+1} \cap (p_{r+k}^{r+k+1})^{-1}(x)$, $x \in \mathscr{R}^k$ typically is an *affine* subspace in the fiber $X_x^{(r+k+1)} = (p_{r+k}^{r+k+1})^{-1}(x)$, (since the top derivatives appear linearly as one differentiates a differential equation) but the dimension $\dim \mathscr{R}_x^{k+1}$ need not be constant in x. The stability of $\mathscr{R}$ amounts to the constancy of $\dim \mathscr{R}_x^{k+1}$ and to the continuity of the subspace $\mathscr{R}_x^{k+1} \subset X_x^{(r+k+1)}$ in $x \in \mathscr{R}^k$.

If $\mathscr{R}$ is stable, then every section $\varphi\colon V \to \mathscr{R}$ lifts to a section $\varphi\colon V \to \mathscr{R}^k$, for all $k = 1, \ldots,$ since the projection $\mathscr{R}^k \to \mathscr{R}$ is a fibration with *contractible* fibers, and the lift is unique up to a homotopy of sections $V \to \mathscr{R}^k$.

The h-Principle for $C^\infty(C^{\mathrm{an}})$-Solutions of $\mathscr{R}$. This h-principle claims, by definition, the stability of $\mathscr{R}^k$ for some integer $k \geq 0$, as well as the possibility to homotope an arbitrary continuous section $V \to \mathscr{R}^k$ to a holonomic $C^\infty(C^{\mathrm{an}})$-section, where C^{an} stands for the real analyticity (which makes sense if the fibration $X \to V$ is C^{an}).

Examples. (a) If $\mathscr{R} \subset X^{(r)}$ is an *open* subset (relation), then it is obviously stable since $\mathscr{R}^k = (p_r^{r+k})^{-1}(\mathscr{R})$ for all $k = 1, 2, \ldots$. In fact, the content of the h-principle for an *open* relation $\mathscr{R}$ is independent of the smoothness class $C^{s \geq r}$ of solutions. Indeed, every C^r-solution $f_0\colon V \to X$ can be finely (see 1.2.2) C^r-approximated by C^∞-sections $f\colon V \to X$. These necessarily satisfy $\mathscr{R}$ when $\mathscr{R} \subset X^{(r)}$ is open.

(b) If $\mathscr{R}$ is stable and if every C^r-solution of $\mathscr{R}$ is (known to be) C^∞-smooth, then, of course, there is no need to specify the smoothness of solutions for stating the h-principle.

In general, however, the stability of $\mathscr{R}$ insures no immediate passage from C^r to C^{r+1}.

1.1.2 The Cauchy-Riemann Relation, Oka's Principle and the Theorem of Grauert

Let the manifolds V and X be endowed with complex analytic structures and let $p: X \to V$ be a complex analytic fibration. The *Cauchy-Riemann* relation is the subset $\mathscr{R} \subset X^{(1)}$ which consists of the *complex* linear maps $T_v(V) \to T_x(X)$ in $X^{(1)}$. The solutions of $\mathscr{R}$ are (well known to be) exactly the *holomorphic* (or *complex analytic*) sections $V \to X$. The projection $\mathscr{R} \to X$ is clearly an affine bundle (whose fibers $\mathscr{R}_x$, $x \in X$, have $\dim_{\mathbb{R}} \mathscr{R}_x = \frac{1}{2} \dim_{\mathbb{R}} X_x^{(1)} = 2nq$ for $2n = \dim_{\mathbb{R}} V$ and $2q = \dim_{\mathbb{R}} X_v, v \in V$). Hence every continuous section $V \to X$ lifts to a section $V \to \mathscr{R}$ which is unique up to a homotopy. Therefore the h-principle for $\mathscr{R}$ reduces to the following.

Oka's Principle. Every continuous section $V \to X$ is homotopic to a holomorphic one.

Remarks. (a) One can easily see the Cauchy-Riemann relation to be stable. In fact, the projection $p_{k+1}^{k+2}: \mathscr{R}^{k+1} \to \mathscr{R}^k$ is an affine bundle (with c_k-dimensional fibers for $c_k = 2q\left(\frac{(n+k)!}{(n-1)!(k+1)!}\right)$. Thus, passing from $\mathscr{R}$ to $\mathscr{R}^k$ does not change the content of the h-principle.

(b) Oka's principle is older than the h-principle and it is rarely stated in a precise form. Furthermore, one does not speak of Oka's principle unless the underlying manifold V is *Stein.* That is (according to one of several equivalent definitions) V is biholomorphic to a complex analytic submanifold in $\mathbb{C}^N$ for some $N = N(V)$. [In fact, the greatest integer $N \leq \frac{3}{2}(\dim_{\mathbb{C}} V + 1)$ will do, see 2.1.5.]

Examples. (a) Every non-singular complex algebraic subvariety $V \subset \mathbb{C}^N$ clearly is Stein.

(b) A Riemann surface V (i.e. $\dim_{\mathbb{C}} V = 1$) is Stein if and only if it is *open* (i.e. it contains no compact connected component). For instance, open subsets $V \subset \mathbb{C}^1$ are Stein (see Gunning and Rossi 1965).

(c) The tangent space $T(V_0)$ of an arbitrary smooth manifold V_0 admits a complex analytic structure which is Stein (Cartan 1957).

(d) Submanifolds in Stein manifolds and Cartesian products of Stein manifolds, are obviously Stein.

(e) No *compact* manifolds of positive dimension is Stein by Liouville's theorem. The complement of a non-empty compact subset in $\mathbb{C}^n$ is non-Stein for $n \geq 2$ by the following

(A) **Lefschetz Theorem.** *Every Stein manifold V admits a homotopy retraction onto a cell subcomplex $K \subset V$ which has* $\dim K = \frac{1}{2} \dim_{\mathbb{R}} V = \dim_{\mathbb{C}} V$.

Proof. Holomorphically embed $V \subset \mathbb{C}^N$ and observe that the Euclidean distance $v \mapsto \operatorname{dist}(z_0, v)$ for a fixed *generic* (see 1.3.2) point $z_0 \in \mathbb{C}^N$, is a *Morse function* on $V \subset \mathbb{C}^N$ with non-degenerate critical points. [This is, in fact, true for all smooth properly embedded submanifolds $V \subset \mathbb{R}^{2N}$, see Milnor (1963).] Since the restriction of $\operatorname{dist}(x_0, v)$ to every complex submanifold $V' \subset V$ of $\dim_{\mathbb{C}} V' = 1$ assumes no local maximum on V'_0 by Liouville's theorem, the critical points of $\operatorname{dist}(x_0, v)$ on V have Morse indices $\leq \dim_{\mathbb{C}} V$. Hence, the Morse subcomplex $K \subset V$, which is a homotopy retract of V, satisfies $\dim K \leq \dim_{\mathbb{C}} V$. Q.E.D. [See Narasimhan (1967) for a similar theorem for *singular* varieties.]

(A′) **Exercises.** (a) Generalize (A) to complex submanifolds $V \subset Z$ where Z is a complete simply connected Kähler manifold of *non-positive* sectional curvature. [In fact, V and Z are known to be Stein under these assumptions, see Green-Wu (1979).]

(b) Take complex submanifolds V and V' in $\mathbb{C}^N$, such that the subset $\{(v, v') \in V \times V' | \operatorname{dist}(v, v') \leq \text{const}\}$ is compact for all const ≥ 0. Assume $\dim_{\mathbb{C}} V + \dim_{\mathbb{C}} V' \geq N$ and show the intersection $V \cap V'$ to be non-empty.

(b′) Generalize (b) to submanifolds in a complete Kähler manifold Z of *non-negative* sectional curvature for $N = \dim_{\mathbb{C}} Z$.

(c) Take an arbitrary (Stein or not) complex manifold V and let $f\colon V \to \mathbb{C}^N$ be a holomorphic map. Show for a fixed *generic* point $z_0 \in \mathbb{C}^n$ that the critical set of the function $v \mapsto [\operatorname{dist}(z_0, f(v))]^2$ on V lies in a *disjoint* union of "fibers" $f^{-1}(z_i) \subset V$ for some points $z_1, \ldots, z_i, \ldots$ in $\mathbb{C}^N$. Let $m = \sup_{z \in \mathbb{C}^N} \dim_{\mathbb{C}} f^{-1}(z)$ and approximate the function $[\operatorname{dist}(z_0, f(v))]^2$ by a function d' on V whose critical points are non-degenerate with the indices $\leq 2m + \dim_{\mathbb{C}} V$. Assume the map f is *proper* and contract V onto a subcomplex $K \subset V$, such that $\dim K \leq 2m + \dim_{\mathbb{C}} V$.

(d) Holomorphically map two complex manifolds into $\mathbb{C}^N$ by $f\colon V \to \mathbb{C}^N$ and $f'\colon V \to \mathbb{C}^N$. Then, consider the function $E(v, v') = [\operatorname{dist}(f(v), f'(v'))]^2$ on $V \times V'$ and let $u_0 = (v_0, v'_0)$ be a non-degenerate critical point of E. Show that $\operatorname{index}(u_0) \leq \dim_{\mathbb{C}} V \times V'$. Assume $E(u_0) > 0$ and prove that

$$(*) \qquad \operatorname{index}(u_0) \geq \dim_{\mathbb{C}} V \times V' - N + 1.$$

(e) Let Z be a complete Kähler manifold of *non-negative* sectional curvature and let $f\colon V \to Z$ be a holomorphic map. Denote by $\mathscr{U}$ the space of triples $u = (v, v', \alpha)$ where $v, v' \in V$ and α is a *smooth path* in Z between $f(v)$ and $f(v')$. That is $\alpha\colon [0,1] \to Z$ is a smooth map for which $\alpha(0) = f(v)$ and $\alpha(1) = f(v')$. Define the (energy) functional $E\colon \mathscr{U} \to \mathbb{R}_+$ by

$$E(u) = E(\alpha) = \int_0^1 \left\| \frac{d\alpha}{dt} \right\|^2 dt,$$

for the Riemann-Kähler norm $\| \; \|$ in $T(z)$, and let $u_0 \in \mathscr{U}$ be a non-degenerate critical point of E, such that $E(u_0) > 0$. Show that

$$\operatorname{index}(u_0) \geq 2 \dim_{\mathbb{C}} V - \dim_{\mathbb{C}} Z + 1,$$

by combining $(*)$ with the Morse index comparison theorem (see Milnor 1963).

(e′) Let the sectional curvature of Z be *strictly positive*. Assume V to be compact and the map f to be finite-to-one. Show, by applying Morse theory to a small perturbation E' of E, that the homotopy groups of the pair $(\mathscr{U}, \mathscr{U}_0)$, for $\mathscr{U}_0 = E^{-1}(0)$, satisfy $\pi_i(\mathscr{U}, \mathscr{U}_0) = 0$ for $i = 0, 1, \ldots, k = 2 \dim_{\mathbb{C}} V - \dim_{\mathbb{C}} Z$. Identify $\mathscr{U}_0$ with the subset $\{(v, v') \in V \times V | f(v) = f(v')\}$, and prove the following *vanishing theorem of*

Lefschetz-Barth-Larsen (see Barth 1975) for holomorphic *embeddings* $V \subsetneq Z = \mathbb{C}P^N$,

$$\pi_i(Z, V) = 0, \quad \text{for } 0 \leq i \leq 2 \dim_{\mathbb{C}} V - N + 1.$$

(B) *The h-Principle of Grauert.* Let G be a complex Lie group, take a complex analytic subgroup $H \subset G$ and consider a complex analytic fibration $X \to V$ with the structure group G and with the (homogeneous) fiber G/H.

(B′) **Theorem** (Grauert 1957). *If V is Stein, then every continuous section $V \to X$ can be homotoped to a holomorphic section. In particular, every continuous map $V \to G/H$ is homotopic to a holomorphic one.*

[See Cartan (1958) and Ramspot (1962) for the proof.]

Exercises. (a) Let M be the multiplicative group of *non-vanishing* holomorphic functions on V and let $E \subset M$ consist of the functions e^f for all holomorphic function $f\colon V \to \mathbb{C}$. Show the factor group M/E to be isomorphic to the cohomology group $H^1(V; \mathbb{Z})$ by applying (B′) to maps $V \to \mathbb{C} \setminus \{0\}$.

(b) Apply (B′) to holomorphic maps $V \to \mathbb{C}P^N$ and construct a non-singular complex subvariety $V_0 \subset V$ of $\operatorname{codim}_{\mathbb{C}} V_0 = 1$ whose fundamental class is the Poincaré dual of a given cohomology class $h_0 \in H^2(V; \mathbb{Z})$.

(c) Construct with (A) and (B′) k linearly independent holomorphic tangent vector fields on V, provided $2k \leq n + 1$ for $n = \dim_{\mathbb{C}} V$.

(d) Let W be a *Riemann surface* and let V be an arbitrary (Stein or not) complex manifold. Assume the fundamental group $\pi_1(W)$ to be *non-Abelian* and show holomorphic maps $V \to W$ to *violate* the h-principle, unless $H^1(V; \mathbb{Z}) = 0$.

(e) Let V be a *compact* (hence, *non-Stein*) Riemann surface. Show holomorphic maps $V \to \mathbb{C}P^q$, $q \geq 1$, to satisfy the h-principle if and only if $\pi_1(V) = 0$ (i.e. $V \approx \mathbb{C}P^1$).

(f) Consider a holomorphic map f of a *compact* complex manifold V to the complex torus $\mathbb{C}^q/\mathbb{Z}^{2q}$. Prove the induced homomorphism $H^1(\mathbb{C}^q/\mathbb{Z}^{2q}; \mathbb{R}) \to H^i(V, \mathbb{R})$ has *even* rank. [This contradicts the h-principle if $q \geq 1$ and $H^1(V; \mathbb{R}) \neq 0$.]

We shall return to holomorphic maps in 2.1.5.

1.1.3 Differentiable Immersions and the *h*-Principle of Smale-Hirsch

A C^1-map $f\colon V \to W$ is called an *immersion* if $\operatorname{rank} f \overset{\text{def}}{=} \operatorname{rank} D_f = \dim V$ everywhere on V. For example, if $\dim W = \dim V$, then immersions $V \to W$ are exactly *locally diffeomorphic* maps.

The pertinent jet space $X^{(1)}$, for $X = V \times W \to V$, consists of the linear maps $T_v(V) \to T_w(W)$ for all $(v, w) \in X$. The *immersion relation* $\mathscr{I} \subset X^{(1)}$ is fibered over X by the projection $X^{(1)} \to X$ and the fiber $\mathscr{I}_x$, $x = (v, w) \in X = V \times W$ consists of the *injective* linear maps in $X_x^{(1)} = \operatorname{Hom}(T_v(V) \to T_w(W))$. Now, sections $V \to \mathscr{I}$ correspond to *fiberwise injective* homomorphisms $\varphi: T(V) \to T(W)$, while *holonomic* sections are *differentials* $D_f: T(V) \to T(W)$ *of immersions* $f: V \to W$.

The subset $\mathscr{I} \subset X^{(1)}$ is empty for $\dim W < \dim V$ and thus trivially satisfies the h-principle. If $\dim W \geq \dim V$, then $\mathscr{I}$ is an open dense subset in $X^{(1)}$. This does not, however, insure the existence of immersions or (and) the validity of the h-principle. In fact, if V is a *closed* (i.e. compact without boundary) manifold and $\dim W = \dim V$ then the h-principle may fail to be true. Moreover, if W is an *open* manifold (i.e. no connected component of W is a closed manifold), then no (equi-dimensional!) immersion $V \to W$ exists at all (see 2.1.3) and the relation $\mathscr{I}$ violates the h-principle in so far as it admits a section $V \to \mathscr{I}$.

The h-principle (in the parametric form, see 1.2.1) for immersions $S^n \to \mathbb{R}^q$, $q \geq n + 1$, was discovered by Smale (1958, 1959) and the theory was completed by the following

(A) **Theorem of Hirsch.** *Immersions $V \to W$ satisfy the h-principle in the following two cases*:

(i) *Extra dimension*: $\dim W > \dim V$ (Hirsch 1959).
(ii) *Critical dimension*: $\dim W = \dim V$ *and the manifold V is open* (Hirsch 1961).

See 2.1.1, 2.2.2, and 2.4.3 for three different proofs.

(B) **Examples and Corollaries.** A manifold V is called *parallelizable* if its tangent bundle is trivial, $T(V) = V \times \mathbb{R}^n \to V$ for $n = \dim V$.

(B_1) *If the manifolds V and W are parallelizable, then in cases* (i) *and* (ii), *every continuous map $V \to W$ is homotopic to an immersion. In particular, every open parallelizable manifold V admits an immersion $V \to \mathbb{R}^n$ for $n = \dim V$.*

Proof. Since the bundles $T(V)$ and $T(W)$ are trivial, the fibration $\mathscr{I} \to X = V \times W$ is also trivial. Hence, every section $V \to X$ lifts to $\mathscr{I}$ and the h-principle applies.

(B_1') Let G be a Lie group and let $\Gamma \subset G$ be a discrete subgroup. The manifold $V = G/\Gamma$ *obviously* is parallelizable (and it can be made open, if necessary, by deleting a point from each component of V or by multiplying V by $\mathbb{R}^1$). Yet, it seems unlikely that anybody can immerse these manifolds into $\mathbb{R}^{n+1}$ (or into $\mathbb{R}^n$ if they are open) bypassing Hirsch's theorem. (See p. 43 in Kirby-Siebenmann for a geometric immersion of the punctured n-torus into $\mathbb{R}^n$.)

(B_1'') Let V be an open manifold which admits a possibly non-complete Riemannian metric of *constant negative curvature*. Then there exists a finite *parallelizable* covering $\tilde{V} \to V$ (see Deligne-Sullivan 1975) which admits by (B_1) an immersion $\tilde{V} \to \mathbb{R}^n$, $n = \dim V = \dim \tilde{V}$.

(B_2) A manifold V is called *stably parallelizable* if $V \times \mathbb{R}$ is a parallelizable manifold. For example, every *orientable hypersurface* $V \subset \mathbb{R}^{n+1}$ is stably parallelizable since a small (tubular) neighborhood $U \subset \mathbb{R}^{n+1}$ of V is diffeomorphic to $V \times \mathbb{R}$ and since open subsets U in parallelizable manifolds (like $\mathbb{R}^{n+1}$) obviously, are parallelizable. Moreover, a manifold V (obviously) is stably parallelizable if and only if there is a continuous map $g\colon V \to S^n$, such that the *induced* bundle $g^*(T(S^n))$ over V is isomorphic to $T(V)$. (One may use the tangential Gauss map for hypersurfaces in $\mathbb{R}^{n+1}$.)

(B_2') *Every n-dimensional stably parallelizable manifold V admits an immersion* $V \to \mathbb{R}^{n+1}$.

Proof. Construct a fiberwise homomorphism $T(V) \to T(S^n) \subset T(\mathbb{R}^{n+1})$ and apply (i). Alternatively, apply (ii) to the (open!) manifold $V \times \mathbb{R}$.

(B_2'') **Exercises.** (a) Assume V and W to be stably parallelizable and find with (i) an immersion in a given homotopy class of maps $V \to W$ for $\dim W > \dim V$.

(b) Derive (i) of Hirsch's theorem from (ii) which applies to the total space of the quotient bundle $f^*(T(W))/T(V)$ for the map $f\colon V \to W$ underlying a given fiberwise injective homomorphism $T(V) \to T(W)$.

(B_3) (Whitney 1944). *Every n-dimensional manifold V admits an immersion* $V \to \mathbb{R}^{2n-1}$.

Proof. The obstruction to the existence of a section $V \to \mathscr{I}$ can be easily identified in this case with the *normal Stiefel-Whitney class* w_n of V. This is zero by the Thom-Wu formulae for w_i.

Whitney's result is sharp for $n = 2^k$, $k = 1, 2, \ldots$. For example, the projective space $\mathbb{R}P^n$, $n = 2^k$, admits no immersion into $\mathbb{R}^{2n-2}$, since the normal class $w_{n-1}(\mathbb{R}P^n) \neq 0$ for $n = 2^k$. [See the surveys by Lannes (1982) and Cohen (1984) for the contemporary state of art.]

1.1.4 Osculating Spaces and Free Maps

Consider a C^2-map $f\colon V \to \mathbb{R}^q$ and fix local coordinates $u_1, \ldots, u_n$ in V near a given point $v \in V$. Denote by $T_f^2(V, v) \subset T_w(\mathbb{R}^q) = \mathbb{R}^q$, $w = f(v)$, the subspace spanned by the vectors $\frac{\partial f}{\partial u_i}(v)$ and $\frac{\partial^2 f}{\partial u_i\, \partial u_j}(v)$ in $\mathbb{R}^q$ for $1 \leq i, j \leq n$. This subspace (obviously) is independent of the choice of u_i and it is called the (second) *osculating* space of the map f. [The first osculating space is $D_f(T_v(V)) \subset T_w(\mathbb{R}^q)$.] The dimension of $T_f^2(V, v)$ can vary between zero and $\frac{1}{2}n(n+3)$. Call f *free* if $\dim T_f^2(V, v) = \frac{1}{2}n(n+3)$ for all $v \in V$.

The freedom relation $\mathscr{F} \subset X^{(2)}$, for $X = V \times \mathbb{R}^q \to V$, is fibered over the immersion relation $\mathscr{I} \subset X^{(1)}$ by the projection $X^{(2)} \to X^{(1)}$. The fiber over each point is (naturally isomorphic to) *the Stiefel manifold* $\mathrm{St}_m \mathbb{R}^q$, $m = n(n+1)/2$, of m-frames of independent vectors in $\mathbb{R}^{q-n}$. If $q < n(n+3)/2$ then $\mathscr{F}$ is empty; otherwise it is an open dense subset in $X^{(2)}$ which is invariant under the natural actions of diffeomorphisms of V and of affine transformations of $\mathbb{R}^q$. Moreover, the group $\mathrm{Diff}\, V \times \mathrm{Aff}\, \mathbb{R}^q$ acts transitively on $\mathscr{F}$. In fact, $\mathscr{F}$ is the only subset in $X^{(2)}$ with these properties.

Exercise. Classify open subsets in $X^{(2)}$ for $X = V \times \mathbb{R}^{n+1} \to V$, which lie over $\mathscr{I} \subset X^{(1)}$ and which are invariant under $\mathrm{Diff}\, V \times \mathrm{Aff}\, \mathbb{R}^{n+1}$.

There are relatively few natural examples of free maps. The simplest one is the map $f\colon \mathbb{R}^n \to \mathbb{R}^q$ for $q = \frac{1}{2}n(n+3)$ given by the q monomials x_i and $x_i x_j$ on $\mathbb{R}^n$ for $1 \le i \le j \le n$. A more interesting example is provided by the *Veronese map* $f\colon S^n \to \mathbb{R}^{q+1}$, $q = \frac{1}{2}n(n+3)$, defined by the monomials $x_i x_j$ on $\mathbb{R}^{n+1} \supset S^n$, $1 \le i \le j \le n+1$. The image $f(S^n)$ is diffeomorphic to $\mathbb{R}P^n$ lying in a *hyperplane* $H \approx \mathbb{R}^q \subset \mathbb{R}^{q+1}$, since $\sum_{i=1}^{n+1} x_i^2 = 1$ on $S^n \subset \mathbb{R}^{n+1}$. Thus one obtains a *free* (Veronese) *embedding* of $\mathbb{R}P^n$ into $\mathbb{R}^q$ for $q = \frac{1}{2}n(n+3)$. [One does not know of a single closed manifold V besides S^n and $\mathbb{R}P^n$ which can be *freely* mapped into $\mathbb{R}^q$ for $q = \frac{1}{2}n(n+3)$. Yet, see Wintgen (1978) for interesting specific examples.]

Exercise. Find a linear transformation of $\mathbb{R}^q \supset f(S^n)$ which makes the Veronese map $f\colon S^n \to \mathbb{R}^q$ *isometric* (compare 1.1.5) and which sends $f(S^n)$ into a round sphere $S^{q-1} \subset \mathbb{R}^q$.

(A) **The h-Principle for Free Maps.** *If either $q \ge \frac{1}{2}n(n+3) + 1$, (compare the extra dimension case of Hirsch's theorem) or $q = \frac{1}{2}n(n+3)$ and V is an open manifold (the critical dimension), then free maps $V \to \mathbb{R}^q$ satisfy the h-principle. In particular, every stably parallelizable manifold admits a free map $V \to \mathbb{R}^q$ for $q = \frac{1}{2}n(n+3) + 1$, where* $n = \dim V$.

See 2.2.2 for the proof for open manifolds V and 2.2.1 and 2.4.3 for the extra dimension case.

Questions. Do free maps of *closed* manifolds $V \to \mathbb{R}^q$ for $q = \frac{1}{2}n(n+3)$ and $n \ge 2$ satisfy the h-principle? Does every parallelizable manifold (e.g. the torus T^n, $n \ge 2$) admit a free map into this $\mathbb{R}^q$?

Generalizations. The notion of freedom can be extended to higher order jets. For example, *third order free* maps $S^1 \to \mathbb{R}^3$ are curves with nowhere vanishing curvature and torsion. Pohl conjectured the h-principle for the kth-order free maps $V \to \mathbb{R}^q$ and the answer is positive for open manifolds V and in the extra dimensional case $\left(q \ge 1 + n + \dfrac{n(n+1)}{2} + \cdots + \dfrac{(n+k-1)!}{k!(n-1)!}\right)$. The h-principle in one critical case, namely concerning third order free maps of the circle into $\mathbb{R}^3$, is also true by the work of Little (1971) (compare Hamenstädt (1986)].

Freedom makes sense also for maps into general Riemannian manifolds, where it is expressed by independence of *covariant* derivatives of various orders. Our state of knowledge here is approximately the same as for maps into $\mathbb{R}^q$. The extradimensional second order case for $n = 1$ is due to Feldman (1968, 1971). Closed free curves in S^2 were classified by Little (1970). The techniques of 2.2.2 and 2.4.3 yield the h-principle for open and extra dimensional cases.

Exercises. (a) Study free maps $S^1 \to \mathbb{R}^2$ (that are closed immersed curves in $\mathbb{R}^2$ with nowhere vanishing curvature) and decide which sections $S^1 \to \mathscr{F}$ are homotopic to holonomic sections (that are free maps $S^1 \to \mathbb{R}^2$).

(b) Show with (A) that every (orientable or not) surface admits a free map into $\mathbb{R}^6$.

(c) Show every n-dimensional manifold V to admit a free map $V \to \mathbb{R}^q$ for $q = \frac{1}{2}n(n+5)$ and find topological obstructions for $q < \frac{1}{2}n(n+5)$.

1.1.5 Isometric Immersions of Riemannian Manifolds and the Theorems of Nash and Kuiper

Let g be a *Riemannian C^0-metric* on V, that is a continuous field of Euclidean structures g_v in $T_v(V)$, $v \in V$, and let $\mathbb{R}^q$ always have its standard Euclidean metric. The *isometric immersion relation* $\mathcal{I}_{\mathcal{J}} = \mathcal{I}_{\mathcal{J}_g} \subset X^{(1)}$ for $X = V \times \mathbb{R}^q \to V$ consists of the isometric linear embeddings $(T_v(V), g_v) \to T_w(\mathbb{R}^q) = \mathbb{R}^q$ for all $(v, w) \in V \times \mathbb{R}^q$. Solutions of $\mathcal{I}_{\mathcal{J}}$ are called *isometric immersions* $V \to \mathbb{R}^q$. These are indeed (differential) immersions, since the immersion condition $\mathcal{I} \subset X^{(1)}$ (properly) contains $\mathcal{I}_{\mathcal{J}}$. In fact, $\mathcal{I}_{\mathcal{J}}$ is a subfibration of $\mathcal{I} \to X$ whose fiber $\mathcal{I}_{\mathcal{J}_x} \subset \mathcal{I}_x$, $x = (v, w) \in V \times \mathbb{R}^q = X$, can be identified with the Stiefel manifold $\mathrm{St}_n^{\mathcal{O}}\, \mathbb{R}^q \subset \mathrm{St}_n\, \mathbb{R}^q$ of *orthonormal* n-frames in $\mathbb{R}^q$ for $n = \dim V$, where $\mathrm{St}_n\, \mathbb{R}^q \approx \mathcal{I}_x$ is the manifold of independent frames. Hence, $\mathcal{I}_{\mathcal{J}}$ is a fiberwise homotopy retract of $\mathcal{I}$ and every section $V \to \mathcal{I}$ can be homotoped to a (unique up to a homotopy) section $V \to \mathcal{I}_{\mathcal{J}}$.

The condition $\mathcal{I}_{\mathcal{J}}$ can be described geometrically by observing that a C^1-map $f\colon V \to \mathbb{R}^q$ is an isometric immersion if and only if it preserves the Riemannian length of the smooth curves in V. Alternatively, one can express the isometry property of $D_f\colon T(V) \to T(\mathbb{R}^q)$, and of f itself, in local coordinates $u_1, \ldots, u_n$ in V by the following system of $\frac{1}{2}n(n+1)$ (that is the codimension of $\mathcal{I}_{\mathcal{J}} \subset X^{(1)}$) non-linear partial differential equations (P.D.E.) of the first order

$$(\mathcal{I}_{\mathcal{J}}) \qquad \left\langle \frac{\partial f}{\partial u_i}, \frac{\partial f}{\partial u_j} \right\rangle = g_{ij}, \quad 1 \le i \le j \le n.$$

The unknown vector-function f has q components $f_1, \ldots, f_q$ and $\langle\ ,\ \rangle$ denotes the scalar product

$$\sum_{k=1}^{q} \frac{\partial f_k}{\partial u_i} \frac{\partial f_k}{\partial u_j}.$$

The functions g_{ij} are the components of the metric tensor of V. Namely $g_{ij} = \left\langle \frac{\partial}{\partial u_i}, \frac{\partial}{\partial u_j} \right\rangle_g$. Analytically speaking, our problem is to solve $(\mathcal{I}_{\mathcal{J}})$ for a given positive symmetric matrix of functions $g_{ij} = g_{ij}(v)$ on V. This is achieved in 2.4.9 with the following

(A) **Theorem.** *Isometric C^1-immersions $V \to \mathbb{R}^q$ satisfy the h-principle for all Riemannian manifolds $V = (V, g)$ and for all $q >$* dim V.

This fact is an immediate corollary of the h-principle of Hirsch for the extra dimension and of the following striking theorem discovered by Nash (1954) for $q \ge n + 2$ and extended to $q = n + 1$ by Kuiper (1955).

(A′) *An arbitrary differentiable immersion* $f_0\colon V \to \mathbb{R}^q$ *admits a* C^1*-continuous homotopy of immersions* $f_t\colon V \to \mathbb{R}^q$, $t \in [0, 1]$, *to an isometric immersion* $f_1\colon V \to \mathbb{R}^q$.

Remarks and Corollaries. (a) One obtains with the examples in (B) of 1.1.3 and with either (A) or (A′) an impressive list of isometric C^1-immersions $V \to \mathbb{R}^q$.

(b) The Theorems (A) and (A′) remains true for isometric C^1-immersions of V into a given convex open subset $W \subset \mathbb{R}^q$. Thus, for example, *the unit sphere* $S^n \subset \mathbb{R}^{n+1}$ *admits an isometric* C^1*-immersion into an arbitrary small ball in* $\mathbb{R}^{n+1}$ for all $n = 1, 2, \ldots$. This result of Kuiper (1955) disproved a long standing conjecture which had claimed every isometric C^1-immersion $f\colon S^n \to \mathbb{R}^{n+1}$ for $n \geq 2$ to be *congruent* to the unit sphere. (If f is C^2-smooth, then it is congruent to the unit sphere $S^n \subset \mathbb{R}^{n+1}$ by the classical rigidity theory).

Remark. An interesting (non-differential) functional equation whose solution is also sensible to the smoothness assumption is presented in Hilbert's 13th problem on superpositions of functions. A simple (counting parameters, compare 2.3.8) argument shows a generic C^∞-function in k variables to be *no* superposition of C^∞-functions in $k - 1$ variables. Yet, *every* C^0-function in $k \geq 3$ variables is a superposition of C^0-functions in two variables (Kolmogorov 1956; Arnold 1957).

(c) The existence of an isometric C^1-immersion of an arbitrarily small (yet, non-empty!) neighborhood $U \subset V$ into $\mathbb{R}^{n+1}$ is a *non-trivial* geometric phenomenon (discovered by Nash and Kuiper). This has no counterpart in the Smale-Hirsch theory where the emphasis is laid upon *global* immersions as the local (non-isometric) ones exist by the very definition of a smooth manifold. In fact, the local solution of an arbitrary *open* differential relation $\mathscr{R} \subset X^{(r)}$ (e.g. $\mathscr{I} \subset X^{(1)}$) is a trivial matter. But if $\operatorname{codim} \mathscr{R} = k > 0$ [e.g. $\operatorname{codim} \mathscr{I}_\sigma = \frac{1}{2}n(n+1)$], then the solutions f of $\mathscr{R}$ satisfy a certain P.D.E. system $\Psi_i(J_f^r) = 0$, $i = 1, \ldots, k$ (where the functions Ψ_i on $X^{(r)}$ have $\mathscr{R}$ for their zero set) and the local solvability of $\mathscr{R}$ becomes a non-trivial analytic problem. However, the relations $\mathscr{I}$ and $\mathscr{I}_\sigma$ share some common geometric features (concerning their convex hulls in $X^{(1)}$) which allow us to treat them on an equal footing in 2.4.

(d) There is no meaningful h-principle for equidimensional isometric immersions. However, *every n-dimensional Riemannian manifold admits a continuous (non-smooth!) map* $V \to \mathbb{R}^n$ *which preserves the Riemannian length of the smooth curves in* V (see 2.4.11).

Exercise. Let $W \subset \mathbb{R}^q$ be an open subset bounded by a smooth closed hypersurface in $\mathbb{R}^q$, and let S^1 be the circle with the standard metric $d\sigma^2$ on S^1. Prove the h-principle for isometric immersions $S^1 \to W$, provided W is *simply connected* and $q \geq 2$. Find counterexamples to this h-principle for $\pi_1(W) \neq 0$.

(B) C^∞*-Immersions* $(V, g) \to \mathbb{R}^q$. Let the metric g on V be C^∞-smooth and consider the lift (see 1.1.1) $\mathscr{I}_\sigma^r \subset X^{(r+1)}$ of the relation $\mathscr{I}_\sigma = \mathscr{I}_{\sigma_g} \subset X^{(1)}$. If $q < \frac{1}{2}n(n+1)$ and if $r \geq r_0$ for some $r_0 = r_0(n)$, then $\mathscr{I}_\sigma^r$ is *empty* for *generic* (see 1.3.2) C^∞-metrics g (see 2.3.8). Therefore, most C^∞-manifolds V admit no isometric C^∞-immersions into

$\mathbb{R}^q$ for $q < \frac{1}{2}n(n+1)$. It is unknown if a small neighborhood of each point $v \in V$ admits an isometric C^∞-immersion into $\mathbb{R}^q$, for $q = \frac{1}{2}n(n+1)$; however, if (V, g) is *real analytic*, such a local C^{an}-immersion does exist by a theorem of Janet (1926) (see 3.1.6).

The relation $\mathscr{I}_\mathscr{J} \subset X^{(1)}$ is never stable (see 1.1.1) for $n \geq 2$ and $q \geq n+1$. However, the lift $\mathscr{I}_\mathscr{J}^r$ may (or may not) become stable for large r.

Exercises. (a) Prove the above mentioned non-stability of $\mathscr{I}_\mathscr{J}$ by invoking Gauss theorema egregium (see 3.1.5).

(b) Prove the stability of $\mathscr{I}_\mathscr{J}$ for $n = 1$ and for all $q \geq 1$. Then study $q = n$ and $n = 1, 2, \ldots$.

(c) Let $n = 2$, $q = 3$ and show $\mathscr{I}_{\mathscr{J}_g}^2 \subset X^{(3)}$ to be stable for metrics g on V whose Gauss curvature is either everywhere positive or everywhere negative on V. Find a C^∞-metric g on V for which $\mathscr{I}_{\mathscr{J}_g}^r$ is non-stable for all $r = 1, 2, \ldots$.

(B_1) *Free Isometric Immersions.* Lift $\mathscr{I}_\mathscr{J}$ to $\mathscr{I}_\mathscr{J}^1 \subset X^{(2)}$, intersect it with the freedom relation (see 1.1.4), put $\mathscr{F}\mathscr{I}_\mathscr{J} = \mathscr{F} \cap \mathscr{I}_\mathscr{J}^1 \subset X^{(2)}$ and observe that solutions of $\mathscr{F}\mathscr{I}_\mathscr{J}$ are those maps $f\colon V \to \mathbb{R}^q$ which are both free and isometric. This $\mathscr{F}\mathscr{I}_\mathscr{J}$ is yet unstable; however, the lift $(\mathscr{F}\mathscr{I}_\mathscr{J})^1 \subset X^{(3)}$ is stable for all Riemannian metrics g (see 3.1.6). Moreover, for every continuous section $\varphi_0\colon V \to \mathscr{F}$ there exists a unique up to homotopy continuous section $\varphi_1\colon V \to (\mathscr{F}\mathscr{I}_\mathscr{J})^1$ whose projection to $\mathscr{F}$ is homotopic to φ_0 (see 3.1.6).

Question. Do free isometric C^∞-immersions satisfy the h-principle for $n \geq 2$? (If $n = 1$ then the answer is, obviously, "no", for $q = 2$, and it is an easy "yes" for $q \geq 3$.)

The positive answer for $q \geq \frac{1}{2}(n+2)(n+3)$ is given in the following

(B_2) **Theorem.** *Let (V, g) be a C^k-smooth Riemannian manifold for* $k = 5, 6, \ldots, \infty$, an. *Then free isometric C^k-immersions $V \to \mathbb{R}^q$ satisfy the h-principle for* $q \geq \frac{1}{2}(n+2)(n+3)$, *where* $n = \dim V$.

This is proven in 3.1.7 by deforming a free immersion $f_0\colon V \to \mathbb{R}^q$ into a free *isometric* one [by using auxiliary isometric immersions $V \times \mathbb{R}^2 \to \mathbb{R}^q$ which require $q \geq \frac{1}{2}(n+2)(n+3)$].

Remarks and Corollaries. (a) The relation $\mathscr{F} \subset X^{(2)}$ fibers over $X = V \times \mathbb{R}^q$ with the fiber $\approx \mathrm{St}_m \mathbb{R}^q$, $m = \frac{1}{2}n(n+3)$. Hence, $\mathscr{F}$ admits a section $V \to \mathscr{F}$ for $q \geq \frac{1}{2}n(n+5)$ and every two sections are homotopic for $q \geq \frac{1}{2}n(n+5)+1$. This extends with the above discussion to $(\mathscr{F}\mathscr{I}_\mathscr{J})^r$, $r \geq 1$, thus showing the h-principle (B_2) to be equivalent to the following

(B_2') **Existence Theorem.** *Every C^k-manifold (V, g) admits a free isometric C^k-immersion into $\mathbb{R}^q$, $q = \frac{1}{2}(n+2)(n+3)$, for all* $k = 5, 6, \ldots, \infty$, an.

(b) The principle possibility of the realization of Riemannian manifolds in $\mathbb{R}^q$ was established by Nash (1956). He proved that a compact manifold of class C^k,

$k = 3, 4, \ldots, \infty$, can be isometrically and freely immersed into $\mathbb{R}^q$ for $q = \frac{3}{2}n(n+9)$ and for $q = \frac{3}{2}n(n+1)(n+9)$ in the non-compact case. Ten years later Nash (1966) solved the problem for compact C^{an}-manifolds.

The methods invented by Nash have not lost their importance although his values for q have been improved since then. In fact, a slight modification of those provides C^k-immersions into $\mathbb{R}^q$ for $q = n^2 + 5n + 3$ [including $k = 3, 4$ not covered by (B'_2)] of all (compact or not) C^k-manifolds V. Yet, one does not know whether every Riemannian C^2-manifold admits an isometric C^2-immersion into some $\mathbb{R}^q$.

1.2 Homotopy and Approximation

1.2.1 Classification of Solutions by Homotopy and the Parametric *h*-Principle

Two solutions, f_0 and f_1 of a relation $\mathscr{R} \subset X^{(r)}$, are called *$C^r$-homotopic* if they can be joined by a *C^r-continuous* homotopy of solutions $f_t: V \to X$, $0 \le t \le 1$, where a homotopy of sections $f_t: V \to X$ is, by definition, C^r-continuous, if and only if the homotopy of the jets $J^r_{f_t}: V \to X^{(r)}$ is continuous. [This is equivalent to the continuity of the map $t \mapsto f_t$ of $[0, 1]$ into the space of C^r-sections $V \to X$ with the ordinary C^r-topology, see (C) below.] We say that $\mathscr{R}$ and (or) solutions of $\mathscr{R}$ satisfy the *one-parametric h-principle* if the existence of a continuous homotopy of sections $\varphi_t: V \to \mathscr{R}$, $0 \le t \le 1$, between $\varphi_0 = J^r_{f_0}$ and $\varphi_1 = J^r_{f_1}$ is sufficient (it is, obviously, necessary) for a C^r-homotopy of solutions between f_0 and f_1 for all pairs of solutions f_0 and f_1 of $\mathscr{R}$.

The one-parametric h-principle holds true for the relations studied in 1.1.2–1.1.5 under the same assumptions that are used for the ordinary h-principle (with the same references to the proofs). For instance, [compare (B) in 1.1.2] *holomorphic maps f_0 and f_1 of a Stein manifold V into a complex homogeneous space G/H can be joined by a homotopy of holomorphic maps $f_t: V \to G/H$ if and only if there is a homotopy of continuous maps between f_0 and f_1.*

(A) *Regular Homotopies.* This is a name for C^1-continuous homotopies of immersions $f_t: V \to W$ (immersions are often called *regular maps*). Let us assume V to be an n-dimensional stably parallelizable manifold and let $W = \mathbb{R}^{n+1}$. Then any splitting of the tangent bundle $T(V \times \mathbb{R}^1) = (V \times \mathbb{R}^1) \times \mathbb{R}^{n+1}$ (obviously) induces some splitting of the immersion relation, $\mathscr{I} = V \times \mathbb{R}^{n+1} \times \mathrm{St}_n \mathbb{R}^{n+1}$, which provides (with the one-parametric h-principle!) a one-to-one correspondence between the regular homotopy classes of immersions $V \to \mathbb{R}^{n+1}$ and the homotopy classes of maps $V \to \mathrm{St}^{\mathscr{C}}_n \mathbb{R}^{n+1} = \mathrm{SO}(n+1)$.

Examples. (A_1) *Immersions* $S^1 \to \mathbb{R}^2$. Here $\mathrm{SO}(2) = S^1$ and maps $S^1 \to S^1$ are classified by degree $d = \ldots, -2, -1, 0, -1, \ldots$. This implies the following

Theorem of Whitney (1937). *Every immersion $S^1 \to \mathbb{R}^2$ is regularly homotopic either to the figure* ∞ *(which corresponds to $d = 0$), or to the immersion $z \mapsto z^d$ of the unit circle $S^1 \subset \mathbb{C}$ into $\mathbb{C} = \mathbb{R}^2$ if $d \neq 0$.*

Exercise. Give a direct geometric proof of this Theorem.

(A_2) *Immersions $S^2 \to \mathbb{R}^3$.* The orthogonal group SO(3) is doubly covered by S^3 and therefore $\pi_2(\mathrm{SO}(3)) = \pi_2(S^3) = 0$. Hence, every immersion $S^2 \to \mathbb{R}^3$ is regularly homotopic to the standard embedding $S^2 \subsetneq \mathbb{R}^3$. In particular, the reflection in the center of the round sphere $S^2 \subsetneq \mathbb{R}^3$ can be homotoped to the original embedding by a regular homotopy of immersions $S^2 \to \mathbb{R}^3$, and in the course of such a homotopy the inward looking normal field on S^2 turns outward. This is the famous

Smale's "Paradox". *The sphere $S^2 \subset \mathbb{R}^3$ (unlike $S^1 \subset \mathbb{R}^2$) can be turned inside out.*

(A_2') *Immersions $V^2 \to \mathbb{R}^3$.* Since $\pi_1(\mathrm{SO}(3)) = \mathbb{Z}/2\mathbb{Z}$ and $\pi_2(\mathrm{SO}(3)) = 0$, immersions of an oriented surface $V \to \mathbb{R}^3$ are classified by the cohomology group $H^1(V; \mathbb{Z}/2\mathbb{Z})$. This gives 4^m classes of immersions for closed surfaces of genus m.

(A_3) *The Degree of an Immersed Hypersurface $V \to \mathbb{R}^{n+1}$.* Let $G_f: V \to S^n$ be the tangential (*Gauss*) map associated to an immersion $f: V \to \mathbb{R}^{n+1}$ of a connected *oriented* n-dimensional manifold V. Since G_f induces the bundle $T(V)$ from $T(S^n)$, it sends the Euler class $\chi(T(S^n)) \in H^n(S^n; \mathbb{Z}) \approx \mathbb{Z}$ to $\chi(T(V) \in H^n(V; \mathbb{Z}) \approx \mathbb{Z}$. If n is even, then $\chi(T(S^n)) = 2 \in \mathbb{Z}$ (for the standard orientation of S^n), which implies the following relation [due to Hopf (1925)] between the topological degree $\deg G_f$ and the Euler characteristic of V.

If $f: V \to \mathbb{R}^{n+1}$ is an immersed closed oriented hypersurface, then $2 \deg G_f = \chi(V)$ for n even.

Now, let n be odd. Then $\deg G_f$ may depend on (the regular homotopy class of) the immersion f as well as on the topology of V. For instance, let V bound a compact $(n + 1)$-dimensional manifold V' and let an immersion $f': V' \to \mathbb{R}^{n+1}$ extend f from the boundary $\partial V' = V$ to V'. *Then*

$$\deg G_f = \chi(V') \quad \textit{for } n \textit{ odd.}$$

Proof. Double the immersion f' and smooth (in an obvious way) the resulting map of the double $V' \cup_V V' \to \mathbb{R}^{n+1}$ to an *immersion* $f'': V' \cup V' \to \mathbb{R}^{n+2} \supset \mathbb{R}^{n+1}$. Clearly, $\deg G_f = \deg G_{f''}$ and $\chi(V' \cup_V V') = 2\chi(V')$; hence, the above even-dimensional formula applies.

(A_3') **Subexample.** Let V' be obtained by deleting an open ball from a closed stably parallelizable manifold V''. Then V', being parallelizable, admits an immersion $f': V' \to \mathbb{R}^{n+1}$, which gives the degree $\deg G_f = \chi(V'') - 1$ to the boundary sphere $S^n = \partial V'$ for $f = f' | \partial V'$. If $\chi(V'') \neq 2$, this degree $\neq 1 = \deg G_{f_0}$ of the standard sphere $f_0: S^n \subsetneq \mathbb{R}^{n+1}$. In particular, for $V'' = S^m \times S^m$, $2m = n + 1$, the immersion

$f\colon S^n \to \mathbb{R}^{n+1}$ has $\deg G_f = 3$ for m *even* and $\deg G_f = -1$ for m odd. In fact, one does not need Hirsch's theorem to immerse (the parallelizable) manifold $V' = S^m \times S^m \setminus B^{n+1}$ into $\mathbb{R}^{n+1}$. Indeed, this V' can be isotoped to an arbitrarily small neighborhood of the wedge $S^m \vee S^m \subset S^m \times S^m$. Then V' obviously goes to $\mathbb{R}^{n+1}$ with the standard map $S^m \vee S^m \to \mathbb{R}^{n+1}$ which sends each copy of S^m onto a round sphere in $\mathbb{R}^{n+1}$, where the two round spheres meet at two points in $\mathbb{R}^{n+1}$ with one of them receiving the joint point of the wedge. The resulting immersion $f\colon S^n \to \mathbb{R}^{n+1}$ (with $\deg G_f = 3$) has no triple self intersection point, while no (known) general theory insures immersions $V^n \to \mathbb{R}^{n+1}$ without triple points (compare 2.1.1).

(A_4) *The Signature of a Hypersurface* $V \to \mathbb{R}^{n+1}$. An immersion $f\colon V \to \mathbb{R}^{n+1}$ is called *null-cobordant* if V bounds an oriented $(n+1)$-dimensional manifold V' which goes into the half-space $\mathbb{R}_+^{n+2} \supset \mathbb{R}^{n+1}$ by an immersion $f'\colon V' \to \mathbb{R}_+^{n+2}$, such that $f = f'|\partial V' = V$ and which is orthogonal to $\mathbb{R}^{n+1}$ along the boundary $\partial V'$. For example, the above immersions $V' \to \mathbb{R}^{n+1}$ can be (obviously) pushed to such positions in $\mathbb{R}_+^{n+2} \supset \mathbb{R}^{n+1}$. Let I^n denote the Abelian group with the generating set $\{(V, f)\}$, for all closed oriented (connected or not) n-dimensional manifolds V and for all immersions $f\colon V \to \mathbb{R}^{n+1}$, and with the relations

$$\{(V_1 \cup V_2, f_1 \cup f_2) = (V_1, f_1) + (V_2, f_2)\},$$

for all (V_i, f_i), $i = 1, 2$, where $V_1 \cup V_2$ denotes the disjoint union of V_i and $f_1 \cup f_2 | V_i = f_i$ for $i = 1, 2$. Denote by $I_0^n \subset I^n$ the subgroup generated by null-cobordant immersions. Then the factor group $\Pi_n = I_0^n/I_0^n$ is called the *cobordism group* of (oriented one-codimensional) *immersions*. It is isomorphic [Wells (1966), compare (G) in 2.2.7] to the (*stable*) *homotopy group*,

$$\Pi_n \approx \pi_{N+n}(S^N) = \pi_{N+n+1}(S^{N+1}) = \ldots, \quad \text{for } N \geq n+2.$$

Hence, by the fundamental theorem of Serre (1953) the group Π_n is finite for $n \geq 1$. [In fact, $\Pi_0 \approx \mathbb{Z}$, $\Pi_1 \approx \mathbb{Z}_2 = \mathbb{Z}/2\mathbb{Z}$, $\Pi_2 \approx \mathbb{Z}_2$, $\Pi_3 \approx \mathbb{Z}_{24}$, $\Pi_4 \approx \Pi_5 = 0$, $\Pi_6 \approx \mathbb{Z}_2$, $\Pi_7 \approx \mathbb{Z}_{240}$, $\Pi_8 = \mathbb{Z}_2 \oplus \mathbb{Z}_2$, $\Pi_9 = \mathbb{Z}_2 \oplus \mathbb{Z}_2 \oplus \mathbb{Z}_2$, $\Pi_{10} = \mathbb{Z}_6$, $\Pi_{11} = \mathbb{Z}_{504}$, $\Pi_{12} = 0$, $\Pi_{13} = \mathbb{Z}_3, \ldots$, see Toda (1962).] The Serre finiteness theorem insures for every immersion $f\colon V \to \mathbb{R}^{n+1}$, $n \geq 1$, the existence of a manifold V' whose boundary consists of $d = \operatorname{ord} \Pi_n$ copies of V and of an immersion $f'\colon V' \to \mathbb{R}_+^{n+2}$ which equals f on each copy of V in $\partial V'$. Now, for $n = 4k - 1$, we define the *signature* of f by $\sigma(f) = d^{-1}\sigma(V')$ where $\sigma(V')$ stands for the signature (or index) of the (quadratic) intersection (of $2k$-cycles) form on the homology $H_{2k}(V')$ (The intersection form may be degenerate since V' has a boundary, but the signature is defined, as usual, by the difference between the plus and minus signs in the diagonalized form.) If $f_0'\colon V_0' \to \mathbb{R}_+^{n+2}$ is another immersion bounded by d copies of V, then the reflection of f_0' in the hyperplane completes V' to a closed immersed hypersurface $V' \cup V_0' \to \mathbb{R}^{n+2}$ which has zero signature being stably parallelizable (see 2.1.3). Hence,

$$\sigma(V') - \sigma(V_0') = \sigma(V' \cup V_0') = 0$$

by the (obvious) additivity of the signature [Novikov (1965); the minus sign comes from the reversal of the orientation by the reflection of V_0'], which shows $\sigma(f)$ to be independent of V'. In fact, $\sigma(f)$ is a regular homotopy invariant of f, since every regular homotopy $f_t\colon V \to \mathbb{R}^{n+1}$, $t \in [0,1]$, can be perturbed to an immersed cylinder $\hat{f}\colon V \times [0,1] \to \mathbb{R}^{n+2}_{+}$ with $\hat{f}|V \times 0 = f_0$ and $\hat{f}|V \times 1 = f_1$ and since adding the "collar" $V \times [0,1]$ to V' does not change $\sigma(V')$.

Example. Let Y be a closed oriented manifold of dimension $2k$ and let $V' \subset Y \times Y$ be a closed tubular neighborhood of the diagonal $\Delta \approx Y$ in $Y \times Y$. The homology $H_{2k}(V')$ is generated by the fundamental class $[\Delta]$ whose self-intersections (for the natural orientation in V') equals $\chi(Y)$ by the Hopf-Lefschetz fixed point formula. Thus $\sigma(V') = \operatorname{sign}\chi(Y)$ (that is zero for $\chi = 0$ and ± 1 for $\chi \gtrless 0$). If Y is stably parallelizable, then, clearly, V' is parallelizable. This allows an immersion $f'\colon V' \to \mathbb{R}^{n+1}$, for which $\sigma(f) = \operatorname{sign}\chi(Y)$. In particular, we obtain an immersion $f\colon \mathrm{SO}(3) \to \mathbb{R}^4$ with $\sigma(f) = 1$ for $Y = S^2$.

Exercises. (a) *Immersions* $S^3 \to \mathbb{R}^4$. Relate immersions $S^3 \to \mathbb{R}^4$ to the group $\pi_3(\mathrm{SO}(4)) = \pi_3(S^3 \times \mathrm{SO}(3)) = \mathbb{Z} \oplus \mathbb{Z}$ and define with the above deg and σ homomorphisms of $\mathbb{Z} \oplus \mathbb{Z}$ onto $\mathbb{Z}$ and into $\mathbb{Q}$ respectively. [In fact, the values of σ lie in $\frac{2}{3}\mathbb{Z} \subset \mathbb{Q}$ by a theorem of Rohlin, see Kervaire-Milnor (1960).] Show immersions f_0 and f_1 to be regularly homotopic if and only if $\deg G_{f_0} = \deg G_{f_1}$ and $\sigma(f_0) = \sigma(f_1)$.

(b) *Immersions* $S^n \to \mathbb{R}^q$ *for* $q \geq n+1$. Observe that the immersion relation $\mathcal{I} \to S^n$ has simply connected fibers ($\approx \mathrm{St}_n \mathbb{R}^q$) for $q \geq n+2$ and that it admits a section $S^n \to \mathcal{I}$. Establish with this a one-to-one correspondence between the homotopy classes of sections $S^n \to \mathcal{I}$ with those of maps $S^n \to \mathcal{I}_v \approx \mathrm{St}_n \mathbb{R}^q$, $v \in S^n$, and conclude

Theorem of Smale. *The regular homotopy classes of immersions* $S^n \to \mathbb{R}^q$ *are classified for* $q \geq n+1$ *by the homotopy group* $\pi_n(\mathrm{St}_n \mathbb{R}^q)$.

(B) *Isometric Immersions.* Since the isometric immersion relation $\mathcal{I}_{\mathrm{iso}} \subset \mathcal{I} \to X = V \times \mathbb{R}^q$ is fiberwise homotopy equivalent to $\mathcal{I}$, the h-principle of Nash-Kuiper (see 1.1.5) allows the regular homotopies in the above examples to be C^1-isometric. For instance, *the sphere* $S^2 \subset \mathbb{R}^3$ *can be turned inside out by a regular homotopy of isometric* C^1*-immersions* $S^2 \to \mathbb{R}^3$.

Exercise. *The Signature and the* η*-Invariant.* Recall the invariant $\eta(V,g)$ of a C^2*-smooth* Riemannian manifold (V,g) (see Atiyah-Patodi-Singer 1975) and show every isometric C^2-smooth immersion $f\colon (V,g) \to \mathbb{R}^{n+1}$, $n = \dim V = 4k-1$, to satisfy $\sigma(f) = \eta(V,g)$. Then express the degree of G_f for all $n \geq 2$ by the integral over V of a certain curvature function of g. Apply this to an arbitrary Riemannian C^2-metric g on S^2 and show the regular homotopy class of any isometric C^2-immersion $f\colon (S^3,g) \to \mathbb{R}^4$ to be uniquely determined by g [compare (C) in 3.3.4]. Observe that invariants of g impose no restriction on the regular homotopy class of an isometric C^1*-smooth* immersion $(V,g) \to \mathbb{R}^{n+1}$.

(C) *The (Multi-)Parametric h-Principle.* Recall that a continuous map between topological spaces, say $\mu\colon A \to A'$, is called a *weak homotopy equivalence* if either of the two following (obviously) equivalent conditions is satisfied.

(i) The map μ is *bijective* on the homotopy groups, $\mu_i\colon \pi_i(A) \xrightarrow{\sim} \pi_i(A')$, $i = 0, 1, \ldots$.
(ii) Let P be an arbitrary cell complex, let $P_0 \subset P$ be a subcomplex and let $\alpha_0\colon P_0 \to A$ be an arbitrary continuous map. Then α_0 extends to a continuous map $\alpha\colon P \to A$ if and only if $\alpha_0' = \mu \circ \alpha_0\colon P_0 \to A'$ extends to a continuous map $\alpha'\colon P \to A'$.

Next, invoke the space $C^r(X)$ of C^r-sections $V \to X$ (of a fibration $X \to V$, see 1.1.1) with the topology of uniform convergence of sections along with their partial derivatives (jets) of order $\leq r$ on compact subsets in V, and let $\mathrm{Sol}^r \mathcal{R} \subset C^r(X)$ be the subspace of the solution of a given relation $\mathcal{R} \subset X^{(r)}$. Say that *$\mathcal{R}$ and (or) solutions of $\mathcal{R}$ satisfy the parametric h-principle* if the map $J^r\colon \mathrm{Sol}^r \mathcal{R} \to C^0(\mathcal{R})$ for $f \mapsto J_f^r$ is a weak homotopy equivalence. This amounts to the vanishing of the relative homotopy groups $\pi_i(C^0(\mathcal{R}), \mathrm{Hol}^0)$, $i = 0, 1, \ldots$, for the subspace $\mathrm{Hol}^0 \subset C^0(\mathcal{R})$ of holonomic sections $V \to \mathcal{R}$ (which is homeomorphic to $\mathrm{Sol}^r \mathcal{R}$ by $f \mapsto J_f^r$).

Example. The surjectivity of J^r on π_0 expresses the ordinary h-principle, while the injectivity on π_0 is the one-parametric h-principle, which refers to continuous maps (C^r-homotopies) $P = [0, 1] \to \mathrm{Sol}^r \mathcal{R}$.

The h-principles of 1.1.2–1.1.5 hold true in the parametric form with the same reference to the proofs. For example.

(1) (Compare 1.1.5.) *The space of isometric C^1-immersions $\mathbb{R}^n \to \mathbb{R}^q$, $q \geq n + 1$, has the same homotopy groups as the Stiefel manifold* $\mathrm{St}_n \mathbb{R}^q$.
(2) *The space of free isometric C^k-immersions $\mathbb{R}^n \to \mathbb{R}^q$ for $k = 5, 6, \ldots, \infty$*, an. *has the same homotopy groups as* $\mathrm{St}_m \mathbb{R}^q$ *for $m = \frac{1}{2}n(n + 3)$ and for $q \geq \frac{1}{2}(n + 2)(n + 3)$.* [Even in this example of the Euclidean metric on $\mathbb{R}^n$ one does not know how to remove the assumptions $k \geq 5$ and (or) $q \geq \frac{1}{2}(n + 2)(n + 3)$ for $n \geq 2$.]

Exercises. (a) Prove (2) for $n = 1$ and for all $k \geq 2$ and $q \geq 3$.

(b) Let V be a compact simply connected Riemannian C^∞-manifold. Prove with the parametric h-principle the space F of free isometric C^∞-immersions $V \to \mathbb{R}^q$, $q \geq \frac{1}{2}(n + 2)(n + 3)$, to have *finitely generated* homotopy groups [compare Serre (1953)]. Show that $\pi_i(F) = 0$ for $1 \leq i < j = q - \frac{1}{2}n(n + 5)$ and that the group $\pi_j(F)$ is cyclic.

(C′) **Remark.** The parametric h-principle refers to continuous maps $\varphi\colon P \to \mathrm{Sol}^r \mathcal{R}$ which are, in fact, sections of the fibration $X \times P \to V \times P$ satisfying $\mathcal{R}$ over each submanifold $V \times p \subset V \times P$, $p \in P$. If P is a smooth manifold and if the sections $\varphi\colon V \times P \to X \times P$ in question are C^r-smooth, then the relations $\mathcal{R} | V \times p$, for all $p \in P$ are equivalent to a single relation $\mathcal{R}' \subset (X \times P)^{(r)}$. That is the pull-back of $\mathcal{R}$ under the natural projection $(X \times P)^{(r)} \to X^{(r)}$. Thus, the parametric h-principle for $\mathcal{R}$ can be reduced in most cases to the ordinary h-principle for $\mathcal{R}' = \mathcal{R}_P$.In fact, one needs $P = S^i$ to show the surjectivity of the map J^r on $\pi_i(\mathrm{Sol}^r \mathcal{R})$; the ball B^{i+1} is used for the injectivity of J^r on π_i. In the latter case one needs the *h-principle for extensions* (see 1.4.2, 1.4.4) of solutions from $V \times S^i$ to $V \times B^{i+1}$ for $S^i = \partial B^i$.

Strictly speaking, the h-principle for $\mathcal{R}'$ is not equivalent to the h-principle for $\mathcal{R}$ due to somewhat different smoothness assumptions on pertinent sections $V \times P \to V \times X$ used in the definitions of the respective h-principles. However, if

$\mathscr{R} \subset X^{(r)}$ is an *open* subset, then any section $\varphi: V \times P \to X \times P$ which is (only) C^r-continuous in p can be easily approximated by C^r-smooth sections which satisfy $\mathscr{R}'$ in so far as φ satisfies $\mathscr{R} | V \times p$ for all $p \in P$. Hence, the parametric h-principle for *open* relations $\mathscr{R}$ does follow from the ordinary h-principle for extensions of solutions of $\mathscr{R}'$. A similar approximation argument applies to many non-open $\mathscr{R}$ (see 2.3.2). Alternatively, one could slightly modify the notion of the h-principle for $\mathscr{R}'$ by starting with C^r-continuous solutions. In any case, the parametric h-principle for most relations is not harder to prove than the ordinary h-principle. In fact, Smale (1959) originally obtained the one-parametric h-principle for immersions $S^n \to \mathbb{R}^q$ by first proving the *multi-parametric* h-principle by induction on n [compare (C) in 1.4.2; no such induction is possible if one restricts the number of parameters.] We adopt a similar approach in §2.2. On the contrary, the techniques in 2.1 and 2.4 directly yield the ordinary h-principle for $\mathscr{R}$ (as well as for $\mathscr{R}'$), without ever mentioning any homotopy groups.

1.2.2 Density of the *h*-Principle in the Fine Topologies

Define *the fine C^0-topology* in the space $C^0(X)$ of continuous sections $V \to X$ by taking the subsets $C^0(U) \subset C^0(X)$, for all open $U \subset X$, for a base of this topology. Then the fine *C^r-topology* in $C^r(X)$ is induced from the fine C^0-topology in $C^0(X^{(r)})$ by the embedding $f \mapsto J_f^r$ of $C^r(X)$ onto the subspace of holonomic sections in $C^r(X)$.

Example. A family F of C^1-functions $f: \mathbb{R} \to \mathbb{R}$ gives a fine C^1-approximation to a given C^1-function f_0 iff for every positive continuous function $\varepsilon(t)$ on $\mathbb{R}$ there exists an element $f \in F$ for which $|f_0(t) - f(t)| + \left|\frac{df_0(t)}{dt} - \frac{df(t)}{dt}\right| \leq \varepsilon(t)$ for all $t \in \mathbb{R}$.

Definitions. Let $f_0: V \to X$ be a continuous section and consider a neighborhood $U \subset X$ of the image $f_0(V) \subset X$. Intersect the pull-back of U under the projection $p_0^r: X^{(r)} \to X$ with a given relation $\mathscr{R} \subset X^{(r)}$ and write $\mathscr{R}_U = (p_0^r)^{-1}(U) \cap \mathscr{R} \subset X^{(r)}$. Say that $\mathscr{R}$ satisfies *the h-principle C^0-near f_0* if every section $\varphi_0: V \to \mathscr{R}$ which lies over f_0 (i.e. $p_0^r \circ \varphi_0 = f_0$) can be brought to a *holonomic* section φ_1 by a homotopy of sections $\varphi_t: V \to \mathscr{R}_U$, $t \in [0, 1]$, for an arbitrary (small) neighborhood $U \subset X$ of $f_0(V)$. The h-principle is called *everywhere dense* [in $C^0(X)$] if it is holds true C^0-near every section $f_0: V \to X$. Clearly, the h-principle is everywhere dense if and only if the relation $\mathscr{R}_U$ satisfies the ordinary h-principle for all open subsets $U \subset X$.

(A) **Examples.** (1) The h-principle for the Cauchy-Riemann relation (see 1.1.2) is *nowhere dense* since C^0-limit of holomorphic sections is holomorphic.

(2) *Immersions $V \to W$ enjoy the C^0-dense h-principle for* $\dim V < \dim W$. [Hirsch (1959), see 2.1.2, 2.2.2, 2.4.3.]

Corollary. *If a continuous map $f_0: V \to W$ is homotopic to an immersion, then it admits a fine C^0-approximation by immersions $V \to W$. In particular, every continuous map $f_0: \mathbb{R}^n \to \mathbb{R}^{n+1}$ can be C^0-approximated by immersions $f: \mathbb{R}^n \to \mathbb{R}^{n+1}$.*

Proof. Since the projection $\mathscr{I} \to X = V \times W$ is a fibration, the existence of a lift of a section $f_0: V \to X$ to $\mathscr{I}$ depends only on the homotopy class of f_0.

Exercises. (a) Let $f_0: \mathbb{R}^2 \to \mathbb{R}^3$ be given in the Euclidean coordinates by

$$f_0: (u_1, u_2) \mapsto (u_1^2, u_1 u_2, u_2).$$

Observe the map f_0 to be a C^∞-immersion outside the origin $(0,0) \in \mathbb{R}^2$ and show that *no immersion* $f: \mathbb{R}^2 \to \mathbb{R}^3$ has *bounded C^1-distance* from f_0, which means $\left\| \frac{\partial(f - f_0)}{\partial u_1} \right\| + \left\| \frac{\partial(f - f_0)}{\partial u_2} \right\| \leq \text{const} < \infty$.

(b) Show that neither one of the two immersions $\mathbb{R}^2 \to \mathbb{R}^2$ given by $(u_1, u_2) \mapsto (u_1, u_2^2)$ and by $(u_1, u_2) \mapsto (u_1 u_2, u_1^2 - u_2^2)$ respectively admit C^0-approximations by immersions $\mathbb{R}^2 \to \mathbb{R}^2$.

(c) Consider the manifold V of pairs $v = (l, x)$, where l is a straight line in $\mathbb{R}^3$ through the origin and $x \in l$. Map V to $\mathbb{R}^3$ by $v = (l, x) \mapsto x \in \mathbb{R}^3$ and show this map to admit no C^0-approximation by immersions $V \to \mathbb{R}^3$. In contrast, prove the existence of some immersion $V \to \mathbb{R}^3$. [See Siebenmann (1972) for general results on limits of equidimensional immersions.]

(B) *Short Maps and Isometric Immersions Between Riemannian Manifolds.* Let (V, g) and (W, h) be Riemannian manifolds. A continuous map $f: V \to W$ is called *short* if it does not increase the Riemannian length of smooth curves in V. This is equivalent, for C^1-smooth maps f, to the inequality $\langle \tau, \tau \rangle_g \geq \langle D_f(\tau), D_f(\tau) \rangle_h$ for all tangent vectors $\tau \in T(V)$. Call f_0 *strictly short* if there is a strictly positive continuous function $\varepsilon = \varepsilon(v)$ on V such that

$$\operatorname{dist}_h(f(v_1), f(v_2)) \leq (1 - \varepsilon(v_1)) \operatorname{dist}_g(v_1, v_2)$$

for those pairs of points $v_1, v_2 \in V$ which have $\operatorname{dist}_g(v_1, v_2) \leq \varepsilon(v_1)$. If f_0 is C^1-smooth, then this is equivalent to $\langle \tau, \tau \rangle_g > \langle D_f(\tau), D_f(\tau) \rangle_h$ for all $\tau \neq 0$ in $T(V)$.

(B_1) **Theorem.** *If* $\dim W > \dim V$ *then strictly short differentiable immersions* $V \to W$ *satisfy the h-principle near every strictly short* C^0*-map* $f_0: V \to W$. (See 2.4.5.)

Remarks. (a) Strictly short differentiable immersions are (the only) solutions of the differential relation in $X^{(1)}$ which consists of strictly short injective linear maps $(T_v(V), g_v) \to (T_w(W), h_w)$ for all $(v, w) \in X = V \times W$. Thus, one may speak of the h-principle.

(b) Since every C^1-map betweem smooth manifolds becomes strictly short with appropriatly chosen Riemannian metrics, (B_1) sharpens the above C^0-dense h-principle of Hirsch.

(c) $W^{i,p}$*-approximation.* Another refinement of Hirsch's approximation can be achieved for all $q \geq n = \dim V$ with the (Sobolev) $W^{i,p}$-metric in the space of C^∞-maps $f: V \to \mathbb{R}^q$, defined for $i = 0, 1, \ldots$ and $p \geq 1$ by

$$\operatorname{dist}_{i,p}(f_1, f_2) = \left(\int_V \| J_{f_1}^i - J_{f_2}^i \|^p d \right)^{1/p},$$

where $\| \ \|$ is some norm in the vector bundle $X^{(r)} \to V$ (associated to $X = V \times \mathbb{R}^q \to V$) and $d\mu$ is some smooth positive measure on V.

(B_1') **Theorem.** *Let $q - n > p(i-1)$ and let V admit an immersion $V \to \mathbb{R}^q$. Then, for every C^∞-map $f_0: V \to \mathbb{R}^q$ and for each $\varepsilon > 0$ there exists a C^∞-immersion $f: V \to \mathbb{R}^q$ for which* $\operatorname{dist}_{i,p}(f, f_0) \leq \varepsilon$.

See Gromov-Eliashberg (1970, 1970_1) and 2.2.1 for the proof.

Exercise. (Gromov-Eliashberg 1970). Find, for given numbers $n \geq 2$, $q > n$, $i \geq 2$ and for all $p \geq (q-n)/(i-1)$ a C^∞-immersion $f_0: \mathbb{R}^n \to \mathbb{R}^q$ which admits no $W^{i,p}$-approximation by C^∞-immersions $\mathbb{R}^n \to \mathbb{R}^q$.

(B_2) **Theorem.** *If* $\dim W > \dim V$, *then the h-principle for isometric C^1-immersions $(V, g) \to (W, h)$ is C^0-dense near every strictly short continuous map $f_0: V \to W$.*

Remarks and Corollaries. (a) One could start with an approximation of f_0 by a strictly short *immersion* [see (B_2)] and then deform to an *isometric immersion* by applying the geometric method of Nash and Kuiper. But the intermediate approximation is suppressed by the technique of 2.4.9, where the theorem is proved.

(b) *If V and W are parallelizable (for example contractible) Riemannian manifolds, then every strictly short map $V \to W$ can be C^0-approximated by isometric C^1-immersions $V \to W$, provided* $\dim W > \dim V$.

Proof. The parallelizability makes the fibration $\mathscr{I}_\partial \to X = V \times W$ split. Then every section $V \to X$ lifts to $\mathscr{I}_\partial$ and the dense h-principle applies.

Exercises. (i) Let V admit an immersion into $\mathbb{R}^q$ for $q > \dim V$. Construct an isometric C^1-immersion $f: (V, g) \to \mathbb{R}^q$ such that the diameter of the image abides $\operatorname{Diam} f(V) \geq \operatorname{Diam}_g V - \varepsilon$ for a given $\varepsilon > 0$.

(ii) Let $f_0: \mathbb{R}^n \to \mathbb{R}^q$ be a *linear* (non-strictly) short map which is isometric on a straight line $\mathbb{R}^1 \subset \mathbb{R}^n$. Show that f_0 admits no *fine* C^0-approximation by isometric C^1-immersions unless f_0 is isometric to start with.

(iii) Let a *compact* manifold V admit some immersion into $\mathbb{R}^q$ for $q > \dim V$. Show that a continuous map $f_0: V \to \mathbb{R}^q$ admits a C^0-approximation by isometric C^1-immersions $V \to \mathbb{R}^q$ if and only if it is short.

(C) *The C^i-Dense h-Principle for $i \geq 1$.* Let $i \leq r$, and consider a neighborhood $U \subset X^{(i)}$ of the image $J^i_{f_0}(V) \subset X^{(i)}$ for a given C^i-section $f_0: V \to X$. Set $\mathscr{R}_U = (p^r_i)^{-1} \cap \mathscr{R} \subset X^{(r)}$ for the projection $p^r_i: X^{(r)} \to X^{(i)}$. We say $\mathscr{R}$ satisfies the *h-principle C^i-near* f_0 if for every neighborhood $U \subset X^{(i)}$ of $J^i_{f_0}(V)$ and for every section $\varphi_0: V \to \mathscr{R}$ which lies over the jet $J^i_{f_0}$ (i.e. $p^r_i \circ \varphi_0 = J^i_{f_0}$) there exists a homotopy of sections $\varphi_t: V \to \mathscr{R}_U$, $t \in [0, 1]$, for which φ_1 is holonomic. The h-principle is called *C^i-dense in a subspace* of $C^i(X)$ if it holds true C^i-near every section in this subspace.

Warning. The C^i-density of some subset of sections in $C^r(X)$ is, of course, a stronger property than the C^{i-1}-density. Yet the C^i-dense h-principle does not necessarily yield the C^{i-1}-dense one, because the former applies only to those sections φ_0: $V \to X^{(r)}$ whose projections to $X^{(i)}$ are holonomic. In fact, a relation $\mathscr{R}$ may satisfy the everywhere C^i-dense h-principle for some $i \geq 1$ and violate the ordinary h-principle at the same time. A (trivial) example is provided by $\mathscr{R} = (p_i^r)^{-1}(\mathscr{R}')$ for any $\mathscr{R}' \subset X^{(i)}$ which violates the h-principle.

Examples. (C_1) *Free Maps* f: $V \to (W, h)$. Define *the first (covariant) derivatives* of f in local coordinates u_i on V, $i = 1, \dots, n$, by $\nabla_i f = D_f(\partial/\partial u_i)$. These are vector fields in W *along* (the coordinate chart in) V, as $\nabla_i f(v) \in T_w(W)$ for $w = f(v)$. Next we invoke the *actual* covariant derivative ∇ in the *Riemannian* manifold (W, h), which applies to tangent fields in W, and put $\nabla_{ij} f = \nabla_{u_i} \nabla_j f$. The subspace $\operatorname{Span}(\nabla_i f(v), \nabla_{ij} f(v)) \subset T_w(W)$, $1 \leq i, j \leq n$, does not depend on the choice of coordinates around $v \in V$ and it is called the (second) *osculating* space $T_f^2(V, v) \subset T_w(W)$ (compare 1.1.4).

Exercise. Let (W, h) be isometrically realized by a C^2-submanifold $W \subset \mathbb{R}^N$. Show that $\nabla_i f(v) = \dfrac{\partial f}{\partial u_i}(v) \in T_w(W) \subset T_w(\mathbb{R}^N)$ and that $\nabla_{ij} f(v) = P\left(\dfrac{\partial^2 f}{\partial u_i \, \partial u_j}(v)\right) \in T_w(W) \subset T_w(\mathbb{R}^N)$ for the orthogonal projection P: $T_w(\mathbb{R}^N) \to T_w(W)$.

Call a C^2-map f: $V \to (W, h)$ *free* if $\dim T_f^2(V, v) = \frac{1}{2}n(n+3)$ for all $v \in V$.

Theorem. *If* $\dim W \geq \frac{1}{2}n(n+3) + 1$, *then free maps* $V \to (W, h)$ *satisfy the everywhere* C^1*-dense h-principle as well as the everywhere* C^0*-dense h-principle* (see 2.4.3).

Remark. The C^0-dense h-principle is not very interesting here since every C^1-map f_0: $V \to W$ for $\dim W \geq 2n$ admits a fine C^1-approximation by an immersion $V \to W$ according to a (quite simple) theorem by Whitney (see 1.3.2).

Corollary. *If a* C^1*-map* f_0: $V \to W$ *is homotopic to a free map, then it also admits a fine* C^1*-approximation by free maps* $V \to W$, *provided* $\dim W \geq \frac{1}{2}n(n+3) + 1$. (This is unknown for $\dim W = \frac{1}{2}n(n+3)$ and $n \geq 2$.)

Proof. We may assume, with the above remark, the map f_0 is an immersion. Since $\dim W \geq \frac{1}{2}n(n+3) + 1 \geq 2n + 1$, the homotopy class of the jet $J_{f_0}^1$: $V \to \mathscr{I}$ into the immersion relation $\mathscr{I} \subset X^{(1)} \to X \to V$ is determined by the homotopy class of f_0 alone, for the fiber $\mathscr{I}_x \approx \operatorname{St}_n \mathbb{R}^q$, $x = (v, w) \in X = V \times W$, $q = \dim W$, is n-connected. This insures a lift of $J_{f_0}^1$ to a section φ_0: $V \to \mathscr{F}$ for the freedom relation $\mathscr{F} \subset X^{(2)}$ and the C^1-dense h-principle applies.

(C_2) *Free Isometric Immersions* $(V, g) \to (W, h)$. Let the manifold (W, h) be C^∞-smooth (or C^{an} if C^{an}-immersions are under consideration) and let (V, g) be C^k-smooth.

Theorem. *If* $\dim W \geq \frac{1}{2}(n+2)(n+3)$ *and if* $k \geq 5$ *then the h-principle for free isometric* C^k*-immerions* $V \to W$ *is* C^0*-dense in the space of those continuous maps* $V \to W$ *which are fine* C^0*-limits of* (*i.e. can be finely* C^0*-approximated by*) *strictly short* C^1*-maps* $V \to W$. *Furthermore, this h-principle is* C^1*-dense in the space of those* *isometric C^1-immersions which are fine C^1-limits of strictly short immersions.* (See 3.1.7.)

Exercises. (a) Derive from this h-principle the following

(C_2') **Approximation Theorem.** *A C^0-smooth (C^1-smooth) map $f_0: V \to W$ admits under the above assumptions on k and on* dim W *a fine C^0-approximation (respectively, C^1-approximation) by free isometric C^k-immersions $V \to W$ if and only if f_0 is a uniform C^0-limit of strictly short C^1-maps (f_0 is an isometric C^1-immersion which is a fine C^1-limit of strictly short immersions).*

(b) Let (W, g) have a non-positive sectional curvature and let V be compact. Show a short C^1-map f_0: $V \to W$ to be a C^1-limit of strictly short maps if and only if f_0 is a C^0-limit of such maps. (Relations between different strictly short approximations are unknown in the complete generality.)

(C_2'') *C^i-Approximation for $i \geq 2$.* Let f_0: $V \to W$ be a *free isometric* immersion in some *Hölder class $C^{j,\alpha}$* (see 2.3.4; recall that $C^{j,0} = C^j$ and that $C^j \supset C^{j,\alpha} \supset C^{j+1}$ for $0 > \alpha \geq 1$).

If $j + \alpha > 3$ and if $k \geq 5$, then f_0 admits a fine C^2-approximation by free isometric C^k-immersions $V \to W$ (without any restriction on dim W).

This is shown by a purely analytic method [extending the implicit function theorem of Nash (1956)] in 2.2 where we also study C^i-approximation for $i > 2$.

Corollary. *If a C^∞-smooth Riemannian manifold (V, g) admits a free isometric C^4-immersion (or, even, $C^{3,\alpha}$-immersion for $\alpha > 0$) $f_0(V, g) \to \mathbb{R}^q$ then it also admits an isometric C^∞-immersion f: $V \to \mathbb{R}^q$ which can be chosen arbitrarily C^2-close to f_0. (One does not know whether the existence of a free isometric C^3-immersion insures an isometric C^∞-immersion into the same space $\mathbb{R}^q$. Nor does one know whether the freedom assumption is essential for C^4-immersions.)*

1.2.3 Functionally Closed Relations

A relation $\mathcal{R} \subset X^{(r)}$ is called (*functionally*) *C^i-closed* for some $i \leq r$ if every C^i-limit f of C^r-solutions $V \to X$ of $\mathcal{R}$ also satisfies $\mathcal{R}$ in so far as this limit f: $V \to X$ is C^r-smooth, where the limits are understood for the ordinary (non-fine) C^i-topology.

One usually establishes this property by reducing (or "integrating") $\mathcal{R}$ to an equivalent possibly (non-differential) relation which involves no jet of order $\geq i$. For example, the shortness relation, $\langle D_f(\tau), D_f(\tau)\rangle_h \leq \langle \tau, \tau\rangle_g$, $\tau \in T(V)$, for maps f: $(V, g) \to (W, h)$, is equivalent to the non-differential relation $\operatorname{dist}_h(f(v_1), f(v_2)) \leq \operatorname{dist}_g(v_1, v_2)$, $v_1, v_2 \in V$, which shows the shortness relation to be C^0-closed. Another example is provided by the Cauchy-Riemann relation which can be expressed with the Cauchy integral formula by a non-differential relation, showing C^0-limits of holomorphic maps to be holomorphic. On the other hand, the isometric immersion relation is quite far from being C^0-closed as the C^0-dense h-principle allows all strictly short maps for C^0-limits of isometric maps.

(A) *Infinitesimally Enlarging Maps.* A map $f: V \to W$ between complete Riemannian manifolds is called *enlarging* if the image under f of a Riemannian ball $B(v, \rho) \subset V$ with center v and radius ρ, for any v and ρ, contains the ball $B(f(v), \rho)$ with the same radius ρ in W around $f(v)$. A C^1-map f is called *infinitesimally enlarging* if the differential $D_f: T_v(V) \to T_w(W)$, $w = f(v)$ is enlarging on every tangent space $T_v(V)$.

If f is infinitesimally enlarging, then the differential D_f is a surjective map $T_v(V) \to T_w(W)$ for all $v \in V$ and $w = f(v) \in W$ whose kernel $\operatorname{Ker} D_f \subset T(V)$ is a $(q-n)$-dimensional subbundle in $T(V)$ for $q = \dim W$ and $n = \dim V$. Furthermore, if a smooth curve $\tilde{C} \subset V$ is *horizontal* (that is everywhere normal to $\operatorname{Ker} D_f$), then the length of the image satisfies length $f(\tilde{C}) \geq$ length $\tilde{C}$. It follows that for every curve $C \subset W$ and for each point $v \in V$ over C, there exists a (unique if f is C^2-smooth) horizontal curve $\tilde{C} \subset V$ over C which passes through v. By applying this to geodesic segments C of length ρ issuing from $w = f(v)$, one shows the ball $B(w, \rho) \subset W$ to be completely covered by the image $f(B(v, \rho)) \subset W$. Therefore,

$$\textit{infinitesimally enlarging} \Rightarrow \textit{enlarging}.$$

With this one immediately sees the "*infinitesimally enlarging*" relation to be C^0-closed.

(B) **Exercises.** (B_1) Fill in the detail in the above argument and remove the completeness assumption in the final "C^0-closed" conclusion.
(B'_1) Let $f: V \to W$ be an infinitesimally enlarging map of a *complete* manifold V into a *connected* manifold W. Prove that, in fact, W is *complete*, and that f is a *locally trivial fibration* of V *onto* W.
(B_2) Consider a C^1-function $f: \mathbb{R}^n \to \mathbb{R}$ and denote by $\mathrm{Gr}_f: \mathbb{R}^n \to \mathbb{R}^n$ the map given by the functions $\partial f / \partial u_i$ on $\mathbb{R}^n$, $i = 1, \ldots, n$. Show the differential relation $\mathrm{Gr}_f(\mathbb{R}^n) \subset A$ to be C^0-closed for every closed subset $A \subset \mathbb{R}^n$. [*Hint*: Reduce to the special case, $A = \{u \in \mathbb{R}^n \mid \|u\| \geq 1\}$, where solutions are infinitesimally enlarging maps $\mathbb{R}^n \to \mathbb{R}$.]
(B'_2) Consider the trivial line bundle $X = V \times \mathbb{R} \to V$ and show an arbitrary *closed* subset $\mathscr{R} \subset X^{(1)}$ to be a (*functionally*!) C^0-*closed* relation.
(B_3) Consider C^1-maps $V \to W$ which do not increase the k-dimensional Riemannian volume of k-dimensional submanifolds in V. Show the corresponding relation is C^0-closed for all $k = 1, 2, \ldots$.
(B_4) Study the C^0-closure of C^1-maps $f: \mathbb{R}^n \to \mathbb{R}^n$ whose Jacobians $J = \det(\partial f_i / \partial u_j)$ satisfy $J \geq 1$. Show every C^1-immersion in this closure has $J \geq 1$. Then prove the relation $\det(\partial f_i / \partial u_j) \geq 0$ to be C^0-closed.
(B'_4) Show the *quasi-conformality relation* $\|D_f(v)\|^n \leq K \operatorname{Det}\left(\dfrac{\partial f_i(v)}{\partial u_j}\right)$, $v \in \mathbb{R}^n$, is C^0-close for all $K \geq 1$.
(B_5) Show the relation $\mathscr{R}_k$ imposed on C^2-functions $f: \mathbb{R}^n \to \mathbb{R}$ by requiring the Hessian matrix $\left(\dfrac{\partial^2 f(v)}{\partial u_i \, \partial u_j}\right)$ to have at most k positive eigenvalues at all points $v \in \mathbb{R}^n$ to be C^0-closed for all $k = 0, 1, \ldots, n$.

(B_6) Show, for every real α, the following three relations imposed on C^2-smooth Riemannian metrics g on V to be C^0-closed.

(i) $K(g) \geq \alpha$; (ii) $K(g) \leq \alpha$; (iii) $\mathrm{Ricci}(\tau, \tau) \geq \alpha\langle\tau, \tau\rangle_g$ for all $\tau \in T(V)$, where K denotes the sectional curvature and Ricci stands for the Ricci tensor. (One does not know if $\mathrm{Ricci} \leq \alpha\langle\tau, \tau\rangle_g$ is C^0-closed.)

(C) *Convex Relations.* Let $X \to V$ be a vector bundle and let $X^{(r)} \to V$ have the associated vector bundle structure. A relation $\mathscr{R} \subset X^{(r)}$ is called C^i-*convex* if for each point in the complement, $x_0 \in X^{(r)} \setminus \mathscr{R}$, there exists a fiberwise linear C^i-function $\Delta\colon X^{(r)} \to \mathbb{R}$ such that $\Delta(x) \leq \Delta(x_0)$ for all points $x \in \mathscr{R}$ which project to a sufficiently small neighborhood $U \subset V$ of the point $p^r(x_0) \in V$. For example, a *compact* subset $\mathscr{R} \subset X^{(r)}$ is C^∞-convex if and only if $\mathscr{R}_v$ is a convex subset in $X_V^{(r)}$ for all $v \in V$. Furthermore, every C^i-smooth subbundle in $X^{(r)}$ is C^i-convex.

Proposition. *If $\mathscr{R}$ is C^i-convex for some $i \leq r$ then it is functionally C^{r-i}-closed.*

Proof. Let $\mathscr{D}(f) = \Delta(J_f^r)$ for $f\colon V \to X$ and express the relation $\mathscr{D}(f) \leq C_0 = \Delta(x_0)$ over $U \subset V$ by infinitely many inequalities

$$(*) \qquad \int_U \mathscr{D}(f)\varphi\, du \leq C_0 \int_U \varphi\, du,$$

for all C^∞-smooth non-negative functions φ on U with compact supports, where $du = du_1\, du_2, \ldots, du_n$ for some local coordinates in U. Since the coefficients of the differential operator $\mathscr{D}$ are C^i-smooth, one can integrate by parts i times and thus reduce $(*)$ to

$$(**) \qquad \int_U \mathscr{D}'(f)\mathscr{D}''(\varphi)\, du \leq C_0 \int_U \varphi\, du$$

for some operators $\mathscr{D}'(f) = \Delta'(J_f^{r-i})$ and $\mathscr{D}''(\varphi) = \Delta''(J_\varphi^i)$. Since $(**)$ depends only on J_f^{r-i} it is C^{r-i}-closed. Q.E.D.

(C_1) **Example.** Consider the equation $\dfrac{df}{dt} - a\dfrac{dg}{dt} = 0$ in the functions f and g on $\mathbb{R}$, assume $a = a(t)$ is C^1-smooth and integrate by parts. The resulting equation

$$f(t) - a(t)g(t) + \int_0^t \frac{da(\tau)}{dt} g(\tau)\, d\tau = \text{const},$$

is obviously functionally C^0-closed.

(C_1') **Counterexample.** Let $a(t)$ be a continuous *nowhere differentiable* function in $t \in [0, 1]$. That is, for every $\varepsilon > 0$, there are disjoint intervals $I_i = [t_i, t_i'] \subset [0, 1]$ for $i = 1, \ldots, N = N(\varepsilon)$, such that $\sum_{i=1}^N t_i' - t_i \geq 1 - \varepsilon$ and such that $\mathrm{Osc}(a|I_i) > \varepsilon^{-1}(t_i' - t_i)$, for all $i = 1, \ldots, N$, which means $|a(x) - a(y)| > \varepsilon^{-1}(t_i' - t_i)$ for some points x and y in I_i. Then there obviously exist smooth non-negative functions φ and φ' on $[0, 1]$ whose supports lie in $\bigcup_i I_i$, such that

$$\int_{t_i}^{t'_i} \varphi(t)\,dt = \int_{t_i}^{t'_i} \varphi'(t)\,dt = 1$$

and

$$\int_{t_i}^{t'_i} a(t)(\varphi'(t) - \varphi(t))\,dt = \varepsilon^{-1}(t'_i - t_i)$$

for all $i = 1, \ldots, N$.

Next, for a given C^1-function $f_0(t)$, $t \in [0, 1]$, define

$$g(t) = \begin{cases} \dfrac{\varepsilon(f_0(t'_i) - f_0(t_i))}{t'_i - t_i} \displaystyle\int_{t_i}^{t} (\varphi'(\tau) - \varphi(\tau))\,d\tau, & \text{for } t \in I_i,\ i = 1, \ldots, N, \\ 0, \quad \text{for } t \in I \setminus \bigcup_i I_i, & \end{cases}$$

and

$$f(t) = f_0(0) + \int_0^t a(\tau)\frac{dg(\tau)}{d\tau}\,d\tau.$$

The functions f and g clearly are C^1-smooth and

$$\frac{df}{dt} - a\frac{dg}{dt} = 0. \qquad (+)$$

Furthermore, $\| f - f_0 \|_{C^0} \to 0$ and $\| g \|_{C^0} \to 0$ for $\varepsilon \to 0$ as a straightforward computation reveals. Therefore, *the relation* $(+)$ *is not functionally* C^0*-closed.*

Exercises. (a) Show the C^1-solutions of $(+)$ to be C^0-dense in the space of pairs of continuous functions $f_0(t)$ and $g_0(t)$, for the above nowhere differentiable $a(t)$, and express this as a density of the *graph* of the operator $g(t) \mapsto \int_0^t a(\tau)\frac{dg(\tau)}{d\tau}\,d\tau$.

(b) Find further examples of integro-differential operators on functions on $\mathbb{R}^n$ whose graphs are C^0-dense.

(c) Identify triples of C^1-functions f, g and a on $\mathbb{R}$ which satisfy $(+)$ with *Legendre* maps (compare 3.4.3) $\mathbb{R} \to \mathbb{R}^3$, which are by definition, everywhere tangent to the plane field $\operatorname{Ker} \eta \subset T(\mathbb{R}^3)$ for the 1-form $\eta = dz - x\,dy$ on $\mathbb{R}^3$. Prove Legendre C^1*-immersions* $\mathbb{R} \to \mathbb{R}^3$ to be C^0-dense in the space of continuous map $\mathbb{R} \to \mathbb{R}^3$.

(d) Consider a one-dimensional subbundle $l \subset T(\mathbb{R}^q)$ and study C^1-maps f: $\mathbb{R} \to \mathbb{R}^q$ which are everywhere tangent to l [i.e. $D_f T(\mathbb{R}) \subset l$]. Prove the pertinent differential relation to be functionally C^0-closed for all C^1*-smooth* subbundles. Show for "sufficiently non-differentiable" C^0-subbundles the maps $\mathbb{R} \to \mathbb{R}^q$, $q \geq 2$, tangent to l to be C^0-dense in the space of continuous maps $\mathbb{R} \to \mathbb{R}^q$. Study the C^0-closure of C^1*-immersions* $\mathbb{R} \to \mathbb{R}^q$ tangent to l.

(e) Consider C^2-maps f: $\mathbb{R}^n \to \mathbb{R}^q$, such that the *mean curvature* M of the graph $\Gamma_f \subset \mathbb{R}^{n+q}$ (which is a vector field M: $\Gamma_f \to T(\mathbb{R}^{n+q})|\Gamma_f$ normal to Γ_f) satisfies $\| M(\gamma) \| \leq \varepsilon$, for all $\gamma \in \Gamma_f$. Show this condition to be functionally C^0-closed for all $\varepsilon \geq 0$. Generalize this for maps between non-flat Riemannian manifolds f: $V \to W$.

1.3 Singularities and Non-singular Maps

1.3.1 Singularities as Differential Relations

Let $M = M(n, q)$ denote the space of linear maps $\mathbb{R}^n \to \mathbb{R}^q$ and let $\Sigma^i \subset M$, $i = 0, 1, \ldots, m = \min(n, q)$ consist of the linear maps of rank $m - i$. These are the orbits of the natural action of the group $GL_n \times GL_q$ on M; hence, they are C^{an}-smooth (in fact real algebraic) submanifolds in M, and a straightforward computation shows $\operatorname{codim} \Sigma^i = (n - m + i)(q - m + i)$.

Next, we turn to the jet space $X^{(1)}$ for $X = V \times W \to V$ and we let $\Sigma^i \subset X^{(1)}$ consist of the linear maps $T_v(V) \to T_w(W)$ of rank $m - i$ for $m = \min(n, q)$ where $n = \dim V$ and $q = \dim W$. These Σ^i clearly are the orbits of the natural action of the group Diff $V \times$ Diff W on $X^{(1)}$ and $\operatorname{codim} \Sigma^i = (n - m + i)(q - m + i)$. The partition $X^{(1)} = \bigcup_{i=0}^{m} \Sigma^i$ is called *the stratification of Whitney-Thom* of the space $X^{(1)}$. Observe that the topological boundary $\partial \Sigma^i = \operatorname{Cl} \Sigma^i \backslash \Sigma^i$ equals the union $\bigcup_{j>i} \Sigma^j$.

Consider a C^1-map $f: V \to W$ and denote by $\Sigma_f^i \subset V$ the pull-back $(J_f^1)^{-1}(\Sigma^i)$ of the jet $J_f^1: V \to X^{(1)}$. The equivalent definition is $\Sigma_f^i = \{v \in V | \operatorname{rank}_v f = m - i\}$. If the map f is C^k-smooth for $k \geq 2$ and if the jet $J_f^1: V \to X^{(1)}$ is *transversal* to Σ^i, then Σ_f^i is a C^{k-1}-submanifold of codimension $(n - m + i)(q - m + i)$ in V by the implicit function theorem. Recall that a C^1-map between smooth manifolds, say $\varphi: A \to B$, is *transversal* to a submanifold $C \subset B$ if

$$\operatorname{codim}[D_f^{-1}(T_c(C)) \subset T_a(A)] = \operatorname{codim}[T_c(C) \subset T_c(B)]$$

for all $c \in C$ and for all $a \in A$, such that $f(a) = c$.

The subset $\Sigma_f^i \subset V$ for $i > 0$ is often called the Σ^i-*singularity* of f. If $\Sigma_f^i = \varnothing$ for all $i > 0$, then the map f is called *regular*. If $\dim W \geq \dim V$, then the regular maps $f: V \to W$ are immersions defined by $\operatorname{rank} f \equiv n = \dim V$ (see 1.1.3). If $\dim W \leq \dim V$, then regular maps satisfy $\operatorname{rank} f \equiv q = \dim W$ and are called *submersions*. The following result by Phillips (1967) generalizes the h-principle for equidimensional immersions.

(A) **Theorem.** *If V is an open manifold, then submersions $V \to W$ satisfy the parametric h-principle.*

Corollary. *An open manifold V admits a submersion $V \to \mathbb{R}^q$ if and only if there are q linearly independent vector fields on V.*

Proof. The independent fields define a fiberwise *surjective* homomorphism $T(V) \to T(\mathbb{R}^q)$ which can be homotoped, according to (A), to the differential of a submersion $V \to \mathbb{R}^q$.

(A_1) A C^1-map $f: V \to W$ is called a *k-mersion* if $\operatorname{rank}_v f \geq k$ for all $v \in V$. This can be expressed with the differential relation $\mathscr{I}_k = X^{(1)}$ which is the union of

Σ^i for $i \leq m - k$. Clearly, every *open* subset in $X^{(1)}$ which is invariant under Diff $V \times$ Diff W equals the k-mersion relation for some $k = 0, 1, \ldots$.

(A'_1) **Theorem** (Feit 1969). *If $k < \dim W$, then k-mersions $V \to W$ satisfy the parametric h-principle; moreover, the h-principle is C^0-dense.*

This generalizes Hirsch's immersion theorem in the extra dimension case. We prove (A'_1) and (A) along with Hirsch's theorem in Sects. 2.1.2, 2.2.3, and in 2.4.3.

Remark. Regular maps $f: V \to W$ can be (obviously) characterized geometrically by the existence of local coordinates $u_1, \ldots, u_n$ in V around every point $v \in V$ and of some coordinates $u'_1, \ldots, u'_q$ in W around $W = f(v)$, such that $f: (u_1, \ldots, u_n) \mapsto (u_1, \ldots, u_m, 0, \ldots, 0)$ in these coordinates. This is equivalent to the transitivity of the action of the group Diff $V \times$ Diff W on the germs of regular maps $V \to W$. But this action is far from transitive on germs of k-mersions.

Example. Here are five 1-mersions $\mathbb{R}^2 \to \mathbb{R}^2$ with completely different local behaviour.
(1) (*Immersion*) $f: (u_1, u_2) \mapsto (u_1, u_2)$.
(2) (*Folding*) $f: (u_1, u_2) \mapsto (u_1^2, u_2)$. This map f *folds* along the line $u_1 = 0$ in $\mathbb{R}^2$; it is an immersion on this line as well as on the complement to this line.
(3) (*Whitney's cusp*) $f: (u_1, u_2) \mapsto (u_1^3 + u_1 u_2, u_2)$. The singularity Σ_f^1 is the parabola $3u_1^2 + u_2 = 0$, where the map $f|\Sigma_f^1: \Sigma_f^1 \to \mathbb{R}^2$ is not regular at the point $u_1 = 0$, $u_2 = 0$. The singularity $\Sigma_g^1 \subset \Sigma_f^1$ of the restricted map $g = f|\Sigma_f^1$ is called the *cusp* Σ_f^{11} of the map f.
(4) (*Blow-down*) $f: (u_1, u_2) \mapsto (u_1 u_2, u_2)$. This map is regular outside the line $u_2 = 0$ and it collapses this line to $(0, 0) \in \mathbb{R}^2$.
(5) (*Totally degenerate map*) $f: (u_1, u_2) \mapsto (u_1, 0)$. Here $\Sigma_f^1 = \mathbb{R}^2$.

Observe the maps (1), (2) and (3) to be stable under small C^∞-perturbations of maps, while the geometry of (4) and (5) can be completely destroyed by small perturbations.

(B) *The h-Principle for Folded Maps.* A C^2-map $f: V \to W$ is said to *fold* along a submanifold $V_0 \subset V$ if f is regular on $V \setminus V_0$, the map $f|V_0: V_0 \to W$ is an immersion, the jet: $J_f^1: V \to X^{(1)}$ is transversal to $\Sigma^1 \subset X^{(1)}$ and $\Sigma_f^1 = V_0$. This description refers to the second jets of f and, hence, it is expressed by certain open relation $\mathcal{F}l \subset X^{(2)}$ which is invariant under Diff W as well as under the diffeomorphisms of the pair (V, V_0). Furthermore, the implicit function theorem allows, for $\dim V = \dim W = n$, local coordinates $u_1, \ldots, u_n$ near each point $v \in V_0$ and some coordinates in W near $f(v) \in W$ such that the map f becomes $f: (u_1, u_2, \ldots, u_n) \mapsto (u_1^2, u_2, \ldots, u_n)$ for V_0 given by $u_1 = 0$.

(B_1) **The Folding Theorem** (Eliashberg 1970). *If* $\dim W = \dim V \geq 2$ *and if each connected component of V contains a component of V_0 then C^2-maps $f: V \to W$ folded along V_0 (and only along V_0!) satisfy the everywhere C^0-dense h-principle.* (See 2.1.3.)

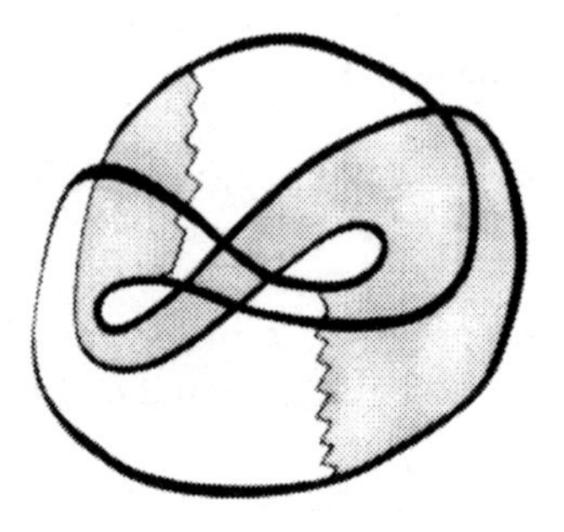
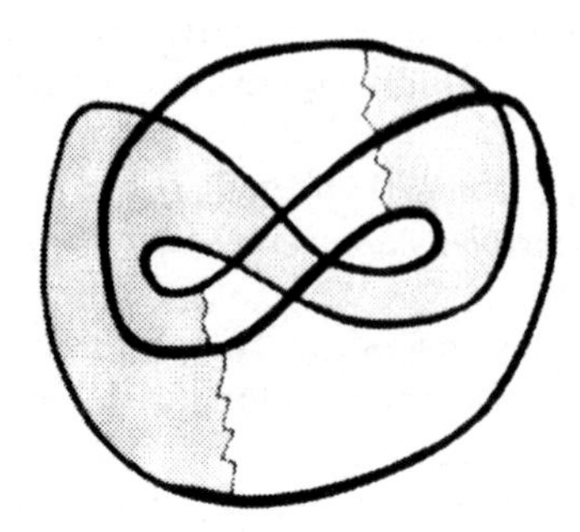

Fig. 1

Remarks and Corollaries. (a) The parametric h-principle may fail to be true for folded maps. For instance, there exists an "exotic" map $f: S^2 \to \mathbb{R}^2$ folded along the equator $S^1 \subset S^2$ which admits no folded C^2-homotopy to the standard folded map f_0: $S^2 \to \mathbb{R}^2$ obtained by the linear projection $S^2 \subset \mathbb{R}^3 \to \mathbb{R}^2$. This is seen in the following amazing picture (Fig. 1) discovered by Milnor and featuring two different (!) immersed discs $\mathscr{D}^2 \to \mathbb{R}^2$ which are bounded by the same immersed circle.

In fact, the folding theorem predicts a great variety of such examples as the dense h-principle yields a map $S^2 \to \mathbb{R}^2$ folded along S^1 and C^0-approximating a given continuous map $S^2 \to \mathbb{R}^2$. Moreover, any two *exotic* folded maps $S^2 \to \mathbb{R}^2$ can be joined by a C^2-homotopy of folded maps (Eliashberg 1972).

(b) The parallelizability of the sphere S^3 implies (an exercise left to the reader) the existence of a section $S^3 \to \mathscr{F}\ell = \mathscr{F}\ell_{V_0}$ for an arbitrary closed surface $V_0 \subset S^3$. It follows, for example, that *there exists a C^2-map $f: S^3 \to \mathbb{R}^3$ which folds along a pair of concentric 2-spheres in S^3 (and has no singularity anywhere else!).*

(c) Consider two immersions of a closed oriented n-dimensional manifold into $\mathbb{R}^{n+1}$, say f_0 and f_1: $V \to \mathbb{R}^{n+1}$. Write $f_0 \prec f_1$ in case there is an *immersion* of the cylinder, $\hat{f}: V \times [0,1] \to \mathbb{R}^{n+1}$ such that $\hat{f}|V \times 0 = f_0$ and $\hat{f}|V \times 1 = f_1$. For example, two round embedded spheres f_0 and f_1: $S^n \to \mathbb{R}^{n+1}$ satisfy $f_0 \prec f_1$, if and only if $f_0(S^n)$ lies in the open ball bounded by $f_1(S^n) \subset \mathbb{R}^{n+1}$. Thus, we obtain a partial order in the set of immersions $f: V \to \mathbb{R}^{n+1}$.

The relative version (see 1.4.2) of Eliashberg's theorem allows cylinders $\hat{f}$: $V \times [0,1] \to \mathbb{R}^{n+1}$ between given *regularly homotopic* immersions f_0 and f_1: $V \to \mathbb{R}^{n+1}$, such that $\hat{f}$ folds along the hypersurface $V \times \frac{1}{2} \subset V \times [0,1]$. Therefore,

there exists an immersion f_2: $V \to \mathbb{R}^{n+1}$ such that $f_2 \succ f_0$ and $f_2 \succ f_1$ for any given pair of regularly homotopic immersions f_0 and f_1: $V \to \mathbb{R}^{n+1}$.

Little is known about this order besides Eliashberg's theorem. A happy exception is an effective condition [due to Blank (1967)] for an immersed circle $S^1 \to \mathbb{R}^2$ to bound an immersed disk (see Poenaru 1968).

(C) *Totally Degenerate Maps.* A map $f: V \to W$ is called *totally degenerate* if $\operatorname{rank}_v f < m = \min(\dim V, \dim W)$ for all $v \in V$. The corresponding relation $\mathscr{R} \subset X^{(1)}$ is the union $\bigcup_{i>0} \Sigma^i$.

Exercise. Show the relation $\bigcup_{i \geq k} \Sigma^i \subset X^{(1)}$ to be functionally C^0-closed for all $k = 0, 1, \ldots$, and find counterexamples to the h-principle for all $k \geq 1$.

(C′) One does not know how to refine the h-principle (in order to make it valid) for maps $f\colon V \to W$ with $k_1 \leq \operatorname{rank}_v f \leq k_2$, $v \in V$, where $k_1 < k_2 < m$. However, such a refinement is indicated in (B) of 2.2.7 for maps of *open* manifolds, $f\colon V \to W$ of *constant* rank $= k < m$. For example

If V is an open n-dimensional manifold, then C^∞-maps $f\colon V \to W$ of $\operatorname{rank} f \equiv n - 1$ *satisfy the h-principle. In particular, there always exists a C^∞-map $f\colon V \to V$ homotopic to the identity, such that* $\operatorname{rank} f \equiv n - 1$.

(D) *Invariant Relations $\mathscr{R} \subset X^{(1)}$.* Every subset $\mathscr{R} \subset X^{(1)}$ invariant under the action of $\operatorname{Diff} V \times \operatorname{Diff} W$ is the union of some subsets $\Sigma^i \subset X^{(1)}$. No h-principle is known for such an $\mathscr{R}$ except for the above-mentioned results.

Exercises. (a) Study C^∞-maps $f\colon S^n \to S^n$ which collapse the equator $S^{n-1} \subset S^n$ to a single point and are regular outside the equator. Show that the equality $\Sigma_f^{n-1} = S^{n-1}$ for such a map f implies $\deg f = \pm(1 + (-1)^{n-1})$. Then provide examples of C^{an}-maps with $\Sigma_f^{n-1} = S^{n-1}$. Finally, construct C^{an}-maps f of degrees $d = 0$ and $d = 2$, such that $\Sigma_f^n = S^{n-1}$.

(b) Find a C^{an}-map f of the projective space P^n onto S^n whose only singularity is the subspace $P^{n-1} = \Sigma_f^{n-1}$.

(c) Let V be a closed oriented manifold. Construct a C^∞-map $f\colon V \to S^n$, $n = \dim V$, of a given degree $\deg_f = d$, and such that Σ_f^i is empty for $0 < i < n$.

(d) Let V and W be connected manifolds without boundary, such that $\dim W \leq \dim V$, and let $f\colon V \to W$ be a C^∞-map for which Σ_f^1 is empty (e.g. a holomorphic map between complex manifolds). Assume f to be a *proper* map (e.g. V is compact) and show that either f is *onto* or is totally degenerate. In particular, if V is compact and the subgroup $f_*(\pi_1(V)) \subset \pi_1(W)$ has infinite index, then $\Sigma_f^1 = \varnothing$ implies the total degeneracy of f.

(e) Let V and W be closed connected orientable n-dimensional manifolds and let $f\colon V \to W$ be a C^∞-map which is not totally degenerate. Assume Σ_f^i to be empty for $i = 1, \ldots, k$, and show the homomorphism $f_*\colon H_i(V; \mathbb{R}) \to H_i(W; \mathbb{R})$ to be surjective for $i = 1, \ldots, k$. Moreover, assume $\pi_1(W) = 0$ and prove $f_*\colon H_i(V; \mathbb{Z}) \to H_i(W; \mathbb{Z})$ to be surjective for $i = 1, \ldots, k$.

Question. Under what assumptions does there exist a C^∞-map f in a given homotopy class of maps $V \to W$, such that $\Sigma_f^{i_j}$ is empty for a given subset of indices $(i_1, i_2, \ldots, i_j, \ldots, i_k)$? For example, when can one find f (looking like a holomorphic map) for which Σ_f^i is empty for all odd i?

(E) *Ramified Maps.* A map between equidimensional manifolds, $f\colon V \to W$, is said to *ramify* along a codimension 2 submanifold $V_0 \subset V$ if in suitable local coordinates $u_1, \ldots, u_n$ near each point $v_0 \in V_0$ and for some coordinates in W near $f(v_0)$ the map f becomes

$$f\colon (u_1, u_2, u_3, \ldots, u_n) \mapsto (\operatorname{Re} z^s, \operatorname{Im} z^s, u_3, \ldots, u_n),$$

for $z = u_1 + \sqrt{-1}u_2$ and for some integer $s \geq 1$. (This s may be different for different components of V_0.)

Example. Take a codimension 2 submanifold $V_0' \subset W$ and let $\tilde{W}'$ be a finite covering of the complement $W \setminus V_0'$. Then the metric completion V of $\tilde{W}'$ (for the metric induced from some Riemannian metric in W) admits a smooth structure, such that the covering map $\tilde{W}' \to W \setminus V_0'$ extends to a C^∞-map $V \to W$ which ramifies along $V \setminus \tilde{W}' \subset V$, and has no singularity on $\tilde{W}'$.

Question. Does every closed n-dimensional parallelizable manifold V admit a C^∞-map $f\colon V \to S^n$ which ramifies along some $V_0 \subset V$ and has no singularity outside V_0? The positive answer is known for $n = 3$ [Alexander (1920), see Rolfsen (1976)].

(F) *Invariant Relations in $X^{(r)}$ for $r \geq 2$.* The action of the group Diff $V \times$ Diff W on $X^{(r)}$ for $X = V \times W \to V$ may have infinitely many orbits for $r \geq 2$. In fact, the number of orbits is infinite for $r \geq 2$ and $\dim V \geq 2$, unless $r = 2$ and $\dim W = 1$. (This is an easy exercise for the reader.) Any union of such orbits gives an invariant relation $\mathscr{R} \subset X^{(r)}$ which may be tested for the h-principle. According to Levin (1965) and Poenaru (1966) maps $f\colon V \to W$ *without* cusps Σ^{11} (i.e. $\Sigma_f^i = \varnothing$ for $i \geq 2$ and the map $f | \Sigma_f^1\colon \Sigma_f^1 \to W$ is regular) satisfy the h-principle.

The h-principle for a general class of open invariant relations $\mathscr{R} \subset X^{(r)}$ is due to Du Plessis (1976). Apparently, all known h-principles for open invariant relations $\mathscr{R} \subset X^{(r)}$ follow from the generalized folding theorem (Eliashberg 1972) which applies to non-equidimensional maps (see 2.1.3).

1.3.2 Genericity, Transversality and Thom's Equisingularity Theorem

Define the C^∞-dimension $\dim S$ of an arbitrary subset S in a smooth manifold Y to be the lower bound of the integers m, such that S is contained in a countable union of C^∞-submanifolds of dimension m in Y.

Examples. (A_1) Take a C^∞-function $f\colon \mathbb{R}^n \to \mathbb{R}$ and consider the set $\Sigma = \Sigma_f^1 \subset \mathbb{R}^n$ where the differential (or the gradient) df of f vanishes. Let $\Sigma_k \subset \Sigma$, $k = 1, 2, \ldots, \infty$, be the subset where the partial derivatives $\partial^I f = \dfrac{\partial^l f}{\partial u_{i_1} \partial u_{i_2} \ldots \partial u_{i_l}}$ vanish for $1 \leq |I| \overset{\text{def}}{=} i_1 + i_2 + \cdots + i_l = l \leq k$ [here I stands for the multi-index $(i_1, \ldots, i_l)$]. Then $S_k = \Sigma_k \setminus \Sigma_{k+1}$ lies in the union of $d = (n + k - 1)!/(n - 1)!k!$ C^∞-hypersurfaces $H_I \subset \mathbb{R}^n$, for $|I| = k$, where each H_I is defined by the equation $\partial^I f(v) = 0$ and by the non-equality $d\partial^I f(v) \neq 0$ for $v \in \mathbb{R}^n$. Hence, $\dim(\Sigma \setminus \Sigma_k) \leq n - 1$ for all $k = 1, 2, \ldots, \infty$.

(A_2) *Stratified Sets.* A stratification of S is a partition of S into finitely many locally closed C^∞-submanifolds (*Strata*) $S^i \subset Y$, $i = 0, \ldots, k$, such that the topological boundary $\partial S^i = (S \cap \mathrm{Cl}\, S^i) \setminus S^i$ lies in the union $\bigcup_{j>i} S^j \subset S$ for all $i = 0, 1, \ldots, k$. Thus, S^0 is an open subset in S, that is $S^0 = S \cap U$ for some open $U \subset Y$, and $S^0 \subset U$ is a C^∞-submanifold which is a closed subset in U. Next, S^1 is open in $S \setminus S^0$, S^2 is open in $S \setminus (S^0 \cup S^1)$ and so on. Clearly, $\dim S \leq \sup_i \dim S^i$. In fact, the C^∞-dimension of S (obviously) equals the topological dimension.

(A'_2) *Semi-algebraic Sets.* A subset $S \subset \mathbb{R}^n$ is called *semi-algebraic* if it is a finite union, $\bigcup_j S(j)$, $j = 1, \ldots, m$, where ech $S(j)$ is defined by a finite system of polynomial equations and (strict or not) inequalities in $\mathbb{R}^n$. Every such S, as well as the image of S under an arbitrary C^∞-diffeomorphism $\mathbb{R}^n \to \mathbb{R}^n$, can be *canonically* stratified as follows. Define $S^0 = \mathrm{Reg}\, S \subset S$ to be the maximal open subset in S which is a (locally closed) C^∞-submanifold in $\mathbb{R}^n$. Then take $S^1 = \mathrm{Reg}(S \setminus S^0)$, $S^2 = \mathrm{Reg}(S \setminus (S^0 \cup S^1))$ and so on. It is not hard to show that S^i is empty for $i > \dim S$ (See Levin 1971; Wall 1971; Whitney 1957; Lojasievicz 1965), which makes the partition $\bigcup_i S^i$ a stratification of S. Moreover, an arbitrary partition of S into semi algebraic subsets $S_\mu \subset S$, $\mu = 1, \ldots, M$, can be *canonically refined* to a stratification of S by first taking the minimal $\mu = \mu_0$ for which $\dim S_\mu = \dim S$ and then by defining S^0 to be the maximal open subset in S which is contained in S_{μ_0} and is a C^∞-submanifold in $\mathbb{R}^n$. The same applies to the complement $S \setminus S_0$ partitioned by $(S \setminus S_0) \cap S_\mu$, thus giving the stratum $S^1 \subset S \setminus S_0$; then one gets S^2 in the subset $S \setminus (S^0 \cup S^1)$ partitioned by $[S \setminus (S^0 \cup S^1)] \cap S_\mu$ and so on. Eventually, one stratifies S by at most $M \dim S$ strata S^i.

(B) *Generic Points, Maps and Sections.* One says that a property of a point x (e.g. of a function or a section) in a Bair space $\mathscr{X}$ is *generic* if it holds for $x \in \mathscr{X}' \subset \mathscr{X}$, where $\mathscr{X}'$ is a *residual* subset (i.e. a countable intersection of open dense subsets in $\mathscr{X}$). This $\mathscr{X}'$ can be defined any time you like in a discussion. It can become smaller in the course of an argument when we need it, but it must remain residual. The expression: "*a property A is satisfied by a generic C^k-section* (*or sections*) $V \to X$", is often used instead of "the property$A$ is generic in the space $\mathscr{X} = C^k(X)$ with the fine (or the ordinary if so indicated) C^k-topology".

Remark. The notion of genericity could be based on any class of "large" subsets in a given space (like subsets of full measure in a Lebesgue space) which is stable under unions and countable intersections of subsets. In fact, a deeper analysis of this notion belongs with the mathematical logic which is not discussed here.

Example. Let $\mathscr{X}$ be an *infinite dimensional* Banach space. Call a subset $\mathscr{Y} \subset \mathscr{X}$ *thin* if, for every $\varepsilon > 0$, there is a sequence of balls $B_i \subset \mathscr{X}$, $i = 1, 2, \ldots$ whose radii satisfy $r_i < \varepsilon$ and $r_i \to 0$ for $i \mapsto \infty$, such that $\bigcup_i B_i \supset \mathscr{Y}$. Show countable unions of thin subsets to be thin and prove that no residual subset is thin. (Thus, one could associate a genericity notion to the complements of thin subsets.)

(C) **Theorem** (A.P. Morse 1939). *Let S be an m-dimensional manifold and let $f: S \to \mathbb{R}$ be a C^∞-function. Then f is transversal to a generic point $x \in \mathbb{R}$.*

Proof. Since S is a *countable* union of balls, we may assume S to be such a ball $B \subset \mathbb{R}^m$ to start with. The function $f|\Sigma_f^0: \Sigma_f^0 \to \mathbb{R}$ for $\Sigma_f^0 = S \setminus \Sigma_f^1$ is transversal [see the definitions in (A_1) and in 1.3.1] to all $x \in \mathbb{R}$. Furthermore, we may assume by induction in m the map $f|H_I: H_I \to \mathbb{R}$ for $|I| < m$ to be transversal to generic points $x \in \mathbb{R}$, which implies the desired transversality on $S \setminus \Sigma_m$. Finally, the subset $\Sigma_m \subset S = B \subset \mathbb{R}^m$ can be covered (as any other subset in $B \subset \mathbb{R}^m$) by $N \leq \text{const}\, \varepsilon^{-m}$ balls with the centers in Σ_m and of radii $\leq \varepsilon$, for any given $\varepsilon > 0$. The f-image of such a ball is an interval in $\mathbb{R}$ of length $\leq \text{const}'\, \varepsilon^{m+1}$ by the Taylor remainder theorem. Hence, the (compact!) image $f(\Sigma_m) \subset \mathbb{R}$ can be covered by N intervals of total length $\leq \text{const}''\, \varepsilon$ for all $\varepsilon > 0$, and so no *generic* point $x \in \mathbb{R}$ lies in $f(\Sigma_m)$. Thus, f has the required transversality on all of S as well as on $S \setminus \Sigma_m$. Q.E.D.

(C′) **Corollary** (Sard 1942). *Take a C^∞-submanifold in a product, $S \subset U \times \mathbb{R}^d$. Then S is transversal to the submanifold $U \times z \subset U \times \mathbb{R}^d$ for generic points $z \in \mathbb{R}^d$.*

Proof. If $d = 1$ this is equivalent to (C) applied to the projection $S \to \mathbb{R}^1$. Then, for $d \geq 2$, we split $\mathbb{R}^d = \mathbb{R}^{d-1} \times \mathbb{R}$ and establish with (C) the transversality of S to $U \times \mathbb{R}^{d-1} \times x \subset U \times \mathbb{R}^{d-1} \times \mathbb{R}$ for generic points $x \in \mathbb{R}$. For these, the intersection $S_x = S \cap (U \times \mathbb{R}^{d-1} \times x)$ is an $(m-1)$-dimensional manifold, which may be assumed, by induction in $m = \dim S$, to be transversal to $U \times y \times x \subset U \times \mathbb{R}^{d-1} \times x$ for generic $y \in \mathbb{R}^{d-1}$. Hence, the subset $Z' \subset \mathbb{R}^{d-1} \times \mathbb{R}$ of the pairs $z = (y, x)$ for which S is transversal to $U \times y \times x$ is dense in $\mathbb{R}^d = \mathbb{R}^{d-1} \times \mathbb{R}$. Since the complement $\mathbb{R}^d \setminus Z'$ (obviously) is a countable union of compact subsets in $\mathbb{R}^d$, the subset Z' is residual as well as dense. Q.E.D.

Exercise. Prove Sard's theorem for C^k-submanifolds S for $k \geq \min(1, m-d)$.

(D) **Transversality Theorem** (Thom 1955). *Let $S \subset X^{(r)}$ be a C^∞-submanifold. Then the jet $J_f^r: V \to X^{(r)}$ is transversal to S for generic C^∞-sections $f: V \to X$. In particular,* $\operatorname{codim} S_f = \operatorname{codim} S$ *for* $S_f = (J_f^r)^{-1}(S) \subset V$.

Proof. Since S can be covered by countably many small compact balls, one may assume S to be such a ball, which projects to a small (coordinate) neighborhood $U \subset V$.

The subspace $\mathscr{X}' \subset C^\infty(X)$ of sections $f: V \to X$ whose jets $J_f^r: V \to X^{(r)}$ are transversal to such a (compact) S is obviously open. To approximate a given C^∞-section $f_0: V \to X$ by those in $\mathscr{X}'$, take a holonomic splitting $Y^{(r)} = U \times \mathbb{R}^{d_r} \subset X^{(r)}$ around $J_{f_0}^r(U) \subset X^{(r)}$ [see (c) in 1.1.1] and apply (C′) to $S_0 = S \subset Y^{(r)} \subset U \times \mathbb{R}^{d_r}$ and to $U \times z = J_{f_z}^r(U) \subset Y^{(r)}$. This yields the transversality of $J_{f_z}^r$ to S for generic $z \in \mathbb{R}^{d_r}$, which allows such a "transversal" f_z arbitrary C^∞-close to $f_0|U$. Finally, the section $f_z: U \to Y \subset X$ extends to all of V by a small perturbation outside $p^r(S) \subset U$. [Take a C^∞-function $\varphi: U \to \mathbb{R}$ with a compact support, which equals one near $p^r(S)$, and extend f_z by $f_0 + \varphi(f_z - f_0)$ outside $p^r(S)$.] Hence, $\mathscr{X}' \subset C^\infty(X)$ is dense as well as open and Thom's theorem follows.

(D′) **Corollary.** *If $S \subset X^{(r)}$ is an arbitrary (e.g. stratified) subset whose C^∞-codimension abides* $\operatorname{codim} S \geq \dim V + 1$, *then the jet $J_f^r\colon V \to X^{(r)}$ misses S for generic C^∞-sections $f\colon V \to X$.*

Remark. In fact one does not need Sard's theorem for the proof of (D′) but rather the following obvious fact: *Every C^∞-map $S \to \mathbb{R}^d$ misses a generic point $z \in \mathbb{R}^d$, provided* $\dim S < d$. Furthermore, in many cases one can reduce (D) to (D′) which applies to the subset $S' \subset X^{(r+1)}$ of the 1-jets of those holonomic C^∞-sections $V \to X^{(r)}$ which are not transversal to $S \subset X^{(r)}$. This works, for example, for $S \subset X^{(0)} = X$ and also for real analytic subsets $S \subset X^{(r)}$ for all $r = 0, 1, \ldots$.

Exercises. (a) Apply (D′) to the Thom-Whitney strata $\Sigma^i \subset X^{(1)}$, for $X = V \times W$ and $i \geq 1$ (see 1.3.1) and prove

Whitney's Theorem (1936). *A generic C^∞-map $V \to W$ is an immersion for* $\dim W \geq 2n$, *where* $n = \dim V$.

(b) (Nash 1956). Show generic C^∞-maps $V \to \mathbb{R}^q$ to be free for $q \geq \frac{1}{2}n(n+5)$.
(c) Generalize (D) and (D′) in order to obtain the following theorem.

(Whitney 1936). *A generic C^∞-map $f\colon V \to W$ has not double points for* $\dim W \geq 2n+1$. *Furthermore, the subset of k-multiple points, $M_k = \{(v_1, \ldots, v_k) \in V \times V \times \cdots \times V \mid f(v_1) = f(v_2) = \cdots = f(v_k)$, for $v_1 \neq v_2 \neq \cdots \neq v_k\}$ has* $\dim M_k \leq n - (k-1)(\dim W - n)$ *for generic maps $f\colon V \to W$.*

(E) *Equisingular Maps and Sections.* Consider a point $z \in X^{(1)}$ and let $y = p_0^1(z) \in X$ and $v = p^1(z) = p(y)$. Take a linear subspace $\tau \in T_v(V)$ and denote by $A(z, \tau) \subset X_y^{(1)}$ the (affine) subspace of those linear maps $x\colon T_v(V) \to T_y(X)$ for which $x|\tau = z|\tau$. Notice that $\dim A(x, \tau) = n(q - \dim \tau)$ for $n = \dim V$ and $q = \dim X_v$. Next, take a stratified subset $\Sigma = \bigcup_i \Sigma^i \subset X^{(1)}$, $i = 0, \ldots, k$, and call a C^∞-section $f\colon V \to X$ *Σ-equisingular* along a submanifold $S \subset V$ if

(i) the image $J_f^1(S)$ lies in a single stratum Σ^i for some $i = i(S, f)$.
(ii) The C^∞-dimension of the intersection $\Sigma^j \cap A(J_f^1(s), T_s(S))$ is constant in $s \in S$ for all $j = 0, 1, \ldots, k$.

Example. If $\Sigma = \bigcup_i \Sigma^i \subset X^{(1)}$ is the Whitney-Thom stratification for $X = V \times W$, then a C^∞-map $f\colon V \to W$ is equisingular along some $S \subset V$ if and only if the ranks $\operatorname{rank}_s f$ and $\operatorname{rank}_s(f|S)$ for $f|S\colon S \to W$ are constant in $s \in S$.

Remark. It would be natural to strengthen (ii) by requiring all topological invariants of the pertinent intersection to be constant in s, but for our applications the weak "equidimensional" definition suffices.

Next, call a stratification $\bigcup_j S_j = S$, of the subset $S = \Sigma_f = (J_f^1)^{-1}(\Sigma) \subset V$ *equisingular* if the section f is Σ-equisingular along S_j for all $j = 0, 1, \ldots, m$.

Finally, a stratified subset $\Sigma = \bigcup_i \Sigma^i \subset X^{(1)}$ is called *locally semi-algebraic* if there exists, for each $x \in \Sigma$, a split neighborhood $Y = U \times W_0 \subset X$ with local coordinates $u_1, \dots, u_n$ and $y_1, \dots, y_n$ [see (c) in 1.1.1] such that $Y^{(1)} \ni x$ and the subset $\Sigma^i \cap Y^{(1)}$ is semi-algebraic in $Y^{(1)}$ for all $i = 0, 1, \dots, k$, where $Y^{(1)}$ is identified with $\mathbb{R}^{n+qn}$ by means of the coordinates $u_1, \dots, u_n, y_1, \dots, y_q$ [compare (c) in 1.1.1]. For example, the Whitney-Thom stratification (obviously) is locally semi-algebraic.

(E_1) **Theorem.** *If $\Sigma = \bigcup_i \Sigma^i \subset X^{(1)}$ is locally semi-algebraic, then the subset $\Sigma_f \subset V$ admits an equisingular stratification for generic C^∞-sections $f: V \to X$.*

This fact (and the proof which follows) is an abstract version of the

(E_2) **Thom Equisingularity Theorem** (1955). *If $f: V \to W$ is a generic C^∞-map, then there exists a stratification $\bigcup_j S^j = V$, such that the ranks of the maps f and $f|S_j: S_j \to W$ is constant on each stratum S_j.*

Proof. (E_2) follows from (E_1) which applies to the Whitney-Thom stratification. To prove (E_1) we need the following

Canonical Partition. Consider a submanifold $Z \subset X^{(r)}$ and let $\tilde{Z} = (p_r^{r+1})^{-1}(Z) \subset X^{(r+1)}$. Take a point $\tilde{z} \in \tilde{Z}$ represented by the differential $D_\varphi: T_v(V) \to T_z(X^{(r)})$ of a germ of a holonomic C^∞-section $\varphi: V \to X^{(r)}$ for $v = p^{r+1}(\tilde{z})$ and $z = p_r^{r+1}(\tilde{z}) \in Z \subset X^{(r)}$ (see 1.1.1). Set $\tau(\tilde{z}) = D_\varphi^{-1}(T_z(Z)) \subset T_v(V)$ and let $d_i(\tilde{z}) = \dim(\Sigma^i \cap A(p_1^r(z), \tau(\tilde{z}))$. Then partition $\tilde{Z}$ into the subsets $\tilde{Z}(i, \delta) = \{\tilde{z} \in \tilde{Z} | d_i(z) = \delta\}$, for $i = 0, \dots, k$, $\delta = 0, 1, \dots, \dim \Sigma^i$ and canonically refine the partition $\{\tilde{Z}(i, \delta)\}$ [see (A_2')]. This gives us a stratification of Z in case the partition $\{\tilde{Z}(i, \delta)\}$ is semi-algebraic in some local coordinates around each point $\tilde{z} \in \tilde{Z}$.

Now, we apply the partition and the refinement procedures to each stratum $\Sigma^i \subset X^{(1)}$ in place of Z in order to stratify the pull-back $(p_1^2)^{-1}(\Sigma^i) \subset X^{(2)}$ for all $i = 1, \dots, k$. Thus, we obtain some stratification, say $\bigcup_j \Sigma^j(2) \subset X^{(2)}$ of $(p_1^2)^{-1}(\Sigma)$ which refines the lift to $X^{(2)}$ of the stratification $\bigcup_i \Sigma^i = \Sigma$. We do the same to each stratum $\Sigma^j(2)$, thus getting a stratification of $(p_1^3)^{-1}(\Sigma) \subset X^{(3)}$, then we pass to $X^{(4)}$ and so on. The resulting tower of stratifications of $(p_1^r)^{-1}(\Sigma) \subset X^{(r)}$ for $r = 1, 2, \dots$, enjoys the following *stability* which is immediate with the definition of $\tilde{\tau}(\tilde{z})$: if some stratum $Z' \subset Z^{(r)}$ lies over $Z \subset X^{(r-1)}$ and if codim Z' = codim Z, then $d_i(\tilde{z})$ is constant in $\tilde{z} \in \tilde{Z}' = (p_r^{r+1})^{-1}(Z') \subset X^{(r+1)}$. Hence, no stratum in $X^{(r)}$ of codimension $\leq n$ is stratified further when lifted to $X^{(r+1)}$ for $r \geq n + 1$.

Now the pull-back of the stratification in $X^{(n+1)}$ for $n = \dim V$ under the jet J_f^{n+1} of a generic section $f: V \to X$ clearly is the required equisingular stratification of Σ_f. Q.E.D.

Remark. Return to a generic map $f: V \to W$, assume $\dim W \geq \dim V$ and show that no smooth curve $C \subset V$ goes to a single point in W. In fact, the vanishing of the first r derivatives of f along C at a fixed point $c \in C$ is given by rq independent equations for $q = \dim W$. On the other hand, the r-jets of germs of non-parametrized curves C in V form an $n + r(n - 1)$ dimensional manifold for $n = \dim V$. Hence, the condition $\mathscr{R}_r \subset X^{(r)}$ which expresses the vanishing of the first r-derivative of f on *some* curve C satisfies codim $\mathscr{R}_r = rq - nr(n-1) \geq r - n$. Therefore, the r-jet of a generic f misses $\mathscr{R}_r$ for $r > 2n$. Q.E.D.

(E_2') **Corollary.** *A generic map f is an immerion on each stratum S^j of the equisingular stratification insured by* (E_2). *That is the subbundle* Ker $D_f|S^j$ *is transversal to S^j for all strata S^j.*

Exercises. (a) Show for dim $W <$ dim V as well as for dim $W \geq$ dim V the subbundle Ker $D_f|S^j \subset T(V)|S^j$ is transversal to S^j for codim $S^j > 0$ (assuming f is generic, of course).

(b) Generalize (E_1) to stratified subsets $\Sigma \subset X^{(r)}$ for $r \geq 2$, and apply this to $\Sigma = X^{(2)} \setminus \mathscr{F}$ for the freedom relation $\mathscr{F} \subset X^{(2)}$.

(c) Observe with (E_2) that the C^∞-dimension is monotone non-increasing under *generic* C^∞-maps (This is unlikely for non-generic C^∞-maps. Probably, there exists a C^∞-map $f\colon \mathbb{R} \to \mathbb{R}^2$ whose image is not contained in a countable union of C^∞-smooth curves in $\mathbb{R}^2$).[1]

(d) Generalize (E_1) and (E_2) to generic C^k-sections (and maps) for $k \geq n + 2$. Observe the set Σ_f^1 for a generic C^2-function $f\colon V \to \mathbb{R}$ to be *discrete*. Show Σ_f^1 for a generic C^1-function f is a Cantor set (i.e. $\dim_{\text{top}} \Sigma_f^1 = 0$ and no point in Σ_f^1 is isolated). Study generic C^1- and C^2-maps $\mathbb{R}^2 \to \mathbb{R}^2$.

References. See Boardman (1967) for a detailed study of equisingular stratifications of generic maps $f\colon V \to W$. See Golubitsky-Guillemin (1973) for an introduction to singularities of smooth maps.

1.4 Localization and Extension of Solutions

1.4.1 Local Solutions of Differential Relations

Take a subset $C \subset V$ and study solutions of a given relation $\mathscr{R} \subset X^{(r)}$ over a small neighborhood $U \subset V$ of C. These are holonomic sections $U \to \mathscr{R}$. In the following consideration we often use an arbitrarily small but non-specified neighborhood of C, denoted by $\mathcal{O}p\, C \subset V$ (*opening of C in V*). This is a small neighborhood of C which may become even smaller in the course of the argument. The following dictionary helps to avoid any ambiguity in dealing with these openings.

The space of C^k-sections $\mathcal{O}p\, C \to X$, by definition, is the *direct* (*inductive*) *limit* of the spaces of C^k-sections $U \to X$ over all neighborhoods $U \subset V$ of C. This is also called the *space of germs* of sections (defined) near C. There is no useful natural topology in this space; however, there is a weaker structure, called *quasi-topology*, which nicely behaves under direct limits.

[1] Such a map f [with $f(\mathbb{R}) \supset A \times A$ for some Cantor set $A \subset \mathbb{R}$] was constructed by A.G. D'Farrell (1986).

A function or a section f on $\mathcal{O}p C \subset V$.	Such a function or a section on some neighborhood $U \subset V$ of C.
$f_1 \| \mathcal{O}p C = f_2 \| \mathcal{O}p C$.	There exists a neighborhood $U' \subset V$ of C such that both functions are defined and equal on U'.
An extension of f from $\mathcal{O}p C_1$ to $\mathcal{O}p C_2$ for $C_1 \subset C_2 \subset V$.	This is a function f' on some neighborhood $U' \subset V$ of C_2, which equals f on a sufficiently small neighborhood $U'' \subset V$ of C_1 where f and f' are simultaneously defined.
Two sections f_1, f_2: $\mathcal{O}p C \to \mathcal{R}$ are homotopic.	There is a neighborhood U of C on which f_1 and f_2 are defined and homotopic.
A C^k-continuous family of C^k-section f_p: $\mathcal{O}p C \to X$ for $p \in P$.	such a family on some neighborhood of C.

(A) **Definition** (Spanier-Whithead 1957). A *quasi-topological structure* in a set A is given by distinguishing a subset in the set of all point-set maps of topological spaces P into A, such that these distinguished maps, called "continuous" for the moment, enjoy the following formal properties of ordinary continuous maps.

(i) If $\mu: P \to A$ is "continuous" and if $\varphi: Q \to P$ is an ordinary continuous map, then the composed map $\mu \circ \varphi: Q \to A$ is "continuous".
(ii) If a map $\mu: P \to A$ is locally "continuous", then it is "continuous" where the local "continuity" requires a neighborhood $U \subset P$ of every point in P such that the map $\mu | U: U \to A$ is "continuous".
(iii) Let P be covered by two *closed* subsets P_1 and P_2 in P. If a map μ is "continuous" on P_1 and on P_2, then it is continuous on all of P. Therefore, if $\bigcup_{i=1}^k P_i = P$ is a covering of P by finitely many closed subsets, then a map $\mu: P \to A$ is "continuous" if and only if $\mu | P_i: P_i \to A$ is "continuous" for all $i = 1, \ldots, k$. [The above (ii) implies a similar property for finite a well as for infinite coverings of P by *open* subsets.]

Next, a map between quasi-topological spaces, $\alpha: A \to B$, is called *continuous* if $\alpha \circ \mu: P \to B$ is "continuous" for all continuous maps $\mu: P \to A$ and for all topological spaces P.

We will from now on write continuous instead of "continuous" if the meaning is clear from the context.

Now, the space of C^k-sections $\mathcal{O}p C \to X$ is endowed with the quasi-topology which is the direct limit of the quasi-topologies associated to the C^k-topologies in the spaces of C^k-sections $U \to X$ for all neighborhoods $U \subset V$ of C. Thus, the notion of continuity for maps $\mu: p \mapsto f_p: \mathcal{O}p C \to X$, $p \in P$, agrees with the above C^k-continuity of families.

The standard definitions of homotopy theory (e.g. the weak homotopy equivalence) obviously generalize to quasi-topological spaces. With this we formulate the following

(B) *Local h-Principles.* A relation $\mathscr{R} \subset X^{(r)}$ is said to satisfy *the h-principle near a subset* $V_0 \subset V$ (or the h-principle on $\mathcal{O}\!p V_0 \subset V$) if every section $\varphi: \mathcal{O}\!p V_0 \to \mathscr{R}$ is homotopic to a holonomic section. That is, according to the dictionary, for every neighborhood $U \subset V$ of V_0 and for every section $\varphi: U \to \mathscr{R}$ there exists a neighborhood $U' \subset U$ of V such that $\varphi | U'$ is homotopic to a holonomic section $U' \to \mathscr{R}$. Furthermore, *the parametric h-principle near* V_0 claims that the map $f \mapsto J_f^r$ of the space of solutions of $\mathscr{R}$ on $\mathcal{O}\!p V_0$ to the space of sections $\mathcal{O}\!p V_0 \to \mathscr{R}$ a weak homotopy equivalence. Finally, the local (near V_0) h-principle is called C^0*-dense* in a subspace $\mathscr{Y} \subset C^0(X|V_0)$ (compare 1.2.2) if for every section $f_0 \in \mathscr{Y}$, for every neighborhood $U \subset X$ of $f_0(V_0) \subset X | V_0 = p^{-1}(V_0) \subset X$, and for every section φ_0: $\mathcal{O}\!p V_0 \to \mathscr{R}$, such that $p_0^r \circ \varphi_0 | V_0 = f_0$, there exists a homotopy of φ_0 to a holonomic section φ_1 by a homotopy of sections $\varphi_t: \mathcal{O}\!p V_0 \to \mathscr{R}, t \in [0, 1]$, such that $p_0^r \circ \varphi_t(V_0) \subset U$ for all $t \in [0, 1]$.

Exercise. Give a consistent definition of the C^i-dense local h-principle for $i \geq 1$.

Remarks and Examples. (B_1) *Localization of* (Diff V)*-Invariant Relations.* Suppose there is a natural action of diffeomorphisms of V on X, and let $\mathscr{R}$ be invariant under the associated action of Diff V on $X^{(r)}$. For example, immersions and free maps $V \to W$ are defined by (Diff V)-invariant relations while isometric maps are not. If $U \subset V$ is a *regular* neighborhood of a piecewise smooth subpolyhedron $V_0 \subset V$ (e.g. a tubular neighborhood of a submanifold $V_0 \subset V$), then U can be brought to $\mathcal{O}\!p V_0$ by a diffeotopy in U which is constant on V_0. Therefore, the local h-principles on $\mathcal{O}\!p V_0$ are equivalent to the corresponding global h-principles on U. If, for instance, V is diffeomorphic to $\mathbb{R}^n$, then the global h-principle on V reduces to the local one near a single point $v_0 \in V$. Furthermore, let V be an open manifold. Then there exists a codimension one subpolyhedron $V_0 \subset V$, such that the h-principle on V localizes to $\mathcal{O}\!p V_0 \subset V$ by a diffeotopy of V which brings V arbitrarily close to V_0.

Indeed, every open manifold admits a positive Moore function μ without local maxima. The Morse complex $V_\mu \subset V$ of such a μ clearly has codim $V_\mu \geq 1$ and the gradient flow of μ brings V to $\mathcal{O}\!p(V_\mu)$.

Exercises. (a) Prove the existence of the above μ.

(b) Construct $V_0 \subset V$ by using some triangulation of μ. *Hint.* Since V is open it embeds to the complement of the set of barycenters of the n-simplices.

(c) Show every open manifolds V admits a (non-Morse) C^∞-function without critical points.

(B_2) *Solutions of* $\mathscr{R}$ *Near a Point* $v_0 \in V$. If $\mathscr{R} \subset X^{(r)}$ is an *open* subset, then the local h-principle on $\mathcal{O}\!p v_0 \subset V$ obviously holds true. Indeed, every jet $x \in X^{(r)}$ over v_0 is represented by a holonomic germ $f = f_x: \mathcal{O}\!p v_0 \to X^{(r)}$, such that $f(v_0) = x$. Therefore, $f(\mathcal{O}\!p v_0) \subset \mathscr{R}$ for any fixed open subset $\mathscr{R} \ni x$. Moreover, the germ f_x can be easily made *continuous* in $x \in X^{(r)}$ which insures the *parametric* h-principle near v_0.

Now, let $\mathscr{R}$ be a locally *closed* submanifold of codimension s in $X^{(r)}$ which is given as the zero set, $\mathscr{R} = \{x \in U \subset X^{(r)} | \Psi(x) = 0\}$, for some C^k-map, $\Psi: U \to \mathbb{R}^s$ transversal to $0 \in \mathbb{R}^s$, for an open subset $U \subset X^{(r)}$. If s equals the dimension q of the fiber $X_v \subset X$, $v \in V$, then the h-principle may easily fail to be true. This is seen, for instance, in the well-known examples (see Hörmander 1963) of locally non-solvable linear differential equations. However, if $s \neq q$ [which makes the corresponding P.D.E. system $\Psi(J_f^r) = 0$ overdetermined for $s > q$ and underdetermined for $s < q$] then the local h-principle *generically* holds true near each point $v \in V$. The following two theorems make this claim meaningful.

(B_2') **Frobenius' Integrability Theorem.** *Let* $\Delta\colon X^{(r)} \to \mathbb{R}^s$ *be a generic* C^∞*-map for* $s > q$. *Then* C^∞*-solutions of the (over-determined) system* $\Delta(J_f^r) = g$ *satisfy the h-principle on* $\mathcal{O}\!p\, v \subset V$ *for all* $v \in V$ *and for all* C^∞*-maps* $g\colon V \to \mathbb{R}^s$.

Remark. This h-principle insures local solutions f of $\Delta(J_f^r) = g$ for the maps g which satisfy necessary *consistency conditions* [see (F) in 2.3.8].

(B_2'') *Let* $\Delta\colon X^{(r)} \to \mathbb{R}^s$ *be a generic* C^∞*-map for* $s < q$. *Then there exists a stratified subset* $\Sigma \subset X^{(r')}$ *for some* $r' = r'(\dim X^{(r)}) \geq r$ *of positive codimension such that* C^∞*-sections* $f\colon V \to X$ *for which*

$$(*) \qquad \Delta(J_f^r) = g \quad \text{and} \quad J_f^{r'}(V) \subset X^{r'} \backslash \Sigma$$

satisfy the h-principle on $\mathcal{O}\!p\, v \in V$ *for all* $v \in V$ *and all* C^∞*-maps* $g\colon V \to \mathbb{R}^s$.

This is shown in (E) of 2.3.8 for generic fiberwise *linear* maps Δ by a purely algebraic argument. Then the non-linear case follows by Nash's implicit function theorem (see 2.3.2).

Remarks. (a) The analytic techniques in 2.3 also delivers C^k-solutions of $(*)$ for $k \geq k_0 = k_0(\dim X^{(r)})$, provided the map g is C^{k+k_0}-smooth. But the analytic method does not apply to maps g of low smoothness. However, the geometric method of *convex integration* yields C^r-solutions of $(*)$ for all *continuous* maps g (see 2.4.6). Yet, the local h-principle for C^{r+1}-solutions of $(*)$ is unknown unless g is C^{k_0}-smooth for a sufficiently large k_0.

(b) If Δ and g are real analytic, then the local h-principle for $(*)$ holds true for all s (including $s \geq q$) by the classical *Cauchy-Kovalevskaya* theorem.

(B_3) *Local Isometric Immersions.* The local h-principle is especially interesting when the global one is not available. For example, the h-principle is unknown for (free) isometric C^∞-immersions $f\colon (V, g) \to \mathbb{R}^q$ for $q \leq \frac{1}{2}(n+2)(n+3)$; however, the local h-principle does hold true by the following

(B_3') **Theorem.** *If a* C^∞*-submanifold* $V_0 \subset V$ *has* codim $V_0 \geq 2$, *then free isometric* C^∞*-immersions* $(\mathcal{O}\!p\, V_0, g \,|\, \mathcal{O}\!p\, V_0) \to \mathbb{R}^q$ *satisfy the parametric h-principle for all* q. *This h-principle is also valid for* codim $V_0 = 1$ *unless* $q = \frac{1}{2}n(n+3)$ *(where the h-principle is unknown). Moreover, the h-principle is* C^0*-dense in the space of strictly short maps* $V_0 \to \mathbb{R}^q$ *for such* codim V_0 *and* q.

This is derived in 3.1.6 and 3.1.7 from the global h-principle for isometric immersions $V_0 \to \mathbb{R}^q$.

Observe that (B'_3) agrees with (B''_2) when V_0 reduces to a single point $v_0 \in V$. The pertinent map Δ is given in local coordinates by $s = \frac{1}{2}n(n+1)$ functions, $\left\langle \frac{\partial f}{\partial u_i}, \frac{\partial f}{\partial u_j} \right\rangle$, for $1 \leq i \leq j \leq n = \dim V$ and $\Sigma = X^{(2)} \backslash \mathscr{F}$, where $\mathscr{F}$ is the freedom relation. In this case, the existence of a free C^∞-immersion $(\mathcal{O}p\, v_0, g | \mathcal{O}p\, v_0) \to \mathbb{R}^q$ for $q = \frac{1}{2}n(n+3)$ is due to R. Green (1970); moreover, one has with (B'_3) *a weak homotopy equivalence between the space of free isometric immersions* $\mathcal{O}p\, v_0 \to \mathbb{R}^q$ *and the Stiefel manifold* $\mathrm{St}_m \mathbb{R}^q$ *for* $m = \frac{1}{2}n(n+3)$. Yet, for no small (but fixed) ball B in V around $v \in V$ one knows how to construct a C^∞-homotopy of isometric immersions between two given free isometric C^∞-immersions of B into $\mathbb{R}^q$, unless $q \geq \frac{1}{2}(n+2)(n+3)$, where the global h-principle of 1.1.5 applies.

Exercise. Assume V to be a parallelizable manifold with a C^∞-metric g and derive from (B'_3) the following

Local C^∞-Immersion Theorem. *A sufficiently small neighborhood* $U \subset V$ *of every* C^∞*-submanifold* $V_0 \subset V$ *of positive codimension admits a free isometric* C^∞*-immersion into* $\mathbb{R}^q$ *for* $q = \frac{1}{2}n(n+3) + 1$, *and for* $q = \frac{1}{2}n(n+3)$, *provided* codim $V_0 \geq 2$.

Then find examples of non-parallelizable manifolds V for which no free map $\mathcal{O}p\, V_0 \to \mathbb{R}^q$, $q = \frac{1}{2}n(n+3) + 1$ exists.

(A''_3) Let (V, g) be a C^{an}-manifold. Then the Cauchy-Kovalevskaya theorem allows a non-free isometric extension of immersions from V_0 to $\mathcal{O}p\, V_0 \subset V$ (see 3.1.6). This yields the following generalization of Janet's theorem (which applies to $V_0 = v_0 \in V$, (see 1.1.5).

The C^{an}-Immersion Theorem. *If* V *is parallelizable and if a* C^{an}*-submanifold* $V_0 \subset V$ *has the trivial normal bundle, then, for* codim $V_0 \geq 2$, *some sufficiently small neighborhood* $U \subset V$ *of* V_0 *admits an isometric* C^{an}*-immersion into* $\mathbb{R}^q$ for $q = \frac{1}{2}n(n+1)$. [Compare Gromov (1970).]

Exercise. Let V_0 be the projective line in the projective plane $P^2\mathbb{R}$ with the metric of constant curvature $+1$. Show that no neighborhood $U \subset P^2\mathbb{R}$ of V_0 admits an isometric C^2-immersion into $\mathbb{R}^3$.

1.4.2 The *h*-Principle for Extensions; Flexibility and Micro-flexibility

The h-principle for *extensions of* C^k*-solutions of* $\mathscr{R} \subset X^{(r)}$, *for some* $k \geq r$, *from a subset* $C' \subset V$ *to a subset* $C \supset C'$ in V claims, for every C^k-section $\varphi_0 \colon \mathcal{O}p\, C \to \mathscr{R}$ which is holonomic on $\mathcal{O}p\, C'$, there exists a C^k-homotopy to a holonomic C^k-section φ_1 by a homotopy of sections $\varphi_t \colon \mathcal{O}p\, C \to \mathscr{R}$, $t \in [0,1]$, such that $\varphi_t | \mathcal{O}p\, C'$ is constant in t. This is also called *the h-principle over* $\mathcal{O}p\, C$ *relative to* $\mathcal{O}p\, C'$, *or the h-principle over the pair* (C, C').

Exercise. Define the parametric and the dense h-principles over (C, C').

The relative h-principle allows one to build global solutions of a relation with the following

(A) Lemma. *Let V be a triangulated manifold and let $\mathscr{R}$ satisfy the h-principle over $(S, \partial S)$ for all simplices S of the triangulation. Then $\mathscr{R}$ satisfies the ordinary h-principle (over all of V).*

Proof. Use the standard induction by skeletons along with the

(A′) Flexibility Sublemma. *Consider a pair (C, C') of compact subsets in V for $C' \subset C$. Let $\varphi_0\colon \mathcal{O}p\,C \to \mathscr{R}$ be a section and $\varphi_t'\colon \mathcal{O}p\,C \to \mathscr{R}$, $t \in [0, 1]$, be a homotopy of $\varphi_0' = \varphi | \mathcal{O}p\,C'$. Then φ_t' extends to a homotopy $\varphi_t\colon \mathcal{O}p\,C \to \mathscr{R}$* of φ_0 (i.e. $\varphi_t | \mathcal{O}p\,C' = \varphi_t'$).

Proof. Take a continuous function $\delta\colon \mathcal{O}p\,C \to \mathbb{R}_+$ with a compact support, such that $\delta \equiv 1$ in a smaller neighborhood $U' \subset \mathcal{O}p\,C'$ of $C' \subset \mathcal{O}p\,C'$. Let $\tau = t\delta(v)$ and set: $\varphi_t(v) = \varphi_\tau'(v)$ for $v \in \mathcal{O}p\,C'$ and $\varphi_t(v) = \varphi_0(v)$ for $v \in \mathcal{O}p\,C \backslash \mathcal{O}p\,C'$.

Exercise. Let $\mathscr{R}$ satisfy the h-principle over $(B^k, \partial B^k)$ for all smooth embedded balls $B^k \subset V$ with smooth boundaries $\partial B^k = S^{k-1} \subset V$ for $k = 0, 1, \ldots, \dim V - 1$. Assume V to be an *open* manifold and let $\mathscr{R}$ be (Diff V)-invariant [compare ($\mathbf{B}_1$) in 1.4.1]. Show $\mathscr{R}$ satisfies the h-principle.

The *flexibility* of sections $V \to \mathscr{R}$ expressed by (A′) can be strengthened with the following

(B) Definitions. Let $\alpha\colon A \to A'$ be a continuous map between quasi-topological spaces. Consider a continuous map of a compact polyhedron into A, say $\varphi\colon P \to A$, and let $\Phi'\colon P \times [0, 1] \to A'$ satisfy $\Phi' | P \times 0 = \varphi'$ for $\varphi' = \alpha \circ \varphi\colon P \to A'$.

The map α is called a (*Serre*) *fibration* if Φ' lifts to a map $\Phi\colon P \times [0, 1] \to A$ such that $\Phi | P \times 0 = \varphi$ and $\alpha \circ \Phi = \Phi'$, for all polyhedra P, maps $\varphi\colon P \to A$ and homotopies Φ' of φ'.

Call α a *micro-fibration* if for all P, φ and Φ' there exists a *positive* $\varepsilon \le 1$ and a map $\Phi\colon P \times [0, \varepsilon] \to A$ (where ε may depend on P, φ and Φ') such that $\Phi | P \times 0 = \varphi$ and $\alpha \circ \Phi = \Phi' | P \times [0, \varepsilon]$.

Examples. (a) A submersion between smooth manifolds, $\alpha\colon A \to A'$, is a micro-fibration. This α, if a proper map, is necessarily a fibration. Another condition which insures the Serre fibration property of a submersion α is the contractibility of $\alpha^{-1}(a) \subset A$ for all $a \in A'$, where $\alpha^{-1}(a)$ is assumed non-contractible in case it is empty (see 3.3.1).

(b) If A is a topological space and $\alpha\colon A \to A'$ is a micro-fibration, then $\alpha | B\colon B \to A$ (obviously) is a micro-fibration for all *open* subsets $B \subset A$. However, the restriction of a *fibration* α to B, is usually *not* a fibration, but only a micro-fibration $B \to A'$. For example, the restriction of the identity map $A \to A$ on B is a fibration if and only if B is a path component of A.

(B′) The fibration property of a map $\alpha: A \to A'$ allows *lifts of polyhedral homotopies* Φ' *from* A' to A, while for a micro-fibration the initial phase of Φ' lifts to A. If a homotopy Φ' is constant in t on some subpolyhedron $P_0 \subset P$, then the lift to A can be chosen constant on P_0 as well. Moreover, let P be a subpolyhedron in another polyhedron $Q \supset P$ and let $\varphi: P \to A$ and $\Phi': Q \to A'$ be continuous maps, such that $\alpha \circ \varphi = \Phi' | P$. Then

(i) *if* α *is a micro-fibration there exists a lift* $\Phi: \mathcal{O}p\, P \to A$, *such that* $\Phi | P = \varphi$ *and* $\alpha \circ \Phi = \Phi' | \mathcal{O}p\, P$, *for* $\mathcal{O}p\, P \subset Q$;
(ii) *if* α *is a fibration and* P *is a homotopy retract in* Q [*or, equivalently*, $\pi_i(Q, P) = 0$, $i = 0, 1, \ldots$], *then there exists a lift* $\Phi: Q \to A$, *such that* $\Phi | P = \varphi$ *and* $\varphi \circ \Phi = \Phi'$.

The proof (which is easy and well known) is left to the reader.

Now, we observe that the proof of (A′) applies to continuous families of sections φ_p for $p \in P$, thus showing the restriction map $\varphi \to \varphi | \mathcal{O}p\, C'$ for $\varphi \in C^0(\mathcal{R} | \mathcal{O}p\, C)$ to be a Serre fibration $C^0(\mathcal{R} | \mathcal{O}p\, C) \to C^0(\mathcal{R} | \mathcal{O}p\, C')$. We express this by calling C^0-sections $V \to \mathcal{R}$ *flexible* (*over all pairs of compact subsets* C *and* $C' \subset C$ *in* V). A similar flexibility property is satisfied by C^k-sections of an arbitrary C^k-fibration $X \to V$, as the restriction map $C^k(X | \mathcal{O}p\, C) \to C^k(X | \mathcal{O}p\, C')$ is obviously [compare the proof of (A′)] a Serre fibration for $k = 0, 1, \ldots, \infty$ (but not for C^{an}-sections).

Next, we call C^k-*solutions* $V \to X$ *of* $\mathcal{R}$ *flexible* (*micro-flexible*) over (C, C') if the restriction map of the space of C^k-solutions $\mathcal{O}p\, C \to X$ to the space of those over $\mathcal{O}p\, C'$ is a fibration (micro-fibration).

Examples. (B_1) *Open Relations.* If $\mathcal{R} \subset X^{(r)}$ is an open subset, then, clearly, [compare the above (b)] C^k-solutions of $\mathcal{R}$ are micro-flexible over all pairs of compact subsets in V for $k = r, r + 1, \ldots, \infty$. However, the flexibility may fail to be true. This happens, for instance, to immersions of the disk $D^2 \to \mathbb{R}^2$, as a regular homotopy near the boundary $\partial D^2 = S^1$ brings an embedded circle $S^1 \subset \mathbb{R}^2$ to the figure in $\mathbb{R}^2$ which extends to no immersion $D^2 \to \mathbb{R}^2$. Yet, immersions $V \to W$ are flexible over all pairs of compact subsets in V in the *extra dimension* case, that is for $\dim W > \dim V$. [See (C) below.]

(B_2) *Generic Underdetermined Systems of P.D.E.* If a relation $\mathcal{R} \subset X^{(r)}$ of codim $\mathcal{R} < q = \dim X_v$ is given by the system ($*$) in (B_2'') of 1.4.1, then C^∞-solutions, as well as C^r-solutions of $\mathcal{R}$ are micro-flexible over all compact pairs in V. This is proven along with the local h-principle [see (E) in 2.3.8 and 2.4.6].

(B_2') *Linear Equations.* Let X and Y be vector bundles over V and let $\Delta: X^{(r)} \to Y$ be a homomorphism. Then the flexibility of solutions f of the *linear* P.D.E. system $\mathcal{D} f = g$ for $\mathcal{D}: f \mapsto \Delta(J_f^r)$ and for any given section $g: V \to Y$ is equivalent to the micro-flexibility. In fact, both properties are obviously equivalent to the existence of an extension of every solution of the homogeneous equation $\mathcal{D} f = 0$ from $\mathcal{O}p\, C$ to all of V for all closed subsets $C \subset V$. Such an extension is, as a rule, impossible for determined and overdetermined systems, where $\dim Y \geq \dim X$. But if $\dim Y < \dim X$, and if Δ is a *generic* C^∞-homomorphism, then these extensions do exist [see (E) in 2.3.8] and insure the flexibility of C^∞-solutions f to $\mathscr{D}f = g$ for all C^∞-sections $g: V \to Y$.

(B_2'') *Free Isometric C^∞-Immersions* $(V, g) \to (W, h)$. These are micro-flexible over all pairs of compact subsets in V (see 2.3.2).

(C) *Flexibility and the h-Principle.* Suppose a relation $\mathscr{R} \subset X^{(r)}$ satisfies the *parametric* h-principle over a compact set $C \subset V$ and also over a smaller compact subset $C' \subset C$. That is the map $f \mapsto J_f^r$ is a weak homotopy equivalence of the space A of C^r-solutions $\mathcal{O}p\,C \to X$ to the space $B = C^0(\mathscr{R}|C)$ and the space A' of solutions over $\mathcal{O}p\,C$ is w.h. equivalent to $B' = C^0(\mathscr{R}|C')$.

(C_1) **Lemma** (Smale 1959). *If C^r-solutions of $\mathscr{R}$ are flexible over (C, C'), then they satisfy the (parametric) h-principle over (C, C').*

Proof. Consider the commutative diagram of continuous maps

$$\begin{array}{ccc} A & \xrightarrow{J} & B \\ {\scriptstyle\alpha}\downarrow & & \downarrow{\scriptstyle\beta} \\ A' & \xrightarrow[J']{} & B' \end{array}$$

where the horizontal arrows are $f \mapsto J_f^r$ and where α and β are restrictions of sections from C to C'. Take an arbitrary point $a' \in A'$, let $b' = J'(a') \in B'$, and consider the fibers $\alpha^{-1}(a') \subset A$ and $\beta^{-1}(b') \subset B$. Then the exact homotopy sequences of the *fibrations* α and β form the commutative diagram

$$\begin{array}{ccccccccccc} \cdots \to & \pi_i(A) & \to & \pi_i(A') & \to & \pi_{i-1}(\alpha^{-1}(a')) & \to & \pi_{i-1}(A) & \to & \pi_{i-1}(A') & \to \cdots \\ & \downarrow & & \downarrow & & \downarrow & & \downarrow & & \downarrow & \\ \cdots \to & \pi_i(B) & \to & \pi_i(B') & \to & \pi_{i-1}(\beta^{-1}(b')) & \to & \pi_{i-1}(B) & \to & \pi_{i-1}(B') & \to \cdots \end{array}$$

where the four non-central vertical arrows are isomorphisms. Hence, by the five homomorphism lemma, the vertical arrow in the middle also is an isomorphism for all $a' \in A'$. This is clearly equivalent to the (parametric) h-principle over (C, C').

(C_2) **Corollary** [Smale-Hirsch, compare (B_1) in 1.4.1]. *Let $\mathscr{R} \subset X^{(r)}$ be a* (Diff V)-*invariant relation which satisfies the parametric h-principle over $\mathcal{O}p\,v \subset V$ for all $v \in V$ and which is flexible over $(B^k, \partial B^k)$ for all balls $B^k \subset V$ with smooth boundaries $B^k = S^{k-1} \subset V$ for $k = 0, 1, \ldots,$* dim V. *Then $\mathscr{R}$ satisfies the h-principle (over V). Furthermore if V is an open manifold, then the flexibility is only needed for $k = 0, 1, \ldots,$* dim $V - 1$.

Proof. Since $\mathscr{R}$ is (Diff V)-invariant the local h-principle implies that over $\mathcal{O}p\,B^k \subset V$ for all balls $B^k \subset V$. Hence, by (C_1), the parametric h-principle over $\mathcal{O}p\,S^{k-1}$ implies that over $(B^k, \partial B^k)$. Since the sphere S^k is covered by two balls B^k which meet over the equator $S^{k-1} \subset S^k$, we conclude with the proof of the above (A) (which generalizes to families of sections) to the parametric h-principle over $\mathcal{O}p\,S^k$ for all $S^k \subset V$. Thus, by induction, we obtain the h-principle over the pairs $(B^k, \partial B^k)$. Finally, we triangulate V and apply (A) to the balls obtained by smoothing the boundaries of the simplices of the triangulation. The details of this proof and the study of an open V are left to the reader.

(C_2') **Remark.** We shall later generalize this corollary (see 2.2.2) to all relations $\mathcal{R}$ with no Diff-invariance assumption.

(C_3') In order to prove the h-principle with (C_2) one needs the flexibility of $\mathcal{R}$ over $(B^k, \partial B^k)$. Smale (1958, 1959) proved this for the immersion relation in the extra dimension case by a direct geometric argument which was generalized to submersions by Phillips (1967) and to k-mersions by Feit (1969). Since these relations are open, the local h-principle is obvious, and so the flexibility does imply the h-principle. The following general fact reduces the flexibility to the micro-flexibility for relations $\mathcal{R} \subset X^{(r)}$ which are invariant under "sufficently many" diffeomorphisms of V. Namely, we assume $V = V_0 \times \mathbb{R}$ and we suppose that $\mathcal{R}$ is invariant under some (natural) action on $X^{(r)}$ of the group of those diffeomorphisms of $V \times \mathbb{R}$ which preserve each line $v \times \mathbb{R} \subset V \times \mathbb{R}$, $v \in V$.

(C_3') **Theorem** (see 2.2.2). *If C^k-solutions of $\mathcal{R}$, for some $k \geq r$, are micro-flexible over all pairs of compact subsets in V, then these solutions are flexible over (C, C') for all compact subsets $C \subset V_0 \times 0 \subset V$ and $C' \subset C$.*

(C_3'') **Corollary.** *If $\mathcal{R} \subset X^{(r)}$ is open and* (Diff V)*-invariant, then it satisfies the h-principle over $\mathcal{O}\!p\, V_0 \subset V$ for $V_0 = V_0 \times 0 \subset V$.*

Proof. Since $\mathcal{R}$ is open, the solutions of $\mathcal{R}$ are micro-flexible, and hence, flexible over the pairs in $V_0 \times 0$. Furthermore, the openness of $\mathcal{R}$ also yields the local h-principle over $\mathcal{O}\!p\, v \in V$, $v \in V$, and (the proof of) (C_2) applies.

Exercise. Prove with (C_3'') the h-principle for k-mersions (in particular for immersions) $V \to W$ for open manifolds V and for all $k = 0, 1, \ldots$. Then derive the extra dimensional case ($\dim V < \dim W$) of Hirsch's immersion theorem for *non-open* manifolds V. [Feit's k-mersion theorem for non-open manifolds V and $k < \dim W$ also follows from the case of V open, but the argument is more difficult than the one for immersions. See Feit (1969) and 2.2.4. A short proof of Feit's theorem is given in 2.1.1.]

(C_4) *Isometric Immersions into Pseudo-Euclidean Spaces.* Let h be the (indefinite) form $\sum_{i=1}^{q_+} dx_i^2 - \sum_{q_++1}^{q} dx_i^2$ on $\mathbb{R}^q$ and let g be an arbitrary quadratic differential C^∞-form on a manifold V_0. A C^1-map $f\colon V_0 \to \mathbb{R}^q$ is called *isometric* if $f^*(h) = g$, which is expressed in local coordinates u_i, $i = 1, \ldots, n = \dim V_0$ on V_0 by the system $\left\langle \frac{\partial f}{\partial u_i}, \frac{\partial f}{\partial u_j} \right\rangle_h = g_{ij}$ [compare (Is) in 1.1.5]. The pertinent differential relation is invariant under the isometry group of (V_0, g) which is a very small (in fact, generically trivial) subgroup in Diff V_0. But the manifold $V = (V_0 \times \mathbb{R}, g \oplus 0)$ has all $v \times \mathbb{R}$ preserving diffeomorphisms for isometries! Furthermore, *free* isometric C^∞-immersions $V \to (\mathbb{R}^q, h)$ are micro-flexible and they satisfy the parametric

h-principle over $\mathcal{O}p\,v \subset V$ for all $v \in V$, as we shall see in 2.2.2 with Nash's implicit function theorem. This implies flexibility and hence, the h-principle [see (C'_2)] for free isometric C^∞-immersions $\mathcal{O}p\,V_0 \to \mathbb{R}$ for $V_0 = V_0 \times 0 \subset V_0 \times \mathbb{R}$. Finall, we construct a section of our (free isometric) relation over $\mathcal{O}p\,V_0$ (see 3.3.1) for

$$(*) \qquad q_+ \geq 2n + 1, q_- = q - q_+ \geq 2n + 1, q \geq \tfrac{1}{2}(n + 1)(n + 8)$$

and thus conclude to the following

(C'_4) **Isometric Immersion Theorem.** *The inequalities* (*) *insure an isometric* C^∞*-immersion* $(V_0, g) \to (\mathbb{R}^q, h)$ *for all* C^∞*-forms* g *on* V_0. (See 3.3 for sharper results.)

Remark. A similar approach applies to more general "isometric" immersions f: $V \to W$ which induce given tensors (e.g. symplectic and contact forms, see 3.4) on V.

1.4.3 Ordinary Differential Equations and "Zero-Dimensional" Relations

An ordinary equation on V is associated to a vector field L on V and it is expressed with the *Lie derivative* Lf by $\Psi(v, f(v), \mathrm{Lf}(v)) = g(v)$, where f is the unknown C^1-map $V \to \mathbb{R}^q$, and where Ψ: $V \times \mathbb{R}^q \times \mathbb{R}^q \to \mathbb{R}^s$ and g: $V \to \mathbb{R}^s$ are given maps.

Example. The simplest linear equation $Lf = g$ for functions f, g: $V \to \mathbb{R}$ may easily violate the h-principle. In fact, if V is a closed manifold, then every $g = Lf$ necessarily satisfies $\int_V g\, d\mu = 0$ for all probability measures $d\mu$ on V which are invariant under the flow on V generated by L. This provides a (non-vacuous!) condition for solvability of $Lf = g$ which is not accountable for by the h-principle. For example, the inequality $Lf > 0$ is never solvable on closed manifolds V (which, of course, is obvious without any $d\mu$), while the h-principle predicts such an f for non-vanishing fields L.

Exercise. Let V be a compact connected manifold with a non-empty boundary and take a smooth non-vanishing field L on V whose every integral curve is a segment in V ending in ∂V. Prove the h-principle for the inequality $Lf > 0$ and for the equation $Lf = g$ for all smooth functions g on V.

We shall see in 2.3 that many non-linear under-determined (i.e. $q > s$) ordinary equations do satisfy the h-principle, and then we shall derive the h-principle for a class of partial differential equations on V (e.g. for isometric C^1-immersions $V \to \mathbb{R}^q$). Now, we concentrate on easier relations concerning the behaviour of f on subsets $V_0 \subset V$ whose intersections with the integral curves of L are *zero-dimensional.*

Example. (A_1) Let L be a non-vanishing field on V and let $V_0 \subset V$ be a submanifold transversal (i.e. nowhere tangent) to L. *Then the relation* $Lf \mid V_0 \neq 0$, *which requires* $Lf(v) \neq 0$, *for all* $v \in V_0$ *(and claims nothing what-so-ever outside* V_0*), satisfies the* h*-principle.*

Proof. This is obvious with functions f on V which satisfy $f|V_0 = f_0$ and $Lf|V_0 = f_1$ for given functions f_0 and f_1 on V_0.

(A'_1) **Remark.** The relation $Lf|V_0 \neq 0$ obviously satisfies the h-principles we have met so far. In particular, the h-principle holds true over each pair of closed subsets $C \subset V$ and $C' \subset C$.

(A_2) *Semi-Transversality.* A field L is called *semi-transversal* to a closed subset $V_0 \subset V$ if there is a stratification of V_0 by smooth submanifolds $\Sigma_i \subset V_0$ (see 1.3.2) $i = 0, \ldots, k$, which are transversal to L.

If L is semi-transversal to V_0 then the relation $Lf|V_0 \neq 0$ satisfies the h-principle.

Proof. Let $V_{k-1} = \bigcup_{j \geq i} \Sigma_j$, assume, by induction in $i = 0, 1, \ldots$, the h-principle for the relation $Lf|V_{k-i+1} \neq 0$ and observe that $V_{k-i} = \Sigma_i \cup V_{k-i+1}$ and that the topological boundary $\partial \Sigma_i = Cl\Sigma_i \setminus \Sigma_i$ lies in V_{k-i+1}. Then (A'_2) insures the h-principle over the pair $(\Sigma_i, \Sigma_i \cap Cl(\mathcal{O}p V_{k-i+1}))$ for all $\mathcal{O}p \mathcal{V}_{k-i+1} \subset V$. This allows an extension of the h-principle from $\mathcal{O}p V_{k-i+1}$ to $\mathcal{O}p V_{k+1} \subset V$. Q.E.D.

(A'_2) **Remark.** The definition of the semi-transversality and the above proof obviously generalizes to an arbitrary line subbundle $L \subset T(V)|V_0$, for which our relation can be expressed by $df|L \neq 0$ for the differential df of f. Moreover, one easily obtains the parametric and the dense h-principles as well as the h-principle over the pairs of closed subsets in V.

(A''_2) **Corollary.** *Let $F'\colon V \to \mathbb{R}^{q-1}$ be a C^∞-map. Impose a differential condition $\mathcal{R}$ on $f\colon V \to \mathbb{R}$ by requiring the map $F = F' \oplus f\colon V \to \mathbb{R}^q$ to be an immersion. If F' is generic and $q - 1 \geq \dim V$ then $\mathcal{R}$ satisfies the h-principle.*

Proof. Let $V_0 = \Sigma_{F'} = \{v \in V | \operatorname{rank}_v F' < \dim V\}$. Then the relation $\mathcal{R} \to V$ reads $df|\operatorname{Ker} D_{F'} \neq 0$ (i.e. df should not vanish on the non-zero vectors in $\operatorname{Ker} D_{F'} \subset T(V)|V_0$). If $\mathcal{R}$ admits a section $V \to \mathcal{R}$, then, obviously, $\operatorname{Ker} D_{F'}$ is a *line* subbundle in $T(V)|V_0$. Since F' is generic, this subbundle is semi-transversal to V_0 [see (E'_2) in 1.3.2] and (A'_2) applies.

Remark. We shall see in 2.1.1 how (a slight generalization of) (A''_2) yields the h-principle for immersions, submersions, k-mersions and for free maps into $\mathbb{R}^q$.

Exercises. (a) Let L be a non-vanishing vector field on V and let a closed subset $V_0 \subset V$ contain no open interval in any integral curve C of L [i.e. $\dim_{\text{top}}(V_0 \cap C) = 0$]. Show the relation $Lf|V_0 = g$ satisfies the C^0-dense h-principle for all C^0-functions g on V.

(b) (Hirsch 1961). Let V be an open manifold. Prove, by localizing to some $V_0 \subset V$ of codim $V_0 = 1$ [compare (B_1) in 1.4.1] that C^∞-functions $V \to \mathbb{R}$ without critical points (i.e. submersions $V \to \mathbb{R}$) satisfy the parametric h-principle.

(B) Remarks and Open Questions. One usually proves the h-principle for a relation $\mathcal{R}$ by reducing the problem to an auxiliary elementary relation like those considered in (A_1). The reduction process uses standard (often tedious but straightforward) methods of soft analysis like the partition of unity or an induction by strata of some stratification, etc. But the resulting elementary problem may require a specific geometric construction for its solution. In many cases, however, the elementary relation becomes very simple, like the "zero-dimensional" relation in (A_1) or like an ordinary "one-dimensional" differential equation (see 2.4), and no geometry is needed at all. Yet, there are many interesting relations which do not (seem to) reduce to anything "one-dimensional" an where the geometric intuition fails to provide a 2-dimensional construction.

(B')$\cdot$**Example.** *Directed immersions.* Let V be an oriented n-dimensional manifold and let $A \subset S^n \subset \mathbb{R}^{n+1}$ be an arbitrary subset. An immersion $f: V \to \mathbb{R}^{n+1}$ is, by definition, *directed by* A if the tangential (Gauss) map $G_f: V \to S^n$ sends V into A. If V is an *open manifold* and if A is an *open subset* in S^n, then the h-principle for directed immersions is immediate with (C_2) in 1.4.2. Now, let the manifold V be closed. Then, for every vector $s \in S^n \subset \mathbb{R}^{n+1}$ there obviously exists a hyperplane $H \subset \mathbb{R}^{n+1}$ orthogonal to s which is tangent to a given *closed* immersed hypersurface $f: V \to \mathbb{R}^{n+1}$ at some point $f(v) \in f(V) \subset \mathbb{R}^{n+1}$. Therefore, $G_f(V) \cup (-G_f(V)) = S^n$, for the reflection $-G_f(V)$ of the image $G_f(V) \subset S^n$ in the center. [If $f: V \to \mathbb{R}^{n+1}$ is an *embedding* then, clearly, $G_f(V) = S^n$.]

Question. Let $A \subset S^n$ be an open connected subset, such that $A \cup (-A) = S^n$. Do immersions directed by A satisfy the h-principle? In particular, does there exist a single closed immersed hypersurface in $\mathbb{R}^{n+1}$ directed by A? (If $A \neq S^n$, such a hypersurface is, necessarily, parallelizable.)

If $n = 1$, then the answer is an easy "yes" (an exercise for the reader). But for $n = 2$ and for $V = T^2$ (the torus T^2 is the only parallelizable closed surface) the answer is unknown for most subsets $A \subset S^2$. For example, the "yes" is obtained (with the convex integration, see 2.4.4) for those $A \subset S^n$, whose complement is a finite subset, or, more generally, a disjoint union of small balls. Yet, the question is open for $A = S^2 \backslash C$, where C is a simple arc in S^2 which is ε-dense in S^2 for a small $\varepsilon > 0$, say for $\varepsilon = 0.1$.

1.4.4 The *h*-Principle for the Cauchy Extension Problem

Consider a relation $\mathcal{R} \subset X^{(r)}$ and let $V_0 \subset V$ be a closed subset. Call the Cauchy (*initial value*) data of order i a section $\varphi_0: V_0 \to X^{(i)}$ such that there exists a *holonomic* section $\varphi_2': \mathcal{O}p\, V_0 \to X^{(i)}$ for which $\varphi_0' | V_0 = \varphi_0$. We always assume $i \leq r$; otherwise, we lift $\mathcal{R}$ to $\mathcal{R}^{i-r} \to X^{(i)}$ (see 1.1.1). Furthermore, when dealing with C^k-solutions of $\mathcal{R}$, we assume $\varphi_0' = J^i_{f_0'}$ for a C^k-section $f_0': \mathcal{O}p\, V_0 \to X$. We associate with given Cauchy data φ_0 on V_0 the *Cauchy relation* $\mathcal{R}_{\varphi_0} \subset \mathcal{R}$ which equals $\mathcal{R}$ over $V \backslash V_0$, while $\mathcal{R}_{\varphi_0} | V_0$ equals $(p^r_i)^{-1}(\varphi_0(V)) \cap \mathcal{R}$. Thus, solutions f of $\mathcal{R}_{\varphi_0}$ are those solutions of $\mathscr{R}$ which satisfy $J_f^i | V_0 = \varphi_0$. *The h-principle for the Cauchy problem* for given $\mathscr{R}$ and φ_0 is defined as that for the Cauchy relation $\mathscr{R}_{\varphi_0}$. We distinguish the *local* Cauchy problem, which concerns the h-principle for $\mathscr{R}_{\varphi_0}$ over $\mathcal{O}\!p\, V_0 \subset V$, and the *global* one, where we extend φ_0 to a solution f of $\mathscr{R}_{\varphi_0}$ on all of V. If the h-principle for extensions of solutions of $\mathscr{R}$ from $\mathcal{O}\!p\, V_0$ to V is available, then the local h-principle (for $\mathscr{R}_{\varphi_0}$ on $\mathcal{O}\!p\, V_0$) yields the global one.

Examples. (A) If $\mathscr{R}$ is an *open* relation of the *first* order, $\mathscr{R} \subset X^{(1)}$, then $\mathscr{R}_{\varphi_0}$ *obviously* satisfies the local h-principle for all initial data φ_0 of order zero or one. However, the local extension may run into a non-trivial global problem on V_0 for relations of order ≥ 2. Consider, for instance, the freedom relation $\mathscr{F} \subset X^{(2)}$ over $V = V_0 \times \mathbb{R}$, and let φ_0: $V_0 \to \mathbb{R}^p$, for $p = \frac{1}{2}(n-1)(n+2)$ and $n - 1 = \dim V_0$, be a free map that we want to extend to a free map f: $V \to \mathbb{R}^p \times \mathbb{R}^q \supset \mathbb{R}^p \times 0 = \mathbb{R}^p$ for $V_0 = V_0 \times 0 \subset V$. Denote by f': $V \to \mathbb{R}^q$ the orthogonal projection to $\mathbb{R}^q$ of the derivative (df/dt): $V \to \mathbb{R}^p \times \mathbb{R}^q$, $t \in \mathbb{R}$. If the map f is free on $\mathcal{O}\!p\, V_0 \subset V$, then a straightforward computation shows $f' | V_0$: $V_0 \to \mathbb{R}^q$ to be an immersion. Thus, the extension of free maps from V_0 to $\mathcal{O}\!p\, V_0$ is at least as difficult as producing immersions $V_0 \to \mathbb{R}^q$. In fact, the h-principle for free extensions from V_0 to $\mathcal{O}\!p\, V$ is valid by the techniques of 2.2.

(B) *Isometric C^1-Immersions.* Any extension of such an immersion from a submanifold V_0 must be preceded by an extension which is strictly short outside V_0, in order to apply the Nash-Kuiper techniques. For example, if $V_0 \subset V$ is a nongeodesic line and if f_0: $V_0 \to \mathbb{R}^q$ is an isometric map onto a straight line in $\mathbb{R}^q$, then no short (in particular isometric) extension to $\mathcal{O}\!p\, V_0$ exists.

(C) *Free Isometric C^∞-Immersions.* The local Cauchy problem for these satisfies the h-principle (see 3.1.6). This is used in the proof of the h-principle for free isometric immersions $\mathcal{O}\!p\, V_0 \to \mathbb{R}^q$ [see (B_3') in 1.4.1].

(D) *The Cauchy Problem with C^{an}-Data.* The local Cauchy problem is easier in the C^{an}-case, as the solution can be often reduced to the Cauchy-Kovalevskaya theorem [compare (A_3'') in 1.4.1]. On the other hand, a C^{an}-extension of a C^{an}-section f_0: $V_0 \to X$ to a (global!) section f: $V \to X$, which is not a subject to any relation, is a non-trivial problem solved by H. Cartan (1957). In fact, Cartan's result goes along with the analytic techniques of 2.3 which imply, in particular, the following

Approximation Theorem. *Let f: $(V, g) \to \mathbb{R}^q$ be a free isometric C^∞-immersion which is real analytic on some C^{an}-submanifold $V_0 \subset V$. If the Riemannian manifold (V, g) is real analytic, then f admits a fine C^∞-approximation by isometric C^{an}-immersions f': $(V, g) \to \mathbb{R}^q$, such that $f' | V_0 = f | V_0$.*

Part 2. Methods to Prove the h-Principle

2.1 Removal of Singularities

Consider a differential relation $\mathcal{R} \subset X^{(r)}$ whose complement $\Sigma = X^{(r)} \backslash \mathcal{R}$ is a closed stratified subset in $X^{(r)}$ of codimension $m \geq 1$ and take a generic holonomic C^∞-section $f: V \to X^{(r)}$ whose singularity $\Sigma_f = f^{-1}(\Sigma) \subset V$ may be non-empty (compare 1.3). Let us try to solve $\mathcal{R}$ by deforming f to a holonomic *Σ-non-singular* section $\bar{f}: V \to X^{(r)}$. Such a deformation can not be, in general, localized near Σ_f [see Exercise (a) below] but one can find in some cases an auxiliary subset $\Sigma' = \Sigma'(f) \supset \Sigma_f$ in V of codimension $m - 1$, such that the desired deformation does exist in an arbitrarily small neighbourhood of Σ'. The major difficulty in the construction of $\bar{f}$ comes from the holonomy condition. In fact, the problem becomes quite easy without this condition, as one can see in the following

Exercise. (a) Let Σ be an m-codimensional *submanifold* in $X^{(r)}$ such that the projection $\Sigma \to V$ is a proper map. Let a (holonomic or not) C^∞-section $f_0: V \to X^{(r)}$ be transversal to Σ and let Σ_{f_0} be a *non-empty* submanifold (of codimension m) in V. Consider an arbitrary homotopy of continuous sections $f_t: V \to X^{(r)}$ which are equal to f_0 outside a sufficiently small neighbourhood of Σ_{f_0} in V and show the subsets $\Sigma_{f_t} = f_t^{-1}(\Sigma) \subset V$ to be non-empty for all $t \geq 0$.

(b) Assume the existence of a continuous section $\bar{\varphi}: V \to \mathcal{R} = X^{(r)} \backslash \Sigma$ and construct a smooth family of (possibly non-holonomic) sections $f_t: V \to X^{(r)}$, such that Σ_{f_1} is empty and the sections f_t are equal to f_0 outside an arbitrarily small neighbourhood of some stratified subset $\Sigma' = \Sigma'(f_0, \varphi_1)$ in V of codimension $m - 1$.

We are interested, of course, in *holonomic* Σ-non-singular sections and these cannot be constructed without additional assumptions on Σ. The method of removal of singularities which we present in this section applies, roughly speaking, to those differential relations $\mathcal{R} = X^{(r)} \backslash \Sigma$ which are "semi-transversal" (compare 1.4.3) to sufficiently many subsets Σ' in V.

2.1.1 Immersions and k-Mersions $V \to \mathbb{R}^q$ for $q > k$

Start with a C^∞-map $F: V \to \mathbb{R}^q = \mathbb{R}^{q-1} \oplus \mathbb{R}$ which is split into the orthogonal sum, $F = F' \oplus f$ for the projections F' and f of V to $\mathbb{R}^{q-1}$ and $\mathbb{R}$ respectively. Our first objective is a C^∞-function $\bar{f}: V \to \mathbb{R}$ for which the map $\bar{F} = F' \oplus \bar{f}: V \to \mathbb{R}^q$

is an immersion. The pertinent singularity here is the subset $\Sigma_F \subset V$ where rank $F < n = \dim V$. Clearly codim $\Sigma_F = q - n + 1$ for generic maps F (see 1.3.1). A natural candidate for Σ' is the subset $\Sigma_{F'} \subset V$ where rank $F' < n$. Indeed, the sum $\bar{F} = F' \oplus \bar{f}$ is an immersion outside this Σ' for any $\bar{f}$, since $\Sigma' \supset \Sigma_{\bar{F}}$.

Now, recall (A_2'') of 1.4.3.

(A) **Lemma.** *If $q > n$ and if the map F' is generic then C^∞-functions $\bar{f}$ for which the sum $\bar{F} = F' \oplus \bar{f}$ is an immersion satisfy (all forms of) the h-principle. In particular the following conditions* (a) *and* (a′) *are necessary and sufficient for the existence of $\bar{f}$.*
(a) *The map F' has rank $\geq n - 1$ everywhere on V. Hence the kernel of the differential $D_{F'}$ is a one dimensional subbundle of the tangent bundle $T(V)|\Sigma'$.*
(a′) *The bundle* Ker $D_{F'}$ *on Σ' is trivial. This is equivalent to the existence of a non-vanishing vector field L on V near Σ', such that $D_{F'}(L)|\Sigma' = 0$.*

(A′) **Remark.** Let $\Phi = \Phi_{F'}$ denote the space of those C^1-smooth 1-forms $\bar{\varphi}$ on V which do not vanish on Ker $D_{F'} \subset T(V)$. Then an arbitrary form $\bar{\varphi}_0 \in \Phi$ admits a C^1-homotopy $\bar{\varphi}_t$ in Φ, $t \in [0, 1]$, to an *exact* form $\bar{\varphi}_1$ in Φ. In fact, this is equivalent to the above h-principle, since $L\bar{f} = \bar{\varphi}(L)$ for the (exact) form $\bar{\varphi} = d\bar{f}$.

Furthermore, a map $F = (f_1, \ldots, f_q)\colon V \to \mathbb{R}^q$ is an immersion if and only if the differentials of the coordinate functions, say $\varphi_i = df_i$, $i = 1, \ldots, q$, span the cotangent space $T_v^*(V)$ for all $v \in V$. This suggests the following generalization of (A). Denote by Φ^q the space of the q-tuples of 1-forms $(\varphi_1, \ldots, \varphi_q)$ which span $T_v^*(V)$ for all $v \in V$. Fix a $(q - 1)$-tuple $\psi' = (\varphi_1, \ldots, \varphi_{j-1}, \varphi_{j+1}, \ldots, \varphi_q)$ for some j between 1 and q and let $\Phi_{\psi'}$ consist of those C^1-smooth forms $\bar{\varphi}$ on V for which the q-tuple $(\varphi_1, \ldots, \varphi_{j-1}, \bar{\varphi}, \varphi_{j+1}, \ldots, \varphi_q)$ is contained in Φ^q.

(B) *Let φ_i, $i = 1, \ldots, j - 1$, be generic exact C^∞-smooth 1-forms on V* (i.e. $\varphi_i = df_i$ *for generic C^∞-functions f_i on V) and let φ_i for $i = j + 1, \ldots, q$ be generic (non-exact) C^∞-smooth 1-forms. If $q > n$, then an arbitrary form $\bar{\varphi}_0 \in \Phi_{\psi'}$ admits a homotopy $\bar{\varphi}_t$ in $\Phi_{\psi'}$, $t \in [0, 1]$, to an exact form $\bar{\varphi}_1 = d\bar{f} \in \Phi_{\psi'}$.*

Proof. The pertinent differential relation (on $\bar{f}$) is concentrated on the subset $\Sigma' \subset V$, where the forms $\varphi_1, \ldots, \varphi_{j-1}, \varphi_{j+1}, \ldots, \varphi_q$ fail to span the cotangent bundle $T_v^*(V)$. Then there is a vector field L near Σ', such that $\varphi_i(L)|\Sigma' = 0$ for $i \neq j$ and $\bar{\varphi}_o(L)$ does not vanish on Σ'. The proof of (E_2') in 1.3.2 shows that field L is semitransversal to Σ' under our genericity condition on $\psi' = (\varphi_1, \ldots, \varphi_{j-1}, \varphi_{j+1}, \ldots, \varphi_q)$, for $q - 1 \geq n$. Hence, the proof of (A) applies.

(C) **The *h*-Principle of Smale-Hirsch for Immersions** $V \to \mathbb{R}^q$. *If $q > n$ then an arbitrary q-tuple $\psi = (\varphi_1, \ldots, \varphi_q) \in \Phi^q$ admits a homotopy in Φ^q to a q-tuple $\bar{\psi} \in \Phi^q$ of exact forms $\bar{\varphi}_i = d\bar{f}_i$, $(i = 1, \ldots, q)$.*

Proof. The q-tuple ψ is made component-wise exact in q steps with lemma (B). The genericity assumption in (B) is satisfied at each step with a small generic perturbation of the pertinent function $f_1, \ldots, f_{j-1}$ and the forms $\varphi_{j+1}, \ldots, \varphi_q$. Q.E.D.

Remark. This h-principle is equivalent to the one stated in 1.1.3, since q-tuples of 1-forms on V define homomorphisms $T(V) \to T(\mathbb{R}^q)$ modulo translations in $\mathbb{R}^q$.

(C′) **The Parametric h-Principle and the h-Principle for Extensions.** The proof of (C) equally applies to families of immersions (compare 1.2.1) thus providing the parametric h-principle. Furthermore, the same argument delivers the h-principle for extensions (see 1.4.2) as well as the C^0-dense h-principle. In fact, the reduction to the zero-dimensional case allows the $W^{i,p}$-approximation (see 1.2.2) which is left as an exercise to the reader.

(D) **Further Exercises.** (a) Prove the h-principle for k-mersions $V \to \mathbb{R}^q$ for $q > k$. Namely, take an arbitrary q-tuple ψ of 1-forms φ_i, $i = 1, \ldots, q$, which span a subspace of dimension $\geq k$ in the cotangent space $T_v^*(V)$ for all $v \in V$ and then deform this ψ to a q-tuple of exact forms with the same k-spanning property.

(a′) State and prove a similar result for q-tuples of exterior forms of degree $d \geq 1$ on V. Then derive the following

Corollary. *Denote by $\Lambda_v^d(V)$ the space of exterior d-forms on $T_v(V)$ [which has $\dim \Lambda^d = \binom{n}{d} = n!/d!(n-d)!$] and show that every parallelizable manifold V admits $\binom{n}{d} + 1$ exact d-forms which span $\Lambda_v^d(V)$ at all points $v \in V$.*

Remark. If $2 \leq d \leq n-1$ then there are $\binom{n}{d}$ forms which span $\Lambda_v^d(V)$, $v \in V$ but the proof (see 2.4.3) exploits one-dimensional (rather than zero-dimensional) techniques.

(b) Consider two vector bundles Y and Z over V and let $\mathscr{D}$ be a C^∞-smooth linear differential operator on sections, $\mathscr{D}\colon \Gamma^\infty(Y) \to \Gamma^\infty(Z)$. Fix two integers, $k \geq 0$ and $q > k$ and prove the h-principle for those q-tuples of sections $f_1, \ldots, f_q\colon V \to Y$ for which the sections $\mathscr{D}f\colon V \to Z$, $i = 1, \ldots, q$, span a subspace of dimension $\geq k$ in each fiber $Z_v \in Z$, $v \in V$. [Compare Gromov-Eliasberg (1971), Burlet (1976). See 2.4.3 for a stronger result.]

(b′) Derive (a) and (a′) from (b). Then prove the h-principle for free maps $V \to \mathbb{R}^q$ for $q \geq [n(n-3)/2] + 1$, by considering the operator $\mathscr{D}\colon f \to J_f^2/\text{const}$ for the 2-jet J^2 on functions $f\colon V \to \mathbb{R}$.

(c) Consider a manifold V with a given k-dimensional C^∞-subbundle $\tau \subset T(V)$ and prove the h-principle for those C^∞-maps $f\colon V \to \mathbb{R}^q$, $q > k$, whose differential $\mathscr{D}_f\colon V \to T(\mathbb{R}^q)$ is injective on τ. Then prove this h-principle for an arbitrary *continuous* subbundle $\tau \subset T(V)$.

Hint. Study the pertinent singularities of generic C^1-maps $V \to \mathbb{R}^{q-1}$.

(E) *Embeddings* $V \to W$. If V is compact, then embeddings $f\colon V \to W$ are characterized by the absence of double points. That is $v_1 \neq v_2$ implies $f(v_1) \neq f(v_2)$. This is not a differential relation, of course, but even so a suitably chosen h-principle makes good sense. Denote by $S = S(V, W)$ the space of those continuous maps $s\colon V \times V \to W \times W$ commuting with the involutions $(v_1, v_2) \to (v_2, v_1)$ on $V \times V$ and $(w_1, w_2) \to (w_2, w_1)$ on $W \times W$ and having $s(v_1, v_2) \neq s(v_2, v_1)$ for all pairs of distinct points v_1 and $v_2 \neq v_1$ in V. The Cartesian square $s_f\colon V \times V \to W \times W$ of any map $f: V \to W$ commutes with the above infolutions, while the second property of $s = s_f$ expresses the "no double point" condition. Now we state the h-principle as the possibility to deform an arbitrary continuous map in S inside S to a smooth map of the form s_f. Haefliger (1962) proved this h-principle for smooth proper embeddings under the assumption $q > \frac{3}{2}(n+1)$ for $q = \dim W$ and $n = \dim V$. He also found counterexamples below this dimension. He removes double points by a more geometric (and more complicated) method than our step by step argument [see the proof of (C)]. In fact the step by step removal procedure also yields Haefliger's h-principle, at least for embeddings $V \to \mathbb{R}^q$ (see Szücs 1980). Moreover, this method applies to other comparable classes of maps, such as maps without triple points (see Szücs 1982, 1983, 1984).

Exercises. (a) Express the "no double point" condition for maps $F = (f_1, \dots, f_2): V \to \mathbb{R}^q$ in terms of the (linear difference) operator which sends functions f on V to antisymmetric function on $V \times V$ by the rule $f \to f(v_1) - f(v_2)$.

(b) Show a C^1-map $F: V \to W$ to be an immersion if and only if every sufficiently small C^1-perturbation of F is a *local* embedding, which means no double points which are close in V.

(c) Derive from Haefliger's h-principle the following corollaries (see Haefliger 1962; Haefliger-Hirsch 1962).

(c′) *Every topological embedding $V \to W$ for $q > \frac{3}{2}(n+1)$ admits a C^0-approximation by C^∞-embeddings.*

(c″) *If V is a k-connected manifold* [i.e. $\pi_i(V) = 0$, $i = 1, \dots, k$] *for $k < n/2$, then it admits a C^∞-embedding into $\mathbb{R}^{2n-k}$.*

Question. Consider a vector bundle $Y \to V$ and take its mth Cartesian power $(Y)^m \to (V)^m$. The power $(f)^m: (V)^m \to (Y)^m$ of a section $f: V \to Y$ is defined by the rule

$$(f)^m(v_1, v_2, \dots, v_m) = f(v_1) \oplus f(v_2) \oplus \cdots \oplus f(v_m).$$

Observe that the permutation group S_m naturally acts on $(Y)^m$ and on $(V)^m$. Take another bundle $Z \to (V)^m$ with an action of S_m which covers the action on $(V)^m$ and consider a differential operator on sections, $\mathscr{D}: \Gamma^\infty(Y)^m \to \Gamma^\infty(Z)$ which commutes with S_m. We are interested in those q-tuples of sections $f_1, \dots, f_2: V \to Y$ for which the sections $\mathscr{D}(f_1)^m, \dots, \mathscr{D}(f_q)^m: (V)^m \to Z$ span in each fiber of Z a subspace of dimension $\geq k$. What conditions on n, q, m, k and $\mathscr{D}$ would imply the h-principle for such q-tuples $(f_1, \dots, f_q)$?

Example. Let $m = 2$ and let Y and Z be trivial line bundles, say $Y = V \times \mathbb{R} \to V$ and $Z = V \times V \times \mathbb{R} \to V \times V$, where (v_1, v_2, t) goes to $(v_2, v_1, -t)$ under the generator of $S_2 \approx \mathbb{Z}/2\mathbb{Z}$. Let $\mathscr{D}$ be the zero order operator which acts on sections $(f_1, f_2): V \times V \to (V \times V) \times \mathbb{R} \times \mathbb{R}$ by $\mathscr{D}: (f_1(v_1, v_2), f_2(v_1, v_2)) \to f_1(v_1, v_2) - f_2(v_1, v_2)$. Then the q-tuples in question (for $k = 1$) are *embeddings* $F = (f_1, \dots, f_q)$: $V \to \mathbb{R}^q$ [compare the above (a)].

Remark. To make the general situation completely consistent with the case of embeddings one must augment the requirement dimension $\geq k$ by some condition at the diagonals of the manifold $(V)^m = V \times V \times \ldots \times V$. For example, the "no m-multiple points" requirement for C^∞-maps $F\colon V \to \mathbb{R}^q$ brings forth (among others) the *differential* condition (at the principal diagonal $\Delta \approx V$ in $V^{(m)}$) which claims "no *local* m-multiple points" for small C^∞-perturbations of F [compare the above (b)].

Another Example. Consider those C^∞-immersions F of $V \approx S^1$ into $\mathbb{R}^q$ for which the vectors $\frac{dF}{ds}(v_1), \ldots, \frac{dF}{ds}(v_m)$ in $\mathbb{R}^q$ are linearly independent for all m-tuples of *distinct* points $v_1, \ldots, v_m$ in V. Then the associated differential condition is the mth order freedom of F (see 1.1.4).

(E′) **Further Questions and Exercises.** The conception of "multiplicity" generalizes to maps $f\colon V^n \to W^q$, for $q \le n$, by introducing some measure of the "topological complexity" of the pull-backs $f^{-1}(w) \in V^n$, $w \in W$. For example, let $\beta(X)$ denote the sum of the Betti numbers of a space X with the coefficients $\mathbb{Z}/2\mathbb{Z}$ and let $\beta(f) = \sup_{w \in W} \beta(f^{-1}(w))$. Then one would like to evaluate $\inf \beta(f)$ over some class (e.g. a homotopy class) of maps $f\colon V \to W$.

(a) Find a map $f\colon V \to \mathbb{R}^q$ which minimizes $\beta(f)$ among all smooth maps for a given *surface* V and a given integer $q \ge 1$.

(b) Let V be a closed n-dimensional manifold. Find a (generic) C^∞-map $f\colon V \to \mathbb{R}^n$ for which $\beta(f) \le N$ for some universal constant $N = N(n)$.

Hint. Construct a cobordism W^{n+1} between V and a disjoint union V' of certain standard n-dimensional manifolds such that W^{n+1} is obtained from V' by attaching [$(n+1)$-thickenings of] k-handles for $k \le n-1$. [Compare Gromov-Lawson (1979)].

(b′) Construct, for a given $q \ge n$, a C^∞-map $f\colon V \to \mathbb{R}^q$ which meets each affine $(q-n)$-dimensional subspace A in $\mathbb{R}^q$ at no more than $N = N(n, q)$ points [which means $\beta(f^{-1}(A)) \le N$ for generic maps f].

(b″) Let $\pi_1(V) = 0$ and $n = \dim V \ge 5$. Find a (generic) C^∞-map $f\colon V \to \mathbb{R}^{n-1}$ for which $\beta(f) \le N = N(n)$.

2.1.2 Immersions and Submersions $V \to W$

Fix a submanifold $V_0 \subset V$ of positive codimension, take a generic C^∞-map $F'\colon V \to \mathbb{R}^{q-1}$ and look for a C^∞-function $\bar{f}\colon V \to \mathbb{R}$ for which the map $F' \oplus \bar{f}\colon V \to \mathbb{R}^q$ is a submersion near V_0.

(A) **Lemma.** *The above functions $\bar{f}$ satisfy all forms of the h-principle.*

Proof. An induction by strata of some equisingular stratification of F' (see 1.3.2, 1.4.1, 1.4.2) reduces the lemma to the equisingular case where the map $F'|V_0\colon V_0 \to \mathbb{R}^{q-1}$ has constant rank r_0. Furthermore, the kernel bundle $\operatorname{Ker} D_{F'}|V_0$ is not contained in the tangent bundle $T(V_0) \subset T(V)|V_0$ for *generic* maps F'. [In fact, the inclusion relation $\operatorname{Ker} D_{F'} \subset T(V_0)$ imposes a differential condition on F' which has infinite codimension in the space of jets of infinite order, compare 1.3.2.]

Take an arbitrary 1-form $\bar{\varphi}_0$ on V which does not vanish on $\operatorname{Ker} D_{F'}$. Then there exists a vector field L on V near V_0 such that $L|V_0$ is contained in $\operatorname{Ker} D_{F'} \subset T(V)|V_0$ and such that the function $\bar{\varphi}_0(L)$ does not vanish. Since $\operatorname{Ker} D_{F'} \not\subset T(V_0)$, *generic* fields L with the above properties are semitransversal to V_0. [This is obtained with an obvious computation of the codimension of the "bad" jets of fields, compare (E_2') in 1.3.2.] Now we may assume L semitransversal to V_0 and then we obtain [see (A_2'') in 1.4.3] a homotopy of forms $\bar{\varphi}_t$, $t \in [0, 1]$, which do not vanish on L, and such that the form $\bar{\varphi}_1$ is exact, say $\bar{\varphi}_1 = d\bar{f}$. But the non-equality $\bar{\varphi}_1(L)|V_0 \neq 0$ implies $\bar{\varphi}_1|\operatorname{Ker} D_{F'} \neq 0$ which is equivalent to the sujectivity of the differential of the map $F' \oplus \bar{f}$. Thus we homotopied $\bar{\varphi}_0$ to the desired exact form as required by the h-principle. Furthermore, this argument automatically gives the C^0-dense h-principle for extensions and it also applies to families of functions which amounts to the parametric h-principle.

(A′) **Corollary to the Proof.** *Submersions $\mathcal{O}p\, V_0 \to \mathbb{R}^q$ satisfy all forms of the h-principle.*

Indeed, an arbitrary q-tuple of 1-forms φ_i on V, $i = 1, \ldots, q$, which are linearly independent on $T(V)|V_0$ can be deformed in q steps [compare (B) and the proof of (C) in 2.1.1] to a q-tuple of *exact* forms $\bar{\varphi}_i = d\bar{f}_i$ which are independent on $T(V)|V_0$ as well. Then the map $F = (\bar{f}_1, \ldots, \bar{f}_q)$: $V \to \mathbb{R}^q$ is a submersion on a small neighborhood $\mathcal{O}p\, V_0 \subset V$. Q.E.D.

(B) The *h*-Principle for Submersions. *If V is an open manifold, then submersions of V into an arbitrary manifold W satisfy the parametric h-principle.*

Proof. The manifold V can be identified with an arbitrarily small regular neighborhood of some smooth subpolyhedron V_0 of positive codimension in V (see 1.4.1). Then the (non-parametric) h-principle reduces to the C^0-dense h-principle for submersions $\mathcal{O}p\, V_0 \to W$. To prove the dense h-principle, subdivide V_0 into small simplices Δ such that the image $f_0(\Delta)$ lies in a small neighborhood $U = U(f_0, \Delta) \approx \mathbb{R}^q$, $q = \dim W$ for all Δ. Then the h-principle for (extensions of) submersions $\mathcal{O}p\, \Delta \to U$ [see (A′)] implies (with the standard induction by skeletons) the C^0-dense h-principle for submersions $\mathcal{O}p\, V_0 \to W$. Furthermore, this argument applies to families of submersions $\mathcal{O}p\, V_0 \to W$, thus yielding the parametric h-principle.

(B′) **Remarks.** The above globalization argument which reduces submersions $V \to W$ to those into $\mathbb{R}^q$ applies to all differential relations which are invariant under diffeomorphisms of the target manifold. In particular, *Hirsch's theorem for immersions* $V^n \to W^q$, $q > n$, follows from (C) in 2.1.1. Furthermore, *the h-principle* [due to A. Phillips (1969)] *for maps $V \to W$ transversal to a given foliation of codimension p in W* reduces to the above (A′) provided V is open. Indeed, the problem localizes to the maps $V \to \mathbb{R}^q$ which project to submersions $V \to \mathbb{R}^p = \mathbb{R}^q/\mathbb{R}^{q-p}$. However, this localization does not work for maps $V \to W$ transversal to a non-integrable subbundle $\tau \in T(V)$, where the techniques of 2.2 are needed.

Exercise. Prove the h-principle for free maps $V \to W$, where W is a Riemannian manifold of *constant* sectional curvature and $\dim W \geq [n(n + 3)/2] + 1$.

Hint. This W is locally projectively isomorphic to $\mathbb{R}^q$ and the freedom is a projective invariant of maps $V \to \mathbb{R}^q$.

2.1.3 Folded Maps $V^n \to W^q$ for $q \leq n$

The h-principle is no longer true for immersions $V \to W$ between *equidimensional* manifolds, if V is a *closed* (i.e. compact without boundary) manifold. In fact, no immersion f of a closed manifold V into an open W is possible for $q = n$. This is, of course, obvious, but the following three proofs illustrate different kinds of obstructions to the h-principle.

(1) Compose f with a C^∞-function $\varphi\colon W \to \mathbb{R}$ without critical points [see (B_1) in 1.4.1]. Then, the map f fails to be an immersion at each *maximum* point of $\varphi \circ f\colon V \to \mathbb{R}$.

(2) Assume V and W to be orientable (if not pass to the oriented double coverings) and let ω be a (non-vanishing!) oriented volume form on W. Since $H^n(W, \mathbb{R}) = 0$, the pull-back form $f^*(\omega)$ has $\int_V f^*(\omega) = 0$ by the De Rham-Stokes theorem and so f fails to be an immersion at the zero set of the induced form $f^*(\omega)$ on V.

(3) Since V has no boundary, every equidimensional immersion $f\colon V \to W$ is an open map, and since V is compact, the image $f(V) \subset W$ is compact as well as open. This is absurd, for W is open.

Exercises. (a) Show that a closed surface with a metric of negative curvature admits no isometric C^2-immersion $f\colon V \to \mathbb{R}^3$.

Hint. Study maximum points of the function $\|f\|$ on V.

(b) Let W be a (possibly closed) simply connected $(n+1)$-dimensional manifold with a fixed volume form ω and let L be a non-vanishing divergence free (i.e. $L\omega = 0$) vector field on W. Show that no closed n-dimensional manifold V admits an immersion $f\colon V \to W$ which is everywhere transversal to L.

(c) Let V and W be connected complex analytic manifolds of the same dimension and let V be closed. Suppose there is a holomorphic map $f\colon V \to W$ whose Jacobian is not identically zero. Show that W is also closed and that the map f is onto.

(d) Let $f_0\colon V \to W$ be a submersion, where $q \leq n$ and where V is a closed manifold. Show (assuming W is connected) that every map $f\colon V \to W$ homotopic to f_0 sends V *onto* W.

(e) Prove for an arbitrary submersion $f\colon \mathbb{R}^n \to W^{n-1}$ that $\sup_{w \in W} \operatorname{length} f^{-1}(w) = \infty$. Generalize this to submersions $\mathbb{R}^n \to W^{n-k}$.

Now we turn to those maps $f\colon V^n \to W^q$, $q \leq n$, whose singularity is as simple as possible. Recall (see 1.3.1) that the singularity $\Sigma_f \subset V$ (where f fails to be a submersion) has $\operatorname{codim} \Sigma_f = q - 1$ for generic maps f.

Basic Example. Let $U_0 = V_0 \times \mathbb{R}^{q-n+1}$ for $\dim V_0 = q - 1$ and let $f_0\colon \mathbb{R}^{q-n+1} \to \mathbb{R}$ be the polynomial $f_0(x) = f_0(x_1, \ldots, x_{q-n+1}) = \sum_{i=1}^{p} x_i^2 - \sum_{i=p+1}^{n-q+1} x_i^2$. Then the singularity of the map $\tilde{f}: U_0 \to V_0 \times \mathbb{R}$ given by $\tilde{f}(v, x) = (v, f_0(x))$ is supported on $V_0 = V_0 \times 0 \subset U$ and this is called *the standard folding* (along V_0) of type $|\sigma| = |n - q + 1 - 2p|$.

An arbitrary smooth map $f: V \to W$ is said *to fold with type* $|\sigma|$ along a $(q-1)$-dimensional submanifold $V' \subset V$ if each point $v \in V'$ admits a (small) neighborhood $V_0 \subset V'$ and a split neighborhood $U_0 = V_0 \times \mathbb{R}^{q-n+1} \subset V$ such that the map $f|U_0$ decomposes into $f = I_0 \circ \tilde{f}$, where $\tilde{f}: U_0 \to V_0 \times \mathbb{R}$ is the standard folding of type $|\sigma|$ and where $I_0: V_0 \times \mathbb{R} \to W$ is a C^∞-immersion.

Remark. One can often take $V_0 = V'$. This is possible, for example, for $q = n$ if the submanifold $V' \subset V$ is normally orientable (compare 1.3.1).

Theorem (Eliashberg 1972). *Let $V_i \subset V$, for $i = 0, \ldots, \sigma_0$, be disjoint (non-empty!) $(q-1)$-dimensional properly embedded submanifolds where σ_0 is the greatest integer $\leq (n - q + 1)/2$. If $q \geq 2$, then the C^∞-maps $f: V \to W$ which fold along V_i with type $n - q + 1 - 2i$ for $i = 0, \ldots, \sigma_0$, and which are non-singular outside $\bigcup_i V_i$ satisfy the C^0-dense h-principle.*

We shall sketch below the proof for $q = n$ and we refer to Eliashberg's 1972-paper for the general case. [Compare Yoshifumi (1982).]

Remark. Consider a smooth family of functions $f_v: \mathbb{R}^{q-n-1} \to \mathbb{R}$, $v \in V_0$, and define $F: U_0 \to V_0 \times \mathbb{R}$ by $F(v, x) = (v, f_v(x))$. Then the singularity of the map F consists of those pairs $(v, x) \in U_0$, for which the point x is critical (i.e. singular) for the function f_v. This connection between singularities of maps and critical points of families of functions plays a crucial role in the study of both. In particular, the equidimensional folding theorem is obtained with a simple (one-dimensional) analysis of functions $f_v: [0, 1] \to \mathbb{R}$ as follows.

(A) **Lemma.** *Let Ω_+ be an open subset in $V_0 \times [0, 1]$ and let $\varphi_0 = \varphi_0(v)$ be an arbitrary non-negative C^∞-function on V_0. If the projection $V_0 \times [0, 1] \to V_0$ sends Ω_+ onto V_0, then there exists a non-negative C^∞-function φ on $V_0 \times [0, 1]$ whose support lies in Ω_+ and such that $\int_0^1 \varphi(v, t)\,dt = \varphi_0(v)$ for all $v \in V_0$.* [Compare (B) in 2.4.1.]

Proof. With a partition of unity in V_0, the lemma reduces to the obvious case where the support $S \subset V_0$ of φ_0 satisfies $S \times t \subset \Omega_+$ for some $t \in [0, 1]$.

Let $U'_0 = V_0 \times [0, 1]$ and let $\Sigma = \Sigma(\psi) \subset U'_0$ be the zero set of a C^∞-function $\psi: U'_0 \to \mathbb{R}$. Denote by $\Omega_+ \subset U'_0$ the subset where $\psi > 0$ and let Ω_- be the subset where $\psi < 0$. Let $f_0: \mathcal{O}p\,\partial U'_0 \to \mathbb{R}$ be a C^∞-function, such that $df_0/dt = \psi|\mathcal{O}p\,\partial U'_0$.

(A') **Lemma.** *If the subset Ω_+ as well as Ω_- goes onto V_0 under the projection $U'_0 = V_0 \times [0, 1] \to V_0$, then there exists a function $f: U'_0 \to \mathbb{R}$ such that*

(a) $f|\mathcal{O}p\,\partial U'_0 = f_0$,

(b) $\left.\frac{df}{dt}\right| \mathcal{O}p\Sigma = \psi$,

(c) *the derivative df/dt is positive on Ω_+ and negative on Ω_-.*

Proof. One easily constructs with (A) a C^∞-function φ on U_0' which equals ψ on $\mathcal{O}p\Sigma$, whose derivative $d\varphi/dt$ equals f_0 on $\mathcal{O}p\partial U_0'$ and such that

(i) φ is positive on Ω_+ and negative on Ω_-,
(ii) $\int_0^1 \varphi(v,t)\,dt = f_0(v,1) - f_0(v,0)$.

Then the function $f(v,t) = f_0(v,0) + \int_0^t \varphi(v,\tau)\,d\tau$ is the required one.

Remark. This lemma establishes, in fact, the h-principle for the functions f in question, since the linear homotopy of ψ to df/dt does not destory the pertinent properties of ψ.

(B) **Exercises.** (a) Consider a non-vanishing C^∞-vector field L on a manifold V. Let ψ be a C^∞-function on V such that every orbit of L meets Ω_+ (where $\psi > 0$) as well as Ω_-. Prove the existence of a C^∞-function f on V whose Lie derivative Lf is positive on Ω_+, negative on Ω_- and $Lf|\mathcal{O}p\Sigma = \psi|\mathcal{O}p\Sigma$ for the zero set Σ of ψ.

(a′) Define by induction $L^kf = L(L^{k-1}f)$ and say that critical points of f are *k-nondegenerate along L* if the equality $Lf(v) = 0$ (which characterizes critical points v of f along L) implies $L^if(v) \neq 0$ for some $i \leq k$. Show that functions f with k-nondegenerate critical points along V satisfy the h-principle for every $k \geq 3$. In particular, prove the existence of an f whose critical points along L are 3-nondegenerate.

(b) Study counterexamples for the above h-principle for $k = 2$. Consider, in particular, $V = S^{2m+1}$ where the vector field L is the infinitesimal generator of the standard S^1-action on S^{2m+1}. Prove for an arbitrary C^∞-function f on S^{2m+1} that the subset $\Sigma^1 \subset S^{2m+1}$ where Lf and L^2f vanish has codimension ≤ 3 in S^{2m+1}. Denote by $K^1 \subset \mathbb{C}P^m = S^{2m+1}/S^1$ the image of Σ^1 under the quotient (Hopf) map $h: S^{2m+1} \to \mathbb{C}P^m$ and prove the restriction homomorphism $H^{2m-2}(\mathbb{C}P^m) \to H^{2m-2}(K^1)$ to be non-zero.

(b′) Denote by $K_i \subset \mathbb{C}P^m$ the subset of those points $x \in \mathbb{C}P^m$ for which the function $f|S_x^1$ has at least $2i$ critical points on the circle $S_x^1 = h^{-1}(x)$, where f is an arbitrary smooth function on S^{2m+1}. Show the homomorphism $H^{2m-2i+2}(\mathbb{C}P^m) \to H^{2m-2i+2}(K_i)$ to be non zero. In particular, the function $f|S_x^1$ has at least $2m+2$ critical points for some circle S_x^1.

(b″) Let $C^0(V)$ be the space of continuous functions on a compact n-dimensional manifold V and let X be a k-dimensional linear subspace in $C^0(V)$. Prove the existence of a non-zero function $\varphi \in X$ whose maximum set $M_\varphi \subset V$ (where φ assumes the *absolute* maximum) contains at least s points, where s is the first integer $\geq (k-1)/(n+1)$.

Hint. Study the convex hull of $F_X(V) \subset \mathbb{R}^k$, where the map $F_X: V \to \mathbb{R}^k$ is given by k independent functions in L; alternatively, study the set valued map $\varphi \mapsto M_\varphi$ for all $\varphi \in L$, whose pertinent property is $M_\varphi \cap M_{-\varphi} = \phi$ for $\varphi \neq 0$.

(C) **Poenaru's Pleating Lemma.** Consider a C^∞-map $F\colon U_0' = V_0 \times [0,1] \to W$, where $\dim W = n = \dim U_0'$, such that $F_t = F|V_t = V_0 \times t$ is an immersion $V_t \to W$ for all $t \in [0,1]$. If the map F is also an immersion on $\mathcal{O}\!p\, V_0 \subset U_0'$ then the (non-vanishing) field $\left.\frac{\partial F}{\partial t}\right| V_0$ induces an orientation in the (one-dimensional) normal bundle of the immersion $F_0\colon V_0 \to W$ which uniquely extends to a normal orientation of $F_t\colon V_t \to W$ for all $t \in [0,1]$. Now we assume the manifold V_0 to be connected and the map F to be an immersion on $\mathcal{O}\!p\, V_1 \subset U_0'$ as well as on $\mathcal{O}\!p\, V_0 \subset U_0'$. Thus we get two normal orientations of the immersions F_t. We call the map F *even* if these orientations coincide; otherwise F is *odd*. For example, if F is an immersion on all of U_0', then it is even. If the only singularity of F is a folding along V_t for a single value $t \in [0,1]$, then F is odd.

Take some points $0 = t_0 < t_1 < \ldots < t_{N+1} = 1$, where the parity of N equals that of F.

If V_0 is compact and if $t_{i+1} - t_i \le \varepsilon$ for $i = 0, \ldots, N$ and for some sufficiently small $\varepsilon = \varepsilon(F) > 0$, then there exists a C^∞-map $F^\varepsilon\colon U_0' \to W$ whose only singularity is a folding along $V_{t_1} \cup V_{t_2} \cup \ldots \cup V_{t_N}$ and such that F^ε equals F on $\mathcal{O}\!p(V_0 \cup V_1) \subset U_0'$. Furthermore, the maps F^ε C^0-converge to F as $\varepsilon \to 0$.

Proof. Let $t_i' = \frac{1}{2}(t_i + t_{i+1})$ for $i = 0, \ldots, N$. There obviously exists a C^∞-map $F'\colon U_0' \to W_0$, such that F_t' is arbitrarily C^∞-close to F_t for all $t \in [0,1]$ and F' is an immersion near $V_{t_0'} \cup V_{t_1'} \cup \ldots V_{t_N'}$ in U_0'. Moreover, this F' can be made *odd* on $V_0 \times [t_i', t_{i+1}']$ for all $i = 0, \ldots, N$, and such that $F'|\mathcal{O}\!p(V_0 \cup V_1) = F|\mathcal{O}\!p(V_0 \cup V_1)$. If ε is small, then the map F_t' is close to $F'_{t_{i+1}}$ for $t \in [t_i', t_{i+1}']$, and using this we perturb F' outside $\mathcal{O}\!p(V_{t_0} \cup \ldots \cup V_{t_N'})$ to the required map F^ε as follows. Since the perturbation we are after *separately* applies to each interval $[t_i', t_{i+1}']$, the problem reduces to the proof of the lemma for $N = 1$, where the maps F_t are assumed C^∞-close to F_0 for all $t \in [0,1]$. Then there obviously exist a C^∞-immersion $\bar{F}\colon V_0 \times \mathbb{R} \to W$ and a C^∞-map $\tilde{F}\colon U_0' \to V_0 \times \mathbb{R}$ such that $\bar{F} \circ \tilde{F} = F$ and $\tilde{F}(v,t) = (v, f_0(v))$ for some function f_0 on U_0'. Now we apply (A') to $\Sigma = V_{t_1}$ and to a suitable function ψ on U_0' whose derivative $d\psi/dt$ does not vanish on Σ. Then the function f insured by (A') gives us the map $\tilde{F}'\colon (v,t) \mapsto (v, f(t))$ which folds along Σ and for which the composition $\bar{F} \circ \tilde{F}'\colon U_0' \to W$ is the required map with the only folding along $V_{t_1} = \Sigma$. Q.E.D.

Remark. The maps F_t^ε obtained by this construction are C^∞-close to F_t for all $t \in [0,1]$. This is stronger than the mere C^0-closeness between F^ε and F (compare $C^\perp$-approximation in 2.4.1).

(C') *The h-Principle for Folded Maps* (Poenaru 1966). Let V be a connected n-dimensional manifold, let V_0 be a normally oriented closed hypersurface in V and let $f_0\colon V \to W$, where $\dim W = n$, be a continuous map which lifts to a homomorphism of the tangent bundles, say $\varphi\colon T(V) \to T(W)$, such that

(a) the homomorphism φ is fiberwise injective outside a small neighborhood $\mathcal{O}p\, V_0 \subset V$;
(b) there is a split tubular neighborhood $U_0 = V_0 \times \mathbb{R} \subset V$, which properly contains the above $\mathcal{O}p\, V_0$, such that φ is injective on the tangent bundle $T(V_t) \subset T(V)|V_t$ for $V_t = V_0 \times t$ and for all $t \in \mathbb{R}$.

Then there exists a C^∞-map $f: V \to W$ homotopic to f_0 whose only singularity is a folding along finitely many submanifolds $V_{t_i} \subset V$ where $t_i = i/N + 1$ for $i = 1, \ldots, N$ and for some $N \leq N_0(f)$. Moreover, the differential $D_f: T(V) \to T(W)$ (for a suitable f) can be deformed to φ by a homotopy of homomorphisms which is injective on $T(V)|V \setminus V_0 \times [0,1]$ and on $T(V_t)$ for all $t \in \mathbb{R}$.

Proof. The h-principle holds true for immersions $V \setminus V_0 \to W$ (since $V \setminus V_0$ is open) and for immersions $V_0 \to W$ (since $\dim V_0 < \dim W$). Thus, we obtain a C^∞-map $F: V \to W$ which is an immersion outside $U_0' = V_0 \times [0,1]$ in V and such that $F_t = F|V_t$ is an immersion for all $t \in \mathbb{R}$. Then this F is modified on U_0' to the desired folded map $f: V \to W$.

Corollary. *If the manifold V is stably parallelizable* (i.e. $V \times \mathbb{R}$ is parallelizable) *then there is a C^∞-map $f: V \to \mathbb{R}^n$ whose only singularity is a folding along some closed (possibly disconnected) normally oriented hypersurface in V.*

An Application to the Signature Theorem. Let V be a closed oriented $4k$-dimensional manifold and let $\sigma(V)$ denote the signature of the intersection form on the homology $H_{2k}(V; \mathbb{R})$. The famous theorem of Thom-Hirzebruch claims the existence of a universal polynomial L in the Pontryagin classes p_i of V, such that $\langle L(p_i), [V] \rangle = \sigma(V)$. In particular, $\sigma(V) = 0$ for *stably parallelizable manifolds* V. Let us reduce this vanishing theorem to the above corollary with the following

(C″) **Lemma.** *Suppose there exists a Riemannian C^∞-metric g on V and an open subset $U \subset V$ which admits an orientation reversing isometric involution $I: (U, g) \to (U, g)$ and such that the complement $V \setminus U$ is a stratified subset of codimension one in V. Then $\sigma(V) = 0$.*

Warning. An involution I on U which preserves $g|U$ may be discontinuous on $V \supset U$.

Proof. Since the signature is the index of some elliptic differential operator on V associated to g, it can be expressed by an integral over the (oriented!) manifold V of some universal polynomial in covariant derivatives of the curvature tensor of g, say $\sigma(V) = \int_V P\, dv$. Then $\int_V P\, dv = \int_U I^*(P)(-du) = -\int_U P\, du = 0$. [See Atiyah (1976) for a conceptional explanation of this "locality" of σ.]

Now, if V is stably parallelizable, it admits a C^∞-map $f: V \to \mathbb{R}^n$ which folds along some normally oriented hypersurface $V' \subset V$. Then there exists a C^∞-function f_0 on V with support in a small neighborhood $U' \subset V$ of V', such that the map $f' = f \oplus f_0: V \to \mathbb{R}^{n+1} = \mathbb{R}^n \times \mathbb{R}$ is an immersion which is symmetric on U' in the hyperplane $\mathbb{R}^n \times 0 \subset \mathbb{R}^{n+1}$. Let $U \subset V$ be the maximal open subset whose image $f(U) \subset \mathbb{R}^n$ does not meet $f(V')$. Then the induced metric g on $V \subset \mathbb{R}^{n+1}$ obviously has the required involution on U. Q.E.D. [See (G) in 2.2.7 for a similar study of non stably paralellizable manifolds.]

(D) *The Equidimensional Folding Theorem of Eliashberg.* Take a normally oriented hypersurface Σ in V and let $\Sigma \times [-1,1] \subset V$ be the tubular neighborhood of Σ. Let the space $\bar{V}$ be obtained from V by identifying the points (σ, t) and $(\sigma, -t)$ for all $(\sigma, t) \in \Sigma \times [-1,1]$ and let $\bar{\rho}$: $V \to \bar{V}$ be the obvious map. There is a unique vector bundle $\bar{T} \to \bar{V}$ such that the induced bundle $\tilde{T} = \rho^*(\bar{T}) \to V$ is canonically isomorphic on $V \backslash \Sigma$ to the tangent bundle $T(V)|V\backslash\Sigma$.

Example. If Σ is an equator in the sphere S^n then $\tilde{T}$ is the trivial bundle which is induced from the tangent bundle of the ball B^n by the obvious map $S^n \to B^n$ which sends Σ onto ∂B^n.

If a smooth map f: $V \to W$, for $\dim V = \dim W$ folds along Σ, then the differential of f naturally defines a homomorphism $\tilde{D}_f$: $\tilde{T} \to T(W)$, which equals D_f outside Σ and which is a fiberwise isomorphism over Σ.

Theorem. *Let a continuous map f_0: $V \to W$ lift to a fiberwise isomorphic homomorphism φ: $\tilde{T} \to T(W)$. Then for $n \geq 2$ there exists a C^∞-map f: $V \to W$ which folds along Σ and such that the homomorphism $\tilde{D}_f$ is homotopic to φ by a homotopy of fiberwise isomorphic homomorphisms.*

Warning. This theorem is false for $n = 1$ and also for $n \geq 2$ if Σ is empty.

Proof. Take an arbitrary normally oriented closed hypersurface $V_0 \subset V$ which transversally meets Σ over a non-empty $(n-1)$-dimensional manifold $\Sigma \cap V_0$. Fix a split tubular neighborhood $U_0 = V_0 \times \mathbb{R}$ of V_0 such that the projection $U_0 \to V_0$ is an *immersion* of $\Sigma \cap U_0$ into V_0. Then we reduce the proof [compare the proofs of (C) and (C′)] to the following special case.

Let f_0 be a smooth function on $U_0' = V_0 \times [0,1]$, such that the map $\tilde{F}$: $U_0' \to U_0$ given by $\tilde{F}(v,t) = (v, f_0(t))$ satisfies the following two conditions,
(1) the only singularity of $\tilde{F}$ near $V_0 \cup V_1 \subset U_0'$ is a folding along $\Sigma \cap \mathcal{O}p(V_0 \cup V_1)$,
(2) there exists a C^∞-function ψ on U_0', whose zero set equals Σ, whose derivative $d\psi/dt$ does not vanish on Σ and such that ψ equals f_0 near $V_0 \cup V_1$ [compare (A′)].

If the projection $U_0' \to V_0$ were surjective on Ω_+ and Ω_- we could prove the theorem by modifying $\tilde{F}$ inside U_0' with an appropriate function f [compare the proof of (C′)]. To achieve this surjectivity we start with a small embedded circle $S^1 \subset V$ which transversally meets Σ at two points. Then we consider a small tubular neighborhood U_1 of S^1 in V and take $V_0 = \partial U_1 \approx S^1 \times S^{n-2}$, such that the intersection $\Sigma \cap U_0'$ becomes the union of two disjoint copies of $S^{n-2} \times [0,1]$ embedded into $U_0' = S^1 \times S^{n-2} \times [0,1]$, whose projections to $V_0 = S^1 \times S^{n-2}$ are immersions. Furthermore, we can (and we do) arrange this V_0 such that each segment $s \times [0,1]$ in $\Sigma \cap U_0'$ goes to S^1 under either projection by an *orientation preserving immersion.*

Now, let $\tau\colon [0,1] \to S^1$ be a (surjective) C^∞-map, such that,

(i) $\tau(t) \equiv e$ for $t \in [0, \frac{1}{4}] \cup [\frac{3}{4}, 1]$, where e is the neutral element of the additive group S^1,

(ii) τ is an orientation preserving immersion on the open interval $(\frac{1}{4}, \frac{3}{4})$.

Define the following diffeomorphism $\tilde{\tau}$ of V which is the identity outside U_0' and which acts on $U_0' = S^1 \times S^{n-2} \times [0,1]$ by $\tilde{\tau}\colon (s, s', t) \mapsto (s + \tau(t), s', t)$. The manifold $\tilde{\Sigma} = \tilde{\tau}(\Sigma) \cap U_0'$ retains the properties (1) and (2) we insisted upon earlier, but now $\tilde{\Sigma}$ goes *onto* V_0 under the projection $U_0' \to V_0$. Hence, the pertinent sets Ω_+ and Ω_- also go onto V_0 and then (A′) applies. Q.E.D.

(D′) **Exercise.** Let V be a connected manifold with a non-empty boundary ∂V. Fix a collar $C = \partial V \times [0,1] \subset V$ and let a finite group G freely act on C preserving $\partial V_0 \times 0 \subset C$. Prove the h-principle for those C^∞-immersions $f\colon V \to W$ for $\dim V \geq 2$, which satisfy $f(gv) = f(v)$ for all pairs $(g, v) \in G \times C$. [See (G) in 2.2.7 for a general result of this kind.]

(D″) **Additional Remarks and Exercises.** The h-principle of Eliashberg fails for $\dim W = 1$ since the Morse theory gives additional restrictions (which are not accounted for by the h-principle) on the critical points of maps of V into $\mathbb{R}$ and into S^1. A similar situation arises for maps $f\colon V \to W^q$, $q \geq 2$, if the topology of the map $f|\Sigma\colon \Sigma \to W^q$ is brought forward. [Compare Serf (1984).]

(a) Consider a C^∞-map f of a closed manifold V into $\mathbb{R}^2$ which folds along a union of circles $\Sigma = \bigcup_{i=1}^k S_i$, and such that the immersion $f|\Sigma\colon \Sigma \to \mathbb{R}^2$ has d transversal double points and no triple point. Establish the following bound on the sum of the Betti numbers $\beta(f^{-1}(w))$

$$\beta(f^{-1}(w)) \leq k + d \qquad \text{for all } w \in \mathbb{R}^2.$$

(b) Prove that $\beta(V) \leq (k + d)(2k + 3d)$.

(E) *Lagrange and Legendre Immersions.* Let α denote the standard (linear differential) form $\sum_{i=1}^n x_i\,dy_i$ on $\mathbb{R}^{2n}$ whose differential $h = d\alpha = \sum_{i=1}^n dx_i \wedge dy_i$ is *the standard symplectic* 2-form. A C^1-map $f\colon V \to \mathbb{R}^{2n}$ is called *Lagrange* if $f^*(h) \equiv 0$ on V, which amounts to the identity $d\sum_{i=1}^n X_i\,dY_i = 0$ for the coordinate functions $X_i, Y_i\colon V \to \mathbb{R}$ of f.

A Lagrange map $f = (X, Y) = (X_1, \ldots, X_n, Y_1, \ldots, Y_n)\colon V \to \mathbb{R}^{2n}$ is called *exact* if the induced form $f^*(\alpha) = \sum_{i=1}^n X_i\,dY_i$ on V is exact. This can be expressed with the *contact form* $\beta = \sum_{i=1}^n x_i\,dy_i + dz$ on $\mathbb{R}^{2n+1}$ by defining the exactness of f as the existence of a lift of f to a *Legendre* map $F = (f, Z)\colon V \to \mathbb{R}^{2n+1}$ for some C^1-function Z on V, where Legendre maps $F\colon V \to \mathbb{R}^{2n+1}$, by definition, are those for which $F^*(\beta) = \sum_{i=1}^n X_i\,dY_i + dZ \equiv 0$ on V.

Examples. An arbitrary smooth map $f = (X, Y)\colon V \to \mathbb{R}^{2n}$ is the sum of two exact Lagrange maps, $f = (0, Y) + (X, 0)$.

Let T^n be the n-torus with the cyclic coordinates t_i, $i = 1, \ldots, n$. Then the map $(X_i = \sin t_i, Y_i = \cos t_i)\colon T^n \to \mathbb{R}^{2n}$ is a Lagrange C^∞-embedding but it is not exact.

Proposition. *Let $Y: V \to \mathbb{R}^n$, $n = \dim V$, be a C^∞-map whose only singularity is a folding along a normally oriented hypersurface $V_0 \subset V$. Then the map Y lifts to an exact Lagrange C^∞-immersion $f = (X, Y): V \to \mathbb{R}^{2n}$ for some C^∞-map $X = (X_1, \ldots, X_n): V \to \mathbb{R}^n$.*

Proof. Let $U = V_0 \times [-1, 1] \subset V$ be a tubular neighborhood of V_0 which admits a C^∞-immersion $\bar{Y}: U \to \mathbb{R}^n$ such that $\bar{Y}(v, t^2) = Y(v, t)$ for all points $(v, t) \in U$, and let Z be a C^∞-function on V, such that $Z|U = t^3$. Since the map Y is an immersion outside V_0, the differentials dY_i, $i = 1, \ldots, n$, span the cotangent bundle $T^*(V)|V \setminus V_0$. Hence, there exist unique C^∞-functions X_i, $i = 1, \ldots, n$ on $V \setminus V_0$, for which $\sum_{i=1}^n X_i \, dY_i = -dZ$. Furthermore, there are unique C^∞-functions $\bar{X}_i$ on U, such that $\sum_{i=1}^n \bar{X}_i \, d\bar{Y}_i = dt$. Then the functions $X_i'(v, t) = -\frac{3}{2} t \bar{X}_i(v, t^2)$ satisfy $\sum_{i=1}^n X_i' \, dY_i = -3t^2 \, dt = dt^3$ on U. Hence, $X_i'|U \setminus V_0 = X_i|U \setminus V_0$, which shows the functions X_i to be C^∞-smooth near V_0 as well as outside V_0.

Since the form $dt = \sum_{i=1}^n \bar{X}_i \, d\bar{Y}_i$ does not vanish, the map $X = (X_1, \ldots, X_n): U \to \mathbb{R}^n$ also does not vanish, and hence, the derivative

$$\frac{\partial X(v, t)}{\partial t} = -\tfrac{3}{2} \bar{X}(v, t^2) - \tfrac{3}{2} t \, d\bar{X}(v, t^2)$$

does not vanish on $V_0 = V_0 \times 0 \subset U$. Therefore, the exact Lagrange C^∞-map $f = (X, Y): V \to \mathbb{R}^{2n}$ is an immersion. Q.E.D.

Corollary. *An arbitrary stably parallelizable manifold V (for example $V = S^n$) admits an exact Lagrange C^∞-immersion $f: V \to \mathbb{R}^{2n}$.*

Indeed, such a V admits the required folded map $Y: V \to \mathbb{R}^n$ by Poenaru's theorem.

Exercises. (a) Show the normal bundle of an arbitrary Lagrange immersion $f: V \to \mathbb{R}^{2n}$ to be isomorphic to the tangent bundle $T(V)$. (This in particular implies the vanishing of the rational Pontryagin classes of V.)

(a′) Show that the existence of a Lagrange *embedding* of a closed orientable manifold V into $\mathbb{R}^{2n}$ implies the vanishing of the Euler characteristic of V. (One knows that no closed manifold $V = V^n$ admits an *exact* Lagrange *embedding* into $\mathbb{R}^{2n}$, compare 3.4.4.)

(b) Prove the h-principle for exact Lagrange immersions $V \to \mathbb{R}^{2n}$ by lifting to $\mathbb{R}^{2n}$ generic C^∞-maps $Y: V \to \mathbb{R}^n$. (See 3.4.2 for a different approach to this h-principle.)

2.1.4 Singularities and the Curvature of Smooth Maps

Consider an oriented n-dimensional manifold V and an immersion $f: V \to \mathbb{R}^{n+1}$. The orientation of V defines a unit normal vector ν at each point $v \in V$ and thus we get the normal (Gauss) map $G_f: V \to S^n \subset \mathbb{R}^{n+1}$. The Jacobian of G_f equals the product of the principal curvatures of the hypersurface $V \to \mathbb{R}^{n+1}$, say

$J(v) = \prod_{i=1}^{n} \alpha_i(v, \nu)$, where the choice of the unit normal ν at v determines the signs of the curvatures $\alpha_i(v)$, such that $\alpha_i(v, -\nu) = -\alpha_i(v, \nu)$. If the manifold is closed, then, obviously, $\int_V J(v)\,dv = (\deg G_f)\operatorname{Vol} S^n$. Furthermore, if n is even then $\deg G_f = \frac{1}{2}\chi(V)$ for the Euler characteristic $\chi(V)$ (compare 1.2.1).

Exercise. Let $V \subset \mathbb{R}^{n+1}$ be a smooth closed *embedded* hypersurface and let $x_1, \ldots, x_{n+1}$ be linearly independent vectors in $\mathbb{R}^{n+1}$. Define

$$C_i = \{v + tx_i, t \mid v \in V \subset \mathbb{R}^{n+1}, t \in [0, \infty)\} \subset \mathbb{R}^{n+1} \times [0, \infty)$$

for $i = 1, \ldots, n+1$ and let $C_0 = V \times [0, \infty) \subset \mathbb{R}^{n+1} \times [0, \infty)$. The cylinders C_i have no common point for small $t > 0$ and they are disjoint for large $t \to \infty$. Hence, small generic perturbations C_i' of C_i transversally meet at finitely many points away from $t = 0$. Prove the algebraic (i.e. counted with properly chosen $\pm$ signs) number of points of the intersection $\bigcap_{i=0}^{n+1} C_i'$ to be equal to the degree d of the normal map $V \to S^n$ (that is the Euler characteristic of the compact domain in $\mathbb{R}^{n+1}$ bounded by V, see 1.2.1). Assume $d \neq 0$ and prove the existence of some points $v_i \in V$, $i = 0, \ldots, n+1$, such that $v_i - v_0 = tx_i$, $i = 1, \ldots, n+1$, for some $t > 0$. Show, in particular, the *existence of a regular* $(n+1)$*-simplex in* $\mathbb{R}^{n+1}$ *whose vertices lie in* V.

Question. Do embeddings $V \to \mathbb{R}^{n+1}$ for which the intersection $\bigcap_{i=0}^{n+1} C_i$ is empty away from $t = 0$ (for fixed vectors x_i) satisfy some h-principle?

(A) *A Lower Bound on the Total Curvature.* Let $f_x\colon V \to \mathbb{R}$ denote the orthogonal projection to the line in $\mathbb{R}^{n+1}$ through a given point $x \in S^n$, that is $f_x(v) = \langle f(v), x\rangle$. Then the singular set of the function f_x (i.e. the set of *critical points* of f_x) equals the pull-back $G_f^{-1}\{x \cup -x\} \subset V$. Furthermore, the critical points of f_x are non-degenerate if and only if the points x and $-x \in S^n$ are *non-critical* values of G_f, that is $\pm x \in S^n \setminus G_f(\Sigma)$ for the zero set Σ of the Jacobian of G_f. Then $2 \deg G_f = \sum_v (-1)^{\operatorname{ind} v}$, where v runs over the critical point of f_x. This yields the equality $\deg G_f = \frac{1}{2}\chi(V)$ for n even.

If the immersion $f\colon V \to \mathbb{R}^{n+1}$ is C^∞-smooth, then the map $G_f\colon V \to S^n$ also is C^r and by Sard's theorem almost all $x \in S^n$ are non-critical values of G_f. Thus, critical points of f_x are discrete and non-degenerate for almost all x and their number $c(x)$ obviously satisfies $\int_{S^n} c(x)\,dx = \int_V |J(v)|\,dv$. Therefore, Morse inequalities bound from below the *total curvature* $\int_V |J(v)|\,dv$ of *closed* manifolds $V \to \mathbb{R}^{n+1}$ (see Chern-Lashof 1957; Kuiper 1958),

$$(*) \qquad \int_V |J(v)|\,dv \geq \tfrac{1}{2}\beta(V)\operatorname{Vol} S^n$$

for the sum of the Betti numbers $\beta(V) = \sum_{i=0}^{n} b_i(V)$ with an arbitrary coefficient field.

Observe that $(*)$ provides an obstruction to the h-principle for immersions $V \to \mathbb{R}^{n+1}$ with a prescribed curvature $J = J(v)$.

(A′) *Non-orientable Manifolds* V. The inequality $(*)$ makes perfect sense without any orientation in V and the above proof immediately extends to non-orientable manifolds V. Furthermore, the normal map is naturally defined (regardless of an orientation in V) on the unit normal bundle of V in $\mathbb{R}^{n+1}$ which is (in the codimension one case) a double covering $\tilde{V} \to V$. Thus Hopf's formula for $\chi(V)$ generalizes to $\int_{\tilde{V}} J(\tilde{v})\, d\tilde{v} = (\operatorname{Vol} S^n)\chi(V)$, where $J(\tilde{v})$ is the Jacobian of the normal map $\tilde{V} \to S^n$ which equals the (signed!) curvature at the point v under $\tilde{v}$.

Exercises. (a) Generalize the above to immersions $V \to \mathbb{R}^q$, $q > n$, by considering the normal map of the unit normal bundle of $V \looparrowright \mathbb{R}^q$ to S^{q-1}.

(b) Suppose some m among $n + 1$ Betti numbers $b_i(V)$ are non-zero. Show that the complement $V \backslash \Sigma$ to the singularity Σ of the normal map G_f: $V \to S^n$ has $k \geq m/2$ components. (This indicates an obstruction to the h-principle for immersions f whose normal map has a prescribed singularity.)

(c) Let v_τ be the number of cusps of the orthogonal projection of $V \subset \mathbb{R}^q$ onto the 2-plane $\tau \subset \mathbb{R}^q$, for all $\tau \in Gr = Gr_2 \mathbb{R}^q$. Find a density function μ on V expressible in terms of the curvature of $V \subset \mathbb{R}^q$ and first derivatives of the curvature, such that $\int_V \mu\, dv = \int_{Gr} v_\tau\, d\tau$. [Consult Morin (1965).]

(B) Let us generalize (A) by allowing an arbitrary smooth map f: $V \to \mathbb{R}^{n+1}$ whose singularities $\Sigma^i = \{v \in V \mid \operatorname{rank}_v f = n - i\}$ are stratified subsets in V. The curvature $J(v)$ is now defined on the non-singular locus Σ^0 of f only.

If $\dim \Sigma^i \leq n - i - 1$ *for* $i = 1, \ldots, n$, *(which is the case for generic maps f) then* $\int_{\Sigma^0} |J(v)|\, dv \geq \frac{1}{2}\beta(V) \operatorname{Vol} S^n$. *Furthermore, if* $\int_{\Sigma^0} |J(v)|\, dv < \infty$ *and n is even, then* $\int_{\Sigma^0} J(v)\, dv = \frac{1}{2}\chi(V) \operatorname{Vol} S^n$. [If Σ_0 is non-orientable then $\frac{1}{2}\int_{\tilde{\Sigma}^0} J(\tilde{v})\, d\tilde{v}$ is used instead of $\int J(v)\, dv$, compare (A′).]

Proof. Since $\dim \Sigma^i \leq n - i - 1$ the critical set of the function f does not meet Σ^i for almost all $x \in S^n$ and the argument in (A) applies.

(B′) *Maps* f: $V \to \mathbb{R}^q$ *for* $2 \leq q \leq n$. Let the singularity $\Sigma^i = \{v \in V \mid \operatorname{rank}_v f \leq q - i\}$ be a stratified subset for all $i = 0, 1, \ldots, q$, and let $\dim \Sigma^i \leq q - i - 2$ (which is a generic condition on f). Let $\Sigma_1^1 \subset \Sigma^1$ be a stratified subset of dimension $\leq q - 2$, such that the complement $\Sigma_0^1 = \Sigma^1 \backslash \Sigma_1^1$ is a smooth $(q - 1)$-dimensional submanifold in V on which the map f is an immersion $\Sigma_0^1 \to \mathbb{R}^q$. The critical set of the function f_x: $V \to \mathbb{R}$ now lies in Σ_0^1 for almost all $x \in S^{q-1}$, which implies the following bound on the curvature J of the hypersurface Σ_0^1 in $\mathbb{R}^q$,

$$(**) \qquad \int_{\Sigma_0^1} |J(v)|\, dv \geq \tfrac{1}{2}\beta(V) \operatorname{Vol} S^{q-1},$$

Hence, the geometry of Σ_f (unlike the topology, see 2.1.3) is not accountable for by the h-principle.

(B″) **Hopf's Formula.** Let $K \to \Sigma_0^1$ denote the kernel $\operatorname{Ker} D_f \subset T(V) | \Sigma_0^1$ (which is a $(n - q + 1)$-dimensional bundle over Σ_0^1) and let $N \to \Sigma_0^1$ be the (one-dimensional) normal bundle of the immersion $f | \Sigma_0^1$: $\Sigma_0^1 \to \mathbb{R}^q$. Assume the second differential D_f^2: $K \to N$ to be non-singular (as a quadratic form) outside a stratified subset Σ' of dimension $\leq q - 2$ in Σ_0^1 (which is a generic condition on f) and let $\Sigma_0 = \Sigma_0^1 \backslash \Sigma'$.

Take a unit normal vector $\tilde{v} \in N_v$, $v \in \Sigma_0$, and define $s(\tilde{v}) = \pm 1$ as follows. The vector $\tilde{v}$ defines an isomorphism $\tilde{\alpha}: N_v \rightleftarrows \mathbb{R}$ for $\tilde{\alpha}(\tilde{v}) = 1$ and then $\tilde{D}_f = \tilde{\alpha} \circ \tilde{D}_f | K_v$ is a quadratic polynomial $K_v \to \mathbb{R}$, that is $\sum_{i=1}^{p} y_i^2 - \sum_{i=p+1}^{n-q+1} y_i^2$ for some basis in K_v. Then we put $s(\tilde{v}) = (-1)^p$, and we have the following integral formula for the curvature $J(\tilde{v})$ of Σ_0 for $n = \dim V$ even.

If $\int_{\Sigma_0} |J(v)|\, dv < \infty$, *then*

$$(+) \qquad \int_{\tilde{\Sigma}_0} s(\tilde{v}) J(\tilde{v})\, d\tilde{v} = \chi(V) \operatorname{Vol} S^{q-1},$$

where $\tilde{\Sigma}_0$ is the (zero-dimensional) unit normal bundle of $\Sigma_0 \to \mathbb{R}^q$ with the (naturally oriented) volume element $d\tilde{v}$.

Proof. The index of each critical point $v \in \Sigma_0 \subset V$ of the function $f_x: V \to \mathbb{R}$ is the sum of $\operatorname{ind}_v(f_x | \Sigma_0)$ and of the index of $\tilde{D}_f: K_v \to \mathbb{R}$ which corresponds in the chosen normal $\tilde{v}$ at v. Hence, the argument in (A) applies.

(C) *On the Convergence of* $\int_{\Sigma_0} |(J(v)|\, dv$. Let $c(x)$ denote the number of critical points of the function $f_x | \Sigma_0$. Then

$$\int_{\Sigma_0} |J(v)|\, dv = \int_{S^{q-1}} c(x)\, dx.$$

Hence, $\int_{\Sigma_0} |J(v)|\, dv < \infty$ provided the function $c(x)$ is bounded on S^{q-1}.

Corollary. *If V is a closed real analytic manifold and the map $f: V \to \mathbb{R}^q$ is C^{an}, then* $\int_{\Sigma_0} |J(v)||\, dv < \infty$.

Proof. This is obvious in the real *algebraic* case as $c(x)$ is bounded (according to Bezout's theorem) by the algebraic degrees of V and X. Moreover, the results of Hironaka (1973) on *subanalytic* sets imply the following general fact which suffices for our purpose.

Let X and Y be compact real analytic spaces and let $h: X \to Y$ be a real analytic map. Then for each $i = 0, 1, \ldots,$ the Betti number $b_i(h^{-1}(y))$ (with $\mathbb{Z}/2\mathbb{Z}$ coefficients) is bounded. Namely, $b_i(h^{-1}(y)) \leq \text{const} = \text{const}(X, Y, h)$, for all $y \in Y$.

Observe that the curvature J is (obviously) expressible in terms of the vector bundles homomorphism $D_f: T(V) \to T(\mathbb{R}^q)$ and that C^∞-automorphisms of the bundles $T(V)$ and $T(\mathbb{R}^q)$ change the integral $\int |J(v)|\, dv$ by a *bounded* factor. According to Mather (1973), a homomorphism $T(V) \to T(\mathbb{R}^q)$ transversal to Whitney-Thom singularities is locally reducible by some automorphisms of the bundles to a canonical C^{an}-form. This implies *the convergence of* $\int |J(v)|\, dv$ *for generic C^∞-maps* $f: V \to \mathbb{R}^q$. [See Burago (1968) and Levin (1971) for another approach to Hopf's formula in the presence of singularities.]

(D) Exercises. (a) Construct a C^∞-map $f\colon S^2 \to \mathbb{R}^2$ which folds along the equator $S^1 \subset S^2$ and such that the curvature J of $f|S^1\colon S^1 \to \mathbb{R}^2$ vanishes at exactly two points in S^1, and $\int_{S^1} |J(v)|\,dv = C$ for a given constant $C > 2\pi$.

(a′) Find a necessary and sufficient condition on a 1-form ω on S^1 for the existence of a C^∞-map $f\colon S^2 \to \mathbb{R}^2$ with the only fold along S^1 and such that $J(v)\,dv$ on S^1 equals ω.

(a″) Divide a closed surface V of genus 2 by a circle $S^1 \subset V$ into two punctured tori. Study the curvature J on S^1 of C^∞-maps $V \to \mathbb{R}^2$ which folds along S^1 and have no other singularities.

(b) Let V be a closed 3-dimensional manifold and let $f\colon V \to \mathbb{R}^3$ be a generic map such that $\operatorname{rank}_v f \geq 2$ for all $v \in V$. Then the singularity $\Sigma_f^1 \subset V$ is a smooth surface, while the map $f|\Sigma_f^1\colon \Sigma_f^1 \to \mathbb{R}^3$ may be singular along some curve $\Sigma_f^{11} \subset \Sigma_f^1$. Bound the Betti numbers of Σ in terms of the curvature of Σ_f^1 and Σ_f^{11} and of the number of singular points $\Sigma_f^{111} \subset \Sigma_f^{11}$ of the map $f|\Sigma_f^{11}$.

(c) Let V be a closed oriented 4-dimensional manifold and let $f\colon V \to \mathbb{R}^5$ be a generic C^∞-map. Then the singularity $\Sigma = \Sigma_f^1$ is a smooth closed surface in V such that $\operatorname{rank}_v f = 3$ for all $v \in \Sigma$. Let $g\colon \Sigma \to Gr_3\mathbb{R}^5$ be the map which assigns the image $D_f(T_v) \subset \mathbb{R}^5$ (which is a 3-dimensional subspace in $\mathbb{R}^5$) to each point $v \in V$. Prove, for properly normalized Euler form ω in $Gr_3\mathbb{R}^5$, [which is a closed SO(5) invariant 2-form on the Grassmann manifold $Gr_3\mathbb{R}^5 = Gr_2\mathbb{R}^5$], the equality $\int_\Sigma g^*(\omega) = p_1(V)$ for a natural orientation in Σ and for the first Pontryagin number $p_1(V)$ of V.

Hint. Express $p_1(V)$ by the Σ^2-singularity of the projection of V onto a hyperplane $\mathbb{R}^4$ in $\mathbb{R}^5$ and then average over the Grassmannian $Gr_4\mathbb{R}^5 = S^4 \subset \mathbb{R}^5$.

(b′) Find similar formulae for the Pontryagin numbers of closed manifolds V mapped into $\mathbb{R}^q$ with generic singularities.

(b″) Generalize (b′) to homomorphisms of vector bundles $X \to Y$ over V, where Y is a trivial bundle. Allow complex vector bundles and express Chern numbers by integrals over pertinent singularities.

Question. Let g be a C^∞-smooth positive *semidefinite* quadratic differential form on a closed manifold V. If g is definite then the Euler characteristic equals the Gauss-Bonnet integral of a certain polynomial in curvature of g. Furthermore, the Gauss-Bonnet formula holds true according to (B) for forms g induced by generic smooth maps $V \to \mathbb{R}^q$, $q \geq n + 1$. However, these forms g have no (?) simple intrinsic description. For example, one does not know under what condition the Gauss-Bonnet integral absolutely converges and whether the convergence implies [under suitable assumptions, compare (B)] the Gauss-Bonnet theorem for singular metrics g on V.

2.1.5 Holomorphic Immersions of Stein Manifolds

Let V be a complex analytic manifold and let $X \to V$ be a holomorphic fiber bundle. Then one defines in a natural way the bundle $X^{(r)} \to V$ of rth order jets of germs of *holomorphic* sections $V \to X^{(r)}$. The *holomorphic h-principle* for a locally closed complex analytic subset (relation) $\mathscr{R} \subset X^{(r)}$ is defined as the possibility to deform an arbitrary holomorphic section $\varphi_0 \colon V \to \mathscr{R}$ to a holonomic section $\varphi_1 = J_f^r \colon V \to \mathscr{R}$ by a homotopy of holomorphic sections $V \to \mathscr{R}$. Similarly one defines the parametric h-principle and the h-principle for extensions from analytic subsets in V. One must be careful, however, with the approximation problem since the (ordinary) C^i-topologies are equivalent for all $i = 0, 1, \ldots, \infty$, for holomorphic sections, while the fine C^0-topology is discrete for non-compact connected manifolds V.

The holomorphic h-principle, when it holds true for a given relation $\mathscr{R}$, does not immediately yield holomorphic solutions of $\mathscr{R}$, but rather reduces the problem to Oka's principle that is the ordinary h-principle for the Cauchy-Riemann relation (compare 1.1.2).

(A) *Immersions of Stein Manifolds V into $\mathbb{C}^q$.* Let V be a *Stein* manifold of complex dimension n. Recall that the Stein property is equivalent to the existence of a proper holomorphic embedding $V \to \mathbb{C}^N$ for some sufficiently large N.

Theorem (Gromov-Eliashberg 1971). *If $q > n = \dim_{\mathbb{C}} V$, then holomorphic immersions $f \colon V \to \mathbb{C}^q$ satisfy the holomorphic h-principle.*

Proof. We mimic the C^∞-argument in 2.1.1 with the following

(A′) "*Zero-Dimensional*" *h-Principle.* Consider a closed analytic subset V_0 in V and let L be a non-vanishing holomorphic vector field along V_0 that is a holomorphic section $V_0 \to T(V)|V_0$. Fix a point $v_0 \in V_0$ and let $u_1, \ldots, u_n$ be local coordinates near v_0 such that $L = \partial/\partial u_1$ near v_0. We say that L has *finite tangency to V_0 near v_0 of order $\leq N$*, if there exists a holomorphic function $F = F(u_1, \ldots, u_n)$ whose zero set contains the germ of V_0 near v_0 and such that the restriction of F to each line $\{u_2 = \mathrm{const}_2, u_3 = \mathrm{const}_3, \ldots, u_n = \mathrm{const}_n\}$ in the coordinate domain has each zero (in the variable u_1) of order $\leq N + 1$. This definition of the tangency is clearly independent of the local coordinates.

The field L is called *semi-transversal* to V_0 if it is tangent to V near each point $v \in V$ with order at most N for some integer N which does not depend on v.

Denote by I_0 is ideal of holomorphic functions on V which equal zero on V_0 and let I_i, $i = 1, 2, \ldots$, be the ideal generated by I_{i-1} and the derivatives $\bar{L}_f$ for the functions $f \in I_{i-1}$, where $\bar{L}$ is some holomorphic extension of L to V. It is clear that the ideals I_i do not depend on the choice of this extension. If L is tangent to V_0 with order $\leq N$, then the Weierstrass preparation theorem allows one to choose the above function F of the form $F = \sum_{i=0}^{N} F_i u_1^i$, where F_i are holomorphic functions in $u_2, \ldots, u_n$ and $F_N \equiv 1$. Then the local ideal I_N near v_0 contains all germs of holomorphic functions at v_0. This applies to all points in V_0 and, since V is Stein, the (global) ideal I_N contains all holomorphic functions on V.

Lemma. *Let L be semi-transversal to V_0 and let g be an arbitrary holomorphic function on V. Then there exists a holomorphic function f on V, such that $\bar{L}f | V_0 = g | V_0$.*

Proof. If $g \in I_0$ then the solution is obvious with $\bar{f} \equiv 0$. Furthermore, an arbitrary function $g \in I_k$ by definition is $g = f_0 + \sum_{i=1}^{k-1} \alpha_i L f_i$ for some holomorphic functions α_i on V and for some $f_i \in I_i$. Then the function $\tilde{f} = \sum_{i=1}^{k-1} \alpha_i f_i$ satisfies the equation $L\tilde{f} = g - g'$ for the function $g' = f_0 + \sum_{i=1}^{k-1} (L\alpha_i) f_i$ in I_{k-1}. We assume, by induction in k, the existence of a function $\bar{f}'$, for which $L\bar{f}' | V_0 = g' | V_0$, and then we solve the equation $L\bar{f} = g$ with $\bar{f} = \tilde{f} + \bar{f}'$. Since each function g lies in I_N, the proof follows with $k = N$.

Now let $\mathscr{L}$ be a line subbundle of the tangent bundle of V restricted to V_0, that is $\mathscr{L} \subset T(V) | V_0$, and suppose there is a holomorphic 1-form φ on V which does not vanish on $\mathscr{L}$. Then there exists a unique holomorphic section L: $V_0 \to \mathscr{L}$, such that $\varphi(L) \equiv 1$ on V_0. We call $\mathscr{L}$ *semi-transversal* to V_0 if the field (section) L is semi-transversal to V_0 and we observe this property of $\mathscr{L}$ to be independent of a particular form φ. The above lemma immediately implies the following

Corollary. *If $\mathscr{L}$ is semi-transversal to V_0, then holomorphic functions f on V whose differentials df do not vanish on $\mathscr{L}$ satisfy the holomorphic h-principle.*

Exercise. Generalize the above to the fields L which have finite tangency of order $N = N(v)$ near each point $v \in V_0$, where $N(v)$ is unbounded for $v \to \infty$.

(A″) *Generic Properties of Holomorphic Maps and Sections*. Start with an arbitrary holomorphic fibration $X \to V$ and let Φ be an arbitrary subspace in the space $\Gamma(X)$, of holomorphic sections $V \to X$, with the topology of uniform convergence on the compact subsets in V. A *p-dimensional holomorphic homotopy* in Φ, by definition, is a map ψ: $\mathbb{C}^p \to \Phi \subset \Gamma(X)$ for which the associated map $V \times \mathbb{C}^q \to X$ is holomorphic. A set Ψ of such homotopies is called *composible* if for every two homotopies in Ψ, say ψ_1: $\mathbb{C}^{p_1} \to \Phi$ and ψ_2: $\mathbb{C}^{p_2} \to \Phi$ there exists a homotopy ψ: $\mathbb{C}^{p_1} \times \mathbb{C}^{p_2} \to \Phi$ in Ψ, called a *composition* of ψ_1 and ψ_2, such that $\psi | \mathbb{C}^{p_1} \times 0 = \psi_1$ and $\psi | 0 \times \mathbb{C}^{p_2} = \psi_2$.

(A_1'') **Example.** Let X be the trivial fibration $X = V \times \mathbb{C}^q \to V$ and let $\Sigma'(i) \subset X^{(1)}$, $i = 1, \ldots, q$ consist of the 1-jets of (germs of) those sections φ: $V \to X$, represented by q-tuples of holomorphic functions $f_1, \ldots, f_q$ on V, for which the map $F_i' = (f_1, \ldots, f_{i-1}, f_{i+1}, \ldots, f_q)$: $V \to \mathbb{C}^{q-1}$ fails to be an immersion. That is $J_\varphi^1(v) \in \Sigma'(i)$ if and only if $\operatorname{rank}_v F_i' < n = \dim V$. Denote by $I'(i)$ the sheaf of ideals of holomorphic functions on $X^{(1)}$ which vanish on $\Sigma'(i)$ and fix an integer $k \geq 0$. A homotopy ψ: $\mathbb{C}^1 \to \Gamma(X)$ represented by functions $f_i(v, z)$ on V for $i = 1, \ldots, q$ and $z \in \mathbb{C}^1$ is called $(I'(i))^k$-*stable* if the functions $f_1, \ldots, f_{i-1}, f_{i+1}, \ldots, f_q$ are constant in the variable z and if the function $f_i(v, z) - (f_i(v, 0)$ is contained in the pull-back of the ideal $(I'(i))^k$ under the map $V \times \mathbb{C}^1 \to X^{(1)}$ given by the 1-jets of sections $\psi(z)$: $V \to X$, $z \in \mathbb{C}^1$. Next, we by induction in $p = 1, 2, \ldots$, define a homotopy ψ: $\mathbb{C}^p \times \mathbb{C}^1 \to \Gamma(X)$ to be *k-regular* if the homotopy $\psi | \mathbb{C}^p \times 0$: $\mathbb{C}^p \to \Gamma(X)$ is k-regular and if the homotopy $\psi | z \times \mathbb{C}^1$: $\mathbb{C}^1 \to \Gamma(X)$ is $(I'(i))^k$-stable for some $i = 1, \ldots, q$ and for all $z \in \mathbb{C}^p$ where every "homotopy" for $p = 0$, by definition, is regular. The k-regular homotopies for $k \geq 2$ preserve *immersions* $V \to \mathbb{C}^q$. Namely, if $\Phi \subset \Gamma(X)$ is the subset of those sections which correspond to immersions $V \to \mathbb{C}^q$ and if a k-regular homotopy $\psi: \mathbb{C}^p \to \Gamma(X)$ for $k \geq 2$ has $\psi(z_0) \in \Phi$ for some $z_0 \in \mathbb{C}^p$, then $\psi(z) \in \Phi$ for all $z \in \mathbb{C}^p$. Furthermore, if V is a *Stein* manifold, then k-regular homotopies constitute a *composible* set for each k. In fact, this remains true for an arbitrary sheaf of ideals in the space of jets of any order, say for $I(i) \in X^{(r)}$, $i = 1, \ldots, q$, since the basic theorems A and B of H. Cartan (see Gunning and Rossi 1965) allow extensions of holomorphic functions on $V \times \mathbb{C}^p$ which lie in a given ideal.

(A_2'') **Lemma.** *Let Ψ be a composible set of holomorphic homotopies in Φ and let Σ be an analytic subset in the jet space $X^{(r)}$. Suppose that for each section $\varphi \in \Phi$ and for each point $v \in V$ there exists a neighborhood $U \subset V$ of v and a homotopy $\psi_v: \mathbb{C} \to \Phi$ in Ψ such that $\psi(0) = \varphi$ and such that the jet $J^r_{\psi(z)}$ sends U to the complement $X^{(r)} \setminus \Sigma$ for all $z \neq 0$ in $\mathbb{C}$ which are close to zero. Then each section, say $\varphi_0 \in \Phi$, admits a continuous homotopy $\varphi_t \in \Phi$, $t \in [0, 1]$, such that the jet $J^r_{\varphi_1}$ sends V to the complement $X^{(r)} \setminus \Sigma$.*

Proof. Compose the homotopies ψ_v for all points $v \in V$ and restrict the composition to an appropriate continuous path in the resulting space $\mathbb{C}^\infty$. This is straightforward and the detail is left to the reader.

(A_3'') **Corollary.** *Let V be Stein and let, for the above $\varphi \in \Phi$ and $v \in V$, there exist a holomorphic vector bundle $Y \to V$, a fiber preserving holomorphic map $\alpha: Y \to X$ and a sheaf of ideals I on V (i.e. I is a subsheaf of the structure sheaf of V), such that*

(i) *the zero section of Y is sent by α to φ;*
(ii) *the zero set of I does not contain the point v;*
(iii) *if $\beta \in \Gamma(Y)$ and $\gamma \in I$, then the map α sends the one-dimensional homotopy $z \mapsto z\gamma\beta$, $z \in \mathbb{C}$, in $\Gamma(Y)$ to Ψ;*
(iv) *let $\alpha_r: Y^{(r)} \to X^{(r)}$ denote the natural map induced by α on the jet spaces. Then the codimension of the pull-back $\alpha^{-1}(\Sigma) \subset Y^{(r)}$ near $v \in V = V \times 0 \subset Y^{(r)}$ satisfies* $\operatorname{codim}_v \alpha^{-1}(\Sigma) > n = \dim V$.

Then, each section $\varphi_0 \in \Phi$ admits the above desingularizing homotopy $\varphi_t \in \Phi$.

Proof. If β' is a generic holomorphic section of Y over a small neighborhood $U \subset V$ of v, then by (a simple local analytic version of) Thom's transversality theorem the rth order jet of the section $z\beta': U \to Y$, $z \in \mathbb{C}$, misses $\alpha^{-1}(\Sigma)$ for all small $z \neq 0$. Since V is Stein, there exists a *global* section $\beta: V \to Y$ and some $\gamma \in I$ for which the section $\gamma\beta$ approximates β' on U, such that the jet of $z\gamma\beta$ misses Σ as well. The image ψ_v of the homotopy $z \to z\gamma\beta$ under the map $\alpha: Y \to X$ meets the assumptions of (A_2'') which insure the homotopy φ_t.

(A_4'') Let holomorphic 1-forms $\varphi_1, \ldots, \varphi_{n-1}$ on V be linearly independent at a given point $v \in V$ and let $u_1, \ldots, u_n$ be local coordinates in a small domain $U \subset V$ around v, such that $\varphi_i(\partial/\partial u_1) \equiv 0$ for $i = 1, \ldots, n-1$. Denote by Y_U the cotangent bundle of U and let $\Sigma^{(r)} \subset Y_U^{(r)}$ be the analytic subset which controls the tangency of order $\geq r$ of the field $\partial/\partial u_1$ to the zero set of the functions $F_\varphi = \varphi(\partial/\partial u_1)$ for forms $\varphi\colon U \to Y_U$ [compare (A′)]. Namely, $J^r_\varphi(v) \in \Sigma^{(r)}$ if and only if $\frac{\partial^k F_\varphi}{\partial u_1^k}(v) = 0$ for $k = 0, 1, \ldots, r$. It is clear that $\operatorname{codim} \Sigma^{(r)} = r + 1$. Furthermore, if $\bar{Y}^{(r)}_U \subset Y^{(r)}_U$ denotes the subspace of the jets of *exact* forms φ on U, then $\operatorname{codim}(\Sigma^{(r)} \cap \bar{Y}^{(r)}_U \subset \bar{Y}^{(r)}_U) = r + 1$ as well.

Fix a form $\bar{\varphi}$ on V and some $i = 1, \ldots, q$. Denote by $\Phi = \Phi^{q-1}_i(\bar{\varphi})$ the space of those $(q-1)$-tuples of holomorphic forms $\varphi_1, \ldots, \varphi_{i-1}, \varphi_{i+1}, \ldots, \varphi_q$ on V, where $\varphi_1, \ldots, \varphi_{i-1}$ are exact and such that the forms $\varphi_1, \ldots, \varphi_{i-1}, \bar{\varphi}, \varphi_{i+1}, \ldots, \varphi_q$ span the cotangent bundle of V (compare 2.1.1). Denote by $\Sigma' \subset V$, $j = 1, \ldots, i-1, i+1, \ldots, q$ the subset where the forms φ_j fail to span the cotangent bundle $T^*(V)$ and let $\mathscr{L} \subset T(V)|\Sigma' \subset T(V)|\Sigma'$ be the line subbundle on which these forms vanish.

Lemma. *If V is Stein and $q > n = \dim V$, then an arbitrary $(q-1)$-tuple in Φ admits a continuous homotopy in Φ to a (generic) $(q-1)$-tuple for which the corresponding bundle $\mathscr{L}$ is semi-transversal to the new Σ'.*

Proof. For an arbitrary $(q-1)$-tuple $\{\varphi_j\}$ in Φ and for each point $v \in \Sigma' \subset V$ there are $n-1$ forms among φ_j, which span together with $\bar{\varphi}$ the cotangent space $T^*_v(V)$. To avoid a mess in the notations, we assume these to be $\varphi_1, \ldots, \varphi_{i-1}, \varphi_{i+1}, \ldots, \varphi_n$. Let $\Sigma'' \subset V$ be the subset, where the forms $\varphi_1, \ldots, \varphi_{i-1}, \bar{\varphi}, \varphi_{i+1}, \ldots, \varphi_n$ fail to span the cotangent bundle $T^*(V)$. This Σ'' is the pullback of some subset Σ_1 in the sum of n copies of the cotangent bundle $T^*(V)$ under the section $(\varphi_1, \ldots, \bar{\varphi}, \ldots, \varphi_n)\colon V \to \bigoplus_1^n T^*(V)$. Denote by I_1 the pull-back to V of the ideal of (functions vanishing on) Σ_1. (Observe that the zero set of I_1 equals Σ'', but, yet, I_1 may be *properly* contained in the ideal of Σ''.) Now, we have a distinguished set of holomorphic homotopies in Φ which only move the last component φ_q (which is *not* among the chosen forms) by adding to φ_q forms $z\gamma\beta$, $z \in \mathbb{C}$, where $\gamma \in I_1$ and where β is an arbitrary form on V. [If φ_q happens to be among the chosen forms and the extra form φ was among the exact ones, say $\varphi = \varphi_1 = df_1$, then we would use the homotopies $\varphi_1 + d(z\gamma\beta)$ for $\gamma \in (I_1)^2$ and for functions β on V.] These homotopies, for all $(q-1)$-tuples in Φ and all $v \in V$, do not form a composible set. However, since V is Stein, there exists a composible set Ψ which contains all these homotopies, and which is built up by generalizing (A$''_1$) in an obvious way. Now the lemma follows from (A$''_3$) with the above relation $\operatorname{codim} \Sigma^{(r)} = r + 1$ for $r = n$.

Finally, this lemma reduces the holomorphic h-principle for immersions $V \to \mathbb{C}^q$ to the 1-dimensional h-principle (A′) (compare 2.1.1) and the theorem follows.

(B) Immersions $V \to \mathbb{C}^q$ (Continuation). *If V is an n-dimensional Stein manifold and if $q > n$ then holomorphic immersions $f\colon V \to \mathbb{C}^q$ satisfy the (ordinary) h-principle.*

Proof. The holomorphic h-principle in (A) reduces this h-principle to Oka's principle for holomorphic sections of the bundle $\mathscr{I} \to V$ whose fiber $\mathscr{I}_v$, $v \in V$, consists of injective complex linear maps $T_v(V) \to \mathbb{C}^q$. Since $\mathscr{I}_v$ is transitively acted upon by the linear group $GL_q\mathbb{C}$, Grauert's theorem [see 1.1.2 and (C) below] applies.

Corollary. *If $2q \geq 3n - 1$, then every n-dimensional Stein manifold V admits a holomorphic immersion into $\mathbb{C}^q$.*

Proof. Since the fiber $\mathscr{I}_v$ is $(2q - 2n)$-connected and since V is homotopy equivalent to a polyhedron of real dimension n (see 1.1.2), the fibration $\mathscr{I} \to V$ admits a continuous section $V \to \mathscr{I}$ for $2q \geq 3n - 1$. Then the above h-principle yields immersions $V \to \mathbb{C}^q$ unless $n = q = 1$. If $n = q = 1$, then our proof of the h-principle does not apply, but a holomorphic immersion $V \to \mathbb{C}^1$ does exist for an arbitrary open one-dimensional manifold V. [See Gunning and Narasimhan (1967) and 3.2.4.]

Questions. Does the h-principle hold true for holomorphic immersions of n-dimensional Stein manifolds into $\mathbb{C}^n$? Does every Stein manifold admit a holomorphic function f whose differential does not vanish?

The key difficulty can be seen if one takes a non-vanishing holomorphic vector field $L = \sum_{i=1}^n \varphi_i(\partial/\partial z_i)$ on $\mathbb{C}^n$ and looks for a holomorphic function f whose derivative $Lf = \sum_{i=1}^n \varphi_i(\partial f/\partial z_i)$ does not vanish. Does such an f exist for every field L?

Exercises. (a) Show holomorphic *immersions* $V \to \mathbb{C}^q$, $q > n = \dim V$, to enjoy Thom's transversality theorem. Namely, let $X^{(r)}$ be the space of rth order jets of holomorphic maps $f: V \to \mathbb{C}^q$ and let $\Sigma \subset X^{(r)}$ be an analytic subset of codimension $\geq n + 1$ for $n = \dim V$. Then *immersions* $f: V \to \mathbb{C}^q$ whose r^{th} order jets miss Σ constitute a residual subset in the *space of immersions* (with the topology of uniform convergence on compact subsets in V).

Hint. Prove the transversality theorem for functions on f whose differentials do not vanish on a subbundle $\mathscr{L} \subset T(V)|V_0$ semi-transversal to V_0.

(b) Prove the *parametric* h-principle for holomorphic immersions $V \to \mathbb{C}^q$, $q \geq n + 1$.

(c) Prove the holomorphic h-principle for holomorphic k-mersions $V \to \mathbb{C}^q$ for $k < q$ (compare 2.1.1).

Remark. The ordinary h-principle for these k-mersions in unknown, unless $k = n$ or $k = 1$. The difficulty arises in Oka's principle for the fibration $\mathscr{R} \to V$, whose fiber $\mathscr{R}_v$ consists of linear maps $T_v(V) \to \mathbb{C}^q$ of rank $\geq k$. This $\mathscr{R}_v$ is not homogeneous (at least in an obvious way) for $2 \leq k \leq n - 1$ and so Grauert's theorem does not apply [compare (C) below]. Yet, one can produce holomorphic sections $V \to \mathscr{R}$ under suitable assumptions on n, q and k (see Gromov-Eliashberg 1971) thus making the holomorphic h-principle for k-mersions non-vacuous.

(d) Fix a proper holomorphic map $F_0: V \to \mathbb{C}^{q_0 > n}$ and consider the holomorphic maps $f: V \to \mathbb{C}^p$ for which $F = F_0 \oplus f: V \to \mathbb{C}^q = \mathbb{C}^{q_0} \times \mathbb{C}^p$ is an immersion. Prove the holomorphic h-principle for these maps f, provided the underlying map F_0 is generic [i.e. the jet of F_0 of a sufficiently high order $r = r(n)$ misses a pertinent singularity of codimension $> n$ in the jet space]. Then prove Oka's principle for holomorphic sections of the corresponding differential relation $\mathscr{R} \to V$ and conclude for a generic $F_0: V \to \mathbb{C}^{n+1}$) to the following

Theorem. *If* $2q \geq 3n + 1$ *then every n-dimensional Stein manifold V admits a proper holomorphic immersion into* $\mathbb{C}^q$.

Question. Do proper holomorphic immersions $V \to \mathbb{C}^q, q > n$, satisfy the *h*-principle?

(d′) Study holomorphic maps f for which $F_0 \oplus f: V \to \mathbb{C}^q$ is a holomorphic embedding (i.e. an immersion without double points). Express this property of f in terms of the map $s_f: V \times V \to \mathbb{C}^p$ defined by $s_f(v_1, v_2) = f(v_1) - f(v_2)$ [compare (E) in 2.1.1]. Formulate the holomorphic *h*-principle for these maps f. Prove this *h*-principle, provided F_0 is a proper *immersion* with normal (i.e. transversal) crossings and $2q \geq 3n + 3$. Then prove the pertinent Oka's principle and the (ordinary) *h*-principle for such maps f.

Remark. The above *h*-principle also holds for generic maps F_0 which are not necessarily immersions [This is announced in Gromov-Eliashberg (1971) but the proof has not been published.] This yields *proper holomorphic embeddings* $V \to \mathbb{C}^q$ for all *n-dimensional Stein manifolds V, provided* $3n \leq 2q - 3$. [See Forster (1970) for a direct construction of a proper holomorphic embedding of every V into $\mathbb{C}^q$ for $q = \mathrm{ent}(\frac{5}{3}n) + 2$, compare Schaft (1984).]

(e) Let V be a smooth affine algebraic variety over an algebraically closed field K of characteristic zero. Formulate the *algebraic h-principle* for regular (i.e. given by polynomials on $V \subset K^N$) immersions $V \to K^q$ and prove this *h*-principle for $q > n = \dim V$. Show that every n-dimensional affine group V over K admits a regular immersion $V \to \mathbb{C}^{n+1}$. Give further examples of algebraically parallelizable manifolds V to which this *h*-principle applies.

(f) Prove the holomorphic counterparts of the *h*-principles in (D) of 2.1.1. In particular establish the *h*-principle for free holomorphic maps $V \to \mathbb{C}^q$ for $q \geq (n + 2)(n + 3)/2$. [Compare Gromov-Eliashberg (1971).]

(g) Study holomorphic Lagrange immersions [compare (E) in 2.1.3] of Stein manifolds V into $\mathbb{C}^p \times \mathbb{C}^p$ with the (exact holomorphic) 2-form $h = \sum_{i=1}^{p} dx_i \wedge dy_i$. In particular, construct a holomorphic immersion $f: V \to \mathbb{C}^{2n}$, $n = \dim V$, such that $f^*(h) \equiv 0$ on V for all parallelizable (for example, topologically contractible) Stein manifolds V.

(C) *Holomorphic Maps into Elliptic Spaces.* Consider a complex manifold W which satisfies the following condition

(Ell_1) For each Stein mainfold V and for each holomorphic map $f: V \to W$ there exists a holomorphic map (homotopy) $\varphi: V \times \mathbb{C}^p \to W$ for some $p \geq \dim W$ such that $\varphi | V = V \times 0 = f$ and such that the differential $d\varphi$ sends each tangent space $T_v(\mathbb{C}^p) \approx \mathbb{C}^p \subset T_v(V \times \mathbb{C}^p)$, $v = (v, 0) \in V \times \mathbb{C}^p$ *onto* $T_w(W)$, $w = f(v)$.

Examples and Exercises. (a) Let W be a homogeneous space under some holomorphic action of a complex analytic Lie group G. Then the corresponding action of the Lie algebra $\mathscr{G} \approx \mathbb{C}^p$, $p = \dim G$, on W integrates to the (exponential) map $W \times \mathscr{G} \to W$ which induces the required homotopy for all $f: V \to W$. Hence, W satisfies (Ell_1).

(a′) Let $X \to W_0$ be a holomorphic fibration with the fiber W and the structure group G. If the base W_0 satisfies (Ell_1), then so does X. Indeed, homotopies $V \times \mathbb{C}^p \to W_0$ lift to horizontal homotopies in X with some holomorphic connection in the induced fibration over $V \times \mathbb{C}^p$. These are complemented by vertical (fiberwise) homotopies in X induced by the exponential maps in the fibers.

Observe that the above X may be not homogeneous for homogeneous W_0 and W. For instance, the total space of a *negative* line bundle over $\mathbb{C}P^1$ is not homogeneous.

(b) Let $a_1, \ldots, a_p$ be holomorphic actions of $\mathbb{C}$ on W whose generating (holomorphic) vector fields span the tangent space $T_w(W)$ for all $w \in W$. Then the composition of the actions $a_1, \ldots, a_p$ gives a holomorphic map $W \times \mathbb{C}^p \to W$ which insures (Ell_1).

(b′) Let $A \subset \mathbb{C}^q$ be a Zariski closed subset such that $\operatorname{codim} A \geq 2$. Then the complement $W = \mathbb{C}^q \setminus A$ admits the above actions $a_1, \ldots, a_p$. Indeed let Z be a field $\sum_{i=1}^q C_i(\partial/\partial z_i)$ for some constants c_i. Then there exists a non-zero polynomial P on $\mathbb{C}^q$ which vanishes on A and is constant on the orbits of Z. The field $PZ|W$ clearly integrates to an action of $\mathbb{C}$ in W and there are sufficiently many of such fields to span $T(W)$.

(b″) Take a Zariski closed subset $A \subset \mathbb{C}P^q$, assume $\operatorname{codim} A \geq 2$ and prove (Ell_1) for $\mathbb{C}P^q \setminus A$. Study $W = W_0 \setminus A$ for algebraic *homogeneous* spaces W_0.

(b‴) Let W be covered by open subsets $W_i \subset W$, $i = 1, \ldots, k$, whose complements $A_i = W \setminus W_i$ are analytic subsets in W. Let W_i satisfy (Ell_1) for all $i = 1, \ldots, k$ and show W to satisfy (Ell_1) as well.

Question. Is (Ell_1) a birational invariant of *projective* algebraic manifolds?

(c) Let W admit a geodesically complete holomorphic connection. Then the geodesic spray exp: $T(W) \to W$ insures (Ell_1). For instance, the quotient $W = \mathbb{C}^q/\Gamma$ for an arbitrary discrete group Γ of affine transformations of $\mathbb{C}^q$ satisfies (Ell_1).

(c′) Show the *Hopf* manifold $W = \mathbb{C}^q \setminus \{0\}/\mathbb{Z}$ satisfies (Ell_1). [Here $\mathbb{Z}$ acts on $\mathbb{C}^q$ by $(z, x) \mapsto zx$ for $(z, x) \in \mathbb{Z} \times \mathbb{C}^q$.]

(d) A domain $U \subset V$ is called *Runge* if every holomorphic function on U can be approximated by functions which holomorphically extend to V. Let W satisfy (Ell_1) and let $f_0: V \to W$ and $f_1: U \to W$ be holomorphic maps, where U is Runge and such that f_1 can be joined with $f_0|U$ by a continuous homotopy of holomorphic maps $f_t: U \to W$, $t \in [0, 1]$. Show that f_1 can be approximated by maps which holomorphically extend to V.

(e) Let W satisfy (Ell_1) and let $w_1, \ldots, w_k, \ldots,$ be an arbitrary sequence of points in a connected component of W. Prove the existence of a holomorphic map $f: \mathbb{C} \to W$ for which $f(z_i) = w_i$, $i = 1, \ldots,$ where $z_i \in \mathbb{C}$ are arbitrary points, such that $|z_i| \to \infty$ for $i \to \infty$.

(e′) Let V be an arbitrary Stein manifold and let $f_0: V \to W$ be a holomorphic map. Construct a holomorphic map $f\colon V \to W$ homotopic to f_0, such that $f(z_i) = w_i$ for a given divergent sequence of points $z_i \in V$.

(C′) **The Holomorphic *h*-Principle for Immersions.** *If W satisfies* (Ell_1) *and if V is Stein, then holomorphic maps $V \to W$ satisfy Thom's transversality theorem for analytic subsets Σ in the jet space $X^{(r)}$ for $X = W \times V \to V$ and all $r = 0, 1, \ldots$.*

The proof is straightforward and left to the reader.

Next, the argument in the above (A) and (B) easily generalizes to those W (in place of $\mathbb{C}^q$) which split into products, say $W = W_1 \times W_2 \times \cdots \times W_k$. Namely, if $W_1, \ldots, W_k$ satisfy (Ell_1) and $\dim W - \dim W_i \geq \dim V$ for all $i = 1, \ldots, k$, then immersions of Stein manifolds V to W satisfy the holomorphic h-principle.

Exercise. Let W be a homogeneous space under an action of a complex Lie group G which is the product $G = G_1 \times G_2 \times \cdots \times G_k$. Assume that the orbits of G_i in W have dimension q_i, $i = 1, \ldots, k$, such that $\sum_{i=1}^k q_i = q = \dim W$ and such that $q - q_i \geq n$ for all $i = 1, \ldots, k$. Show holomorphic immersions of n-dimensional Stein manifolds V into W to satisfy the holomorphic h-principle. Apply this to immersions to an arbitrary *commutative* group G which has $\dim G \geq n + 1$.

(C″) *Oka's Principle.* Applications of the holomorphic h-principle depend on the ordinary (Oka's) h-principle which expresses the following property of holomorphic sections of a fibration $X \to V$.

(Ell_2) Let V_0 be an arbitrary analytic subset in V and let $f_0\colon \mathcal{O}\!p\,V_0 \to X$ be a holomorphic section over a small neighborhood $\mathcal{O}\!p\,V_0 \subset V$ of $V_0 \subset V$. Then holomorphic sections $f\colon V \to X$, which have $J_f^r | V_0 = J_{f_0}^r | V_0$ for a given $r = 0, 1, \ldots$, satisfy the h-principle.

A fundamental theorem of Grauert (see 1.1.2) insures (Ell_2) for those fibrations over *Stein* manifolds V which admit a *complex* Lie group for the structure group which *transitively* acts on the fiber. This theorem immediately implies the following

Corollary. *Let $Y \to X$ be a holomorphic fibration whose structure group is a complex Lie group transitive on the fiber. If V is Stein and if a fibration $X \to V$ satisfies* (Ell_2) *then the* (*composed*) *fibration $Y \to V$ also satisfies* (Ell_2).

Next, a manifold W is said to satisfy (Ell_2) iff the trivial fibration $W \times V \to V$ satisfies (Ell_2) for all *Stein* manifolds V.

Remarks. (a) Obviously, (Ell_2) implies (Ell_1) but the implication (Ell_1) $\Rightarrow$ (Ell_2) is unknown.

(b) The above corollary provides many (Ell_2)-manifolds W: these are built out of homogeneous spaces by successive fibrations.

(c) The *ellipticity* axioms (Ell_1) and (Ell_2) allow "many" holomorphic maps $\mathbb{C} \to W$. This is contrary to the *hyperbolicity* of W which prohibits non-constant holomorphic maps $\mathbb{C} \to W$. Observe that every connected Riemannian surface W (i.e. $\dim W = 1$) is either hyperbolic (if it is covered by the disc) or elliptic (if it is covered by $\mathbb{C}$ or by S^2). One may also view the ellipticity as a dual to the Stein property of W which claims "many" holomorphic maps $W \to \mathbb{C}$.

Finally, the above corollary immediately yields *the h-principle for holomorphic fiberwise injective homomorphisms* $T(V) \to T(W)$, *provided V is Stein and W satisfies* (Ell_2). *Hence, the holomorphic h-principle* [see (C′)] *for immersions* $V \to W$ *implies the ordinary h-principle in case W is* (Ell_2).

(D) *Harmonic Immersions.* Many properties of the Cauchy-Riemann system generalize to other systems of linear (and quasi-linear) systems of elliptic P.D. equations. Let, for example, V be an *open Riemannian* manifold. Green and Wu (1975) have pointed out that the *harmonic structure* of V is Stein. Namely, V satisfies Stein's axioms with harmonic functions in place of the holomorphic ones.

Corollary (Green-Wu). *Every open n-dimensional Riemannian manifold V admits a proper harmonic embedding* $f: V \to \mathbb{R}^{2n+1}$.

Questions. Does the removal of singularities yield the h-principle for (proper) harmonic immersions $V \to \mathbb{R}^q$ for $q \geq n + 1$? What happens for $q \leq n$? For example, does every open Riemannian manifold V admit a harmonic function $f: V \to \mathbb{R}$ whose gradient nowhere vanishes? Which Riemannian manifolds W satisfy (Ell_1) and (Ell_2) for harmonic maps of open Riemannian manifolds $V \to W$?

2.2 Continuous Sheaves

Consider a relation $\mathcal{R} \subset X^{(r)}$ for a fibration $X \to V$, fix an integer $k \geq r$ and denote by $\Phi(U)$ the space of C^k-solutions of $\mathcal{R}$ over U for all open subsets $U \subset V$. The collection of the spaces $\Phi(U)$ comes with an additional structure given by the *restriction maps*, called $\Phi(I)$: $\Phi(U) \to \Phi(U')$ for all open subsets U' in U, where $I = I(U', U)$ stands for the inclusion, $I: U' \subset U$, and where $\Phi(I)(\varphi) = \varphi | U'$ for all $\varphi \in \Phi(U)$. The assignment $\{U \mapsto \Phi(U), I \mapsto \Phi(I)\}$ (which is a contravariant functor from the category of open subsets in V to the category of topological spaces) is called *the sheaf of C^k-solutions* of $\mathcal{R}$ over V (compare 1.4.2).

Recall that *an abstract sheaf* Φ over a topological space V (see Godement 1958), by definition, assigns a set $\Phi(U)$ to each open subset $U \subset V$ and a map $\Phi(I)$: $\Phi(U) \to \Phi(U')$ to each inclusion $I: U' \subset U$, such that the following three axioms are satisfied.

(1) *Functoriality.* If $I': U'' \subset U'$ and $I: U' \subset U$, then the value of Φ at the inclusion $I \circ I': U'' \subset U$ abides $\Phi(I \circ I') = \Phi(I') \circ \Phi(I)$.

One also agrees $\Phi(\mathrm{Id}: U \subset U) = \mathrm{Id}$ for all $U \subset V$ and $\Phi(\emptyset) = \emptyset$. One calls elements $\varphi \in \Phi$ *sections of* Φ *over* U and writes $\varphi | U'$ instead of $\Phi(I)$.

(2) *Locality* (*Uniqueness*). If two sections φ_1 and φ_2 of Φ over U are *locally equal*, then they are equal, where the local equality means there exists a neighborhood $U' \subset U$ of every point $u \in U$, such that $\varphi_1 | U' = \varphi_2 | U'$.

(2′) *Locality* (*Existence*). Let open subsets $U_\mu \subset U$, $\mu \in M$, cover U and let sections $\varphi_\mu \in \Phi(U_\mu)$ satisfy $\varphi_\mu | U_\mu \cap U_{\mu'} = \varphi_{\mu'} | U_\mu \cap U_{\mu'}$ for all μ and μ' in M. Then there exists a section $\varphi \in \Phi(U)$ [which is unique by (2)], such that $\varphi | U_\mu = \varphi_\mu$ for all $\mu \in M$.

The axioms (2) and (2′) show every sheaf Φ to be uniquely defined by $\Phi(U_\nu)$ for any *base* of open subsets $U_\nu \subset V$.

Next, one extends Φ to non-open subsets $C \subset V$ by $\Phi(C) = \Phi(\mathcal{O}\!p\, C)$ which denotes the inductive (direct) limit of $\Phi(U)$ over all neighborhoods $U \subset V$ of C (compare 1.4.1). In particular, one defines the *stalk* $\Phi(v) = \Phi(\mathcal{O}\!p\, v)$ for all $v \in V$ and one writes $\varphi(v) \in \Phi(v)$ for $\varphi | \mathcal{O}\!p\, v$. Then one can *restrict* Φ to a sheaf over C, called $\Phi | C$ and defined by $(\Phi | C)(D) = \Phi(\mathcal{O}\!p\, D)$ for all open subsets $D \subset C$ and for $\mathcal{O}\!p\, D \subset V$. Thus, the sheaf $\Phi | C$ has the same stalks over the points $c \in C$ as Φ.

A sheaf Φ is called *continuous* (or quasi-topological) if every set $\Phi(U)$, $U \subset V$, is endowed with a quasi-topology (see 1.4.1), such that the map $\Phi(I)$ is continuous for all inclusions $I: U' \subset U$. In this case the space $\Phi(C)$ is equipped with the inductive limit quasi-topology for all subsets $C \subset V$.

A *homomorphism* between continuous sheaves over V, say $\alpha: \Phi \to \Psi$, is a collection of *continuous* maps $\alpha_U: \Phi(U) \to \Psi(U)$, for all open $U \subset V$ which commute which the restrictions of sections, that is $\alpha_{U'} \circ \Phi(I) = \Psi(I) \circ \alpha_U$ for all $I: U' \subset U$. Finally, one defines a *subsheaf* $\Phi' \subset \Phi$ by giving a subspace $\Phi'(U) \subset \Phi(U)$ for all $U \subset V$, such that Φ' satisfies (2) and (2′).

2.2.1 Flexibility and the *h*-Principle for Continuous Sheaves

Consider a continuous sheaf Φ over V and define the *parametric sheaf* Φ^P over $V \times P$ for an arbitrary topological space P by first claiming that its sections are just the continuous families of sections of Φ parametrized by P. To complete the definition we only need to specify $\Phi^P(U \times R)$ for open sets $U \subset V$ and $R \subset P$. Set $\Phi^P(U \times R)$ equal to $(\Phi(U))^R$, the space of continuous maps $R \to \Phi(U)$ with the following quasi-topology. A map $Q \to (\Phi(U))^R$ is continuous iff the corresponding map $R \times Q \to \Phi(U)$ is continuous.

Next, we apply this construction to $P = V$, and then restrict the parametric sheaf Φ^V over $V \times V$ to the diagonal $\Delta \subset V \times V$. The resulting sheaf over $\Delta = V$ is denoted Φ^*. Intuitively, sections in Φ^* are continuous families of germs $\varphi_v \subset \Phi(v)$, $v \in V$. For example, if Φ the sheaf of locally constant functions $V \to \mathbb{R}$, then Φ^* is canonically isomorphic to the sheaf of continuous functions on V.

Every section of Φ corresponds to a unique constant family of sections with the parameter space V. Thus, we obtain a natural injective homomorphism $\Delta: \Phi \to \Phi^*$ which makes Φ a subsheaf $\Delta(\Phi) = \Phi$ in Φ^*.

(A) **Definitions.** A sheaf Φ satisfies the (sheaf theoretic) *h-principle*, if every section $\varphi \in \Phi^*(U)$ can be homotoped to $\Phi(U) \subset \Phi^*(U)$ for all open subsets $U \subset V$. *The parametric h-principle* for Φ claims the homomorphism Δ to be a weak homotopy equivalence. That is $\Delta_U: \Phi(U) \to \Phi^*(U)$ is a w.h. equivalence for all open $U \subset V$.

(A′) **Remark.** If Φ is the sheaf of solutions of a relation $\mathscr{R} \subset X^{(r)}$ and if Ψ is the sheaf of continuous sections $V \to \mathscr{R}$, then there is a natural (and obvious) homomorphism $J: \Phi^* \to \Psi$, such that $(J \circ \Delta) f = J_f^r: U \to \mathscr{R}$ for all $f \in \Phi(U)$ and all $U \subset V$. If this J is a w.h. equivalence, then the sheaf theoretic h-principle obviously implies the h-principle for $\mathscr{R}$.

(A″) **Examples.** (a) If $\mathscr{R}$ is an *open* subset in $X^{(r)}$, then J is clearly a w.h. equivalence (compare 1.4.1).

(b) If $\mathscr{R}$ is given by (∗) in (B_2'') of 1.4.1, then J is a w.h. equivalence, which is proven in 2.3.2. In particular, *J is a w.h. equivalence for the sheaf of free isometric C^∞-immersions $(V, g) \to (W, h)$.*

(B) *Flexibility and Microflexibility.* A sheaf Φ is called *flexible* (*micro-flexible*) if the restriction map $\Phi(C) \to \Phi(C')$ is a fibration (micro-fibration) for all pairs of compact subsets C and $C' \subset C$ in V (compare 1.4.2).

The main result of this section is the following

Theorem. *If V is a locally compact countable polyhedron* (e.g. *a manifold*), then *every flexible sheaf over V satisfies the parametric h-principle* (compare 1.4.2).

Proof. Start with the following

(B_1) **Definition.** A homomorphism $\alpha: \Phi \to \Psi$ is a *local w.h. equivalence* if $\alpha_v: \Phi(v) \to \Psi(v)$ is a w.h. equivalence for $v \in V$.

(B_2) **Local Lemmas.** *Let Φ be any* (*possibly non-flexible*) *sheaf over a locally finite polyhedron V. Then*
(a) *the sheaf Φ^* is flexible;*
(b) *the inclusion $\Phi = \Delta(\Phi) \subset \Phi^*$ is a local w.h. equivalence.*

Proof. The first claim is obvious by the proof of (A′) in 1.4.2. To prove (b), take a point $v \in V$ and consider a fundamental system of neighborhoods $U_i \subset V$ of v, $U_{i+1} \subset U_i$, $i = 1, 2, \ldots$. The space $\Phi(v)$ is the inductive limit of the sequence $\Phi(U_1) \to \Phi(U_2) \ldots$ and the space $\Phi^*(v)$ is the inductive limit of the sequence $[\Phi(U_1)]^{U_1} \to [\Phi(U_2)]^{U_2} \to \cdots$, where $[\Phi(U_i)]^{U_i}$ denotes the space of maps $U_i \to \Phi(U_i)$. Since V is a polyhedron we can choose all U_i to be contractible, to make the inclusions $\Phi(U_i) \to [\Phi(U_i)]^{U_i}$ w.h. equivalences. By passing to the inductive limit we conclude that the map $\Phi(v) \to \Phi^*(v)$ is also a w.h. equivalence.

Thus the main theorem reduces to the following

(B) Homomorphism Theorem. *Let Φ and Ψ be flexible sheaves over a locally compact countably compact finite dimensional space V. Then every local w.h. equivalence $\alpha: \Phi \to \Psi$ is a weak homotopy equivalence.*

Proof. We start with the following functorial construction, which transforms an arbitrary continuous map $a: X \to Y$ into a fibration $\tilde{a}: \tilde{X} \to Y$. Start with the space $P(Y)$ of all continuous maps $p: [0, 1] \to Y$ and define $\tilde{X}$ as the subset in $X \times P(Y)$ of the pairs (x, p), such that $a(x) = p(0)$. We define the map $\tilde{a}$ by setting $\tilde{a}(x, p) = p(1)$. There is a canonical homotopy equivalence between X and $\tilde{X}$ and the map $\tilde{a}: \tilde{X} \to Y$ is a Serre fibration. Clearly, the map $\tilde{a}$ is a w.h. equivalence iff a is a w.h. equivalence. Now, an arbitrary Serre fibration is a w.h. equivalence iff the fibers are *weakly contractible.* (A space Z is called weakly contractible if for an arbitrary compact polyhedron K the space of continuous maps $K \to Z$ is path connected. For such a Z, the space of maps $K \to Z$, in fact, is weakly contractible.)

Let us apply our fibration construction to the homomorphism $\alpha: \Phi \to \Psi$. We obtain a new sheaf $\tilde{\Phi}$, $\tilde{\Phi}(U) = \widetilde{\Phi(U)}$, and a homomorphism $\tilde{\alpha}$, such that the maps $\tilde{\alpha}_U: \tilde{\Phi}(U) \to \Psi(U)$ are Serre fibrations for all open sets $U \subset V$. In order to get "a fiber" of $\tilde{\alpha}$ we take a section $\Psi \in \Psi(U)$ and define the *fiber sheaf* $\Omega = \Omega_\psi$ over U by setting $\Omega(U') = \tilde{\alpha}_{U'}^{-1}(\psi)$ for all open subsets $U' \subset U$. The flexibility of Φ and Ψ implies (easy to see) the flexibility of Ω, which reduces (B) to the following

(B′) Contractibility Theorem. *Let Ω be a flexible sheaf over V. If Ω is locally contractible [that is the space $\Omega(v) = \Omega(\mathcal{O}\!p(v)$ is weakly contractible for all $v \in V$] then the space $\Omega(V)$ is weakly contractible.*

Proof. We proceed in four steps.

Step 1. If V is a compact subset in $\mathbb{R}$ then the space $\Omega(V)$ is path connected.

Proof. To join sections ω_1 and ω_2 in $\Omega(V)$ by a path $\omega_t \in \Omega(V)$, $t \in [0, 1]$, we cover $V \subset \mathbb{R}$ by two closed subsets V' and V'', such that each of them lies in a *disjoint* union of small ε-intervals in $\mathbb{R}$ and such that the intersection $V' \cap V''$ is a *finite* set. Since the space $\Omega(v) = \Omega(\mathcal{O}\!p\,v)$ is path connected for all $v \in V$, there is a path $\omega'_t \in \Omega(V')$ between $\omega_1 | V''$ and $\omega_2 | V''$ as well as a path $\omega''_t \in \Omega(V'')$ between $\omega_1 | V''$ and $\omega_2 | V''$, provided the above $\varepsilon > 0$ is small enough. Since the space $\Omega(V' \cap V'')$ is simply connected (being the product of the simply connected spaces $\Omega(v)$, $v \in V' \cap V''$), there is a homotopy of the path $\omega'_t | V' \cap V''$ to $\omega''_t | V' \cap V''$, which extends, by the flexibility, to a homotopy of ω'_t to another path $\bar{\omega}'_t \in \Omega(V')$ between $\omega_1 | V'$ and $\omega_2 | V''$, (where the homotopies by definition are fixed at the ends of the paths). Now the paths $\bar{\omega}'_t$ and ω'_t agrees over $\mathcal{O}\!p(V' \cap V'')$ and, hence, they define the required path ω_t over V.

Step 2. The above $\Omega(V)$ is weakly contractible.

Proof. Since Ω is flexible, the sheaf $U \mapsto (\Omega(U))^K$ over V is also flexible and the previous step applies.

Step 3. If V is a compact subset in $\mathbb{R}^N$ then $\Omega(V)$ is weakly contractible.

Proof. Let $p: V \to \mathbb{R}$ be an orthogonal projection and let Ω_* denote the *push-forward sheaf* over $p(V) \subset \mathbb{R}$ defined by $\Omega_*(U_*) = \Omega(p^{-1}(U_*))$ for all $U_* \subset p(V)$. Since Ω is flexible, the sheaf Ω_* is also flexible. Furthermore, $\Omega_*(v) = \Omega(p^{-1}(v))$ for $v \in p(V)$, which implies by induction in N the weak contractibility of $\Omega_*(v)$ for all $v \in p(V)$, as $p^{-1}(v)$ lies in a hyperplane in $\mathbb{R}^N$. Hence, the space $\Omega(V) = \Omega_*(p(V))$ is weakly contractible by Step 2.

Step 4. Exhaust an arbitrary V by compact subsets $V_1 \subset V_2 \subset \cdots \subset V_i \subset \cdots \subset V$.

Since $\dim V < \infty$, every V_i embeds into $\mathbb{R}^N$ and so $\Omega(V_i)$ is weakly contractible for all $i = 1, 2, \ldots$. The space $\Omega(V)$ is the inverse (projective) limit of the sequence of the *Serre fibrations*, $\Omega(V_1) \leftarrow \Omega(V_2) \leftarrow \cdots$, and hence, it is weakly contractible as well. Q.E.D.

Exercise. Assume $\Omega(v)$, $v \in V$, to be $(n + k)$-connected for $n = \dim V$ and a given $k = 0,1, \ldots$. Prove $\Omega(V)$ to be k-connected for an arbitrary flexible sheaf Ω over V.

2.2.2 Flexibility and Micro-flexibility of Equivariant Sheaves

We have already seen in 1.4.2 that many interesting sheaves over V are acted upon by the group of diffeomorphisms Diff(V). In fact, it is customary to use the *pseudogroup* of diffeomorphisms, that is the set of all pairs (U, f), where U is an open set and f is a diffeormorphism of U onto another open set $U' = f(U)$

Examples. Let Φ be the sheaf of immersions $V \to W$ which relates to each open set $U \subset V$ the space $\Phi(U)$ of immersions $U \to W$. A diffeomorphism $f: U \to U'$ sends the space $\Phi(U')$ to $\Phi(U)$ by $\varphi \to \varphi \circ f$ for each immersion $\varphi: U' \to W$.

Further examples of such Diff(V)-invariant sheaves are provided by k-mersions (see 1.3.1), by free maps (see 1.1.4) and by maps transversal to a foliation or to a general subbundle as in 2.1.4. On the other hand the sheaf of isometric immersions $V \to W$ is not Diff(V)-invariant. It is invariant under the (pseudo)-group of isometries of the underlying Riemmanian manifold V. Another important (pseudo) group is attached to a smooth map $\pi: V \to V_0$. This (pseudo) group consists of diffeomorphisms f commuting with π, that is $\pi(f(v)) = \pi(v)$, $v \in V$. It is called the (pseudo) group Diff(V, π) *of fiber preserving diffeomorphisms*.

Main Flexibility Theorem. *Let $V = V_0 \times \mathbb{R}$, and let $\pi: V \to V_0$ denote the projection on the first factor. Let Φ be a microflexible sheaf over V, invariant under* Diff(V, π). *Then the restriction $\Phi | V_0 = V_0 \times 0$ is a flexible sheaf over $V_0 = V_0 \times 0$.*

We prove this theorem in the next section, but first give some applications.

Flexibility for Diff-*invariant Sheaves. Let Φ be a microflexible* Diff(V)-*invariant sheaf over a manifold V. Then the restriction to an arbitrary piecewise smooth polyhedron $K \subset V$ of positive codimension, $\Phi | K$, is a flexible sheaf over K.*

Proof. Start with an arbitrary sheaf Ψ over a locally finite polyhedron K and observe (using the induction by skeletons) the following simple fact

(*) *If the restriction of Ψ to every simplex in K is a flexible (microflexible) sheaf, then Ψ is also flexible (microflexible) over all* of K.

By applying this to sufficiently fine subdivisions of K we conclude to the following

Localization Lemma. *If for each point $k \in K$ the restriction of Ψ to some neighborhood of k is a flexible (microflexible) sheaf then Ψ itself is a flexible (microflexible) sheaf.*

Exercise. Prove the localization lemma for an arbitrary locally compact space K.

Let us return to our Diff(V)-invariant sheaf Φ. For each simplex Δ in K there is a neighborhood $U \subset V$ which splits into the product $U = V_0 \times \mathbb{R}$ such that $\Delta \subset V_0 = V_0 \times O \subset U$. The restricted sheaf $\Phi | U$ over U is Diff(U)-invariant and, in particular, it is Diff(U, π)-invariant for the projection $\pi\colon U \to V_0$. The main flexibility theorem implies that the sheaf $\Phi | V_0$ is flexible, so $\Phi | \Delta$ is also a flexible sheaf, and Lemma (*) applies.

The *h*-Principle for Open Manifolds. *An arbitrary microflexible* Diff(V)*-invariant sheaf Φ over an open manifold V satisfies the parametric h-principle.*

Proof. Take a codimension one polyhedron $K \subset V$ such that V is isotopic to an arbitrarily small neighborhood of K (see 1.4.1). Since that sheaf $\Phi | K$ is flexible it satisfies the h-principle (see 2.2.1); therefore the sheaf $\Phi | \mathcal{O}p(K) = V$ also satisfies the h-principle. [See Bierstone (1973) for a generalization.]

Remark. As we know (see 1.4.2), the sheaf Φ is not flexible in general.

Immersions, Free Maps, etc. *All open* Diff(V) *invariant differential relations over an open manifold V satisfy the parametric h-principle. In particular, maps of rank $> k$, free maps, hyperbolic immersions (see* 3.2.2) *and maps $V \to \mathbb{R}^q$ directed by an open set $A \subset S^{q-1}$ (see* 1.4.4) *satisfy this h-principle as long as the underlying manifold V is open.*

Proof. The openess of a differential condition implies mircoflexibility and it also shows that the sheaf theoretic h-principle is equivalent of the usual one (see 1.4.1).

Isometric Immersions Between Pseudo-Riemannian Manifolds. The sheaf of isometric C^∞-immersions $(V, g) \to (W, h)$ is not, in general, microflexible. However, the sub-sheaf Φ of *free* isometric immersions is microflexible (see 2.3.2). Furthermore, if $(V, g) = (V_0 \times \mathbb{R}, g_0 \oplus 0)$, then Φ is Diff(V, π)-invariant for the projection $\pi\colon V = V_0 \times \mathbb{R} \to V_0$. In this case Φ is a flexible sheaf, which , moreover, satisfies the h-principle. With this, one obtains isometric immersions $V \to W$ under suitable dimension assumptions (see 1.4.2, 3.3).

2.2.3 The Proof of the Main Flexibility Theorem

Let Φ be an arbitrary sheaf over V. We treat a map $\psi: Q \to \Phi(A)$, $A \subset V$, as a function $\psi = \psi(v, q)$, $v \in A$, $q \in Q$, where $\psi(v, q) \in \Phi(v)$ denotes the image of the section $\psi(q) \in \Phi(A)$ under the restriction map $\Phi(A) \to \Phi(v) = \Phi(\mathcal{O}p(v))$. In short, $\psi(v, q)$ is the restriction of $\psi(q)$ to $\mathcal{O}p(v)$.

When Q is split into the product of a compact polyhedron and a closed interval, $Q = P \times [x, y]$, then the maps $\psi: P \times [x, y] \to \Phi(A)$ are called *deformations over* A. When the set A is compact then according to our definitions (see 1.4.1), each deformation ψ over A is actually defined over an open set $U = U(\psi) \supset A$, and we denote this extended deformation $P \times [x, y] \to U$ also by ψ. We say that a deformation ψ over A is *fixed* at a point $v \in U = U(\psi)$ if $\psi(v; p, t) = \psi(v; p, x)$ for all pairs $(p, t) \in P \times [x, y]$. We call the set of non-fixed points of ψ *the support of the deformation*, $\operatorname{supp} \psi \subset U$.

(A) *Compressibility*. A deformation ψ over A is called compressible if for an arbitrarily small neighbourhood $\bar{U}$ of A there exists a deformation $\bar{\psi}: P \times [x, y] \to \Phi(U)$, $U = U(\psi)$, with the following three properties:

(i) $\bar{\psi} | A = \psi | A$, that is $\bar{\psi}(a; p, t) = \psi(a; p, t)$ for all points $a \in A$, $p \in P$, $t \in [x, y]$.
(ii) $\bar{\psi} | P \times x = \psi | P \times x$, that is $\bar{\psi}(v; p, x) = \psi(v; p, x)$ for all $v \in U$ and $p \in P$.
(iii) $\operatorname{supp} \bar{\psi} \subset \bar{U}$.

Observe, that this compressibility property of a deformation ψ, depends only on the behaviour of ψ near A, that is on $\mathcal{O}p(A) \subset U = U(\psi)$, and so the particular choice of the neighborhood U does not affect the content of the notion of compressibility. However, one can not express compressibility in terms of the restricted sheaf $\Phi | A$. In fact, if a deformation ψ is given over the whole space where a sheaf is defined, then this ψ is tautologically compressible. Let us show that compressibility is equivalent to flexibility. More precisely, *a sheaf Φ over a locally compact space V is flexible if for every compact set $A \subset V$ all deformations over A are compressible.*

Proof. Let Φ be a flexible sheaf. The condition (i) prescribes $\bar{\psi}$ on A and the condition (iii) prescribes $\bar{\psi}$ on $U \setminus \bar{U}$ by requiring it to be fixed there. When the neighbourhood $\bar{U} \subset U$ is sufficiently small and its closure in U, $\operatorname{Cl}(\bar{U})$, is compact, then the restriction map $\Phi(A \cup (\operatorname{Cl}(\bar{U}) \setminus \bar{U})) \to \Phi(\operatorname{Cl}(\bar{U}))$ is a fibration, and we can extend this deformation from $A \cup (\operatorname{Cl}(\bar{U}) \setminus \bar{U})$ to $\operatorname{Cl}(\bar{U})$. Such an extension gives us $\bar{\psi}$ on $\operatorname{Cl}(\bar{U})$ and so on U.

Now, let $\psi: P \times [0, 1] \to \Phi(A)$ be a compressible deformation over A and let its restriction to $P \times 0 \subset P \times [0, 1]$, $\psi | P \times 0$, be extended to a larger compact set $B \subset V$, $B \supset A$. The properties (i) and (ii) show that in order to extend ψ, it is sufficient to extend some compression $\bar{\psi}$ of ψ. When the neighbourhood $\bar{U}$ is chosen sufficiently small, the condition (iii) allows us to extend $\bar{\psi}$ to B just by making it fixed on $B \setminus \bar{U}$.

Microcompressibility. A deformation $\psi: P \times [x, y] \to \Phi(A)$ is called *microcompressible* if there is a positive ε, $\varepsilon \in (0, x - y]$, such that the restricted deformation

$\psi | P \times [x, x+\varepsilon] = \psi_\varepsilon \colon P \times [x, x+\varepsilon] \to \Phi(A)$ is compressible. This property is reminiscent of microflexibility but, in fact, is much stronger. If Φ is a microflexible sheaf, then, for a *given* neighborhood $\bar{U}$ of a compact set A, we can find a positive ε, such that the deformation ψ_ε can be compressed to $\bar{\psi}_\varepsilon$ with supp $\psi_\varepsilon \subset \bar{U}$, but this ε here depends on $\bar{U}$. On the other hand, the microcompressibility property provides a *universal* ε for all (arbitrarily small) neighborhoods $\bar{U}$. This difference is crucial as the following lemma shows.

Microcompressibility Lemma. *A sheaf Φ over a locally compact space is flexible iff all deformations over compact sets are microcompressible.*

Proof. We must show that microcompressibility implies compressibility. Take a deformation $\psi \colon P \times [x, y] \to \Phi(A)$ and let us apply the microcompressibility property to the following auxiliary deformation $\eta \colon Q \times [0, 1] \to \Phi(A)$, where $Q = P \times [x, y]$ and η is defined by $\eta(p, t, \tau) = \psi(p, \min(y, t+\tau))$, $p \in P$, $t \in [x, y]$, $\tau \in [0, 1]$. Since η is microcompressible with some $\varepsilon > 0$, the restriction of ψ to an arbitrary interval $[x_1, y_1] \subset [x, y]$ of length $\leq \varepsilon$, that is $\psi | P \times [x_1, y_1] = \psi_1 : P \times [x_1, y_1] \to \Phi(A)$, is a compressible deformation. Now, we subdivide the interval $[x, y]$ into some finite number of intervals of length $\leq \varepsilon$ such that the restriction of ψ to these intervals are compressible. Now an obvious induction reduces the whole matter to the following.

If for a point $z \in [x, y]$ the restricted deformations $\psi_0 = \psi | P \times [x, z]$ and $\psi_1 = \psi | P \times [z, y]$ are compressible then ψ is also compressible.

Indeed, for a given $\bar{U} \supset A$, we first compress ψ_0 to $\bar{\psi}_0$ with supp $\bar{\psi}_0 \subset \bar{U}$. Next, we take a smaller neighbourhood $\bar{U}_1$ of A such that $\psi_0 | \bar{U}_1 = \bar{\psi}_0 | \bar{U}_1$. Such a $\bar{U}_1$ exists by the property (i) above. Finally, we compress ψ_1 to $\bar{\psi}_1$ with supp $\bar{\psi}_1 \subset \bar{U}_1$ and we define $\bar{\psi}$ by "glueing" $\bar{\psi}_0$ and $\bar{\psi}_1$ together:

$$\bar{\psi}(P, t) = \begin{cases} \bar{\psi}_0(P, t) & \text{for } t \in [x, z], \\ \bar{\psi}_1(P, t) & \text{for } t \in [z, y]. \end{cases}$$

The property (ii) shows that this $\bar{\psi}$ is a correctly defined compression.

(B) *Actions of Diffeotopies on Φ.* Take open subsets $U' \subset V$ and $U \subset U'$ and move U in U' by a diffeotopy $\delta_t \colon U \to U'$, $t \in [0, 1]$, for $\delta_0 = \mathrm{Id} \colon U \to U \subset U'$. Let Φ' be a subset in $\Phi(U')$ and let δ_t *act* on Φ' by assigning to each section $\varphi \in \Phi'$ a homotopy of sections in $\Phi(U)$, called $\delta_t^* \varphi \in \Phi(U)$, such that $\delta_0^* \varphi = \varphi | U$ and such that the following four conditions are satisfied.

(i) If two sections in Φ' are equal at some point in U', say $\varphi_1(u_0') = \varphi_2(u_0')$, and if $\delta_{t_0}(u_0) = u_0'$ for some $u_0 \in U$ and $t_0 \in [0, 1]$, then $(\delta_{t_0}^* \varphi_1)(u_0) = (\delta_{t_0}^* \varphi_2)(u_0)$. This allows us to write $\varphi(\delta_t(u))$ for $(\delta_t^* \varphi)(u)$, $u \in U$.

(ii) Let $U_0 \subset U$ be the maximal open subset where δ_t is constant in t, that is $\delta_t(u) = \delta_0(u)$, $u \in U_0$. Then the homotopy $\delta_t^* \varphi$ (whenever defined) is constant in t over U_0.

(iii) If the diffeotopy δ_t is constant in t for $t \geq t_0$ over all of U, then the homotopy $\delta_t^* \varphi$ also is constant in t for $t \geq t_0$.

(iv) If $\varphi_p \in \Phi'$, $p \in P$, is a *continuous* family of sections, then the family $\delta_t^* \varphi_p$ is jointly continuous in p and t.

(B′) **Examples.** (a) If the sheaf Φ is acted upon by the group Diff V, then all diffeotopies obviously act on all of Φ. For instance, the diffeotopies in V act on the sections of the trivial fibration $X = V \times W \to V$. Now, let Y be an open subset in X and consider the sheaf Φ of sections $V \to Y$. Since Y is not assumed invariant under Diff(V), the diffeotopies of V do not preserve Φ. However, for an arbitrary *compact* family of sections $\varphi \in \Phi(U')$, the action $\delta_t^* \varphi_p$ is defined in Φ for the diffeotopies δ_t which are sufficiently C°-close to the identity.

(b) Let a diffeotopy δ_t be smooth in t and let δ_t' denote the vector field $d\delta_t/dt$ on $U_t = \delta_t(U) \subset U'$. Take a continuous function α on $U \times [0,1]$ which vanishes at every point $(u,t) \in U \times [0,1]$, where the vector $\delta_t'(u) \in T_{u'}(U')$, $u' = \delta_t(u)$, vanishes. Then the assignment $\varphi \mapsto \varphi(\delta_t(u)) + \int_0^t \alpha(u,t)\,dt$ defines an action of δ_t on (the sheaf of) continuous functions φ on V.

(C) *Sharp Diffeotopies.* Let V_0 be a closed subset in the above $U' \subset V$. A diffeotopy of V_0 in U' is by definition a diffeotopy $\delta_t\colon U \to U'$ for an arbitrarily small neighborhood $U = \mathcal{O}p\, V_0 \subset U'$. Fix some metric in V and call some set $\mathcal{A}$ of diffeotopies $V_0 \to U'$ *strictly moving* a given subset $S \subset V_0$, if $\operatorname{dist}(\delta_t(S), V_0) \geq \mu > 0$ for $t \geq \frac{1}{2}$ and for all $\delta_t \in \mathcal{A}$.

Call $\mathcal{A}$ *sharp* at S if for every $v > 0$ there exists a diffeotopy $\delta_t \in \mathcal{A}$ such that

(i) $\delta_t | \mathcal{O}p(v) = \delta_0 | \mathcal{O}p(v)$, $t \in [0,1]$, for all points $v \in V_0$ which have $\operatorname{dist}(v, S) \geq v$, where $\mathcal{O}p(v) \subset V$ is an (arbitrarily) small neighborhood of v.

(ii) $\delta_t = \delta_{1/2}$ for $t \geq \frac{1}{2}$.

Call a set of diffeotopies *sharply moving* V_0 at S, if it contains a subset $\mathcal{A}$ which strictly moves S and is sharp at S.

Finally, for a given sheaf Φ on V and for given actions of diffeotopies δ_t on subsets $\Phi' = \Phi'_{\delta_t} \subset \Phi(U')$, we say that *acting diffeotopies sharply move V_0 at S*, if for every compact family of sections $\varphi_p \in \Phi(U')$ there exists the above subset $\mathcal{A}$, such that $\varphi_p \in \Phi'_{\delta_t}$ for all $\delta_t \in \mathcal{A}$.

(C′) Let V_0 be a submanifold in V, let $B \subset V_0$ be a codimension zero submanifold bounded by a closed hypersurface $S = \partial B \subset V_0$ and let $\psi\colon P \times [0,1] \to \Phi(B)$ be a deformation for a given sheaf Φ on V. Denote by $U' = U'(\psi) \subset V$ the neighborhood of $B \subset V$ where ψ is actually defined and let $V_0' = V_0 \cap U'$. Take an arbitrary compact subset $A \subset B$ which does not intersect $S = \partial B$ and let $\bar{U}_0 \subset V_0'$ be an arbitrary neighborhood of B.

Main Lemma. *If the sheaf Φ is microflexible and if acting diffeotopies $\delta_t\colon V_0' \to U'$ sharply move V_0' at S, then the deformation $\psi | A$ admits a microcompression in the sheaf $\Phi_0 = \Phi | V_0$ to a deformation with the support in $\bar{U}_0$.*

Proof. Since Φ is microflexible, there exist $\varepsilon = \varepsilon(\mu) > 0$ for all $\mu > 0$, and a deformation $\psi_\mu\colon P \times [0,\varepsilon] \to \Phi(U')$, which equals $\psi | P \times [0,\varepsilon]$ near B and such that the support $\operatorname{supp}\psi_\mu$ lies in the μ-neighborhood $U_\mu \subset U'$ of B (for a given metric on V). Moreover, there exists $\bar{\varepsilon} = \bar{\varepsilon}(\bar{U}_0) > 0$, such that the above homotopy can be made additionally constant in $t \in [0, \bar{\varepsilon}]$ on $V_0 \setminus \bar{U}_0$, that is

$$\operatorname{supp}(\psi_\mu | P \times [0, \bar{\varepsilon}]) \cap V_0' \subset \bar{U}_0.$$

Next, take a diffeotopy δ_t: $V_0' \to U'$, $t \in [0, 1]$, which does act on ψ_μ and such that

(i) δ_t is constant in t away from a small ν-neighborhood $\bar{U}_\nu \subset \bar{U}_0$ of $S \subset \bar{U}_0$ for some $\nu > 0$, such that

$$\nu < \min(\operatorname{dist}(S, A), \operatorname{dist}(S, \partial \bar{U}_0)).$$

(ii) δ_t is constant in t for $t \geq \frac{1}{2}$.
(iii) δ_1 sends S outside U_μ,

$$\delta_1(S) \subset U' \setminus U_\mu.$$

Let $\bar{\delta}_t = \delta_{\lambda t}$ for $\lambda = \bar{\varepsilon}^{-1}$ where $\bar{\delta}_t$ for $t \geq \bar{\varepsilon}$ is defined by $\bar{\delta}_t = \delta_1$. Now compress $\psi_\mu | V_0'$ to a deformation $\bar{\psi}$ in Φ_0 by putting

$$\bar{\psi}(v, p, t) = \begin{cases} \psi_\mu(\bar{\delta}_t(v), p, t) & \text{for } v \in B \text{ and } t \in [0, \varepsilon]; \\ \psi_\mu(\bar{\delta}_t(v), p, \min(t, \bar{\varepsilon})) & \text{for } v \in V_0 \setminus B. \end{cases}$$

These formulae agree along S for $t \in [\bar{\varepsilon}, \varepsilon]$, as $\bar{\delta}_t$ for $t \geq \bar{\varepsilon}$ sends S outside the support of ψ_μ, and hence, $\bar{\psi}$ is correctly defined. Since the support of $\psi_\mu | V_0'$ lies in $\bar{U}_0$ and since δ_t is constant in t outside $\bar{U}_0$, the deformation $\bar{\psi} | V_0'$ is supported in $\bar{U}_0$, while $\bar{\psi} | A = \psi_\mu | A = \psi | A$. Q.E.D.

(C″) Say that *acting diffeotopies sharply move* a submanifold $V_0 \subset V$ if each point $v \in V_0$ admits a neighborhood $U' \subset V$ of v, such that acting diffeotopies δ_t: $V_0' = V_0 \cap U' \to U'$ sharply move V_0' at any given closed hypersurface $S \subset V_0'$.

Theorem. *Let Φ be a microflexible sheaf over V and let a submanifold $V_0 \subset V$ be sharply movable by acting diffeotopies. Then the sheaf $\Phi_0 = \Phi | X_0$ is flexible and, hence, it satisfies the h-principle.*

Proof. An arbitrary deformation over $A \subset V_0'$ can be microcompressed by applying the main lemma to some hypersurface S around A. Hence [see (A)], $\Phi | V_0$ is locally flexible and therefore (see the localization lemma in 2.2.2) a flexible sheaf.

Corollaries. (a) The main flexibility theorem (see 2.2.2) follows immediately since diffeotopies preserving the fibration $V = V_0 \times \mathbb{R} \to V_0$ (obviously) sharply move $V_0 = V_0 \times 0 \subset V$.

(b) Let Φ' be an open Diff(V)-invariant subsheaf of the sheaf of exterior differential k-forms on V and let $\Phi \subset \Phi'$ be the subsheaf which consists of the *closed* forms in Φ'. Notice that the sheaf Φ (unlike Φ') is not microflexible. Consider the subspace $G_0 \subset \Phi(V)$ of forms g in $\Phi(V)$ cohomologous to a given closed k-form g_0 on V. Then we "parametrize" G_0 with the sheaf Ψ_0 of the $(k-1)$-form ψ on V for which

$g_0 + d\psi \in \Phi$. The sheaf Ψ_0 is clearly microflexible and the map $\psi \mapsto g_0 + d\psi$ is a homotopy equivalence of $\Psi_0(V)$ onto G_0.

If $g_0 \not\equiv 0$, then the sheaf Ψ_0 is not Diff(V)-invariant. However, each smooth diffeotopy $\delta_t\colon U \to U'$ defines an action on Ψ_0 as follows. Take the interior product (see 3.4.1) of the field $\delta'_t = d\delta_t/dt$ on $U_t = \delta_t(U) \subset U'$ with g_0 and then pull-back this product to U by δ_t. Thus we obtain a $(k-1)$-form, say $g_t^* = \delta_t^*(\delta'_t \cdot g_0)$ on U for all $t \in [0,1]$.

Assign $\psi \mapsto \delta_t^*(\psi) + \int_0^t g_t^*\, dt$ for all $\psi \in \Psi_0$. One easily checks this to be an action [compare (b) in (B)] of δ_t on Ψ_0 which yields the parametric h-principle for Ψ_0 and, hence for Φ, provided V is *open*.

(C‴) **Example.** Let Φ be the sheaf of *symplectic* (see 3.4.2) forms on an open manifold V and let forms g_0 and g_1 in $\Phi(V)$ represent a given class $\alpha_0 \in H^2(V;\mathbb{R})$. If there exists a homotopy of *non-singular* (possibly non-closed) 2-forms g_t between g_0 and g_1, then the h-principle for the sheaf Ψ_0 yields a homotopy of 1-forms $\psi_t \in \Psi_0(V)$, $t \in [0,1]$, such that $\psi_0 = 0$ and $d\psi_1 = g_1 - g_0$.

Exercises. Let L be a *completely non-integrable* k-plane field on a C^∞-manifold W, that is a k-dimensional subbundle $L \subset T(W)$, such that successive Poisson brackets of the C^∞-sections $W \to L$ (that are vector fields on W tangent to L) span the tangent bundle $T(W)$.

(a) Show the sheaf of C^∞-immersions $\mathbb{R} \to W$ which are everywhere *tangent* to L to be microflexible.

Hint. This can be done with a straightforward (but lengthy) geometric argument or (much faster) with the analytic techniques of 2.3.8.

(b) Let V be an arbitrary smooth manifold and let $\tilde{\Phi}'$ be the sheaf of C^∞-maps $F\colon V \times \mathbb{R} \to W$, such that

(i) the restriction of F to each line $v \times \mathbb{R} \subset V \times \mathbb{R}$, $v \in V$, is an immersion $\mathbb{R} = v \times \mathbb{R} \to W$ everywhere tangent to L.
(ii) the map F is *transversal* to L. Here it means the fiberwise *surjectivity* of the homomorphism $T(V \times \mathbb{R}) \to T(W)/L$ obtained by composing the differential $D_F\colon T(V \times \mathbb{R}) \to T(W)$ with the quotient homomorphism $T(W) \to T(W)/L$.

Prove the sheaf $\tilde{\Phi}'$ to be microflexible. Then show the sheaf $\tilde{\Phi} = \Psi | V \times 0$ to be a *flexible* sheaf on $V = V \times 0$.

(b′) Prove the sheaf Φ of C^∞-maps $V \to W$ transversal (in the above sense) to L satisfies the C^0-dense parametric h-principle.

2.2.4 Equivariant Microextensions

Consider two sheaves Φ and $\tilde{\Phi}$ over V and a homomorphism $\alpha\colon \tilde{\Phi} \to \Phi$ over V. We say that α is *surjective* if for each point v the corresponding map $\alpha(\mathcal{O}p(v))\colon \tilde{\Phi}(v) \to \Phi(v)$ is onto. Notice that for a surjective homomorphism α the map $\alpha(V)\colon \tilde{\Phi}(V) \to \Phi(V)$ is not, in general, surjective.

Examples. Let $V = V \times 0 \subset V \times \mathbb{R}$, let Φ denote the sheaf (of germs) of immersions $V \to W$ and let $\tilde{\Phi}$ be the restriction to V of the sheaf immersions $V \times \mathbb{R} \to W$. The natural restriction homomorphism $\tilde{\Phi} \to \Phi$ is surjective iff $\dim V < \dim W$, but the corresponding map $\tilde{\Phi}(V) \to \Phi(V)$ may not be surjective for $\dim W < 2(\dim V) - 1$. For instance, if V is the real projective plane and $W = \mathbb{R}^3$, then there is an immersion $V \to \mathbb{R}^3$ but the product $V \times \mathbb{R}$ has no immersions into $\mathbb{R}^3$. However, for a contractible manifold V, each immersion $V \to W$ extends, for $\dim W > \dim V$, to an immersion $V \times \mathbb{R} \to W$. This allows one to apply the main flexibility theorem to $V \times \mathbb{R}$ and to conclude the flexibility of immersions of a contractible manifold V into W for $\dim W > \dim V$. Since any V is covered by balls, one can use the localization lemma (see 2.2.2) and obtain flexibility, and thus the h-principle, for extra-dimensional immersions of *non*-open manifolds V.

Unfortunately, for more general maps this approach does not work. For example, in general, a map $V \times M$ of rank $\geq k$, even for $V = \mathbb{R}^n$, cannot be extended to a map $V \times \mathbb{R} \to W$ of rank $\geq k + 1$, though such extensions may exist at every point $v \in V$.

Let us describe a property of the homomorphism $\alpha: \tilde{\Phi} \to \Phi$ that "descend flexibility" from $\tilde{\Phi}$ to Φ. Take two sets A and $B \subset A$ in V and call two sections $\varphi \in \Phi(A)$ and $\tilde{\varphi} \in \tilde{\Phi}(B)$ *coherent* if the restriction $\varphi | B \in \Phi(B)$ equals to $\alpha(\tilde{\varphi}) \in \Phi(B)$. The space of all coherent pairs $(\varphi, \tilde{\varphi}) \in \Phi(A) \times \tilde{\Phi}(B)$ is denoted by $\Omega = \Omega(A, B)$ and the natural map $\tilde{\Phi}(A) \to \Omega$ is denoted by $\eta = \eta(A, B)$.

Definition. A homomorphism α is called *microextension* if it is surjective and if for every pair of compact subsets A and B in V the map $\eta: \tilde{\Phi} \to \Omega$ is a microfibration.

Example. Let Φ be the sheaf of immersions $V \to W$ and let $\tilde{\Phi}$ be the restriction to $V \subset V \times \mathbb{R}$ of the sheaf of immersions $V \times \mathbb{R} \to W$. For a set $A \subset V$, a section in $\tilde{\Phi}(A)$ is an immersion of $\mathcal{O}p(A) \subset V \times \mathbb{R}$ to W, and the space $\Omega(A, B)$ for $B \subset A$ consists of the maps of the union $\mathcal{O}p(B) \cup (V \cap \mathcal{O}p(A))$, $\mathcal{O}p(B) \subset V \times \mathbb{R}$, to W, such that the restrictions of these maps to $\mathcal{O}p(B)$ and to $V \cap \mathcal{O}p(A)$ are immersions: $\mathcal{O}p(B) \to W$ and $V \cap \mathcal{O}p(A) \to W$ respectively. Here the map $\eta: \tilde{\Phi}(A) \to \Omega$ amounts to the restriction of immersions from $\mathcal{O}p(A)$ to the union of $\mathcal{O}p(B)$ and $V \cap \mathcal{O}p(A)$.

Since the immersion condition is open, we immediately conclude that η is a microfibration (compare 1.4.2). The same conclusion holds for the restriction map of $(k + 1)$-mersions $V \times \mathbb{R} \to W$ to k-mersions $V \to W$, and in general, for all open relations whenever the homomorphism η makes sense.

Microextension Theorem. *If $\tilde{\Phi}$ is a flexible sheaf and $\alpha: \tilde{\Phi} \to \Phi$ is a microextension then the sheaf Φ is also flexible. In other words, if Φ admits a flexible microextension then Φ itself is a flexible sheaf.*

The proof is given in the next section. This theorem provides the h-principle for immersions and k-mersions in the extra-dimensional case. Now try free maps $V = V^n \to \mathbb{R}^q$. The only suitable extension comes from free maps $V \times \mathbb{R} \to \mathbb{R}^q$. These maps only exist for $q > (n + 1)(n + 4)/2 = [n(n + 3)/2] + n + 2$ and only for such a q the corresponding homomorphism $\tilde{\Phi} \to \Phi$ is surjective. Thus, we need $n + 2$ extra dimensions to get the h-principle for the free maps of closed manifolds into $\mathbb{R}^q$. The method of removal of singulatities serves much better for free maps: it only requires one extra dimension.

Invariant Extensions. Let $X \to V$ be an arbitrary smooth fibration and let $Y \to V' = V \times \mathbb{R}$ be the fibration induced by the projection $\pi: V \times \mathbb{R} \to V$. Denote by $\Pi^{(r)}: Y^{(r)} \to X^{(r)}$ the map that assigns to an r-jet represented by a germ $g: V' \to Y$ at $v' = (v, t) \in V'$, the jet of the restriction of g to $V = V \times t \subset V \times \mathbb{R} = V'$.

For a differential relation $\mathcal{R} \subset X^{(r)}$ we call a relation $\mathcal{R}' \subset Y^{(r)}$ an *extension* of $\mathcal{R}$ if the map $\Pi^{(r)}$ sends $\mathcal{R}'$ *onto* $\mathcal{R}$.

Now, we invoke the group $\mathcal{D} = \mathcal{D}(V', \pi)$ of the fiber preserving diffeomorphisms for $\pi: V' \to V$ and we observe that this $\mathcal{D}$ naturally acts on Y and on $Y^{(r)}$. An extension $\mathcal{R}' \subset Y^{(r)}$ of $\mathcal{R} \subset X^{(r)}$ is called *invariant* if it is invariant under this action of $\mathcal{D}$.

Examples. For the immersions relation $\mathcal{R} \subset X^{(1)}$, $X = V \times W \to V$, there is a natural non-invariant extension whose solutions are the maps $V \times \mathbb{R} \to W$ such that their restrictions to the manifolds $V \times t \subset V \times \mathbb{R}$ are immersions for all $t \in \mathbb{R}$. Notice that this extensions is an open condition in $Y^{(1)}$ for $Y = V' \times W \to V'$. There is also an invariant but not open extension of the immersion relation $\mathcal{R} \subset X^{(1)}$. The solutions of this last extension are the maps $V \times \mathbb{R} \to W$) which are immersions on all manifolds $V \times t \subset V \times \mathbb{R}$ and which are constant on all lines $v \times \mathbb{R} \subset V \times \mathbb{R}$, $v \in V$. These maps are not microflexible.

The immersion relation $\mathcal{R} \subset X^{(1)}$ has an extension $\mathcal{R}' \subset Y^{(1)}$ which is both open and invariant only for dim $W >$ dim V, that is the immersion condition for $V' \to W$. (If dim V = dim W this $\mathcal{R}' \subset Y^{(r)}$ is an empty set and so the projection $\Pi^{(1)}: \mathcal{R}' \to \mathcal{R}$ is by no means onto.)

The Open Extension Theorem. *If a relation $\mathcal{R} \subset X^{(r)}$ admits an open invariant extension $\mathcal{R}' \subset Y^{(r)}$ then $\mathcal{R}$ abides by the h-principle and solutions of $\mathcal{R}$ are flexible.*

Proof. Denote by $\tilde{\Phi}'$ the sheaf of solutions of $\mathcal{R}'$. Since $\mathcal{R}'$ is open the sheaf $\tilde{\Phi}'$ is microflexible and by the main flexibility theorem its restriction to $V = V \times 0 \subset V \times \mathbb{R} = V'$ is a flexible sheaf $\tilde{\Phi}$ over V. There is a natural restriction homomorphism of $\tilde{\Phi}$ to the sheaf Φ of solutions of $\mathcal{R}$. Since $\mathcal{R}' \subset Y^{(r)}$ is open, this homomorphism is a microextension; therefore Φ is flexible and it satisfies the h-principle as well (see 2.2.1).

Let us give a canonical method for locating an invariant open extension of $\mathcal{R} \subset X^{(r)}$. Take the pullback $\mathcal{R}^* = (\Pi^r)^{-1}(\mathcal{R}) \subset Y^{(r)}$. This $\mathcal{R}^*$ is an extension of $\mathcal{R}$ which is not invariant. Next, we take the maximal $\mathcal{D}$-invariant subset in $\mathcal{R}^*$, namely $\mathcal{R}' = \bigcap_\delta \delta(\mathcal{R}^*)$, where δ runs over $\mathcal{D}$. The condition $\mathcal{R}'$ is an invariant extension of $\mathcal{R}$, but it may be not open. Finally, we take the interior of $\mathcal{R}'$ and project this interior, Int$(\mathcal{R}')$, by $\Pi^{(r)}$ to $X^{(r)}$. The image of this projection is an open set $\tilde{\mathcal{R}} \subset X^{(r)}$ which is contained in $\mathscr{R}$ and the condition $\mathrm{Int}(\mathscr{R}') \subset Y^{(r)}$ serves as the maximal open invariant extension of $\tilde{\mathscr{R}}$. This immediately leads to the following conclusion.

A condition $\mathscr{R} \subset X^{(r)}$ admits an open invariant extension iff $\tilde{\mathscr{R}} = \mathscr{R}$.

Examples. For the k-mersion condition, $\mathscr{R} \subset X^{(1)}$, $X = V \times W$, the condition $\tilde{\mathscr{R}} \subset X^{(1)}$ is empty for $k > \dim W$, but for $k < \dim W$ we have $\tilde{\mathscr{R}} = \mathscr{R}$. This gives Feit's k-mersion theorem in the extra dimensional case.

Take an arbitrary fibration $X \to V$ and a subbundle τ in the tangent bundle $T(X)$.

Transversality Theorem. *If* $\mathrm{codim}(\tau) > \dim(V)$, *then section $V \to X$ transversal to τ satisfy the h-principle.*

Proof. A straightforward check up shows that the inequality codim $(\tau) > \dim(V)$ implies $\tilde{\mathscr{R}} = \mathscr{R}$.

Here our condition $\mathscr{R}$ is not $\mathrm{Diff}(V)$-invariant and even for an open manifold V we can not drop the dimension restriction. Notice also that for the trivial fibration $V \times W \to V$ this transversality theorem reduces to Hirsch's theorem if τ equals the kernel of the differential of the projection $V \times W \to W$.

2.2.5 Local Compressibility and the Proof of the Microextension Theorem

We start with an abstract version of the "induction by skeletons" procedure. Consider a compact space A covered by compact subsets A_i, $i = 1, \ldots, k$, $\bigcup_{i=1}^k A_i = A$. For a subset $I \subset \{1, \ldots, k\}$ we denote by A_I the intersection $\bigcap_{i \in I} A_i$ and by ∂A_I the union $\bigcup_{j \notin I} (A_I \cap A_j)$. Consider some unions of A_i, say $B = \bigcup_{j \in J} A_j$ for $J \subset \{1, \ldots, k\}$ and $B' = \bigcup_{j \in J'} A_j$ for $J' \subset J$, and observe that the extension of any homotopy from B' to B reduces to extensions from ∂A_I to A_I for all $I \subset J$. Here, the flexibility (microflexibility) of a sheaf Φ over A on the pairs $(A_I, \partial A_I)$ implies flexibility (microflexibility) on (B, B'), where the flexibility (microflexibility) of Φ *on a pair* (C, C') means the restriction map $\Phi(C) \to \Phi(C')$ is a fibration (microfibration).

Lemma. *If A admits an arbitrarily fine finite cover by compact subsets A_i such that Φ is flexible (microflexible) on all pairs $(A_I, \partial A_I)$, then the sheaf Φ is flexible (mirco-flexible).*

Proof. For every pair (C, C') of compact subsets in A and for all neighborhoods $U \supset C$ and $U' \supset C'$ in A there exists a pair (B, B') built from A_i, such that $C \subset B \subset U$ and $C' \subset B' \subset U'$, provided the cover by A_i is sufficiently fine. Hence, the extension of deformations from $\mathcal{O}p\,C'$ to $\mathcal{O}p\,C$ reduces to that from B' to B for sufficiently fine covers $A = \bigcup A_i$. Q.E.D.

Flexibility of Parametric Sheaves. Let Φ be a sheaf over A and let Q be a finite polyhedron.

If Φ is flexible (microflexible) then the parametric sheaf Φ^Q over $A \times Q$ (see 2.2.1) is also flexible (microflexible).

Proof. Cover A by small compact sets A_i, $i = 1, \ldots, k$, and subdivide Q into some small simplices S_j, $j = 1, \ldots, l$. The above argument reduces our problem to the following

Sublemma. *For every simplex $S \subset Q$ and for every compact subset $B \subset A$ the sheaf Φ^S is flexible (microflexible) on the pair $(A \times S, \bar{B})$ for $\bar{B} = (B \times S) \cup (A \times \partial S)$.*

Proof. Let $\varphi\colon P \to \Phi^S(A \times S)$ be an arbitrary map and let $\psi\colon P \times [0,1] \to \Phi^S(\bar{B})$ be a deformation such that $\psi | P \times 0 = \varphi | \bar{B}$. By the definition of Φ^S (see 2.2.1), the map φ is given by a map

$$\varphi'\colon S \times P \to \Phi(A)$$

and ψ is given by two maps

$$\psi'\colon \mathcal{O}p(\partial S) \times P \times [0,1] \to \Phi(A), \quad \mathcal{O}p(\partial S) \subset S,$$

and

$$\psi''\colon S \times P \times [0,1] \to \Phi(B).$$

The maps φ' and ψ' agree on the intersection $(S \times P) \cap (\mathcal{O}p(\partial S) \times P \times [0,1]) = (\mathcal{O}p(\partial S)) \times P \subset S \times P \times [0,1]$, $P = P \times 0 \subset P \times [0,1]$, and so they define a map $(\varphi', \psi')\colon P' \to \Phi(A)$ for $P' = (S \times P) \bigcup (\mathcal{O}p(\partial S) \times P \times [0,1]) \subset S \times P \times [0,1]$. Now for the restriction map $\rho\colon \Phi(A) \to \Phi(B)$, we have

$$\psi'' | P' = \rho \circ (\varphi', \psi'),$$

and in order to extend the deformation ψ to $A \times S$ we must extend the lift (φ', ψ') of ψ'' from P' to $S \times P \times [0,1]$. If we use for $\mathcal{O}p(\partial S)$ a small standard collar of the boundary ∂S in S, we get a compact polyhedral pair (P', P), such that P' is a deformation retract of P. When Φ is flexible, then ρ is a fibration and this lift extends to the entire $S \times P \times [0,1]$. If the sheaf Φ is microflexible, then we only have an extension to $\mathcal{O}p(P') \subset S \times P \times [0,1]$ but this is all we need in this case.

Microextensions of Parametric Sheaves. Let Φ and $\tilde{\Phi}$ be sheaves over A, let $\alpha\colon \tilde{\Phi} \to \Phi$ be a homomorphism and let Q be a finite polyhedron.

If α is a microextension then the corresponding homomorphism of the parametric sheaves, $\alpha^Q\colon \tilde{\Phi}^Q \to \Phi^Q$, is also a microextension.

Proof. The microextension property in particular implies that each map $\alpha(a) = \alpha(\mathcal{O}p(a))\colon \tilde{\Phi}(a) \to \Phi(a)$, $a \in A$, is a surjective *microfibration*. It follows that the maps $\alpha^Q(a,q)\colon \tilde{\Phi}^Q(a,q) \to \Phi^Q(a,q)$ are surjective for all points $(a,q) \in A \times Q$.

Now we must show that for an arbitrary pair of compact sets in $A \times Q$, the corresponding map η^Q is a microfibration (see 2.2.4). We use as earlier some covers $\{A_i\}$ of A and $\{S_j\}$ of Q which allow us to assume that $Q = S$ and that the pair in question is $(A \times S, \bar{B})$ for $\bar{B} = (B \times S) \cup (A \times \partial S)$ and $B \subset A$. A map $\tilde{\varphi}\colon P \to \tilde{\Phi}(A \times S)$ amounts to a map

$$\tilde{\varphi}'\colon S \times P \to \tilde{\Phi}(A).$$

A coherent pair $(\psi, \tilde{\psi})\colon P \times [0,1] \to \Omega^S$, for $\psi\colon P \times [0,1] \to \Phi^S(A)$ and $\tilde{\psi}\colon P \times [0,1] \to \tilde{\Phi}^S(\bar{B})$, which deforms the projection $\eta^S \circ \tilde{\varphi}\colon P \to \Omega^S = \Omega(A \times S, \bar{B})$ (that is $\eta^S \circ \tilde{\varphi} = (\psi, \tilde{\psi})|P$, $P = P \times 0 \subset P \times [0,1]$), is given by the following three maps

$$\psi'\colon S \times P \times [0,1] \to \Phi(A),$$

$$\tilde{\psi}'\colon \mathcal{O}p(\partial S) \times P \times [0,1] \to \tilde{\Phi}(A), \text{ and}$$

$$\tilde{\psi}''\colon S \times P \times [0,1] \to \tilde{\Phi}(B),$$

where the pair $(\psi', \tilde{\psi}'')$ is coherent, that is $\rho \circ \psi' = \alpha \circ \tilde{\psi}''$. Here $\rho\colon \Phi(A) \to \Phi(B)$ is the restriction map and α stands for $\alpha(B)\colon \tilde{\Phi}(B) \to \Phi(B)$. Furthermore, the maps $\tilde{\varphi}'$ and $\tilde{\psi}'$ agree on their common domain of definition, $\mathcal{O}p(\partial S) \times P \subset S \times P \times [0,1]$ thus providing a map

$$(\tilde{\varphi}', \tilde{\psi}')\colon P' \to \tilde{\Phi}(A),$$

for $P' = (S \times P) \cup (\mathcal{O}p(\partial S) \times P \times [0,1]) \subset S \times P \times [0,1]$. The coherent pair $(\psi', \tilde{\psi}'')$ defines a map $(\psi', \tilde{\psi}'')\colon S \times P \times [0,1] \to \Omega(A, B)$ and the pair $(\tilde{\varphi}, \tilde{\psi}')$ lifts the restriction $(\psi', \tilde{\psi}'')|P'$ to $\tilde{\Phi}(A)$ for the map $\eta\colon \tilde{\Phi}(A) \to \Omega(A, B)$. Since η is a microfibration, this lift extends to a lift $\mathcal{O}p(P') \to \tilde{\Phi}(A)$, for $\mathcal{O}p(P') \subset S \times P \times [0,1]$. Thus, for a small $\varepsilon > 0$, we obtain a lift $S \times P \times [0,\varepsilon] \to \tilde{\Phi}(A)$, which gives us the required lift $P \times [0,\varepsilon] \to \tilde{\Phi}^S(A)$ of $(\psi, \tilde{\psi})$.

Diagonal Products. Consider a sheaf Φ over V and fix a compact polyhedron P. A *double-deformation* over a compact subset $A \subset V$ is by definition a continuous map $\psi\colon P \times [0,\varepsilon] \times [0,\varepsilon] \to \Phi(A)$. By restricting ψ to $P \times [0,\varepsilon]$ and to $P \times \Delta$, where $\Delta \subset [0,\varepsilon] \times [0,\varepsilon]$ is the diagonal, we obtain two deformations $P \times [0,\varepsilon] \to \Phi(A)$, denoted by $\psi^0 = \psi(p,t,0)$ and $\psi^* = \psi(p,t,t)$ correspondingly. Furthermore, ψ tautologically defines a deformation $\psi'\colon P' \times [0,\varepsilon] \to \Phi(A)$ for $P' = P \times [0,\varepsilon]$. Namely $\psi'(P',\tau) = \psi(p,t,\tau)$ for $p' = (p,t)$. We say ψ is *compressible* if ψ' is a compressible deformation (see 2.2.3). Call ψ *microcompressible* if the restriction $\psi|P \times [0,\delta] \times [0,\delta]$ is a compressible double-deformation for some positive $\delta \in [0,\varepsilon]$.

Lemma. *If ψ and ψ^0 are compressible (microcompressible) then the deformation ψ^* is also compressible (microcompressible).*

Proof. First (micro) compress ψ^0 to $\bar{\psi}^0$ and then ψ' to $\bar{\psi}'$, such that the support of $\bar{\psi}'$ is contained in a small neighbourhood $U' \supset A$ for which $\bar{\psi}^0|U' = \psi^0|U'$. Then define the required (micro)compression $\bar{\psi}^*$ of ψ^* by

$$\bar{\psi}^*(u,p,t) = \begin{cases} \bar{\psi}'(u,p,t,t) & \text{for} \quad u \in U' \\ \bar{\psi}^0(u,p,t) & \text{for} \quad u \in U \backslash U', \end{cases}$$

where $U = U(\psi^*)$ is the actual domain of definition of ψ^* (compare 2.2.3). Q.E.D.

Call double-deformations $\psi_0, \ldots, \psi_k$ *composable* if $\psi_i^0 = \psi_{i-1}^*$ for $i = 1, \ldots, k$. We define for such ψ_i the (diagonal) *product* by setting $\circ_{i=0}^k \psi_i = \psi_k^*$. The above lemma yields the following

Compression of Products. *If the deformation ψ_0^0 and the double-deformations $\psi_0, \ldots, \psi_k$ are compressible (microcompressible), then the product $\circ_{i=0}^k \psi_i$ is a compressible (microcompressible) deformation.*

Nonlinear Partition of Unity. Call a point $(a,p) \in A \times P$ *fixed* under a double-deformation ψ if $\psi'(a,p,t) = \psi'(a,p,0)$ for all $t \in [0,\varepsilon]$. Define $\operatorname{supp}\psi \subset A \times P$ to be the closure of the non fixed points of ψ. Say a deformation $\varphi\colon P \times [0,\varepsilon] \to \Phi(A)$ admits *a partition of unity* relative to a given cover of the space $A \times P$ by open subsets $U_i \subset A \times P$, $i = 0, \ldots, k$, if $\varphi = \circ_{i=0}^k \psi_i$ for some composable double-deformations ψ_i such that $\operatorname{supp}\psi_i \subset U_i$ for $i = 1, \ldots, k$, and $\operatorname{supp}\psi_0^0 \subset U_0$.

Proposition. *Let $A \times P$ be covered by open subsets U_i, $i = 0, \ldots, k$, and assume the restriction $\Phi|A$ is a flexible sheaf over A. Then every deformation φ admits a partition of unity relative to the cover $\{U_i\}$. Similarly, the microflexibility of $\Phi|A$ insures the partition of unity for the restriction $\varphi|P \times [0,\delta]$, where δ is some positive number $\leq \varepsilon$.*

Proof. Let P consist of a single point, assume $k = 1$, denote $S = \operatorname{supp}\varphi$ and prove the following

Lemma. *If Φ is flexible, then there exists a double-deformation $\psi\colon [0,\varepsilon] \times [0,\varepsilon] \to \Phi(A)$ such that $\psi^* = \varphi$ and*

(i) $$\operatorname{supp}\psi^0 \subset U_0 \cap \mathcal{O}p(S)$$

(ii) $$\operatorname{supp}\psi' \subset U_1 \cap \mathcal{O}p(S).$$

Similarly, if Φ is microflexible, the above ψ exists on $[0,\delta] \times [0,\delta]$ for some positive $\delta \leq \varepsilon$.

Proof. The condition $\psi^* = \varphi$ defines ψ on the diagonal $\Delta \subset [0,\varepsilon] \times [0,\varepsilon]$. Furthermore, (i) defines the restriction $\psi|A\backslash(U_0 \cap \mathcal{O}p S)$ on the segment $[0,\varepsilon] \times 0 \subset [0,\varepsilon] \times [0,\varepsilon]$ while (ii) defines the restriction $\psi|A\backslash(U_1 \cap \mathcal{O}p S)$ on the entire square $[0,\varepsilon] \times [0,\varepsilon]$. Since the union $\Delta \cup ([0,\varepsilon] \times 0)$ is a homotopy retract in the square $[0,\varepsilon] \times [0,\varepsilon]$, this partially defined ψ extends to the required double-deformation $[0,\varepsilon] \times [0,\varepsilon] \to \Phi(A)$ if Φ is flexible. If Φ is micro-flexible, then the extension is possible over $[0,\delta] \times [0,\delta]$ for some $\delta > 0$. Q.E.D.

Now, still assuming $P = \{p\}$, we construct by induction on $i = 0, 1, \ldots, k$, double-deformations $\psi_k, \psi_{k-1}, \ldots, \psi_{k-i}$, such that $\psi_k^* = \varphi$ and

(i) $$\operatorname{supp}\psi_{k-i}^0 = \bigcup_{j=0}^{k-i-1} U_j;$$

(ii) $$\operatorname{supp}\psi'_{k-j} \subset U_{k-j} \quad \text{for} \quad j = 0, \ldots, i.$$

This is done with the above lemma applied to the covering of A by U_{k-i} and $A \backslash \mathcal{O}p(A \backslash U_{k-i})$ at every induction step.

Finally, if P contains more than one point, we apply the above to the parametric sheaf Φ^P over $A \times P$.

Local Compressions. A deformation $\psi\colon P \times [0,\varepsilon] \to \Phi(A)$ is called *S-microcompressible* if for an arbitrary neighborhood $\tilde{U} \subset U = U(\psi)$ (see 2.2) of the support $\operatorname{supp}\psi \subset U$ there exists a smaller positive $\tilde{\varepsilon} \in [0,\varepsilon]$, $\tilde{\varepsilon} = \tilde{\varepsilon}(\tilde{U})$, such that the restricted deformation $\psi | P \times [0,\tilde{\varepsilon}]$ can be compressed to a deformation $\bar{\psi}\colon P \times [0,\tilde{\varepsilon}] \to \Phi(A)$, such that, in addition to properties (i)–(iii) in 2.2 this compression $\bar{\psi}$ satisfies

$$\operatorname{supp}\bar{\psi} \subset \tilde{U}.$$

Recall that the compression procedure involves a neighbourhood $\bar{U}$ of A and the support $\operatorname{supp}\bar{\psi}$ must be contained in this $\bar{U}$. Now, $\operatorname{supp}\psi$ must be contained in the intersection $\tilde{U} \cap \bar{U}$, and $\tilde{\varepsilon}$ may depend on $\tilde{U}$, but not on $\bar{U}$ (compare with the definition of microcompressibility in 2.2.3). It is clear that S-microcompressibility implies microcompressibility. On the other hand, *if the sheaf Φ is flexible, one can construct S-compressions with $\tilde{\varepsilon} = \varepsilon$ for any $\tilde{U}$* in the same way we did it for the usual compressions in 2.2.3.

There is one important difference between compressions and S-compressions. Namely, *the S-microcompressibility of a deformation $\psi\colon P \times [0,\varepsilon] \to \Phi(A)$* only *depends on the behavior of ψ near* the intersection $A \cap \operatorname{supp}\psi$ (but does not on what happens in the interior of $A \backslash \operatorname{supp}\psi$). In other words, if a subset A' in A contains the intersection $A \cap \mathcal{O}p(\operatorname{supp}\psi)$, and if the restricted deformation $\psi | A'\colon P \times [0,\varepsilon] \to \Phi(A')$ is S-microcompressible, then the deformation ψ itself is also S-microcompressible.

Now, take a double-deformation $\psi\colon P \times [0,\varepsilon] \times [0,\varepsilon] \to \Phi(A)$ and consider the corresponding deformation $\psi^P\colon I_\varepsilon \times [0,\varepsilon] \to \Phi^P(A \times P)$, where I_ε is the first interval $[0,\varepsilon]$. We say that the double-deformation ψ is *$\tilde{S}$-microcompressible* if the restriction $\psi^P | I_\delta \times [0,\delta]\colon I_\delta \times [0,\delta] \to \Phi^P(A \times P)$ is an S-microcompressible deformation for some positive $\delta \in (0,\varepsilon]$ where $I_\delta = [0,\delta]$. Clearly, $\tilde{S}$-microcompressibility implies the usual microcompressibility for double-deformations. Again, the property of $\tilde{S}$-microflexibility of a double-deformation ψ only depends on the behaviour of ψ near $\widetilde{\operatorname{supp}\psi} \subset A \times P$.

If Φ is a flexible sheaf, then all double-deformations are $\tilde{S}$-microcompressible since the flexibility of Φ implies the flexibility of the parametric sheaf Φ^P and hence the above remark about deformations applies.

Partial Deformation. Take a map $\varphi_0\colon P \to \Phi(A)$ and a product $A_0 \times P_0 \subset A \times P$, where A_0 is a compact set in A and P_0 is a subpolyhedron in some subdivision of P. We introduce *a partial deformation of* φ_0 as a map $\psi\colon P_0 \times [0,\varepsilon] \times [0,\varepsilon] \to \Phi(A_0)$ such that $\psi(a_0, p_0, 0, 0) = \varphi_0(a_0, p_0)$ for all $(a_0, p_0) \in A_0 \times P_0$. We say that the map φ_0 is *locally stable* if for each point $(a_0, p_0) \in A \times P$ there is a product $A_0 \times P_0$ (as above) which contains a neighbourhood of (a_0, p_0) in $A \times P$, and such that every partial double-deformation $P_0 \times [0,\varepsilon] \times [0,\varepsilon] \to \Phi(A_0)$ of φ_0 is $\tilde{S}$-microcompressible.

Local Criterion for Microcompressibility. *Let Φ be a sheaf over a locally compact space V and let A be a compact subset in V. If the restricted sheaf $\Phi|A$ is microflexible and if all maps $P \to \Phi(A)$ are locally stable, then all deformations Φ: $P \times [0,1] \to \Phi(A)$ are microcompressible.*

Proof. Since $\Phi|A$ is a microflexible sheaf, the deformation $\varphi|P \times [0,\varepsilon]$, for some $\varepsilon \in [0,1]$, can be decomposed into the product of double-deformations with arbitrarily small supports, $\varphi|P \times [0,\varepsilon] = \circ_{i=0}^{k} \psi_i$, ψ_i: $P \times [0,\varepsilon] \times [0,\varepsilon] \to \Phi(A)$. Now, if the support of the double-deformation ψ_i lies inside a "small product", that is $\mathcal{O}p(\widetilde{\operatorname{supp}\psi_i}) \subset A_i \times P_i$, for $A_i \subset A$ and $P_i \subset P$, then the restricted double-deformation $\psi_i|A_i \times P_i$: $P_i \times [0,\varepsilon] \times [0,\varepsilon] \to \Phi(A_i)$ is $\tilde{S}$-microcompressible. Therefore, each double-deformation ψ_i is microcompressible and the product of their microcompressions $\circ_{i=0}^{k} \psi_i$, gives us the required microcompression $\bar{\varphi}$ of φ.

Proof of the Microextension Theorem. We shall prove this theorem for sheaves over a *manifold* V, since this is all we need for the further applications. The general case of an arbitrary locally compact space V is left as an exercise.

Now we have a microextension α: $\tilde{\Phi} \to \Phi$, where $\tilde{\Phi}$ is a flexible sheaf and we have to show that Φ is also flexible. We may assume, by induction on dim V that for each hypersurface $H \subset V$ the restriction $\Phi|H$ is already a flexible sheaf. Actually, we shall only use the microflexibility of $\Phi|H$. (In most examples the microflexibility of Φ is not a problem anyway).

In order to prove flexibility of Φ we only need to establish microcompressibility of deformations φ over a given compact set A in V (see 2.2.3). The compression of a deformation φ to $\bar{\varphi}$ with supp $\bar{\varphi} \subset \bar{U}$ is equivalent to the compression of the restriction of φ to a hypersurface $H \subset \bar{U}$, which separates A and the boundary of $\bar{U}$ in V. (See the main lemma in 2.2.3). So we can work on H, where Φ is microflexible. To save the notations we assume that the sheaf $\Phi|A$ is microflexible and continue to work on A.

Let us enumerate the relevant properties of the microextension α: $\tilde{\Phi} \to \Phi$.

(1) For every section $\varphi_0 \in \Phi(A)$ and for each point $a \in A$ there is a compact subset A_0 in A such that $A_0 \supset A \cap \mathcal{O}p(a)$, and such that the section $\varphi_0|A_0$ lifts (by surjectivity of α) to a section $\tilde{\varphi}_0 \in \tilde{\Phi}(A_0)$. That is $\alpha_0(\varphi_0) = \varphi_0|A_0$, where α_0 is an abbreviation for $\alpha|A_0$.
(2) For every deformation φ: $[0,1] \to \Phi(A)$, $\varphi(0) = \varphi_0$ there exist a positive $\varepsilon \in (0,1]$ and a partial lift $\tilde{\varphi}$: $[0,\varepsilon] \to \tilde{\Phi}(A_0)$. That is $\tilde{\varphi} = \tilde{\varphi}_0$ and $\alpha_0(\tilde{\varphi}(t)) = \varphi(t)|A_0$ for $t \in [0,\varepsilon]$. In fact, for every compact subset B_0 in A_0 the map η: $\tilde{\Phi}(A_0) \to \Omega(A_0, B_0)$ is a microfibration. If B_0 is empty, then $\Omega(A_0, B_0) = \Phi(A_0)$ and the map η reduces to $\alpha_0 = \alpha|A_0$. Therefore the map α_0: $\tilde{\Phi}(A_0) \to \Phi(A_0)$ is a microfibration and the initial deformations can be lifted.
(3) This property refines (2) by giving us a lift $\tilde{\varphi}$ which has almost the same support as φ. Namely, for an arbitrary neighbourhood $\tilde{U} \subset V$ of $A_0 \cap \operatorname{supp}\varphi$, there is a positive $\tilde{\varepsilon} \in (0,\varepsilon]$ for which one can find a lift $\tilde{\varphi}$: $[0,\tilde{\varepsilon}] \to \tilde{\Phi}(A_0)$ such that supp $\tilde{\varphi} \subset \tilde{U}$. We construct this $\tilde{\varphi}$ by first interpreting the restriction $\tilde{\varphi}_0|A_0 \backslash \tilde{U}$ as the constant deformation $\tilde{\varphi}^0$: $[0,1] \to \tilde{\Phi}(A_0 \backslash \tilde{U})$ $\tilde{\varphi}^0(a,t) = \tilde{\varphi}_0(a)$ for $a \in A_0 \backslash \tilde{U}$ and for all $t \in [0,1]$.

Then, for the pair of sets $(A_0, A_0 \backslash \tilde{U}_0)$, we have the coherent pair of deformations, $(\varphi | A_0, \tilde{\varphi}^0)$. Finally, we lift the initial phase of this pair of deformations to $\tilde{\Phi}(A_0)$ for the *microfibration* $\eta\colon \tilde{\Phi}(A_0) \to \Omega(A_0, A_0 U_0)$.
(4) The properties (1)–(3) hold if we pass to the parametric sheaves and to the corresponding homomorphism $\alpha^Q\colon \tilde{\Phi}^Q \to \Phi^Q$, as it was shown in the beginning of this section for all finite polihedra Q.

Now, we prove the microextension theorem by applying the previous local criterion. For a map $\varphi_0\colon P \to \Phi(A)$ and a point $(a, p) \in A \times P$, we find a product $A_0 \times P_0 \subset A \times P$, such that $A_0 \times P_0$ contains $\mathcal{O}p(a, p)$ and such that the restriction $\varphi_0 | A_0 \times P_0$ lifts to a map $\tilde{\varphi}_0\colon P_0 \to \tilde{\Phi}(A_0)$. Next, for a partial double-deformation $\psi\colon P_0 \times [0, 1] \times [0, 1] \to \Phi(A_0)$, we lift its restriction $\psi | P_0 \times [0, \varepsilon] \times 0$ to $\tilde{\Phi}(A_0)$, and then, with a given neighbourhood $\tilde{U}$ in $A_0 \times P_0$ of $\operatorname{supp} \psi \subset A_0 \times P_0$, we find an extension of the last lift to a map $P_0 \times [0, \varepsilon] \times [0, \tilde{\varepsilon}] \to \Phi(A_0)$ for some positive $\tilde{\varepsilon} \in (0, \varepsilon]$, such that the restriction of this map to $P_0 \times [0, \tilde{\varepsilon}] \times [0, \tilde{\varepsilon}]$, called $\tilde{\psi}\colon P_0 \times [0, \tilde{\varepsilon}] \times [0, \tilde{\varepsilon}] \to \Phi(A_0)$ is a double deformation with $\widetilde{\operatorname{supp}}(\tilde{\psi}) \subset \tilde{U}$. Since the sheaf $\tilde{\Phi}$ is flexible, $\tilde{\psi}$ can be $\tilde{S}$-microcompressed (even without making $\tilde{\varepsilon}$ smaller).

Finally, by projecting the $\tilde{S}$-microcompressed $\tilde{\psi}$ back to $\Phi(A_0)$ by α_0 we obtain a $\tilde{S}$-microcompression of ψ.

2.2.6 An Application: Inducing Euclidean Connections

Recall, that a *connection* in a C^∞-smooth vector bundle $E \to V$ is a rule that assignes to each C^∞-smooth vector field L on V a first order linear differential operator in the space $C^\infty(E)$ of smooth sections $V \to E$. This operator, called *the covariant* derivation in the direction L and denoted by $\nabla_L\colon C^\infty(E) \to C^\infty(E)$, must satisfy the formal properties of a derivative. (See Kobayashi-Nomizu 1969.)

If $u_1, \ldots, u_n$ are some local coordinates in V, then each connection in E is uniquely determined over this coordinate chart by the n operators of covariant derivations in the directions $\partial/\partial u_i$, $i = 1, \ldots, n$, denoted by $\nabla_i\colon C^\infty(X) \to C^\infty(X)$.

Examples. Each trivial bundle $V \times \mathbb{R}^m \to V$ carries *the standard flat connection* whose covariant derivatives ∇_i of a section $V \to V \times \mathbb{R}^m$ are just the usual derivatives $\partial/\partial u_i$ of the map $f\colon V \to \mathbb{R}^m$, which corresponds to the section. For a field $L = \sum_{i=1}^n a_i(\partial/\partial u_i)$ the derivatives ∇_L in this case amounts to the Lie derivative $L_f = \sum_{i=1}^n a_i(\partial f/\partial u_i)$.

Take a bundle E with a connection ∇ and a C^∞-subbundle $E' \subset E$. Each linear C^∞-projection $P\colon E \to E'$, $P^2 = \mathrm{Id}$, gives rise to a connection ∇' in E'. Namely, $\nabla' = P \circ \nabla$, that is $\nabla'_L(X) = P \circ \Delta_L(X)$ for all fields L on V and for all sections $X \in C^\infty(E') \subset C^\infty(E)$.

Exercise. Let F be the trivial bundle $V \times \mathbb{R}^q \to V$ with the standard flat connection, denoted by ∇^F and let $E \subset F$ be a subbundle of dimension m. Suppose that $q \geq m(n + 1) + n$, $n = \dim V$, and that $E \subset F$ is a *generic* C^∞-subbundle. Show that for an arbitrary connection ∇ in E there exists a projection $P\colon F \to E$ such that $P \circ \nabla^F = \nabla$.

Induced Connections. Let $E \to V$ be a bundle with a connection ∇ and let $g = V' \to V$ be a C^∞-map of a manifold V' into V. Denote by $E' \to V'$ the induced bundle $g^*(E)$ over V' and let $G: E' \to E$ denote the corresponding fiberwise isomorphic map between the bundles E' and E. The bundle E carries a connection ∇', called the *induced connection* $g^*(\nabla)$, which is uniquely characterized by the following property.

If two sections $X: V \to E$ and $X': V' \to E$ are related by the equality $G \circ X' = X \circ g$, and if two fields $L: V \to T(V)$ and $L': V' \to T(V')$ are related at a point $v' \in V'$ by $D_g(L'(v')) = L(g(v'))$, then $G((\nabla'_L X')(v')) = (\nabla_L X)(g(v'))$.

For example, if g is constant then the induced bundle E' is trivial and $\nabla' = g^*(\nabla)$ is the (standard) flat connection in E'.

Euclidean Connection. We now assume that the bundle $E \to V$ is given a *Euclidean structure* that is a field of Euclidean metrics in the fibers $E_v \subset E$ for v running over V. A connection ∇ in such a bundle is called *Euclidean* if the Lie derivative of the scalar product abides by the Leibniz rule:

$$L\langle X, Y\rangle = \langle \nabla_L X, Y\rangle + \langle X, \nabla_L Y\rangle,$$

for all tangent fields on V and for all section X and Y in $C^\infty(E)$.

Examples. The standard flat connection in a trivial bundle is Euclidean.

Let $E' \subset E$ be a subbundle with the induced Euclidean structure and let $P: E \to E'$ be the *orthogonal* projection. Then for a Euclidean connection ∇ in E the connection $P \circ \nabla$ in E' is also Euclidean.

Let us apply this construction to the canonical m-dimensional bundle over the Grassmann manifold of m-planes in $\mathbb{R}^q$. This bundle, $H \to Gr = Gr_m(\mathbb{R}^q)$, is realized as a subbundle of the trivial bundle $F = Gr \times \mathbb{R}^q \to Gr$. The standard flat connection ∇^F in F and the orthogonal projection $P: F \to H$ yield the connection $P \circ \nabla^F$ in H, called the *canonical connection.*

Let W be a Riemannian C^∞-manifold. The tangent bundle $T(W) \to W$ has a distinguished Euclidean connection, called *the Riemannian connection* ∇^W. If W is isometrically C^∞-immersed into $\mathbb{R}^q$, then the tangential Gauss map $g: W \to Gr_m(\mathbb{R}^q)$, $m = \dim W$, induces ∇^W from the canonical connection in the bundle $H \to Gr_m(\mathbb{R}^q)$.

Now suppose the Riemannian manifold W has dimension $m + n$. Let an n-dimensional submanifold in W and let $E \to V$ be the normal bundle of V in W. By restricting the connection ∇^W to $T(W)|V$ and by applying the orthogonal projection $T(W)|V \to E$ we obtain the so-called *normal* connection in E.

Exercises. (a) Show that for an arbitrary connection ∇ in an abstract bundle $E \to V$, there exists a Riemannian metric in the total space E, such that the normal connection of the zero section $V \subset E$ equals ∇.

(b) Recall the existence of an isometric C^∞-immersion of every $(m + n)$-dimensional manifold into $\mathbb{R}^{q_0}$, for $q_0 = (m + n + 2)(m + n + 3)/2$ (see 3.1.7). Show that an arbitrary Euclidean connection in an m-dimensional bundle over an n-dimensional manifold V can be induced from the canonical connection in $H \to Gr_m(\mathbb{R}^{q_0})$.

The problem of inducing Euclidean connections was first considered by Narasimhan and Ramanan (1961), who also studied non-Euclidean connections. Our aim is to obtain a connection inducing map $V \to Gr_m(\mathbb{R}^q)$ for a relatively small q.

Our major tool is the h-principle for a class of such maps. In fact, we study the following more general problem. Let E and F be two bundles over V with arbitrary Euclidean connections, ∇ in F and ∇' in E. We seek *connection homomorphisms* $X: E \to F$. Namely X must be an isometric isomorphism of E onto a subbundle E', such that $\nabla' = P \circ \nabla$ for the orthogonal projection $P: F = E' = X(E) \approx E$.

We work for a while over a fixed system of local coordinates $u, \ldots, u_n$ in V and we also fix, over this local chart, an orthonormal frame of sections $e_k: V \to E$, $k = 1, \ldots, m$, where m is the dimension of the bundle E. With these fixed data, we can express the connection ∇' in terms of Christoffel's coefficients $\Gamma_i^{kl} = \langle e_k, \nabla_i e_l \rangle$, $1 \leq i \leq n$, $1 \leq k, l \leq m$. Since the vectors e_k are orthonormal, that is $\langle e_k, e_l \rangle = \delta_{kl}$, we have $\frac{\partial}{\partial u_i} \langle e_k, e_l \rangle = 0$ for all i, k and l, and so $\Gamma_i^{kl} = -\Gamma_i^{lk}$.

An isometric homomorphism $X: E \to F$ (over our local chart) now amounts to a system of orthonormal sections $X_k: V \to F$, $k = 1, \ldots, m$,

$$(*) \qquad \langle X_k, X_l \rangle = \delta_{kl}, \qquad k, l = 1, \ldots, m,$$

and the condition $\nabla' = P \circ \nabla$ is expressed by the following system of partial differential equations in unknowns $X_1, \ldots, X_m$,

$$(**) \qquad \langle X_k, \nabla_i X_l \rangle = \Gamma_i^{k,l}, \ 1 \leq i \leq n, \ 1 \leq k, l \leq m.$$

If the bundle F has dimension q, then each section X_k is given by q real functions and the conditions $(*)$ and $(**)$ represent $[m(m+1)/2] + [nm(m-1)/2]$ equations in mq unknown functions on V. We must solve these equations with arbitrary given $nm(m-1)/2$ functions $\Gamma_i^{kl} = -\Gamma_i^{lk}$ on V. The system of these equations is underdetermined for $q > [(m+1)/2] + [n(m-1)/2]$, and so one expects it to be solvable for $q \approx mn/2$. But our method provides solvability only for $q \approx mn$. [See D'Ambra (1985) for the case $q \approx mn/2$.] This restriction is due to the fact that we can only handle some special homomorphisms $X: E \to F$, called *regular*, that are distinguished by the condition of linear independence of the section X_k and $\nabla_i X_k$, $i = 1, \ldots, n$, $k = 1, \ldots, m$, in each fiber $F_v \subset F$, $v \in V$. Observe, that the span of the vectors X_k and $\nabla_i X_k$ only depends on the homomorphism X and not on the choice of the frame $\{e_k\}$ or of the coordinate system $\{u_i\}$. So, the definition of regularity is correct and it is also clear that the connection ∇' in E plays no role in this definition. It is equally clear, that regular maps may exist only for $q \geq m(n+1)$.

Example. Let E be the trivial line bundle. Then isometric homomorphisms $E \to F$ are unitary fields $V \to F$ that are sections of the unit sphere bundle associated to F. If F is the trivial flat bundle, $F = V \times \mathbb{R}^q \to V$, then *regular* unitary fields $V \to F$ correspond to *immersions* $V \to S^{q-1}$.

(A) **Theorem.** *The regular connection C^∞-homomorphisms $X: E \to F$, satisfy the h-principle for $q \geq m(n+2)$, where $n =$ dim V and m and q are the dimensions of the bundles E and F respectively.*

Corollary. *If* $q > q_0 = \max(m(n+2), m(n+1)+n)$, *then there always exists a regular connection* C^∞*-homomorphism* $E \to F$. *In particular, every connection in* E *can be induced by a map* $g: V \to Gr_m(\mathbb{R}^{q_0})$ *from the canonical connection.*

Proof of the Corollary. It will become clear in the proof of the main lemma below that the differential condition $\mathscr{R}$, that governs our homomorphisms $X: E \to F$, fibers over V with fibers $\mathscr{R}_v$, $v \in V$, which are homotopy equivalent to the Stiefel manifold $St_s(\mathbb{R}^q)$ for $s = m(n+1)$. Therefore, we have a section $V \to \mathscr{R}$ for $q \gg m(n+1)+n$. Then the theorem applies for $q \geq m(n+2)$.

Proof of the theorem is based on the following

Main Lemma. *The sheaf of germs of regular connection* C^∞*-homomorphisms* $X: E \to F$, *is microflexible.*

Proof. Since microflexibility is a local property (see 2.2.3) we only need to establish microflexibility for regular solutions of the systems (*) and (**). We proceed inductively for $k = 1, \ldots, m$, and thus reduce the solution of the systems (*) and (**) to the solution of a sequence of m *open* differential conditions, $\mathscr{R}(k)$ imposed on X_k, $k = 1, \ldots, m$, where the vectors $X_1, \ldots, X_l, \ldots, X_{k-1}$ are fixed. So we work with the following system of conditions imposed on $X_1, \ldots, X_k$:

(a) the frame $(X_1, \ldots, X_k)$ is regular, that is the vectors X_l and $\nabla_i x_l$ are linearly independent for $l = 1, \ldots, k$, and $i = 1, \ldots, n$.

(b)
$$\left\{\begin{array}{l} \langle X_\alpha, X_\beta \rangle = \delta_{\alpha\beta}, \\ \langle X_\alpha \nabla_i X_\beta \rangle = \Gamma_i^{\alpha\beta}, \\ \alpha, \beta = 1, \ldots, k,\ i = 1, \ldots, n. \end{array}\right\} \quad (\mathscr{R}_k)$$

We must show, that for any given solution $(X_1^0, \ldots, X_{k-1}^0)$ of the systems $\mathscr{R}_{k-1}$ the conditions imposed by the system $\mathscr{R}_k$ on X_k can be reduced to an open condition. The only *differential* equations in $\mathscr{R}_k$ which involve X_k are

$$\langle X_\alpha^0, \nabla_i X_k \rangle = \Gamma_i^{\alpha k}, \qquad \alpha = 1, \ldots, k-1,\ i = 1, \ldots, n.$$

These equations can be written equivalently as

$$\langle \nabla_i X_\alpha^0, X_k \rangle = -\Gamma_i^{\alpha k},$$

and so *all* equations in (b) imposed on X_k become algebraic. Furthermore, all the equations in (b) are linear in X_k, with the only exception of

$$\langle X_k, X_k \rangle = \delta_{kk} = 1.$$

Since the vectors X_α and $\nabla_i X_\alpha$, $\alpha < k$, are linearly independent, the linear equations, namely

$$\left.\begin{array}{ll} \langle X_\alpha^0, X_k \rangle = \delta_{\alpha,k} = 0 \quad \text{and} & \\ \langle \nabla_i X_\alpha^0, X_k \rangle = -\Gamma^{\alpha,k}, & \alpha = 1, \ldots, k-1,\ i = 1, \ldots, n, \end{array}\right\} \mathscr{R}_0(k)$$

define an affine subbundle A in F of codimension $(n+1)(k-1)$, such that sections $V \to A$ correspond to solutions X_k of these equations. The non-linear equation

$\langle X_k, X_k\rangle = 1$ prescribes a sphere S_v in each fiber $A_v \subset A$, $v\in V$. Observe that the set $S_v \subset A_v$ may be, apriory, empty, or it may consist of one single point. But we could assume from the beginning the existence of at least one solution X_k^0 of $\mathscr{R}_0(k)$, such that the frame $(X_1^0, \ldots, X_{k-1}^0, X_k^0)$ is *regular* and, in particular, X_k^0 is not contained in the span of X_α^0 and $\nabla_i X_\alpha^0$ for $\alpha < k$; otherwise, there is no solution to speak about. This independence of the vectors X_α^0, $\nabla_i X_\alpha^0$ and some vectors X_k^0 implies that our sets $S_v \subset A_v$ are actual spheres and that they form a sphere bundle $S \to V$, such that sections $V \to S$ are exactly solutions X_k satisfying the system (b) in $\mathscr{R}_k$ with fixed $X_\alpha = X_\alpha^0$, $\alpha < k$. In order to satisfy the entire system (a) we must add the regularity condition for the frames $(X_1^0, \ldots, X_{k-1}^0, X_k)$, but this is an *open* differential condition in X_k. In fact, there is a field of n-planes in the total space S, which are transversal to all fibers $S_v \subset S$, and such that the sections $V \to X$ transversal to this field are the solutions of $\mathscr{R}_k$ relative to X_k. Observe that the 1-jets of these transversal sections $V \to S$ at each point $v \in V$ form a manifold of the homotopy type of the Stiefel manifold $\mathrm{St}_{n+1}(\mathbb{R}^p)$, where p denotes the dimension of fibers $A_v \subset A$, $v \in V$. This shows that our original condition $\mathscr{R} \to V$ is fibered in the way we asserted in the proof of the previous corollary.

Notice finally, that by reducing $\mathscr{R}$ to a sequence of open relations, we have not only obtained microflexibility, but also the rest of the nice "open properties". In particular, the h-principle for $\mathscr{R}$ is equivalent to the sheaf theoretic h-principle and we also have the following

Microextension Lemma. *Let V_0 be a submanifold in V and let E_0 and F_0 denote the restrictions $E|V_0$ and $F|V_0$ respectively with induced connections ∇_0' and ∇_0. Let Φ denote the sheaf of connection homomorphisms $E \to F$ and let Φ_0 be the corresponding sheaf of germs of connection homomorphisms $F_0 \to E_0$. If $q \geq m(n+1)$, that is if $\mathscr{R}$ is not the empty set, then the restriction homomorphism $\Phi|V_0 \to \Phi_0$ is a microextension.*

Now, we introduce the bundles $\tilde{E}$ and $\tilde{F}$ over $V \times \mathbb{R}$ which are induced, (together with their connections,) from E and F respecitively by the projection $\pi\colon V \times \mathbb{R} \to V$. The regular connection homomorphisms $\tilde{E} \to \tilde{F}$ are naturally acted upon by $\mathrm{Diff}(V \times \mathbb{R}, \pi)$ and the theory of sheaves (sec 2.2.3) yields the h-principle for connection homomorphisms $\tilde{E} \to \tilde{F}$ without any restriction on the dimension. Finally, the above lemma brings this h-principle back to V for $q \geq m(n+2)$.

Exercises. (a) Prove for $q \geq k(n+2)$ that the natural map of the space of solutions of the system $\mathscr{R}_k$ (over a fixed coordinate chart) to the space of solutions of the system $\mathscr{R}_{k-1}$ is a Serre fibration [*Hint*. Use the group $\mathrm{Diff}(V \times \mathbb{R}, \pi)$].

(b) Describe non-open conditions $\mathscr{R}$ which can be reduced to open conditions. Define an appropriate maximal invariant extension (as in 2.2.4 for the open case) and give a criterion for validity of the h-principle generalizing the theorem on inducing connections. Consider, in particular, the connection inducing problem for pseudo-Euclidean connections.

(B) *Semiregular Homomorphisms*. Let $m = 2$ and let (e_1, e_2) be an orthonormal frame in E which is *parallel* at a given point $v \in V$, that is $\Gamma_i^{k,l}(v) = 0$, $i = 1, \ldots, n$, k, $l = 1, 2$. A connection inducing isometric homomorphism $X\colon E \to F$ is called *semi-*

regular at v if the n vectors $(s\nabla_i X_1 + c\nabla_i X_2)(v) \in F_v$ are linearly independent for all pairs of real numbers s and c such that $s^2 + c^2 = 1$.

Exercise. Show that no homomorphism is semiregular if $q = n + 2$ for n even. However, if n is odd, then semiregular homomorphisms do (locally) exist for all $q \geq n + 2$.

Our analysis of regular homomorphisms equally applies to semiregular ones. In particular, we have the following

Corollary. *Let V be a parallelizable manifold and let the bundles E and F be trivial. If $q \geq n + 4$, or if q is even and $\geq n + 3$, then there exists a connection inducing isometric C^∞-homomorphism $E \to F$ which is semiregular at all points $v \in V$.*

Exercise. Define semiregular homomorphisms for $m \geq 3$ and prove the pertinent h-principle.

(B′) **Remark.** We shall need later (see 3.1.7) the following version of the above corollary.

Let V split into the product, $V = V_0 \times \mathbb{R}$, where V_0 is a parallelizable manifold, let the bundles E and F be trivial and let $e: V \to E$ be a unitary C^∞-section. If $q \geq n + 3$, then there exists a connection inducing isometric C^∞-homomorphism $X: E \to F$ such that

(i) *the section $X(e)$: $V \to F$ is regular;*
(ii) *the restriction of X to the bundle $E_t = E | V_t$ over $V_t = V \times t \subset V \times \mathbb{R}$ is an everywhere semiregular homomorphism of E_t to $F_t = F | V_t$ for all $t \in \mathbb{R}$.*

Proof. The microflexibility of the pertinent homomorphisms X is established as earlier, which yields the existence of X via the h-principle.

2.2.7 Non-flexible Sheaves

The lack of the (micro) flexibility does not necessarily impairs the h-principle. For example, C^{an}-maps $V \to W$ are not microflexible but the h-principle does not suffer due to the following fact which summarizes basic results by Whitney (1934), H. Cartan (1957) and Grauert (1958).

Let $X \to V$ be a C^{an}-submersion and let f_0: $V \to X$ be a C^∞-section whose jet $J^r_{f_0}$: $V \to X^{(r)}$, for a given $r = 0, 1, \ldots,$ is real analytic on some analytic (possibly singular) subvariety $V_0 \subset V$. Then f_0 admits a fine C^∞-approximation by C^{an}-sections f: $V \to X$, such that $J^r_f | V_0 = J^r_{f_0} | V_0$.

This can be refined for vector bundles $X \to V$ with the *cohomology* of the sheaf Φ of C^{an}-sections $V \to X$. Namely, $H^i(\Phi) = 0$ for $i \geq 1$ (Cartan 1957).

For general (non-linear!) continuous sheaves there is no (?) cohomology groups to measure the degree of non-microflexibility.

However, the deviations from the h-principle is reasonably controled for the non-flexible sheaves in the following examples.

(A) *Degenerate Maps.* A continuous map $f: A \to C$ is called *k-contractible* if there exists a k-dimensional polyhedron B and continuous maps $g: A \to B$ and $h: B \to C$, such that $h \circ g$ is homotopic to f.

Exercises. Let $f: V \to W$ be a C^1-map, such that rank $D_f(v) \leq k$ for all $v \in V$.

(a) Show the map f to be k-contractible.

(b) Assume rank $D_f(v) \equiv k$ and let V be a *closed* manifold. Construct a k-dimensional manifold B and two maps of constant rank $= k$, say $g: V \to B$ and $h: B \to W$ such that $f = h \circ g$.

(b') Assume the manifolds V and W as well as the map f to be real analytic and construct the above B, g and h for *open* manifolds V. Find counterexamples in the C^∞-case.

(A') **The *h*-Principle for Maps of Constant Rank** (Gromov 1973; Phillips 1974). *Let V be an open manifold and let $\varphi: T(V) \to T(W)$ be a homomorphism of constant rank k. Then the following condition $(*)$ is necessary and sufficient for the existence of a C^∞-map $f: V \to W$ of constant rank k whose differential is homotopic to φ by a homotopy of homomorphisms of rank k.*

$(*)$ *There exists a k-dimensional bundle over a k-dimensional polyhedron, say $K \to B$, and vector bundle homomorphisms $\varphi_1: T(V) \to K$ and $\varphi_2: K \to T(W)$, of constant rank k, such that $\varphi_2 \circ \varphi_1: T(V) \to T(W)$ is homotopic (via homomorphisms of constant rank) to φ.*

Proof. First, let V be obtained from the ball B^n, $n = \dim V$, by attaching l-handles $H^l = B^l \times \mathbb{R}^{n-l}$ for $1 \leq l \leq k < n/2$. If a map f of rank k is given on a small neighborhood $\mathcal{O}\!\mathit{p}\, S^{l-1} \subset H^l$ of the sphere $S^{l-1} = \partial B \times O \subset H^l$ and if the ball $B^l = B^l \times O \subset H^l$ is transversal to the subbundle $\operatorname{Ker} D_f \subset T(V)$ near S^{l-1}, then one easily extends f to $\mathcal{O}\!\mathit{p}\, B^l \subset H^l$ by applying the h-principle to immersions $B^l \to W$, for $k < \dim W$. (If $k = \dim W$ the h-principle in question amounts to Phillips' submersion theorem.) Furthermore, if B^l is not transversal to $\operatorname{Ker} D_f$, then there exists a C^0-small diffeotopy of the identity map $\mathcal{O}\!\mathit{p}\, S^{l-1} \circlearrowleft$ which achieves the transversality. This follows from the h-principle for immersions $S^{l-1} \to \mathcal{O}\!\mathit{p}\, S^{l-1}$ transversal to $\operatorname{Ker} D_f$ and from the inequality $2k < n$. Thus, we obtain the h-principle for maps $f: V \to W$ of rank k for our special manifold $V = B^n \bigcup_i H_i^{l_i}$ with an obvious induction in $i = 1, 2, \ldots$.

Now, for any V, we construct with $(*)$ an n'-dimensional manifold $V' = B^{n'} \bigcup_i H_i^{l_i}$, $n' = 2k + 3$, a homomorphism $\varphi_2': T(V') \to T(W)$ of rank k and a homomorphism $\varphi_1': T(V) \to T(V')$ such that the composition $\varphi_2' \circ \varphi_1'$ is a homomorphism $T(V) \to T(W)$ of rank k homotopic to φ. Then we have, as earlier, a C^∞-map $f': V' \to W$ of rank k. Furthermore, the h-principle for maps $V \to V'$ transversal to $\operatorname{Ker} D_{f'}$, yields such a map, say $f'': V \to V'$, for which the composition $f' \circ f''$ is the required map $V \to W$ of rank k.

(A″) **Exercises.** (a) Fill in the detail for the above proof.

(b) (Chen 1971). Take two loops γ_1 and γ_2 at some point in a connected manifold V and let ω_1 and ω_2 be closed C^1-smooth 1-forms on V such that $\omega_1 \wedge \omega_2 \equiv 0$ and such that the period matrix $(\int_{\gamma_i} \omega_j)$, $i, j = 1, 2$, has rank 2. Assume $\dim H^1(V; \mathbb{R}) = 2$ and show *Abel's period map*

$$V \to T^2 = H^1(V; \mathbb{R})/H^1(V; \mathbb{Z})$$

to have rank ≤ 1. Then construct a homomorphism of $\pi_1(V)$ into the free group $F_2 = \mathbb{Z} * \mathbb{Z}$, such that the images of $[\gamma_i] \in \pi_1(V)$, $i = 1, 2$, are freely independent in F_2.

(b′) Let $\dim H^1(V; \mathbb{R}) \geq 3$ and apply (b) to the covering $\tilde{V} \to V$ for which the subgroup $\pi_1(\tilde{V}) \subset \pi_1(V)$ is generated by $[\gamma_1]$ and $[\gamma_2]$. Thus show the homotopy classes $[\gamma_i] \in \pi_1(V)$ to be freely independent in $\pi_1(V)$ (with no assumptions on $\dim H^1$).

(b″) Assume $\dim H^1(V; \mathbb{R}) = d < \infty$ and study homomorphisms $\pi_1(V) \to F_2$ by means of Abel's map $V \to T^d$.

(c) *Degenerate 2-Forms.* Take a submanifold $V_0 \subset V$, let $H^2(V_0; \mathbb{R}) \approx H^2(V, \mathbb{R}) = 0$ and consider a closed 2-form ω_0 on V_0 for which $\omega_0 \wedge \omega_0 \equiv 0$. Show that ω_0 does not extend to a closed form ω on V for which $\omega \wedge \omega \equiv 0$ on V, unless $\int_c \alpha_0 \wedge \omega_0 = 0$ for every 3-cycle c in V_0, homologous to zero in V and for every 1-form α_0 on V_0, such that $d\alpha_0 = \omega_0$. Apply this to $V_0 = S^3 \subset B^4 = V$ and to the pullback ω_0 of the area form on S^2 under the Hopf map $S^3 \to S^2$.

(d) *Locally Flat Immersions* $V \to \mathbb{R}^q$. Call an immersion $f: V \to \mathbb{R}^q$ *locally k-flat* if a small neighborhood of each point $v \in V$ is sent by f into an affine k-dimensional subspace in $\mathbb{R}^q$. C^1-Approximate an arbitrary C^1-immersion $f: V \to \mathbb{R}^q$ by k-flat C^∞-immersions for $k \leq k_0 = k_0(\dim V)$, where $k_0(1) = 2$ and $k_0(2) \leq 7$.
Hint. Start with a piecewise linear approximation of f for a suitable triangulation of V.

(d′) Show the normal Pontryagin classes P_i of a k-flat immersion to vanish for $i > 2(k - n)$, $n = \dim V$, and thus obtain a lower bound for $k_0(n)$.

(B) *Integrable Subbundles of* $T(V)$. Let $K \subset T(V)$ be a C^∞-smooth subbundle of codimension k and let $\alpha: T(V) \to N = T(V)/K$ be the quotient homomorphism. There (obviously) exists a unique vector bundle homomorphism $A: K \otimes K \to N$ such that $A(\partial_1 \otimes \partial_2) = \alpha([\partial_1, \partial_2])$ for every pair of sections ∂_1 and $\partial_2: V \to K$, where $[\ ,\ ]$ stands for the Poisson bracket of vector fields on V. The subbundle K is called *integrable* if $A \equiv 0$.

Theorem (Bott 1968). *If K is integrable, then the Pontryagin classes $P_i \in H^{4i}(V, \mathbb{R})$ of N can be represented by closed C^∞-smooth $4i$-forms ω_i on V such that every form ω on V which is a polynomial in ω_i with constant coefficients identically vanishes on V for* $\deg \omega > 2k$.

Proof. Take a small neighborhood $U \subset V$, fix a frame of independent sections $X_j: V \to K$, $j = 1, \ldots, n - k$, and let $Y_1, \ldots, Y_k$ be transversal to K tangent fields on U such that the sections $v_l = \alpha(Y_l): U \to N$, $l = 1, \ldots, k$ are linearly independent. Then there (obviously) exists a unique (affine) connection ∇ in the bundle $N|U$, for which

$$\nabla_{Y_l} v_m = 0, \qquad l, m = 1, \dots, k$$

$$\nabla_{X_j} v_m = \alpha([X_j, Y_m]), \qquad j = 1, \dots, n-k, m = 1, \dots, k.$$

Furthermore, the identity $A \equiv 0$ shows the derivative $\nabla_X v$ is independent of the choices of X_i and Y_l for all fields $X\colon U \to K$ and all sections $v\colon U \to N$. Moreover, the *curvature* $\Omega_{X,Y}\colon N \to N$, defined by $\Omega_{X,Y}(v) = \nabla_X \nabla_Y v - \nabla_Y \nabla_X v - \nabla_{[X,Y]} v$ vanishes on the pairs of vectors X and Y in $K \subset T(V)$, as a straightforward computation shows. By patching these connections on open subsets with a partition of unity on V we arrive at a (Bott) connection $\bar{\nabla}$ on V whose curvature $\bar{\Omega}$ also vanishes on $K \subset T(V)$. Now, there is a unique (Chern-Weil) polynomial of degree $2i$ in $\bar{\Omega}$ (viewed as a matrix valued 2-form on V) which gives us a closed (scalar valued) $4i$-form ω_i on V, such that $[\omega_i] = P_i$. Since $\bar{\Omega}$ vanishes on K and codim $K = k$, every polynomial of degree $> k$ in Ω is zero on V. Q.E.D.

Exercises. (a) Show, by producing examples and using Bott's theorem, that the integrability relation violates the h-principle for $2 \le k \le n-2$. (One does not know if there are obstructions to this h-principle besides Bott's theorem.)

(b) (Shulman 1972). Show the Massey products (as well as ordinary cup-products) of P_i vanish beyond the dimension $2k$ for all integrable subbundles $K \subset T(V)$ of codimension k.

(b′) Let $C_M\colon V \to Gr_{M!}(\mathbb{R}^{M'})$, $M' \ge M! + \dim V$ denote the classifying map for the bundle $(M!)N = \underbrace{N \oplus N \oplus \cdots \oplus N}_{M!}$. Assume V is compact and K is integrable. Show the map C_M to be $2k$-contractible for all sufficiently large integers M.

Hint. Replace the Grassmann manifold by the space G' of operators $P\colon \mathbb{R}^{M'} \to \mathbb{R}^{M'}$, such that $P^2 = P$ and rank $P = M!$. Take a classifying map $C'\colon V \to G'$ (for a large enough M') which induces a Bott connection $\bar{\nabla}$ on V from the (obvious) canonical connection in the natural $M!$-dimensional bundle over G'. Study C' by means of Sullivan's minimal model.

(c) *Secondary Classes* (Godbillon-Vey 1971). Observe that every affine connection in the trivial line bundle $N \to V$ is given by a 1-form, say β on the associated principle bundle which is fiberwise diffeomorphic to $V \times \mathbb{R} \to V$. Assume $N = T(V)/K$ for an integrable subbundle $K \subset T(V)$ of codimension one, let β correspond to a Bott connection and take a section $s\colon V \to V \times \mathbb{R}$. Show that the 3-form $s^*(\beta \wedge d\beta)$ on V *is closed and the cohomology class* $h = [s^*(\beta \wedge d\beta)] \in H^3(V; \mathbb{R})$ depends only on K.

(c′) Let Γ be a cocompact discrete subgroup in $SL_2\mathbb{R}$ and let T be the subgroup of upper triangular matrices. Let $K_0 \subset T(V_0)$, for $V_0 = SL_2\mathbb{R}/\Gamma$, consist of the vectors tangent to the T-orbits in V_0. Show K_0 to be integrable with the Godbillon-Vey class $h \ne 0$. Imbed V_0 into $\mathbb{R}^n$ for $n \ge 6$ and show that there is no integrable subbundle $K \subset T(\mathbb{R}^n)$ for which $K \cap T(V_0) = K_0$. (Thus integrable codimension one subbundles violate the h-principle for extensions.)

(B′) *Foliations*. Let Φ denote the sheaf of C^∞-submersions $V \to \mathbb{R}^k$ and let Ψ be the sheaf of subbundles $K \subset T(V)$ of codimension k. The correspondence

$f \mapsto \operatorname{Ker} D_f \subset T(V)$, $f \in \Phi$, defines a homomorphism of sheaves $\Phi \to \Psi$ whose image, say $\bar{\Phi} \subset \Psi$, consists, by *Frobenius theorem*, of (exactly and only) integrable subbundles. Thus, for every integrable subbundle $K \subset T(V)$, there exist submersions $f_i: U_i \to \mathbb{R}^k$ for some open cover $\bigcup_i U_i = V$, such that $\operatorname{Ker} D_{f_i} = K | U_i$ for all $i = 1, \ldots$. This defines a unique partition, called a *foliation* of V into *connected* $(n-k)$-dimensional submanifolds (which may be non-closed subsets in V) called *leaves* $\mathscr{L} \subset V$, which are everywhere tangent to K, and such that each intersection $\mathscr{L} \cap U_i$ is a countable union of connection components of f_i-pullbacks of some points in $\mathbb{R}^k$.

Example. Let the structure group of a k-dimensional bundle $N \to V$ with the fibers $N_v \approx \mathbb{R}^k$, $v \in V$, reduces to a *discrete* subgroup in $\operatorname{Diff} \mathbb{R}^k$. Then there exists a Galois covering $\tilde{V} \to V$, such that the lift $\tilde{N} \to \tilde{V}$ is *a trivial* fibration over $\tilde{V}$. Moreover, there is a splitting $\tilde{N} = \tilde{V} \times \mathbb{R}^k$, where each submanifold $\tilde{V}_x = \tilde{V} \times x \subset \tilde{V} \times \mathbb{R}^k$, $x \in \mathbb{R}^k$ is the graph of a section $\tilde{V} \to \tilde{N}$, such that the subbundle $\tilde{K} = \bigcup_{x \in \mathbb{R}^k} T(\tilde{V}_x) \subset T(\tilde{N})$ is invariant under the action of the Galois group Γ on $\tilde{N}$. Thus we obtain an integrable subbundle on $N = \tilde{N}/\Gamma$, say $\bar{K} \subset T(N)$, which is transversal to the fibers $N_v \subset N$, $v \in V$, and whose leaves are covered by the manifolds $\tilde{V}_x$.

Corollary (Phillips 1969). *If the structure group of the quotient bundle N of a subbundle $K \subset T(V)$ reduces to a discrete subgroup, then K is homotopic to an integrable subbundle $K' \subset T(V)$, provided V is an open manifold.* [In fact, this is also true, according to Thurston (1974, 1976), for closed manifolds V.]

Proof. The h-principle for maps $V \to N$ transversal to $\bar{K}$ gives us a transversal map $f: V \to N$, for which $K' = D_f^{-1}(\bar{K}) \subset T(V)$ is homotopic to K.Q.E.D.

Exercise. Apply this corollary to subbundles $K \subset T(V)$ of codimension one.

See Lawson (1974), Fuks (1981), and Reinhart (1983) for further information and references on foliations.

(C) *Complex Structures.* An *almost complex* structure on V is an automorphism $J: T(V) \circlearrowleft$ such that $J^2(\tau) = -\tau$ for all $\tau \in T_v(V)$ and $v \in V$. This is equivalent to a reduction of the structure group $GL_n\mathbb{R}$ of the bundle $T(V)$ to $GL_m\mathbb{C}$, where $n = 2m = \dim V$. An almost complex structure is called *complex* (or *integrable*) if it is locally isomorphic to the standard (integrable!) almost complex structure on $\mathbb{C}^m$. This can be expressed with the sheaf Ψ of immersions $V \to \mathbb{C}^m$ and with the obvious homomorphism h of Ψ to the sheaf Φ of almost complex structures on V by defining the sheaf of complex structures on V as the subsheaf $\Phi' = \operatorname{Im} h \subset \Phi$.

Every almost complex structure J on V defines a (unique up to a homotopy) *classifying* map C_J of V to the complex Grassmann manifold $Gr_m\mathbb{C}^N$, for a given $N > 2n$, such that the *complex* vector bundle $(T(V), J)$ [where $(a + \sqrt{-1}b)\tau \overset{def}{=} a + bJ(\tau)$] is induced by C_J from the canonical bundle over $Gr_m\mathbb{C}^N$.

Proposition (Gromov 1973; Landweber 1974). *If the manifold V is open and if the map C_J is $(m+1)$-contractible, then the almost complex structure J is homotopic to a complex one.* [Compare Adachi (1979).]

Proof. Proceed as in the proof of (A′) with the *total reality* condition [see (C) in 2.4.5] for embeddings $S^{l_i-1} \to \mathcal{O}\not{h} S^{l_i-1} \subset H_i^i$ instead of the transversality. The details are left to the reader.

Corollary. *If V is an open manifold of dimension $n = 2m \leq 6$, then the sheaf Φ of complex structures on V satisfies the h-principle.*

Indeed, any map $V \to \mathbb{C}r_m \mathbb{R}^N$ is $(m+1)$-contractible in this case.

(D) *Classifying Space.* Consider a topological space P and a continous sheaf Φ over an arbitrary manifold V. Denote by Φ_0^P the subsheaf of the sheaf Φ^P over $V \times P$ (see 1.5.5) for which $\Phi_0^P(U \times R)$ consists of the *locally constant* maps $R \to \Phi(U)$ for all open subsets $U \times R \subset V \times P$. Next, for a continuous map $f\colon P \to V$ we define the pull-back sheaf $f^*(\Phi)$ over P as the restriction of Φ_0^P to the graph $\Gamma_f = P \subset V \times P$. The contravariant functor $\mathscr{F}$ from topological spaces to sets, which assigns to each P the set of pairs (f, φ), where $f\colon P \to V$ is a continuous map and $\varphi \in f^*(\Phi)(P)$, satisfies:

($*$) The restriction of $\mathscr{F}$ to the category of open subsets in P is a sheaf over P.
($**$) The sheaf axioms are also satisfied for *finite* coverings of P by *closed* subsets $P_i \subset P$, $i = 1, \ldots, k$. Namely, if some elements $\psi_i \in \mathscr{F}(P_i)$ agree (in the obvious sense) on the intersections $P_i \cap P_j$, $i, j = 1, \ldots, k$, then there is a unique $\psi \in \mathscr{F}(P)$ such that $\psi | P_i = \psi_i$.

If X is an arbitrary topological space then the functor

$$P \mapsto \{\text{continuous maps } P \to X\}$$

obviously satisfies ($*$) and ($**$). With this in mind, we define a "space" as an arbitrary contravariant functor from topological spaces to sets which abides ($*$) and ($**$).

Example. The functor which assigns to each P all n-dimensional vector bundles over P is a "space".

Warning. *All* bundles over P do not form a set with the usual meaning of "all". Here and below, we restrict our "alls" to objects to a fixed sufficiently large set theoretic universe. The reader may entertain himself by putting the logic straight.

One obviously extends to "spaces" the usual topological notions, like continuous maps, Serre fibrations, homotopy groups etc. Furthermore, one can assign (in many ways) to each "space" $\mathscr{F}$ an ordinary space, say $[\mathscr{F}]$ and a weak homotopy equivalence $[\mathscr{F}] \to \mathscr{F}$. For example, let Δ^∞ be the infinite dimensional simplex with countably many vertices. Then one constructs a cell complex $[\mathscr{F}]$, such that the set of k-cells in $[\mathscr{F}]$ is the union $\bigcup_\sigma \mathscr{F}(\sigma)$ where σ runs over all k-faces in Δ', and where a $(k-1)$-cell $\alpha' \in \mathscr{F}(\sigma')$ is attached to $\alpha \in \mathscr{F}(\sigma)$ as a face iff σ' is a face of σ and $\alpha|\sigma' = \alpha'$. This $[\mathscr{F}]$ comes with a natural continuous "map" $[\mathscr{F}] \to \mathscr{F}$ which clearly is a weak homotopy equivalence. (For the above vector bundle functor this is the ordinary construction of the classifying space).

Exercises. (a) Define sheaves over V with values in "spaces" and extend the results in 2.2.1–2.2.4 to these sheaves.

(b) Let Φ be a *microflexible* sheave over an arbitrary manifold V and let $\varphi \in \Phi^*(V)$. Construct a continuous map $f\colon V \to V$, which lies in a given C^0-fine neighborhood of the identity map, and a section $\psi \in f^*(\Phi)(V)$ which is (in an obvious sense) "homotopic" to φ. Derive the h-principle for Diff(V)-invariant microflexible sheaves over *open* manifolds V from the C^0-dense h-principle for maps $\mathcal{O}p\,V_0 \to V \times V$ whose projections on the second factor are immersions $\mathcal{O}p\,V_0 \to V$, where $\mathcal{O}p\,V_0 \subset V$ is an (arbitrarily) small neighborhood of a given subpolyhedron $V_0 \subset V$ of positive codimension.

(D′) Fix a Diff-invariant sheaf Φ over $\mathbb{R}^n$ and consider a fibration $T \to P$ with the fiber $\mathbb{R}^n$ and the structure group Diff $\mathbb{R}^n$. Since this group operates on $\Phi(\mathbb{R}^n)$, we can difine the associated fibration, say $T_\Phi \to P$ with the fiber $\Phi(\mathbb{R}^n)$.

Definitions. (a) A Φ-bundle is a fibration $T \to P$ with a given section called a *Φ-structure*, $P \to T_\Phi$, that is a family $\varphi_p \in \Phi(T_p \approx \mathbb{R}^n)$ continuous in p.

(b) A *flat connection* in T is a subsheaf F in the sheaf of sections $P \to T$ such that

(i) for each point $t \in T_p \subset T$, $p \in P$, for each pair of small neighborhoods $S \subset T_p$ of t and $U \subset P$ of p, and for each point $s \in S$ there exists a unique section $f = f_s \in F(U)$ for which $f(p) = s$;

(ii) the map $g_q\colon S \to T_q$ defined by $s \mapsto f_s(q)$ for all $q \in U$, is a diffeomorphism of S onto an open subset in the fiber T_q which C^∞-continuously depends on $q \in U$.

(c) *A flat Φ-bundle* is a pair consisting of a Φ-bundle given by sections $\varphi_p \in \Phi(T_p \approx \mathbb{R}^n)$, $p \in P$, and of a flat connection in T, such that the diffeomorphism $g_q\colon S \to T_q$ sends (for the Diff-action in Φ) $\varphi_p|S$ to $\varphi_q|g_q(S)$ for all $t \in T_p$, and $p \in P$ and all $q \in U$. [Compare Segal (1978).]

(c′) **Examples.** The trivial bundle $T = P \times \mathbb{R}^n \to P$ carries the canonical flat connection given by the sheaf of (the graphs of) locally constant maps $P \to \mathbb{R}^n$. Each section $\varphi \in \Phi(\mathbb{R}^n)$ extends to a unique Φ-structure on this T which is flat for this connection and which is called the *constant* Φ-structure.

Let us project the product $V \times V$ to the diagonal $\Delta \subset V \times V$ by $(v_1, v_2) \mapsto (v_1, v_1)$. Then there is an arbitrarily small neighborhood $T \subset V \times V$ of Δ for which the restricted projection $T \to \Delta$ is fiberwise diffeomorphic to the tangent bundle $T(V) \to V = \Delta$. A natural connection in this T is given by the sheaf of those sections $s\colon \Delta \to T \subset V \times V$ whose projections on the second V-factor are locally constant.

Let Ψ be a Diff-invariant sheaf on V which is locally isomorphic to Φ. Then each section $\psi \in \Psi^*(V)$ defines (by the definition of ψ^*, see 1.5.5) a Φ-structure on small neighborhoods $T \approx T(V)$ of $\Delta = V$. This structure is flat for the above connection if and only if ψ is contained in $\Psi(V) \subset \Psi^*(V)$.

(d) Consider the "space" which relates to each P the set of all Φ-bundles over P and let $[\Phi]^*$ be the corresponding classifying space. Similarly define the (Haefliger-Milnor) classifying space $[\Phi]^0$ for the "space" of flat Φ-bundles. Furthermore, consider a natural (forgetful) map between these spaces, say $H: [\Phi]^0 \to [\Phi]^*$.

Exercises. (1) Show the map H to be a weak homotopy equivalence for *microflexible* sheaves Φ on $\mathbb{R}^n$.

(2) Let Ψ be a Diff-invariant sheaf on a smooth manifold V which is locally isomorphic to Φ. Observe a correspondence between Φ-structures on the tangent bundle $T(V) \to V$ and sections in $\Psi^*(V)$. Let V be an *open* manifold. Show that the existence of a flat Φ-structure in $T(V)$ implies the existence of a section in $\Psi(V)$. Moreover, a section $\psi \in \Psi^*(V) \supset \Psi(V)$ can be homotoped to $\Psi(V)$ if and only if the *classifying map* $C_\psi: V \to [\Phi]^*$ can be "lifted" to a continuous map $C^0: V \to [\Phi]^0$, such that $H \circ C$ is homotopic to C_ψ.

Remark. If $[\mathscr{F}]$ is a classifying space for a "space" $\mathscr{F}$ then each $\psi \in \mathscr{F}(P)$, for all cell complexes P, defines a (unique up to a homotopy) classifying map $C_\psi: P \to [\mathscr{F}]$ which is consistent with the weak homotopy equivalence $[\mathscr{F}] \to \mathscr{F}$.

(3) Reformulate the above (A'), (B) and (C) in terms of the respective maps $H: [\Phi]^0 \to [\Phi]^*$.

(4) Take a subset of tangent n-planes of a manifold W, say $A \subset Gr_n W$, and let Φ be the sheaf of immersions $f: \mathbb{R}^n \to W$, such that $D_f(T_v(\mathbb{R}^n)) \in A$, for all $v \in \mathbb{R}^n$ (compare 2.4.4). Observe the canonical map $[\Phi]^* \to A$ and prove it to be a weak homotopy equivalence for *open* subsets A.

(5) Take a *closed* k-form g on a manifold W and let Φ be the sheaf of *g-isotropic* immersions $f: \mathbb{R}^n \to W$ that is $f^*(g) \equiv 0$. Show the *homotopy fiber* of the map $H: [\Phi]^0 \to [\Phi]^*$ to be the Eilenberg-MacLane space $K(\mathbb{R}, k-1)$. Prove the map H turned into a fibration to be induced from the canonical fibration over $K(\mathbb{R}, k)$ [with the fiber $K(\mathbb{R}, k-1)$] by the composition of the obvious map $[\Phi]^0 \to W$ with the (classifying) map $W \to K(\mathbb{R}, k)$ which represents the cohomology class $[g] \in H^k(W; \mathbb{R})$.

(6) Define the classifying *sheaf* for a given sheaf Φ on V with no (Diff-V)-action. Reduce this sheaf to a single space if Φ is acted upon by a *transitive* (pseudo) group of diffeomorphisms on V. [Compare Brown (1962).]

(D'') *The Weak h-Principle.* Fix a Diff-invariant sheaf Φ over $\mathbb{R}^n$ and consider a locally isomorphic (to Φ) sheaf Ψ (which is also Diff-invariant) over an n-dimensional manifold V. Say that Ψ satisfies the *weak h-principle* on V if for an arbitrary flat Φ-structure φ_0 on $T(V) \to V$ there exists a section $\psi_1 \in \Psi(V)$ whose associated flat Φ-structure φ_1 on $T(V)$ [see the above (c')] is "homotopic" to φ_0. This means the existence of a flat Φ-bundle (T, φ) over $V \times [0,1]$ such that $(T, \varphi)| V \times 0 = (T(V), \varphi_0)$ and $(T, \varphi) V \times 1 = (T(V) \varphi_1)$.

Exercises. Prove the weak h-principle for an arbitrary Diff-invariant sheaf Ψ on an *open* manifold V.

Show the weak h-principle to be equivalent to the ordinary h-principle for all *microflexible* Diff-invariant sheaves Ψ on all manifolds V.

Relate flat Φ-bundles over V to pairs (f, ψ) where $f: V \to V$ is a continuous map and $\psi \in f^*(\Psi)(V)$. Then generalize the above to arbitrary (non-Diff-invariant) sheaves Ψ over V.

(E) *Foliations on Closed Manifolds.* Thurston (1974, 1976) has proved the following

Theorem. *The sheaf of C^i-foliations, $i = 1, 2, \ldots, \infty$, on an arbitrary (possibly closed!) manifold V satisfies the weak h-principle. Moreover, 2-dimensional foliations satisfy the ordinary h-principle.*

Thurston's proof is based on a far-reaching generalization of Reeb's construction [see (c) in (E′)]. Another approach (related to the surgery of singularities) is due to Misharchev and Eliashberg (1977).

Exercise. Derive the following corollaries from Thurston's theorem

(a) If a subbundle $K \subset T(V)$ has a *trivial* normal bundle $N = T(V)/K$, then K is homotopic to an integrable C^∞-subbundle in $T(V)$.

(b) Every codimension one subbundle in $T(V)$ is homotopic to an integrable C^∞-subbundle.

(c) Every 2-dimensional subbundle is homotopic to a C^∞-integrable one.

(d) The sphere S^7 admits a k-dimensional C^∞-foliation of codimension k for all $k = 1, \ldots, 6$. (No simple construction is known for these foliations.)

(E′) C^{an}-*Foliations.* The weak h-principle fails to be true for C^{an}-foliations of codimension one on closed manifolds due to the following

Theorem (Haefliger 1962). *No closed simply connected manifold V admits a real analytic foliation of codimension one.*

Haefliger proves this by applying the classical Poincaré-Bendixson theorem to the intersection of a given C^∞-foliation on V with an appropriately immersed disc $D^2 \to V$. Thus he finds an immersed cylinder $S^1 \times [0, 1] \to V$ transversal to the foliation, such that the closed leaves of the induced foliation on the cylinder are exactly the circles $S^1 \times t$ for $t \in [0, 1/2]$.

Examples of Foliations. (a) Let the fundamental group Γ of a closed manifold V_0 act by C^i-diffeomorphisms on a closed manifold F. Then the group Γ diagonally acts on $\tilde{V}_0 \times F$, where $\tilde{V}_0$ is the universal covering of V, and the closed manifold $V = \tilde{V}_0 \times F/\Gamma$ carries a natural n_0-dimensional C^i-foliation for $n_0 = \dim V_0$. For instance, the unit tangent bundle of an n_0-dimensional manifold V_0 of *constant negative curvature* carries an n_0-dimensional C^{an}-foliation transversal to the fibers of the bundle.

Remark. It is very hard, in general, to find a Γ-action on a given manifold F.

(b) Let Γ be a discrete subgroup in a Lie group G. Then there is an (obvious) one-to-one correspondence between G-invariant foliations on G/Γ and connected subgroups in G.

(c) The action $x \mapsto 2^i x$, $i \in \mathbb{Z}$, $x \in \mathbb{R}^n \setminus \{0\}$, preserves the foliation of $R^n \setminus \{0\}$ into parallel k-planes. Thus one obtains a C^{an}-foliation of dimension k, for any $k = 1, \ldots, n-1$, on the *Hopf manifold* $(\mathbb{R}^n \setminus \{0\})/\mathbb{Z} \approx S^{n-1} \times S^1$. Furthermore, one obtains a C^{an}-foliation of codimension one on $(\mathbb{R}^n_+ \setminus 0)/\mathbb{Z} \approx D^{n-1} \times S^1$, such that the boundary $\partial(D^{n-1} \times S^1) \approx S^{n-2} \times S^1$ is a leaf of this foliation. In particular, one gets such a foliation on the solid torus $D^2 \times S^1$. By gluing two solid tori over the boundary one obtains *Reeb's foliation* on S^3 which is C^∞-smooth (but not C^{an}!) and whose only closed leaf is the (Clifford) torus $T^2 \subset S^3$.

(F) *Compact Complex Manifolds.* Complex structures on closed 4-dimensional manifolds violate the weak h-principle as seen in the following

Example (Yau 1976). *The (parallelizable) manifold $(T^3 \# P^3\mathbb{R}) \times S^1$ admits no complex structure.*

Proof. Every compact complex surface V whose first Betti number is even, $\dim H^1(V;\mathbb{R}) = 2m$, admits (see Kodaira 1964) m linearly independent closed *holomorphic* 1-forms. These forms define the (Abel) holomorphic (!) period map $A\colon V \to T^{2m} = \mathbb{C}^m/\mathscr{L}$ for the period lattice $\mathscr{L} \approx H^1(V;\mathbb{Z}) \approx \mathbb{Z}^{2m}$ in $\mathbb{C}^m$, such that $A^*\colon H^1(T^{2m};\mathbb{Z}) \to H^1(V;\mathbb{Z})$ is an isomorphism. The above manifold $V \approx (T^3 \# p^3\mathbb{R}) \times S^1$ admits a basis $h_i \in H^1(V;\mathbb{Z}) \approx \mathbb{Z}^4$, $i = 1, \ldots, 4$, such that the cup product $h_1 \cup h_2 \cup h_3 \cup h_4$ is a *generator* in $H^4(V;\mathbb{Z}) \approx \mathbb{Z}$. Hence, any map $A\colon V \to T^4$ isomorphic on H^1 must have the topological degree ± 1. The desired contradiction now follows from the (obvious) fact that every proper holomorphic map of degree one between equidimensional complex manifolds induces an *isomorphism* of the respective fundamental groups.

(G) *Φ-Cycles.* Let Φ be a Diff-invariant sheaf over $\mathbb{R}^n$, let $[\Phi]^0$ be the classifying space for the flat Φ-bundles and denote by $\mathscr{T} \to [\Phi]^0$ the canonical fibration (whose flat Φ-structure we forget for the moment) regarded here as an n-dimensional *vector bundle* over $[\Phi]^0$. (The vector bundle structure is unique up to an isomorphism.) Then we consider sheaves Ψ locally isomorphic to Φ over closed oriented manifolds V, dim $= n$. Recall that every section $\psi \in \Psi(V)$ defines a flat Φ-bundle over V and, hence, a classifying map $C_\psi\colon V \to [\Phi]^0$. The maps which arise this way are called *Φ-cycles* in $[\Phi]^0$. Each Φ-cycle C in $[\Phi]^0$ defines an integral homology class $[C] \in H^n([\Phi]^0)$ that is the image of the fundamental class of the underlying manifold V. The following condition $(*)$ is obviously necessary for the representation of a class $h \in H^n([\Phi]^0)$ by a Φ-cycle,

$(*)$ There exists a closed oriented manifold V and a continuous map $B\colon V \to [\Phi]^0$, such that the induced bundle $B^*(\mathscr{T})$ over V is isomorphic to $T(V) \to V$.

Furthermore, if the sheaves Ψ over all closed manifolds V satisfy the weak h-principle, then $(*)$ is clearly sufficient for a representation of h by a Φ-cycle.

Remark. It would be far more interesting to prove (rather than to use) the h-principle by representing the homology of $[\Phi]^0$ by Φ-cycles. But little is known in this direction [compare (H) and (H′)].

Let us relax $(*)$ by allowing a *stable* isomorphism between the bundles $B^*(\mathcal{T})$ and $T(V)$, that is an isomorphism $B^*(\mathcal{T}) \oplus l \approx T(V) \oplus l$, for the trivial bundle $l = V \times \mathbb{R} \to V$.

Lemma. *If $n \geq 3$ then the stable condition $(*)$ implies $(*)$.*

Indeed, for an arbitrary n-dimensional bundle $T \to V$ which is stably isomorphic to $T(V)$ there exists a stably parallelizable manifold ($n \geq 3$) V_0, such that the tangent bundle of the connected sum $V \# V_0$ is isomorphic to the induced bundle $p^*(T)$ over $V \# V_0$ for the obvious (pinching) map $p\colon V \# V_0 \to V$. Q.E.D.

Let us reduce the stable condition $(*)$ to a pure homotopy condition. To do this we fix a countable $(n+1)$-dimnesional subcomplex $K \subset [\Phi]^0$, such that the group $H_n(K)$ injects into $H_n([\Phi]^0)$ and such that a given class $h \in H^n([\Phi]^0)$ comes from some class $h' \in H^n(K)$. Then there (obviously) exist a (unique up to a diffeomorphism) N-dimensional manifold X for a given $N \geq 2n+2$ and a homotopy equivalence $y\colon X \to K$, such that the Whitney sum $T(X) \oplus y^*(\mathcal{T}')$ is a trivial bundle over X for the bundle $\mathcal{T}' = \mathcal{T} | K$ over $K \subset [\Phi]^0$.

Now, the class $h \in H_n([\Phi]^0)$ can be represented by a map $B\colon V \to [\Phi]^0$ satisfying the stable condition $(*)$ if and only if the class $y_*^{-1}(h') \in H_n(X)$ can be represented by a submanifold in X with the *trivial* normal bundle. Such submanifolds form a (cobordism) group which is isomorphic (via Pontryagin-Thom construction) to the (cohomotopy) group of homotopy classes of maps $\bar{X} \to S^{N-n}$ for the one-point compactification $\bar{X}$ of X.

Corollary. *If $n \geq 3$ and if the bundles Ψ locally isomorphic to Φ satisfy the weak h-principle, then an integer multiple of every class $h \in H_n([\Phi]^0)$ can be represented by a Φ-cycle.*

Example. Take a manifold W for $\dim W > n$ and take an open subset of oriented tangent n-planes in W, say $A \subset Gr_n(W)$. Assume smooth immersions $f\colon V \to W$ *directed by A* [which means $D_f(T_v(V)) \in A$, $v \in V$, compare 2.4.4] satisfy the h-principle for all n-dimensional manifolds V. Take a class $h \in H_n(V)$ and let h' be a class in $H_n(A)$ which goes to h under the projection $A \to V$. Then in the following two cases the class $(m!)h$, for all sufficiently large m, can be represented by an immersion directed by A of some closed oriented manifolds V into W,

(1) $n \geq 3$,
(2) there exists a continuous map $\alpha\colon S^n \to A_w = A \cap Gr_n(T_w(W))$ for all $w \in W$, such that the induced n-dimensional bundle over S^n is isomorphic to $T(S^n)$.

Indeed, the case (1) is covered by the above corollary and (2) allows one to derive the non-stable condition (*) from the stable one. [See Eliashberg (1984) for a comprehensive study of cobordisms of differential relations.]

Exercises. Develop cobordism theories for

(a) (Wells 1966) immersions of positive codimensions;

(b) Lagrange and Legendre (see 3.4.2, 3.4.3) immersions into symplectic (respectively contact) manifolds. [See Arnold (1980) for an elementary treatment of low dimensional examples and Audin (1984) for a general approach.]

(c) Free isotropic (see 3.3.4) immersions into pseudo-Riemannian manifolds;

(d) Immersions transversal to a given subbundle in $T(W)$ of codimension $\geq n+1$.

Show, for instance, that a non-zero integer multiple of each class in $H_n(W)$ can be represented by an immersion of parallelizable manifold into W for $3 \leq n < \dim W$.

Remark. The tangent bundle $T(G)$ of the Grassmann manifold $G = Gr_n(W) \to W$ contains a canonical subbundle, say $K \subset T(G)$ of codimension $q - n$ for $q = \dim W$, such that the natural lift of every submanifold $V^n \subset W$ to G is everywhere tangent to K. Next, for a submanifold $A \subset G$, we put $K' \subset T(A) \cap K$ and we view [following Thom (1959)] maps $V \to A$ tangent to K' as generalized maps $V \to W$ directed by A. For example, if $q = n+1$ and if A is an *open* subset in G, then these are just *Legendre* maps $V \to A$ for the *contact* structure $K|A$. Legendre maps unlike the maps $V \to W$ directed by A, *always* satisfy the h-principle (see 3.4.3) which makes the cobordism theory of the *generalized* A-directed cycles quite easy (see the above (b)). In fact, Thom (1959), developed a homology theory for such cycles for *all* $q - n$ without use of the h-principle.

(H) *Curvature Relations.* Consider Riemannian metrics g on a manifold V whose sectional curvatures do not vanish that is either $K(g) > 0$ or $K(g) < 0$ everywhere on V. If V is a closed connected *surface* then the parametric h-principle refined by the Gauss-Bonnet theorem holds true:

If the Euler characteristic $\chi(V)$ is positive, then the metrics g on V with $K(g) > 0$ constitute a non-empty contractible space. If $\chi(V) < 0$, then the metrics with $K(g) < 0$ form such a space.

This is an easy application of the uniformization theory (compare 3.2.4). However, there is no simple refinement of the h-principle for extensions.

Example. Let V be a compact surface bounded by a simple closed curve $S \approx \partial V$. Then, by the Gauss-Bonnet formula, the geodesic curvature $\kappa(s)$, $s \in S$, satisfies $\int_S \kappa(s)\,ds < 2\pi\chi(V)$ if $K(g) > 0$ and $\int_S \kappa(s)\,ds > 2\pi\chi(V)$ for $K(g) < 0$. Furthermore, let $S_+ \subset S$ be a (connected!) segment of length σ_+, such that $\int_{S_+} \kappa(s)\,ds \geq \kappa_+ > 2\pi$, and let every open (possibly disconnected) subset $S_- \subset S$ of total length $\leq \sigma_+$ satisfy $\int_S \kappa(s)\,ds \geq \kappa_-$. *Then, for $K(g) > 0$, the numbers κ_+ and κ_-* satisfy $\kappa_+ + \kappa_- < 2\pi$. *If $K(g) < 0$, then $\kappa_+ + \kappa_- < 2\pi(1 - \chi(V)) + \int_S \kappa(s)\,ds$.* This is proved by deforming

S_+ to the shortest curve γ in V with $\partial\gamma = \partial S_+$ and by applying the above inequalities to the complement $V \backslash \gamma$. [Compare Gromoll-Klingenberg-Meyer (1968).]

Thus, the extension of metrics with non-vanishing curvature from ∂V to V meets an obstruction which is not are accounted for by the h-principle nor by the Gauss-Bonnet theorem. [Compare Gromoll-Meyer (1969).]

If $\dim V \geq 3$, then the relations $K(g) > 0$ and $K(g) < 0$ completely deviate from the h-principle. For instance, *closed* manifolds (V, g) with $K(g) > 0$, have *finite* fundamental group, while the inequality $K(g) < 0$ makes $\pi_1(V)$ *infinite*. This totally desagrees with the parametric h-principle which predicts the space of metrics g satisfying $K(g) > 0$ (or, as well $K(g) < 0$) to be a (non-empty!) contractible space. In fact, this space may be disconnected. See Hitchin (1974) for an example of two metrics of constant positive curvature on S^8 which can not be joined by any homotopy of metrics g with $K(g) > 0$. In fact, Hitchin proves that no homotopy of metrics with *non-negative scalar* curvature exists between these metrics. (No such example is known for $K(g) < 0$.)

The classifying space for the sheaf Φ_+ of metrics on $\mathbb{R}^n$ with $K > 0$ is clearly weakly homotopy equivalent to the Grassmann manifold $Gr_n \mathbb{R}^\infty$, $n = \dim V$, and the same is true for the relations $K \geq 0$, $K < 0$ and $K \leq 0$. It is unknown whether a non-zero integer multiple of every class $h \in H_n(Gr_n \mathbb{R}^\infty)$ can be represented by a Φ_+-cycle. In plain words we ask for a (possibly disconnected) oriented manifold (V, g) with $K(g) > 0$ and with given sufficiently divisible characteristic numbers. If V is a closed *connected* manifold with $K \geq 0$, then the Euler characteristic and the signature are bounded by $|\chi| + |\sigma| \leq \text{const}_n$, (Gromov 1981), which strongly restricts *connected* cycles. However, disjoint unions of spheres and of products of complex projective spaces (which carry obvious metrics with $K \geq 0$) may have arbitrarily large (vectors of) characteristic numbers. Similarly, compact hyperbolic manifolds and products of complex hyperbolic manifolds (which have $K \leq 0$) generate a subgroup of finite index in $H_n(Gr_n \mathbb{R}^\infty)$. On the other hand, the only known restriction on characteristic numbers of a closed manifold (V, g) with $K(g) \leq 0$ is $\chi(V) \geq 0$ for $\dim V = 4$, with the strict inequality $\chi(V) > 0$ for $K > 0$.

(H′) *The Scalar Curvature.* The failure of the h-principle for metrics g with the scalar curvature $S(g) > 0$ less drastic than for $K > 0$. For example, these metrics are amenable to a surgery (Shoen-Yau 1979; Gromov-Lawson 1979) which allows one to produce many closed manifolds with $S > 0$ starting from standard examples (like the above Φ_+-cycles). In particular, *every simply connected manifold V whose second Stiefel class w_2 does not vanish admits a metric g with $S(g) > 0$*, provided $\dim V \geq 5$ (Gromov-Lawson 1979).

On the other hand *if $w_2 = 0$, and if V is homeomorphic to a product $V_1 \times V_2$ where the manifold V_1 admits a metric g_1 with* $K(g_1) \leq 0$ (like $V_1 \approx T^m$) *and V_2 has non-zero $\hat{A}$-genus* (e.g. $\dim V_2 = 0$ or $\dim V_2 = 4$ and $\sigma(V) \neq 0$) *then V admits no metric g with $S(g) > 0$.* [See Gromov-Lawson (1983) and Shoen (1984) for further information and references.]

Exercises. (a) Let $\mathscr{R} \subset X^{(r)} \to X \to V$ be an *open differential relation, let* $V_0 \subset V$ be an arbitrary submanifold and let $f_0 \colon V \to X$ be a C^r-solution of $\mathscr{R}$ [i.e. $J^r_{f_0}(V) \subset \mathscr{R}$]. Let F denote the space of C^r-solutions $f \colon V \to X$ of $\mathscr{R}$, such that $J^{r-1}_f | V_0 = J^{r-1}_{f_0} | V_0$, and let F_0 be the space of jets $\varphi \colon V_0 \to \mathscr{R}$ of such solutions near V_0. That is $\varphi \in F_0$ if and only if there exists a solution $f' \colon \mathcal{O}\!\!\!/\, V_0 \to X$ of $\mathscr{R}$ such that $J^{r-1}_{f'} | V_0 = J^{r-1}_{f_0} | V_0$ and for which $J^r_{f'} | V_0 = \varphi$. Prove the following

Weak Flexibility Lemma. *The map* $f \mapsto J^r_f V_0$ *is a Serre fibration* $F \to F_0$.

Hint. Use the induction in dim V and codim V_0, starting with dim $V = 1$, dim $V_0 = 0$.

(b) Apply (a) to the differential relations $K(g) > 0$, $K(g) < 0$, $S(g) > 0$, and to a closed *geodesic* $V_0 \subset (V, g_0)$. Thus deform a given Riemannian metric g_0 which satisfies one of the above inequalities to a metric g whose sectional curvature is *constant* near V_0, while satisfying the same curvature inequality as g_0 everywhere on V.

(c) Construct a metric of positive scalar curvature on the connected sum of n-dimensional manifolds, $n \geq 3$, with constant sectional curvature > 0. Then, using (b), make this work for all manifolds with (now non-constant) positive scalar curvature.

(d) (Yau). Construct metrics of negative Ricci curvature on the connected sums of manifolds of negative Ricci curvature.

Scalar Curvature $S < 0$. Riemannian metrics g with $S(g) < 0$ on n-dimensional manifolds are likely to satisfy the parametric h-principle for $n \geq 3$. In fact, the non-parametric h-principle is established in the following stronger form by Kazdan-Warner (1975).

Let φ *be a* C^∞*-function on a connected manifold* V, dim $V \geq 3$, *which is somewhere negative at some point in* V. *Then there is a* C^∞*-metric* g *on* V, *such that* $S(g) \equiv \varphi$.

A similar existence theorem (with an easy proof left to the reader) holds for the equation $S(g)\,dg \equiv \omega$, where dg is the Riemannian volume form and ω is a given n-form on V which is negative at some point. In fact, one expects the solvability of the system $dg = \omega_0$, $S(g) = \varphi$ for the above φ. Furthermore, it does not seem to be hard to solve the equation $P(g) = \omega$ for a top-dimensional Pontryagin-Chern-Weil form $P(g)$ and for a given n-form ω, for which $[\omega] = [P(g)] \in H^n(V)$ and which is assumed to change sign [compare (H'')]. Moreover, one hopes to solve *systems* of equations $P_i(g) = \omega_i$, $i = 1, \dots, k$, for k small compared to n. A similar but easier problem consists in finding connections in a given bundle with given Pontryagin forms.

(H'') **Exercises.** (a) (Gromov-Eliashberg 1973). Consider n-dimensional manifolds V and W with given n-forms ω_0 on W and ω on V, where ω_0 nowhere vanishes on W, and prove the following h-principle for maps $f \colon V \to W$, such that $f^*(\omega_0) = \omega$.

Let $f_0 \colon V \to W$ *be a continuous map which lifts to a homomorphism of the tangent bundles* $T(V) \to T(W)$ *of rank* $\geq n-1$ *on every fibre* $T_v(V)$, $v \in V$, *and let*

$f_0^*[\omega_0] = [\omega]$ for the cohomology classes $[\omega_0] \in H^n(W;\mathbb{R})$ and $[\omega] \in H^n(V;\mathbb{R})$. Then there exists, in the following two cases, a C^∞-map $f\colon V \to W$ of rank $\geq n-1$, such that $f^*(\omega_0) = \omega$.

(1) *The manifold V is open.*

(2) *V is connected and the form ω changes the sign somewhere on V* (i.e. neither $\omega \geq 0$ nor $\omega \leq 0$ relative to a fixed volume form).

Show, in particular for an arbitrary C^∞-function φ on a connected *stably parallelizable Riemannian manifold* (V, g), such that $\int_V \varphi\, dg = 0$, there exists a C^∞-map $f\colon V \to \mathbb{R}^n$, $n = \dim V$, whose Jacobian satisfies $J_f \equiv \varphi$.

(a′) Let W be the Grassmann manifold $Gr_m\mathbb{R}^N$ and let ω_0 be a $SO(N)$-invariant closed non-zero form on W. Homotope a given immersion $f_0\colon V \to W$ to a map $f\colon V \to W$, such that $f^*(\omega_0) = \omega$, provided $f_0^*[\omega_0] = [\omega]$ and one of the above (1) and (2) is satisfied. Then construct a connection in an m-dimensional bundle over V whose given Pontryagin form P equals ω, provided $[P] = [\omega]$. Replace (1) and (2) by a suitable condition on ω_0 by using the techniques in 3.4.2.

Remark. The action of Diff V on (scalar curvature) functions has a non-empty (!) open orbit which is equally true for the action on n-forms. This suggests a unified approach to equations like $S(g) = \varphi$ and $f^*(\omega_0) = \omega$.

(G) *Folded Maps and Sections.* We have seen in (H) how the h-principle may fail for microflexible sheaves Ψ over a *closed* manifold V. However, one can save the h-principle by slightly modifying the sheaf Ψ.

The Singular h-Principle. *Let Ψ' be a microflexible sheaf over V' which is acted upon by the* (*pseudo*) group Diff V *and let $f\colon V \to V'$, for* $\dim V = \dim V' = n \geq 2$, *be a C^∞-map whose Jacobian changes sign on every connected component of V* (i.e. *there is a pair of points, say v_+ and v_- in each component of V, such that the local degree of f at v_+ equals $+1$ and* $\deg_{v_-} f = -1$, *where one defines these degrees on the oriented double coverings of V and V' in case the manifolds are non-orientable*). *Then the pull-back sheaf $\Psi = f^*(\Psi')$ over V satisfies the h-principle.* [Compare 6.2.5 in Gromov (1972) and 3.1 in Gromov (1971).]

Proof. First, the formalism in (D), reduces the problem to the sheaf Ψ_0' of continuous maps $V' \to \Omega \subset \mathbb{R}^n \times P$ whose projections to $\mathbb{R}^n$ are C^∞-immersions, where P is a polyhedron and Ω is an open subset in $\mathbb{R}^n \times P$. If the only singularity of f is a folding along a hypersurface $V_0 \subset V$, then the proof is concluded with the equidimensional folding theorem of Eliashberg (see 2.1.3). In general, assuming V is compact, one takes a small open neighborhood $U' \subset V'$ of the set of the critical values of f, such that the (topological) boundary $\partial U'$ is a C^∞-hypersurface in U'. Then one considers the submanifold $V_1 = V \backslash f^{-1}(U) \subset V$ whose boundary $\partial V_1 = f^{-1}(\partial U')$ is *immersed* by f onto $\partial U' \subset V'$. If there are at most two points in the pull-back $f^{-1}(u) \subset \partial V_1$ for all $u \in \partial U'$, then, by identifying these points in pairs, one obtains another manifold out of V_1, say V_2, which is continuously mapped by f into V' with a topological folding (like $x \mapsto |x|$) over those points $u \in U$ where $f^{-1}(u)$ contains exactly two points. In fact, the folding locus is non-empty due to the sign change condition imposed on f. This folding is good enough to apply Eliashberg's theorem to V_2, and then to extend maps $V_2 \to \Omega$ to all of V. If $f^{-1}(u)$ contains many points, then they can be locally organized into pairs to give local folded maps to which the extension version of Eliashberg's theorem applies. Finally, if V is an *open* manifold, then a modification (in fact, a simplification) of the above argument establishes the h-principle for $\Psi = f^*(\Psi')$ with no assumptions on the map f what-so-ever. Filling n the detail in this "proof" is left to the reader.

Remarks. (a) The most interesting (and the easiest) case of the singular h-principle concerns maps $f: V \to V'$ whose only singularity is a folding along some $V_0 \subset V$, as the sheaf $f^*(\Psi')$ for such an f is the nearest to Ψ'.

(b) Saving the h-principle with a folding is also possible for some non-microflexible and non-Diff-invariant sheaves. An instance of that is the case (2) of the equidimensional equation $f^*(\omega_0) = \omega$ [see (H")], where the sign change condition on ω insures "foldings" of maps $f: V \to W$.

Exercises. (a) State and prove the *weak* h-principle for the sheaf $\Psi = f^*(\Psi')$ without assuming Ψ' microflexible [compare (D")].

(b) Let Ψ be a Diff-invariant microflexible sheaf over an n-dimensional manifold V, $n \geq 2$, with a non empty boundary, and let a finite group Γ act freely on an open tubular neighborhood V_0 of the boundary. Show that sections ψ in $\Psi(V)$ which are Γ-equivariant on V_0 (i.e. $\gamma(\psi_0) = \psi_0$ for $\psi_0 = \psi | V_0$, $\gamma \in \Gamma$) satisfy the h-principle. Generalize this to infinite groups Γ whose action on V_0 is discrete. Prove the *weak* version of this h-principle for *non-microflexible* sheaves Ψ.

Locally Split Metrics. A Riemannian C^∞-manifold $V = (V, g)$ is called *locally split*, if a small neighborhood $U \subset V$ of each point $v \in V$ isometrically splits, $(U, g|U) = (U_0 \times (0, \varepsilon), g_0 \oplus dt^2)$.

(c) Show that the Euler and Pontryagin numbers of every locally split manifold vanish.

(c') Construct a locally split C^∞-metric on every manifold V which admits a free S^1-action. (*Simply connected* manifolds carry no *real analytic* locally split metrics.)

(d) Consider the sheaf Φ of locally split C^∞-metrics on $\mathbb{R}^n$ and prove every n-dimensional bundle over an $(n-1)$-dimensional polyhedron to carry a flat Φ-structure. Thus show every open n-dimensional manifold to admit a locally split C^∞-metric. [Compare Pasternak (1975).]

(d') Let V be closed oriented manifold and let $C: V \to G = Gr_{n=1} \mathbb{R}^N$ be the clasifying map for the stabilized tangent bundle, $T(V) \oplus l \to V$ for $l = V \times \mathbb{R} \to V$, where G is the Grassmann manifold of oriented $(n+1)$-planes in $\mathbb{R}^N$ for $N > 2n = 2 \dim V$. Let Ψ be the sheaf of locally split C^∞-metrics on V and assume the map C is $(n-1)$-contractible. Construct a C^∞-map $f: V \to V$ homotopic to the identity whose only singularity is a folding along some closed (possibly disconnected) C^∞-hypersurface $V_0 \subset V$, such that the induced sheaf $f^*(\Psi)$ admits a section $\psi^* \in f^*(\Psi)$.

(d″) Let the Pontryagin numbers of a connected oriented manifold V' vanish. Show the classifying map of the manifold $V = 2V' = V' \# V'$ into $G = Gr_{n+1}\mathbb{R}^N$ to be $(n-1)$-contractible. Construct with the above ψ^* a C^∞-metric g on V which admits an orientation reversing isometric involution $I: (U, g)\circlearrowleft$ for an open subset $U \subset V$ with an $(n-1)$-dimensional complement. Prove with (C″) in 2.1.3 the vanishing of the signature $\sigma(V) = 2\sigma(V') = 0$, and conclude to the following

Thom-Hirzebruch Signature Theorem. *There exists a rational cohomology class $L_n \subset H^n(G; \mathbb{Q})$ (which is a certain polynomial in the Pontryagin classes $p_i \in H^{4i}(G; \mathbb{Z})$ with rational coefficients) such that every closed orientable manifold V, $\dim V = n = 4m$, has $\sigma(V) = L_n(V)$ for $L_n(V) \overset{def}{=} C^*(L_n)$ where C is the classifying map $V \to G$. [The polynomial L_n can be explicitly determined by substituting in the equality $\sigma(V) = L_n(V)$ the products of complex projective spaces for V.]*

2.3 Inversion of Differential Operators

In the previous section we have obtained the h-principle for some P.D.E. systems which are locally solvable by a purely algebraic procedure. Now, we turn to more general non-linear systems that become algebraically solvable only after they have been linearized. In order to come back from (solutions of) the linearized sytems to (solutions of) the non-linear systems themselves one needs an appropriate infinite dimensional *implicit function theorem*. Such a theorem was discovered by Nash (1956) in the course of his solution of the isometric imbedding problem. In the following sections we develop Nash's theory in the context of differential operators and related sheaves of solutions of non-linear P.D.E. systems.

2.3.1 Linearization and the Linear Inversion

Let X denote, as usual, a C^∞-fibration over an n-dimensional manifold V and let $G \to V$ be a C^∞-smooth vector bundle. We denote by $\mathscr{X}^\alpha$ and $\mathscr{G}^\alpha$ respectively the spaces of C^α-sections of the fibrations X and G for all $\alpha = 0, 1, \ldots, \infty$. Let $\mathscr{D}: \mathscr{X}^r \to \mathscr{G}^0$ be a differential operator of order r. The expression "differential of order r" means that for each C^r-section $x: V \to X$ the value of the section $\mathscr{D}(x): V \to G$ at any given point $v \in V$ depends only on the r-jet $J_x^r(v) \in X^{(r)}$. In other words, the operator $\mathscr{D}$ is given by a map $\Delta: X^{(r)} \to G$, namely $\mathscr{D}(x) = \Delta \circ J_x^r$. We say that $\mathscr{D}$ is a *C^α-operator*, $\alpha = 0, 1, \ldots, \infty$, if the map Δ is C^α-smooth. We assume below that $\mathscr{D}$ is a C^∞-operator and so we have continuous maps $\mathscr{D}: \mathscr{X}^{\alpha+r} \to \mathscr{G}^\alpha$ for all $\alpha = 0, 1, \ldots, \infty$. The word "continuous" equally applies here to the usual and to the fine topologies in our spaces of sections.

Example. Let G denote the symmetric square of the cotangent bundle of v, let W be a manifold with a quadratic differential form h, and let $X = W \times V \to V$. We get an operator $\mathscr{D}$ by relating to C^1-maps $x: V \to W$ the induced quadratic forms, $\mathscr{D}(x) = x^*(h)$. This operator $\mathscr{D} = \mathscr{D}_h$ has first order and, for a C^∞-form h, this is a C^∞-operator. In the case when $(W, h) = (\mathbb{R}^q, \sum_1^q dx_i^2)$ we write (compare 1.1.5).

$$\mathscr{D}(x) = g = \{g_{ij}\} = \left\{\left\langle \frac{\partial x}{\partial u_i}, \frac{\partial x}{\partial u_j} \right\rangle\right\}, \qquad i, j = 1, \ldots, n.$$

Linearization. Denote by $T_{\text{vert}}(X) \subset T(X)$ the space of those vectors in $T(X)$ which are tangent to the fibers of the fibration $X \to V$, and for a section $x: V \to X$ we denote by $Y_x \to v$ the induced vector bundle, $x^*(T_{\text{vert}}(X))$. If the section x is C^α-smooth, then $Y_x \to X$ is a C^α-smooth bundle and for each $\beta \leq \alpha$ we denote by $\mathscr{Y}_x^\beta$ the vector space of C^β-sections $V \to Y_x$. The space $\mathscr{Y}_x^\alpha$ can be also defined as the infinite dimensional tangent space $T_x(\mathscr{X}^\alpha)$. If $X \to V$ is a vector bundle, then each bundle Y_x is canonically isomorphic to X and each space $\mathscr{Y}_x^\beta$ is canonically isomorphic to $\mathscr{X}^\beta$.

Now, we assume that the fibers of the fibration $X \to V$ have no boundaries and we observe that for each pair (x, y) of C^r-sections $x: V \to X$ and $y: V \to Y_x$ there exists a C^r-smooth family of sections $x_t: V \to X$, $t \in [0, 1]$, such that $x_0 = x$ and $\partial x_t/\partial t = y$ for $t = 0$. We define *the linearization of $\mathscr{D}$ at x*, called $L_x: \mathscr{Y}_x^r \to \mathscr{G}^0$, by the formula

$$L_x(y) = L(x, y) = \frac{\partial}{\partial t} \mathscr{D}(x_t)|t = 0.$$

This definition does not depend on a specific choice of the family x_t that represents $y \in \mathscr{Y}_x^r$ and one can interpret

$$L_x: T_x(\mathscr{X}^r) = \mathscr{Y}_x^r \to T(\mathscr{G}^0) = \mathscr{G}^0$$

as the differential of our operator $\mathscr{D}: \mathscr{X}^r \to \mathscr{G}^0$ at $x \in \mathscr{X}^r$. It is also clear, that L_x is *a linear differential* operator of order r in y. Furthermore, if x is of class $C^{\alpha+r}$, then L_x is a C^α-operator. Moreover, $L(x, y)$ is a *differential* operator of order r in both variables x and y. This "global" operator L acts on the space $\mathscr{J}^r$ of C^r-sections $V \to T_{\text{vert}}(X)$, and since $\mathscr{D}$ is a C^∞-operator, L is also a C^∞-operator, $L: \mathscr{J}^{\alpha+r} \to \mathscr{G}^\alpha$. If we interpret $\mathscr{J}^{\alpha+r}$ as the tangent bundle, $\mathscr{J}^{\alpha+r} = T(\mathscr{X}^{\alpha+r})$, then $L = L(\mathscr{D})$ amounts to the (global) differential of $\mathscr{D}$.

If $X \to V$ is a vector bundle we have $L_x: \mathscr{X}^r = \mathscr{Y}_x^r \to \mathscr{G}^0$ and $L: \mathscr{X}^r \times \mathscr{X}^r \to \mathscr{G}^0$, namely

$$L(x, y) = \lim_{t \to 0} t^{-1}[\mathscr{D}(x + ty) - \mathscr{D}(x)].$$

Exercise. Extend the definition of L to the case when both, X and G over V, are general (non-vector) fibration.

Infinitesimal Inversions of $\mathscr{D}$. We say that the operator $\mathscr{D}$ is *infinitesimally invertible over a subset $\mathscr{A}$* in the space of sections $x: V \to X$ if there exists a family of *linear differential* operators of a certain *order* s, namely $M_x: \mathscr{G}^s \to \mathscr{Y}_x^0$, for $x \in \mathscr{A}$, such that the following three properties are satisfied.

(1) There is an integer $d \geq r$, called the *defect of the infinitesimal inversion M*, such that $\mathscr{A}$ is contained in $\mathscr{X}^d$, and furthermore, $\mathscr{A} = \mathscr{A}^d$ consists (exactly and only)

of C^d-solutions of an *open* differential relation $A \subset X^{(d)}$. In particular, the sets $\mathscr{A}^{\alpha+d} = \mathscr{A} \cap \mathscr{X}^{\alpha+d}$ are open in $\mathscr{X}^{\alpha+d}$ in the respective *fine* $C^{\alpha+d}$-topologies for all $\alpha = 0, 1, \ldots, \infty$.

(2) The operator $M_x(g) = M(x, g)$ is a (non-linear) *differential* operator in x of order d. Moreover, the "global" operator

$$M: \mathscr{A}^d \times \mathscr{G}^s \to \mathscr{J}^0 = T(\mathscr{X}^0)$$

is a differential operator, that is given by a C^∞-map $A \oplus G^{(s)} \to T_{\text{vert}}(X)$.

(3) $L_x \circ M_x = \text{Id}$, that is

$$L(x, M(x, g)) = g \text{ for all } x \in \mathscr{A}^{d+r} \text{ and } g \in \mathscr{G}^{r+s}.$$

Examples. If $\mathscr{D}$ is a zero order operator and if the corresponding map $\Delta: X \to G$ is a submersion on an open set $A \subset X$, then, obviously, $\mathscr{D}$ is infinitesimally invertible over the corresponding $\mathscr{A} = \mathscr{A}^0$ with order and defect zero.

Our next example is more exciting.

Theorem (Nash 1956). *Let $\mathscr{D}$ be the operator relating to C^1-maps $x: V \to \mathbb{R}^q$ the induced forms $\mathscr{D}(x) = g = \{g_{ij}\} = \left\{\left\langle \frac{\partial x}{\partial u_i}, \frac{\partial x}{\partial u_j} \right\rangle\right\}$. Then, over the space of free maps $V \to \mathbb{R}^q$, this operator $\mathscr{D}$ admits an infinitesimal inversion M of defect $d = 2$ and of order $s = 0$.*

Proof. We clearly have

$$L(x, y) = \left\{\left\langle \frac{\partial x}{\partial u_i}, \frac{\partial y}{\partial u_j} \right\rangle + \left\langle \frac{\partial x}{\partial u_j}, \frac{\partial y}{\partial u_i} \right\rangle\right\}, \qquad i, j = 1, \ldots, n = \dim V.$$

(Accidentally, this L is linear in x as well as in y but this is irrelevant to the subject). In order to construct M we must resolve relative to y the following P.D.E. system

$$\text{(L)} \qquad \left\langle \frac{\partial x}{\partial u_i}, \frac{\partial y}{\partial u_j} \right\rangle + \left\langle \frac{\partial x}{\partial u_j}, \frac{\partial y}{\partial u_i} \right\rangle = g_{ij}, \qquad 1 \le i \le j \le n = \dim V.$$

Following Nash we add to (L) the equations

$$\text{(N)} \qquad \left\langle \frac{\partial x}{\partial u_i}, y \right\rangle = 0, \qquad i = 1, \ldots, n.$$

Then we differentiate (N) and get

$$\text{(N}'\text{)} \qquad \left\langle \frac{\partial x}{\partial u_i}, \frac{\partial y}{\partial u_j} \right\rangle + \left\langle \frac{\partial^2 x}{\partial u_i \partial u_j}, y \right\rangle = 0.$$

Finally we alternate i and j and conclude that under condition (N) the system (L) is equivalent to the following system,

$$\text{(N*)} \qquad \left\langle \frac{\partial^2 x}{\partial u_i \partial u_j}, y \right\rangle = -\tfrac{1}{2} g_{ij}.$$

In particular, every solution y of the *algebraic* (in y) joint system (N) + (N*) also satisfies (L).

As x is a free map, the vectors $\frac{\partial x}{\partial u_i}(v)$, and $\frac{\partial x}{\partial u_i \partial u_j}(v) \in \mathbb{R}^q$ are linearly independent for all $v \in V$ and so the system (N) + (N*) is solvable at every point $v \in V$. In order to get a solution y over the whole manifold V we must specify a canonical procedure for picking up one solution $y(v)$ at each point v. [This problem does not appear for $q = n(n + 3)/2$ when the solution $y(v)$ is unique, but only for $q > n(n + 3)/2$.] We make our choice by taking at each point $v \in V$ the solution y that minimizes the norm $\|y\| = \langle y, y\rangle^{1/2}$, and with this minimal $y = y(x,g)$ we put $M(x,g) = y(x,g)$.

Generalization (Green 1970). *For an arbitrary pseudo-Riemannian manifold* (W, h) *the form inducing operator on free maps* $\mathscr{D}_h(x) = x^*(h)$, *admits an infinitesimal inversion of orders* $s = 0$ *and of defect* $d = 2$ *over free maps* $x\colon V \to W$.

Indeed, the calculation above goes through with the covariant derivatives ∇_i^h and ∇_{ij}^h in place of $\partial/\partial u_i$ and $\partial^2/\partial u_i \partial u_j$, and the minimal y may be taken with any auxiliary Riemannian metric.

See 2.3.8 for further examples of linear P.D.E. systems that are solvable by purely algebraic manipulations.

2.3.2 Basic Properties of Infinitesimally Invertible Operators

In the theorem below we always refer to the respective *fine* topologies in the spaces $\mathscr{X}^\alpha$, $\mathscr{G}^\beta$ and $\mathscr{X}^\alpha \times \mathscr{G}^\beta$. For a subset $\mathscr{A} \subset \mathscr{X}^0$ we denote by $\mathscr{A}^\alpha \subset \mathscr{X}^\alpha$, $\alpha = 0, 1, \ldots, \infty$, the intersection $\mathscr{A} \cap \mathscr{X}^\alpha$ again with the induced *fine* C^α-topology. We apply the same rule to subsets in $\mathscr{G}^0$ and in $\mathscr{X}^0 \times \mathscr{G}^0$, that is for $\mathscr{B} \subset \mathscr{X}^0 \times \mathscr{G}^0$ we put $\mathscr{B}^{\alpha,\beta} = \mathscr{B} \cap (\mathscr{X}^\alpha \times \mathscr{G}^\beta)$ and we deal with the *fine* $C^\alpha \times C^\beta$-topology in $\mathscr{B}^{\alpha,\beta}$.

Let $\mathscr{D}\colon \mathscr{X}^{\alpha+r} \to \mathscr{G}^\alpha$ be a differential C^∞-operator of order r and let $\mathscr{D}$ admit, over an open set $\mathscr{A} = \mathscr{A}^d \subset \mathscr{X}^d$, an infinitesimal inversion M of order s and of defect d. Let us fix an integer σ_0 which satisfies the following inequality.

$$(*) \qquad \sigma_0 > \bar{s} = \max(d, 2r + s).$$

Finally, we fix an arbitrary Riemannian metric in the underlying manifold V.

Main Theorem. *There exists a family of sets* $\mathscr{B}_x \subset \mathscr{G}^{\sigma_0+s}$ *for all* $x \in \mathscr{A}^{\sigma_0+r+s}$, *and a family of operators* $\mathscr{D}_x^{-1}\colon \mathscr{B}_x \to \mathscr{A}$ *with the following five properties.*

(1) *Neighborhood Property: Each set* $\mathscr{B}_x$ *contains a neighborhood of zero in the space* $\mathscr{G}^{\sigma_0+s}$. *Furthermore, the union* $\mathscr{B} = \{x\} \times \mathscr{B}_x$ *where* x *runs over* $\mathscr{A}^{\sigma_0+r+s}$, *is an open subset in the space* $\mathscr{A}^{\sigma_0+r+s} \times \mathscr{G}^{\sigma_0+s}$.

(2) *Normalization Property:* $\mathscr{D}_x^{-1}(0) = x$ *for all* $x \in \mathscr{A}^{\sigma_0+r+s}$.

(3) *Inversion Property:* $\mathscr{D} \circ \mathscr{D}_x^{-1} - \mathscr{D}(x) = \mathrm{Id}$, *for all* $x \in \mathscr{A}^{\sigma_0+r+s}$, *that is*

$$\mathscr{D}(\mathscr{D}_x^{-1}(g)) = \mathscr{D}(x) + g,$$

for all pairs $(x, g) \in \mathscr{B}$.

(4) *Regularity and Continuity: If the section $x \in \mathscr{A}$ is C^{η_1+r+s}-smooth and if $g \in \mathscr{B}_x$ is C^{σ_1+s}-smooth for $\sigma_0 \le \sigma_1 \le \eta_1$, then the section $\mathscr{D}_x^{-1}(g)$ is C^σ-smooth for all $\sigma < \sigma_1$. Moreover the operator $\mathscr{D}^{-1}: \mathscr{B}^{\eta_1+r+s,\sigma_1+s} \to \mathscr{A}^\sigma$, $\mathscr{D}^{-1}(x, g) = \mathscr{D}_x^{-1}(g)$, is jointly continuous in the variables x and g. Furthermore, for $\eta_1 > \sigma_1$, the section $\mathscr{D}^{-1}(x, g)$ is C^{σ_1}-smooth and the map $\mathscr{D}^{-1}: \mathscr{B}^{\eta_1+r+s,\sigma_1+s} \to \mathscr{A}^{\sigma_1}$ is continuous.*

(5) *Locality: The value of the section $\mathscr{D}_x^{-1}(g): V \to X$ at any given point $v \in V$ does not depend on the behavior of x and g outside the unit ball $B_v(1)$ in V with center v, and so the equality $(x, g)|B_v(1) = (x', g')|B_v(1)$ implies $(\mathscr{D}_x^{-1}(g))(v) = (\mathscr{D}_{x'}^{-1}(g'))(v)$.*

The proof is given in 2.3.3–2.3.6, where we also consider Hölder spaces $\mathscr{G}^\sigma$ and $\mathscr{X}^\eta$ for all real σ and η.

Corollaries. (A) *Implicit Function Theorem. For every $x_0 \in \mathscr{A}^\infty$ there exists a fine* $C^{\bar{s}+s+1}$-neighborhood of zero in the space $\mathscr{G}^{\bar{s}+s+1}$, $\bar{s} = \max(d, 2r + s)$, say $\mathscr{B}_0 \subset \mathscr{G}^{\bar{s}+s+1}$, *such that for each $C^{\sigma+s}$-section $g \in \mathscr{B}_0$, $\sigma \ge \bar{s} + 1$, the equation $\mathscr{D}(x) = \mathscr{D}(x_0) + g$ has a C^σ-solution.*

Proof. Take $x = \mathscr{D}_{x_0}^{-1}(g)$.

In the case of isometric immersions this is the famous theorem of Nash.

If a metric g_0 on V can be realized by a free C^∞-immersion $x_0: V \to \mathbb{R}^q$ then the C^σ-metric $g_0 + g$ for $\sigma \ge 3$ can be realized by C^σ-immersions for all g that are C^3-small.

(B) *The operator $\mathscr{D}: \mathscr{A}^\infty \to \mathscr{G}^\infty$ is an open map in the respective fine C^r-topologies.*

Indeed, for every $x_0 \in \mathscr{A}^\infty$, the operator $\mathscr{D}$ is invertible near the point $g_0 = \mathscr{D}(x_0) \in \mathscr{G}^\infty$ by the operator $g_0 + g \mapsto \mathscr{D}_{x_0}^{-1}(g)$ that is C^∞-continuous in g.

(C) *Approximation Theorem. If $g \in \mathscr{G}^\infty$, then every solution $x_0 \in \mathscr{A}^{\eta+r+s}$ of the equation $\mathscr{D}(x) = g$ admits, for $\eta > \bar{s}$, a fine C^σ-approximation by C^∞-solutions for all $\sigma < \eta$.*

Proof. First, we $C^{\eta+r+s}$-approximate x_0 by an arbitrary $x \in \mathscr{A}^\infty$. Then the section $g' = g - \mathscr{D}(x)$ is $C^{\eta+s}$-small. Now, by the neighborhood property, the operator $\mathscr{D}_x(g')$ is defined for g' and

$$\mathscr{D}(\mathscr{D}_x^{-1}(g')) = \mathscr{D}(x) + g' = g,$$

so that $x' = \mathscr{D}_x^{-1}(g')$ satisfies $\mathscr{D}(x') = g$. Finally, by the regularity and continuity of $\mathscr{D}^{-1}$ in x and g, the solution x' is C^∞-smooth and is C^σ-close to $x_0 = \mathscr{D}_{x_0}^{-1}(0)$.

Relation $\mathscr{R} = \mathscr{R}(A, \mathscr{D}, g)$ and the Sheaf Φ of Its Solution. Let us fix a C^r-section $g: v \to G$ and call a C^∞-germ $x: \mathcal{O}p(v) \to X$, $v \in V$, *an infinitesimal solution of order α* of the equation $\mathscr{D}(x) = g$, if at the point v the germ $g' = g - \mathscr{D}(x)$ has zero α-jet, $J_{g'}^\alpha(v) = 0$. This property of x depends only on the jet represented by x, called $j = J_x^{r+\alpha}(v)$, and we denote by $\mathscr{R}^\alpha(\mathscr{D}, g) \subset X^{(r+\alpha)}$ the set of all jets that are represented by these infinitesimal solutions over all points $v \in V$. This relation $\mathscr{R}^\alpha(\mathscr{D}, g)$ has exactly the same $C^{r+\alpha}$-solutions as the equation $\mathscr{D}(x) = g$. Now, we recall the open relation $A \subset X^{(d)}$ defining the set $\mathscr{A} \subset \mathscr{X}^d$ and for $\alpha \geq d - r$ we put

$$\mathscr{R}_\alpha = \mathscr{R}_\alpha(A, \mathscr{D}, g) = A^{r+\alpha-d} \cap \mathscr{R}^\alpha(\mathscr{D}, g) \subset X^{(r+\alpha)},$$

where $A^{r+\alpha-d}$ denotes the pull-back $(p_d^{r+\alpha})^{-1}(A)$ for $p_d^{r+\alpha}\colon X^{(r+\alpha)} \to X^{(d)}$. In other words, the relation $\mathscr{R}_\alpha \subset X^{(r+\alpha)}$ corresponds to those infinitesimal solutions of order α of the equation $\mathscr{D}(x) = g$, which also satisfy the relation $A \subset X^{(d)}$. A $C^{r+\alpha}$-section $x\colon V \to X$ satisfies $\mathscr{R}_\alpha$ iff $\mathscr{D}(x) = g$ and $x \in \mathscr{A}$, and so all relations $\mathscr{R}_\alpha$, $\alpha = d - r, \ldots$, have the same C^∞-solutions. We set $\mathscr{R} = \mathscr{R}_{d-r}$ and denote by $\Phi = \Phi(\mathscr{R}) = \Phi(A, \mathscr{D}, g)$ the sheaf of C^∞-solutions of $\mathscr{R}$.

Local Solution of the Equation $\mathscr{D}(x) = g$. Take a jet $j \in \mathscr{R}_\alpha \subset X^{(r+\alpha)}$ over $v = p^{r+\alpha}(j) \in V$ and represent j by a C^∞-germ $x\colon \mathcal{O}p(v) \to X$, such that $j = J_x^{r+\alpha}(v)$.

(D) *If* $\alpha > \bar{s} + s = \max(d + s, 2r + 2s)$, *then there exists a continuous family of* C^∞-*germs*, $x(t)\colon \mathcal{O}p(v) \to X$, $t \in [1, 0]$, *such that* $x(0) = x$, $x(1) \in \Phi(v)$ *and* $J_{x(t)}^{r+\alpha}(v) \in \mathscr{R}_\alpha$ *for all* $t \in [0, 1]$. *In other words, infinitesimal solutions of* $\mathscr{R}_\alpha$ *can be deformed to local solutions.*

Proof. Take a small neighborhood $V_0 \subset V$ of v such that the section x is defined over this V_0 and there satisfies the relation $A \subset X^{(d)}$. Since $j \in \mathscr{R}_\alpha \subset \mathscr{R}^\alpha(\mathscr{D}, g)$, we have $J_{g'}^\alpha(v) = 0$ for $g' = g - \mathscr{D}(x)$, and so we can find another C^∞-section $g_1\colon V_0 \to G$ that is arbitrarily small on V_0 in the fine C^α-topology and such that $g_1 | \mathcal{O}p(v) = g' | \mathcal{O}p(v)$. Now, we apply the main theorem with V_0 in place of V and obtain our family by setting $x(t) = \mathscr{D}^{-1}(x, tg_1)$.

Local w.h. Equivalence. Denote by $\Gamma(\mathscr{R}_\alpha)$ the sheaf of germs of sections $V \to \mathscr{R}_\alpha$ and consider the homomorphism $J\colon \Phi \to \Gamma(\mathscr{R}_\alpha)$, $J(\varphi) = J_\varphi^{r+\alpha}$.

(D′) *If* $\alpha > \bar{s} + s$, *then* J *is a local w.h. equivalence, that is the maps* $J(v)\colon \Phi(v) \to (\Gamma(\mathscr{R}_\alpha))(v)$ *are weak homotopy equivalences for all* $v \in V$.

Proof. We combine the deformation procedure $x \mapsto x(t)$ above with the following general considerations.

Let $\Gamma = \Gamma^\infty(X)$ be the sheaf of C^∞-sections $V \to X$ and consider the associated sheaf Γ^*. There is a natural (jet) homomorphism $\tilde{J} = \tilde{J}^{r+\alpha}$ of Γ^* to the sheaf of section $\Gamma(X^{(r+\alpha)})$. Indeed, sections $\gamma \in \Gamma^*(U)$, $U \subset V$, are continuous families of germs $x_v \in \Gamma(v)$ where v runs over U. One gets $\varphi = \tilde{J}(\gamma)\colon U \to X^{(r+\alpha)}$ by taking $\varphi(v) = J_{x_v}^{r+\alpha}(v)$, $v \in U$. There also exists a homomorphism $I\colon \Gamma(X^{(r+\alpha)}) \to \Gamma^*$ such that $\tilde{J} \circ I = \mathrm{Id}$. In fact, such an I assigns to every jet $j \in X^{(r+\alpha)}$ a specific germ $x = x(j) \in \Gamma$ over $\mathcal{O}p(v)$, $v = p^{(r+\alpha)}(j)$, that represents this $j = J_x^{r+\alpha}(v)$. Furthermore, for each $x \in \Gamma(v)$ the germ $(I \circ J)(x)$ is homotopic to x and the homotopy can be chosen simultaneously for all $x \in \Gamma(v)$, $v \in V$. In particular, for the pull-back $\tilde{\Gamma}(\mathscr{R}_\alpha) = \tilde{J}^{-1}(\Gamma(\mathscr{R}_\alpha)) \subset \Gamma^*$, the map $\tilde{J}\colon \tilde{\Gamma}(\mathscr{R}_\alpha) \to \Gamma(\mathscr{R}_\alpha)$ is a w.h. equivalence.

Now we turn to the sheaf Φ and we also take $\Phi^* \subset \Gamma^*$. This sheaf Φ^* is contained in $\tilde{\Gamma}(\mathscr{R}_\alpha) \subset \Gamma^*$; because sections in Φ^* are families of *local* solutions of $\mathscr{R}_\alpha$, while sections in $\tilde{\Gamma}(\mathscr{R}_\alpha)$ are families of *infinitesimal* solutions. The deformation procedure $x \mapsto x(t)$ applies to families of infinitesimal solutions, as long as these solutions are defined over a fixed open set $V_0 \subset V$. It follows, that $(\tilde{\Gamma}(\mathscr{R}_\alpha))(v)$ can be deformed to $\Phi^*(v) \subset (\tilde{\Gamma}(\mathscr{R}_\alpha))(v)$ for all $v \in V$, and so the inclusion $i: \Phi^* \to \tilde{\Gamma}(\mathscr{R}_\alpha)$ is a local equivalence.

Finally we invoke the local equivalence $\Delta: \Phi \to \Phi^*$ (see 2.2.1; do not confuse with the map $\Delta: X^{(r)} \to G$ where defines $\mathscr{D}$) and observe that the homomorphism $J: \Phi \to \Gamma(\mathscr{R}_\alpha)$ decomposes into three local equivalences, $J = \tilde{J} \circ i \circ \Delta$. Hence J is a local w.h. equivalence.

(D″) **Remark.** The Theorems (D) and (D′) also hold for $\alpha = \bar{s} + s$. To show this we must transform infinitesimal solutions x of order $\alpha = \bar{s} + s$ to such solutions of order $\alpha + 1$.

Suppose that $X \to V$ is a vector bundle and let $M = M(x, g)$ be an infinitesimal inversion of $\mathscr{D}$ of order s (in g). *Then the section* $x_1 = x + M(x, g')$, *for* $g' = g - \mathscr{D}(x)$, *is an infinitesimal solution of order* $\beta = 2(\alpha - r - s) + 1$. *In particular for* $\alpha \geq \bar{s} + s \geq 2r + 2s$ *we get* $\beta \geq \alpha + 1$.

Proof. By Taylor's formula, $f(x + y) = f(x) + f'(x)y + B(x, y)y^2$, one has $\mathscr{D}(x + y) = \mathscr{D}(x) + L(x, y) + B$, where $B = B(x, y, y, y)$ is a differential operator of order r, and B is *bilinear* in the last two arguments. In particular, the identity $J_y^{k+r}(v) = 0$ implies $J_B^{2k+1}(v) = 0$.

Now, as $J_{g'}^{\alpha}(v) = 0$, we have $J_y^{\alpha - s}(v) = 0$ for $y = M(x, g')$. Furthermore, with this y we have $L(x, y) = g - \mathscr{D}(x)$, since $L_x \circ M_x = \mathrm{Id}$. It follows that $\mathscr{D}(x_1) = \mathscr{D}(x + y) = g + B$ with $J_B^{\beta}(v) = 0$.

Exercise. Prove Taylor's formula for $\mathscr{D}(x + y)$ by applying the usual formula to the map $\Delta: X^{(r)} \to G$ and thus completing the argument above.

Microflexibility of Φ. Take a section $x \in \Phi(V)$ and consider a deformation of x over a compact set $K \subset V$. This deformation, $x_t \in \Phi(\mathcal{O}p(K))$, $t \in [0, 1]$, $x_0 = x | \mathcal{O}p(K)$, extends to a C^∞-continuous deformation $\tilde{x}_t$: $V \to X$, $\tilde{x}_0 = x$. Fix an open set $V_0 \supset K$ in $\mathcal{O}p(K)$, such that $\tilde{x}_t | V_0 = x_t | V_0$, $t \in [0, 1]$, and choose a metric in V such that $\mathrm{dist}(K, V \setminus V_0) > 1$. With this metric we take the operator $\mathscr{D}^{-1}$ and apply it to $(\tilde{x}_t, g_t')$, for $g_t' = \mathscr{D}(x) - \mathscr{D}(\tilde{x}_t) = g - \mathscr{D}(\tilde{x}_t)$, for t in a small interval $[0, \varepsilon] \subset [0, 1]$, $\varepsilon > 0$. The resulting new deformation $x_t' = \mathscr{D}^{-1}(\tilde{x}_t, g_t')$ satisfies the equation $\mathscr{D}(x_t') = g$ and by the locality of $\mathscr{D}^{-1}$ we have $x_t' | K = x_t | K$, $t \in [0, \varepsilon]$. Finally we observe that this construction equally applies to the families of sections $x_{p,t}$ for p running over a polyhedron, and to all open sets $U \supset K$ in place of V. Therefore *all restriction maps* $\Phi(U) \to \Phi(K)$ *are microfibrations*

Example. The sheaf of free isometric C^∞-immersions $(V, g) \to (W, h)$ is microflexible, as it was claimed in 1.4.1.

(E) *The Transversality Theorem for Solutions of* $\mathscr{R}_\alpha$. Let $\alpha \geq d - r$ and observe (compare the above) $\mathscr{R}_\alpha$ to be a (locally closed) C^∞-submanifold in $X^{(r+\alpha)}$. Consider a stratified subset $\Sigma \subset \mathscr{R}_\alpha$ with C^∞-strata.

(E′) *The jet* $J_x^{r+\alpha}\colon V \to \mathscr{R}_\alpha$ *of a generic* C^∞*-solution* $x\colon V \to X$ *of* $\mathscr{R}_\alpha$ *is transversal to each stratum of* Σ.

Proof. The above discussion allows enough deformations of (jets of) solutions of $\mathscr{R}_\alpha$ is order to apply the argument of 1.3.2(D).

Example. A generic free isometric C^∞-immersion $x\colon V \to W$ is transversal to any given (and fixed) C^∞-submanifold $W_0 \subset W$.

2.3.3 The Nash (Newton-Moser) Process

Let us reduce the Main Theorem to the case when $X \to V$ is a vector bundle. To do that, we realize X as a C^∞-subfibration in the trivial bundle, $X \subset \tilde{X} = V \times \mathbb{R}^m \to V$ for m large and then we extend $\Delta\colon X^{(r)} \to G$ to a map $\tilde{\Delta}\colon \tilde{X}^{(r)} \to G$, such that in a normal neighborhood $\tilde{U}$ of X in $\tilde{X}$ we have $\tilde{\Delta} = \Delta \circ P$ for the normal fiberwise projection $P\colon \tilde{U} \to X$. Since P is a submersion, the corresponding zero order differential operator $\Gamma(\tilde{U}) \to \mathscr{X}$ is infinitesimally invertible and so the operator $\tilde{\mathscr{D}}\colon \tilde{\mathscr{X}} \to \mathscr{G}$ associated to $\tilde{\Delta}$ is also infinitesimally invertible over $\mathcal{O}p(\mathscr{A}) \subset \tilde{\mathscr{X}}^d \supset \mathscr{X}^d$, and with $\tilde{\mathscr{D}}^{-1}$ for $\tilde{\mathscr{D}}$ we put $\mathscr{D}^{-1} = P \circ \tilde{\mathscr{D}}^{-1}$.

At this point we assume that X is a vector bundle.

Smoothing Operators. We choose and fix a sequence of linear operators $S_i\colon \mathscr{X}^0 \to \mathscr{X}^\infty$, $i = 0, 1, \ldots$, and another such sequence in $\mathscr{G}$ that is also denoted with a stretch of language, $S_i\colon \mathscr{G}^0 \to \mathscr{G}^\infty$. We call them *local smoothing operators* if the following two properties hold.

Locality. Every S_i, $i = 0, 1, \ldots$, does not enlarge supports of sections $\mathscr{X}$ more than by $\varepsilon_i = (2(i+1))^{-2}$, that is, with our fixed metric in V, the value $(S_i(x))(v)$ only depends on this x *within* the ball $B_v(\varepsilon_i)$, for all x in $\mathscr{X}^0$ or in $\mathscr{G}^0$) and for all $v \in V$.

Convergence. If x is C^α-smooth, $\alpha = 0, 1, \ldots$, then $S_i(x) \underset{C^\alpha}{\to} X$ as $i \to \infty$. Moreover, C^α-convergence $x_i \to x$ implies C^α-convergence $S_i(x_i) \to x$.

Remark. The notion of convergence always refers to the usual (not fine) topologies in the respective spaces.

Exercise. Show that the *fine* convergence $x_i \to x$ makes all x_i for $i \geq i_0$ equal to x outside a fixed compact set in V.

Basic Notations. For a finite or infinite sequence $x(i) \in \mathscr{X}^{(r)}$, we write $\bar{x}(i) = S_i(x(i))$ and with the linearization $L = L(\mathscr{D})$ we consider

$$L^*(x(i), y) = L(\bar{x}(i), y) - \int_0^1 L(x(i) + ty, y)\, dt.$$

Then we put $y(i) = x(i) - x(i-1)$, $i = 1, \ldots$, and we take

$$e(i) = L^*(x(i-1), y(i)),$$

$$E(i) = \sum_1^i e(j)$$

and

$$E^*(i) = \bar{e}(i) + (S_i - S_{i-1})E(i-1), \quad \text{for } \bar{e}(i) = S_i(e_i).$$

Observe that

$$\sum_1^i E^*(j) = \sum_1^i S_j(E(j) - E(j-1)) + \sum_1^i (S_j - S_{j-1})E(j-1) = S_i(E(i)),$$

where we assume $E(0) = 0$.

Now, with a given pair $(x_0, g) \in \mathscr{A} \times \mathscr{G}^0$, $\mathscr{A} \subset \mathscr{X}$, we define *Nash's process directed by* (x_0, g) as a finite or infinite sequence $x(0) = x_0$, $x(1)$, $x(2)$, $\ldots$, $x(i)$, $\ldots \in \mathscr{X}^d$ with the following two properties

(A) $\bar{x}(i) \in \mathscr{A}$, $i = 0, \ldots$, and so the infinitesimal inversion M is defined at $\bar{x}(i)$.
(B) $x(i+1) = x(i) + M(\bar{x}(i), g'(i))$, for $g'(i) = (S_{i+1} - S_i)g + E^*(i)$, $i = 0, 1, \ldots$, where we assume $E^*(0) = 0$.

Observe that the condition (B) inductively defines $x(i)$ for all i as long as (A) holds.

"Raison d'etre" of Nash's Process. Suppose that the sections $x(i)$ are defined for *all* $i = 0, 1, \ldots$, and let the sequence $x(i)$ C^r-converge as $i \to \infty$, to $x = x(\infty)$. In this case we *define* $\mathscr{D}^{-1} = \mathscr{D}^{-1}(x_0, g)$ by setting $\mathscr{D}^{-1}(x_0, g) = x(\infty)$.

If the "total error" sequence $E(i)$ C^0-converges as $i \to \infty$, then this operator $\mathscr{D}^{-1}$ satisfies the normalization, the inversion and the locality properties of the main theorem of 2.3.2. (We now use x_0 in place of x of 2.3.2.)

Proof. If $g = 0$, then $(S_{i+1} - S_i)g = 0$ for all $i = 0, 1, \ldots$, and, by induction, we have all $y(i)$ and $E^*(i)$ equal zero, so that $x(i) = x(0) = x_0$ for all $i = 0, 1, \ldots$, and $x(\infty) = x_0$ as well.

In order to prove the inversion property, $\mathscr{D}(x(\infty)) = \mathscr{D}(x_0) + g$, we first recall that $\frac{d}{dt}\mathscr{D}(x + ty) = L(x + ty, y)$, and then we have

$$\begin{aligned}\mathscr{D}(x(i+1)) - \mathscr{D}(x(i)) &= \int_0^1 L(x(i) + ty(i+1), y(i+1))\,dt \\ &= L(\bar{x}(i), y(i+1)) - L^*(x(i), y(i+1)) \\ &= L(\bar{x}(i), M(\bar{x}(i), g'(i)) - e(i+1) \\ &= g'(i) - e(i+1),\end{aligned}$$

and so we conclude

$$\begin{aligned}\mathscr{D}(x(i+1)) - \mathscr{D}(x_0) &= \sum_0^i g'(j) - E(i+1)\\ &= \sum_0^i (S_{j+1} - S_j)g + \sum_1^i E^*(j) - E(i+1)\\ &= S_{i+1}(g) + S_i(E(i)) - E(i+1),\end{aligned}$$

and for $i \to \infty$ we get

$$\mathscr{D}(x(\infty)) - \mathscr{D}(x_0) = g.$$

To prove the locality, we observe that $x(i+1)$ is obtained from $\bar{x}(j)$, $j = 0, 1, \ldots, i$, and g by some differential operators and by the smoothing operators S_{i-1}, S_i and S_{i+1}. It follows, that value $x(i+1)(v)$ is determined by all $\bar{x}(j)$, $j \leq i$, and g *within* the ball $B_v(\varepsilon_{i-1})$, $\varepsilon_{i-1} = (2i)^{-2}$. Since $\sum_1^\infty (2i)^{-2} < \frac{1}{2}$, the values $x(i)(v)$ are determined by $(x_0, g)|B_v(1)$ for all $v \in V$ and for all $i = 0, 1, \ldots$, and so this is also true for $x(\infty) = \mathscr{D}^{-1}(x_0, g)$.

Remark. For $S_i = \mathrm{Id}$, $i = 0, 1, \ldots$, Nash's process reduces to the usual Newton's method of solving non-linear equations. Unfortunately, Newton's process (without smoothing) diverges in most interesting cases. To make it converge one should use appropriate smoothing operators.

2.3.4 Deep Smoothing Operators

For a function $f\colon \mathbb{R}^n \to \mathbb{R}^q$ we denote by $\|f\|_0$ its C^0-norm, $\|f\|_0 = \sup_{v \in \mathbb{R}^n} \|f\|$, and for $\alpha \in (0, 1)$ we denote by $\|f\|_\alpha$ its Hölder C^α-norm,

$$\|f\|_\alpha = \max(\|f\|_0, \sup_{v,w}(\|w\|^{-\alpha}\|f(v+w) - f(v)\|)),$$

where v runs over $\mathbb{R}^n$ and w runs over all non-zero vectors in the unit ball in $\mathbb{R}^n$ around the origin. For an arbitrary $\alpha = j + \theta$, $j = 0, 1, \ldots$, $\theta \in [0, 1)$, we put $\|f\|_\alpha = \|J_f^j\|_\theta$, and if the map f is *not* C^j-smooth we assume $\|f\|_\alpha = \infty$, for all $\alpha \geq j$.

Digression. The existence of operators $f \mapsto Sf$ satisfying the *smoothing estimates* [see (1)–(3) below] needed for Nash's iteration process depends in a crucial way on Taylor's remainder theorem. A more geometric aspect of this theorem is revealed in

Yomdin's Proof of A.P. Morse' Lemma. Fix a large positive constant $c = c(n, r)$ for $n, r = 1, 2, \ldots$, and consider a C^α-function

$$f\colon \mathbb{R}^n \to \mathbb{R} \text{ for } \alpha \in [r, r+1].$$

Cover the unit ball $B \subset \mathbb{R}^n$ by $N = N_\varepsilon \leq c\varepsilon^{-n}$ balls B_ε of radius ε for some small $\varepsilon > 0$ and let $p\colon B_\varepsilon \to \mathbb{R}$ be the Taylor polynomial of f on B_ε of degree $\deg p = r$.

Let

$$A_\varepsilon = \{x \in B_\varepsilon \,|\, \|\text{grad}\, p\| \le c'\varepsilon^{\alpha-1}\}$$

for $c' = c\|f\|_\alpha$ and observe with Taylor's theorem that the *$c'\varepsilon^\alpha$-neighborhood of the image $p(A_\varepsilon) \subset \mathbb{R}$ contains the set of critical values of $f|B_\varepsilon$*. That is the image $f(\Sigma_\varepsilon) \subset \mathbb{R}$ of the zero set $\Sigma_\varepsilon \subset B_\varepsilon$ of grad f on B_ε. Since the set A_ε is semialgebraic and p is a polynomial *there exists a (semi) algebraic curve $C_\varepsilon \subset A_\varepsilon$ of degree $d \le (n \deg p)^n < c$ which meets the level $A_\varepsilon(y) = A_\varepsilon \cap p^{-1}(y)$ for all $y \in \mathbb{R}$.* To construct C_ε take a generic linear function L on $\mathbb{R}^n$, consider the critical set $C_\varepsilon(y)$ of L on (each stratum of the canonical stratification of) $A_\varepsilon(y)$ and let $C = \bigcup_y A_\varepsilon(y)$ over all $y \in \mathbb{R}$ (compare 1.3.2). The length of C_ε is estimated by *Crofton formula*,

$$\text{length } C_\varepsilon \le cd(\text{Diam } C_\varepsilon) \le 2^n c^2 \varepsilon.$$

Since $\|\text{grad}\, p | C_\varepsilon\| \le c'\varepsilon^{\alpha-1}$ and $p(A_\varepsilon) = p(C_\varepsilon)$, the $c'\varepsilon^\alpha$-neighborhood of $p(A_\varepsilon)$ consists of at most cd intervals of length $\le c''\varepsilon^\alpha$ for $c'' = 2^{n+1}c^2c'$, where we assume $\varepsilon \le 1$. *Therefore, the set of critical values of $f|B$ can be covered by at most $c\varepsilon^{-n}$ intervals of length $\le c''\varepsilon^\alpha$ for all $\varepsilon \in (0, 1)$.* In particular, if $\alpha > n$, then *the set of critical values has measure zero* (compare 1.3.2).

The reader interested in further applications of Yomdin's method is referred to his forthcoming book in Astérisque series and to my Bourbaki talk in June 1986. Here are some samples suggested as exercises.

(a) Let the set of critical values of a C^α-function on a compact n-dimensional manifold contain the subset $\{1, 2^{-\beta}, \ldots, i^{-\beta}, \ldots\} \subset \mathbb{R}$ for some $\beta > 0$. Then $\alpha \le n(\beta + 1)$.

(b) Let f be a C^α-smooth map of a compact n-dimensional manifold V into $\mathbb{R}^q$ and let $\Sigma_0 \subset V$ be the zero set of the differential of f. Then the image $f(\Sigma_0) \subset \mathbb{R}^q$ can be covered by at most $c\varepsilon^{-n}$ balls of radius ε^α for some $c = c(f)$ and all $\varepsilon \in (0, 1)$.

(b′) Let $\Sigma_i \subset V$ be the subset where rank $f \le i$. Then $f(\Sigma_i)$ can be covered by at most $c\varepsilon^{-d}$ balls in $\mathbb{R}^q$ of radius ε for $d = i + \alpha^{-1}(n - i)$ and for all positive $\varepsilon \le 1$.

(c) Let p be a polynomial map of the unit ball $B \subset \mathbb{R}^n$ into $\mathbb{R}^q$ and define $\Delta = \Delta_p \colon B \to \mathbb{R}_+$ by $\Delta = (\det DD^*)^{1/2}$, where $D = D_x \colon \mathbb{R}^n \to \mathbb{R}^q$ is the differential of p at $x \in \mathbb{R}^n$ and D^* is the adjoint operator $\mathbb{R}^q \to \mathbb{R}^n$. Consider the subset

$$A_\varepsilon = \{x \in B \,|\, \Delta(x) \le \varepsilon\}$$

and estimate the q-dimensional measure of the image $p(A_\varepsilon) \subset \mathbb{R}^q$ by $\mu[p(A_\varepsilon)] \le c\varepsilon$ for some constant $c = c(n, q, \deg p)$ and all $\varepsilon \in [0, 1]$.

(d) Consider a vector field $X = \sum_{i=1}^n p_i \frac{\partial}{\partial x_i}$ on $\mathbb{R}^n$ where p_i's are polynomials and construct an algebraic hypersurface $H \subset \mathbb{R}^n$ of degree $\le c = c(n, \deg p_i)$ which meets every orbit of X.

(d′) Consider a q-codimensional foliation on $\mathbb{R}^n$ whose tangent bundle is an algebraic subset in $Gr_{n-q}(\mathbb{R}^q)$ and find an algebraic subset in $\mathbb{R}^n$ of dimension q which meets every leave of the foliation.

Now, for the manifold V, we fix a locally finite cover by relatively compact coordinate neighborhoods, $V = \bigcup_\mu U_\mu$, $\mu = 1, 2, \ldots$, and we take a partition of unity $\{P_\mu\}$ inscribed in $\{U_\mu\}$. Then, we fix some trivializations of the bundles G and X over each U_μ and we divide sections f into the sums, $f = \sum_\mu f^\mu$ for $f^\mu = P_\mu f\colon U_\mu = \mathbb{R}^n \to \mathbb{R}^q$. Finally, for a subset K in V, we select all those neighborhoods among U_μ, which intersect K, called $\{U_\nu\} \subset \{U_\mu\}$, and we define Hölder's (semi)norms $\| \ \|_\alpha(K)$ as $\|f\|_\alpha(K) = \sup_\nu \|f^\nu\|_\alpha$.

Exercise 1. Let a sequence f_i, $i = 1, 2, \ldots$, satisfy the following inequalities

$$\|f_i\|_j(K) \le \mathrm{const}_j(i^{-3} + i^{2(j-\alpha)-1}),$$

for two given successive integer values of j, $j = j_1 \ge 0$ and $j = j_2 = j_1 + 1$, for a fixed α in the open interval (j_1, j_2) and for all $i = 1, 2, \ldots$. Show that $f = \sum_1^\infty f_i$ is a section of class C^α on K, [i.e. $\|f\|_\alpha(K) < \infty$], and that

$$\Big\| \sum_{i=i_0}^{\infty} f_i \Big\|_\beta(K) \to 0, \quad \text{as } i_0 \to \infty,$$

for every $\beta < \alpha$.

Hint. Estimate $\|J_f^{j_1}(v + w) - J_f^{j_1}(v)\|$ for $\|w\| = \varepsilon \le 1$, by using the sum $\sum_{i=1}^{i_0} f_i$ with $i_0^2 \le \varepsilon^{-1} < (i_0 + 1)^2$.

We say that a sequence of smoothing operators $S_0, S_1, \ldots$, has (*Nash's*) *depth* $\bar{d}$, if for every compact set $K \subset V$ there are some constants C_α, $\alpha \in [0, \infty)$, uniformly bounded on every finite interval $[0, a] \subset [0, \infty)$, such that all sections f satisfy the following inequalities with the norms $\| \ \|_\alpha = \| \ \|_\alpha(K)$ for all $\alpha \in [0, \infty)$ and for all $i = 1, 2, \ldots$.

Smoothing Estimates

(1) $$\|S_{i-1}(f)\|_\alpha \le C_\alpha i^{2\beta} \|f\|_{\alpha-\beta}, \qquad \text{for } 0 \le \beta \le \alpha.$$

(2) $$\|(S_i - S_{i-1})f\|_\alpha \le C_\alpha (i^{-2\bar{d}-1} + i^{-2\beta-1}) \|f\|_{\alpha+\beta}, \qquad \text{for } -\alpha \le \beta < \infty.$$

(3) $$\|S_{i-1}(f) - f\|_\alpha \le C_\alpha (i^{-2\bar{d}} + i^{-2\beta}) \|f\|_{\alpha+\beta}, \qquad \text{for } \beta \ge 0.$$

These formulae and the exercise above show that smoothing operators of depth $\bar{d} \ge 1$ naturally interpolate the scale C^j to Hölder's C^α for all real $\alpha \ge 0$. However, in our exposition below and in Sect. 4.3.5 Hölder's spaces are not essential for the proof of the main theorem for *integral* σ_0, σ_1 and η_1, and the reader who does not like Hölder's spaces may assume all α and β below to be integers.

To construct "deep smoothing" we start with a C^∞-function $S\colon \mathbb{R}^n \to \mathbb{R}$ with support in the unit ball $B_0(1) \subset \mathbb{R}^n$, and for a continuous function f on $\mathbb{R}^n$ we write

$$S * f = (S * f)(v) = \int_{\mathbb{R}^n} S(w) f(v + w)\, dw, \qquad v, w \in \mathbb{R}^n.$$

Then we take $S_\lambda(v) = \lambda^n S(\lambda v)$ for $\lambda \ge 1$, and observe that $\mathrm{supp}(S_\lambda) \subset B_0(\lambda^{-1})$ and

$$\|S_\lambda * f\|_\alpha \leq \int_{\mathbb{R}^n} |S_\lambda(w)| \, \|f\|_\alpha \, dw.$$

Therefore,

$$\|S_\lambda * f\|_\alpha \leq C\|f\|_\alpha, \quad \text{for all } \alpha \geq 0,$$

(4)
$$C = C(S) = \int_{\mathbb{R}^n} |S_\lambda(w)| \, dw = \int_{\mathbb{R}^n} |S(w)| \, dw$$

Lemma. *If the function S is orthogonal to the polynomials Q on $\mathbb{R}^n$ of degrees* 0, 1, ..., $\bar{d}$, *then*

$$\|S_\lambda * f\|_0 \leq C\lambda^{-\beta}\|f\|_\beta,$$

for all $\lambda \geq 1$ and $\beta \in [0, \bar{d} + 1]$.

Proof. By Taylor's formula $f(v + w) = Q_v(w) + R_v(w)$, where Q is a polinomial in w of degree $\leq \bar{d}$ and where the remainder R satisfies the following estimate for $\|w\| \leq 1$

$$R_v(w) \leq \|w\|^\beta \|f\|_\beta.$$

since $\int SQ = 0$, we have

$$|(S_\lambda * f)(v) = \left| \int_B S_\lambda(w) R_v(w) \, dw \right|, \qquad \text{for } B = B_0(\lambda^{-1}),$$

and so

$$|(S_\lambda * f)(v)| \leq C\|f\|_\beta \sup_{w \in B} |R_v(w)| \leq C2^{-\beta}\|f\|_\beta.$$

Corollary. For $\alpha \in [j, j + 1]$, $j = 0, 1, \ldots,$ and for $\beta \in [-\alpha, \bar{d}]$ one has

(5)
$$\|S_\lambda * f\|_\alpha \leq C_j\lambda^{-\beta}\|f\|_{\alpha+\beta}.$$

Proof. First show that

(5′)
$$\|S_\lambda * f\|_j \leq C_j\lambda^{-\beta}\|f\|_{j+\beta},$$

this time for $\beta \in [-j, \bar{d} + 1]$. To do that we must estimate $\|\partial^k(S_\lambda * f)\|_0$ for all partial derivatives ∂^k of order $k \leq j$. But $\partial^k(S_\lambda * f) = (-\lambda)^k(\partial^k S)_\lambda * f$, and the partial derivatives $\partial^k S$ are orthogonal to all polynomials of degree $\leq k + \bar{d}$, so we obtain (5′) by applying the lemma to $\partial^k S$.

Now, for a real $\alpha \in [j, j + 1]$, we interpolate (5′) to (5) by means of the following elementary relation

(*)
$$\| \ \|_\alpha \leq \text{const}_j(\| \ \|_\mu)^a(\| \ \|_v)^b,$$

that holds for all μ and v such that $j \leq \mu\alpha \leq v \leq j + 1$ and with those a and b, for which $a + b = 1$ and $a\mu + bv = \alpha$.

Remark. *For an arbitrary function S, with no orthogonality conditions, we have*

(6) $$\|S_\lambda * f\|_\alpha \leq C_j \lambda^\beta \|f\|_{\alpha-\beta}$$

for $\alpha \in [j, j+1]$, $j = 0, 1, \ldots$, *and for* $0 \leq \beta \leq \alpha$.

Proof. For $\alpha = j$ the inequality (6) follows from (5′) with $\bar{d} = -1$, and for $\beta = 0$ it follows from (4) for all α. With these two special cases on hand, we get the remaining α and β by the interpolation inequality (∗).

At this point we normalize S by the condition

$$\int_{\mathbb{R}^n} S(v)\,dv = 1.$$

If S is so normalized, then

(7) $$|\partial^j(S_\lambda * f - f)(v)| \leq C \sup_{w \in B} |(\partial^j f)(v + w) - (\partial^j f)(v)|,$$

for $B = B_0(\lambda^{-1})$, *for all* $v \in \mathbb{R}^n$ *and for all partial derivatives* ∂^j. *In particular the operators* $f \mapsto S_\lambda * f$ *converge, as* $\lambda \to \infty$, *in all* C^j*-topologies to the identity operator.* (Compare with the convergence condition of 2.3.3.)

Proof. Since $\partial^j(S_\lambda * f) = S_\lambda * \partial^j f$, we have with $f_j = \partial^j f$ the following identity

$$\partial^j(S_\lambda * f - f)(v) = \int_B S_\lambda(w)(f_j(v + w) - f_j(v))\,dw,$$

and we get (7) with the constant C of (4).

Warning. For C^α-functions with a non-integer α, there may be no C^α-convergence $S_\lambda * f \to f$ for $\lambda \to \infty$. For example, the C^α-function $f(v) = \|v\|^\alpha$, $0 < \alpha < 1$, admits no C^α-approximation by C^∞-functions at all. However, $\|S_\lambda * f - f\|_\beta \to 0$ for all $\beta < \alpha$.

We say that a *normalized function has depth* $\bar{d}$, if it is orthogonal to the *homogeneous* polynomials of degrees $1, \ldots, \bar{d}$. One produces such functions out of any given normalized function S by taking linear combinations

$$S^{\text{new}} = \sum_{k=0}^{\bar{d}} a_k S_{\lambda_k}.$$

The normalization condition for S^{new} amounts to the identity $\sum_0^{\bar{d}} a_k = 1$, and the depth $\bar{d}$ condition is expressed by the equations $\sum_{k=0}^{\bar{d}} a_k \lambda_k^{-p} = 0$ for $p = 1, \ldots, \bar{d}$.

Notice, that the derivative of S_λ in λ, that is $S'_\lambda = \frac{d}{d\lambda} S_\lambda$, can be written as

$$S'_\lambda = \lambda^{-1} \tilde{S}_\lambda, \qquad \text{for } \tilde{S}_\lambda = (\tilde{S})_\lambda,$$

where

$$\tilde{S} = \tilde{S}(v_1, \dots, v_n) = nS + \sum_1^n v_j \frac{\partial S}{\partial v_j},$$

and if S has depth $\bar{d}$, then this $\tilde{S}$ is orthogonal to all polynomials of degrees 0, 1, ..., $\bar{d}$ *(with* deg = 0 *included this time).*

Theorem. *If S has depth $\bar{d}$, then the operators $S_i(f) = S_{\lambda_i} * f$ for $\lambda_i = (i+1)^2$ satisfy the smoothing estimates* (1), (2), *and* (3).

Proof. The estimate (1) follows from (6). To get (2) we write

$$(S_i - S_{i-1})f = \int_{\lambda_{i-1}}^{\lambda_i} (S'_\lambda * f)\, d\lambda = \int_{\lambda_{i-1}}^{\lambda_i} \lambda^{-1} \tilde{S}_\lambda * f\, d\lambda,$$

and by applying (5) to $\tilde{S}$ we get

$$\|(S_i - S_{i-1})f\|_\alpha \le C_\alpha i^{-2\beta-1} \|f\|_{\alpha+\beta} \qquad \text{for } -\alpha \le \beta \le \bar{d},$$

that is equivalent to (2) for $-\alpha \le \beta < \infty$.

The estimate (3) for $\beta = 0$ follows from (4). If α is an integer, and $\beta \le 1$, then (3) follows from (7). Now we can write

$$f = \sum_{i=1}^{\infty} (S_i - S_{i-1})f,$$

and

$$f - S_{i-1}(f) = \sum_{j=i}^{\infty} (S_j - S_{j-1})f,$$

so that (2) implies (3) for all α and $\beta \ge 1$. For the rest of α and β we interpolate according to (*).

Remark. This construction of smoothing equally applies to vector-functions $f\colon \mathbb{R}^n \to \mathbb{R}^q$ for all $q \ge 1$.

Smoothing on V. With the partitions of unity $\{U_\mu, P_\mu\}$ and with some constants $\lambda_\mu \ge 1$ assigned to each $\mu = 1, 2, \dots$, we obtain smooth sections

$$f = \sum_\mu f^\mu, \qquad f^\mu\colon U_\mu = \mathbb{R}^n \to \mathbb{R}^q,$$

by setting $S_i(f) = \sum_\mu S^{\mu,i} * f^\mu$, for a fixed function $S\colon \mathbb{R}^n = U_\mu \to \mathbb{R}$ of depth d and for $S^{\mu,i} = S_\lambda$, where $\lambda = (i+1)^2 \lambda_\mu$.

These operators S_i satisfy the smoothing estimates (1), (2), and (3) and also the convergence condition of 2.3.3, and by choosing the constants λ_μ sufficiently large we satisfy the locality condition of 2.3.3 as well. This is all we need as long as the space $\mathscr{G}$ is concerned, but for the smoothing in $\mathscr{X}$ we choose the parameters λ_μ more carefully in order to meet the requirement $\bar{x}(i) \in \mathscr{A}$ of Nash's process. Namely, instead of numbers λ_μ now we take *functions* $\lambda_\mu = \lambda_\mu(x_0)$ for $x_0 \in \mathscr{A} = \mathscr{A}^d$, such that every λ_μ, $\mu = 1, 2, \ldots$, is continuous in x_0 relative to the usual C^d-topology in $\mathscr{A}$ and such that with these λ_μ we have $S_i(x_0) \in \mathscr{A}$ for all $x_0 \in \mathscr{A}$ and for all $i = 0, 1, \ldots$. Such a smoothing, or rather a family $S_i^{x_0}: \mathscr{X}^0 \to \mathscr{X}^\infty$, for all $x_0 \in \mathscr{A}$, is called *$\mathscr{A}$-stable*. Now the constants C_α in the estimates (1), (2), and (3) depend on x_0, but with $\lambda_\mu(x_0)$ constinuous in x_0, we can choose all $\bar{C}_\alpha(x_0) = \sup_{0 \le \beta \le \alpha} C_\beta(x_0)$ also continuous in x_0. In fact, we only need these contants to be locally bounded in x_0.

Smoothing on Manifolds with Boundaries. If V has a boundary, then some of the neighborhoods U_μ are modeled by $\mathbb{R}^n_+ = \{v_1, \ldots, v_n, v_1 \ge 0\}$ and the smoothing problem on V is reduced to the smoothing of compactly supported functions f on $\mathbb{R}^n_+$. To smooth such an f, we first extend it to a function $\bar{f}$ on $\mathbb{R}^n$ and then we take $S_i(f) = S_i(\bar{f}) | \mathbb{R}^n_+$. The extension operator $f \to \bar{f}$ must be bounded in all norms $\| \ \|_\alpha$ and it is constructed as follows. Write $f = f(v, w)$, where $v = v_1$, $w = (v_2, \ldots, v_n)$, and seek $\bar{f}$ on $\mathbb{R}^n_-$ in the form $\bar{f}(v, w) = \sum_{i=0}^\infty a_i f(-2^i v, w)$, for $v \in (-\infty, 0]$. The conditions $\partial^k \bar{f} = \partial^k f$ for $v_1 = 0$ amount to the following infinite system of equations in the unknowns a_i,

$$(**) \qquad \sum_{i=0}^\infty a_i(-2^i)^k = 1, \qquad k = 0, 1, \ldots,$$

We satisfy (**) by taking an entire function $p(z) = \sum_0^\infty a_k z^k$ such that $p(z = 2^k) = (-1)^k$ for $k = 0, 1, \ldots$, [see Seely (1964) and Ogradska (1967)].

Non-local Smoothing. All our estimates of the "convolution" integrals $S_\lambda * f$ hold for *rapidly decreasing* functions S, possibly with infinite supports. "Rapidly decreasing" means that the products of all partial derivatives of S by polynomials of all degrees are bounded on $\mathbb{R}^n$, that is $\|J_S^j(v)\| \, \|v\|^j \le \text{const}_j$ for all $v \in \mathbb{R}^n$ and $j = 0, 1, \ldots$. In particular, if one takes a rapidly decreasing function $\hat{S}$ with zero partial derivatives at the identity, $(\partial^i \hat{S})(0)$ for all $j > 0$, then the Fourier transform of $\hat{S}$ has infinite depth and the corresponding smoothing operators have no "$i^{-2\bar{d}}$" term in the estimates (2) and (3) (see Nash 1956). Of course such an infinitely deep smoothing is never local in our sense and so it is less convenient for non-compact manifolds.

Analytic Smoothing. An open set $U \subset \mathbb{C}^n = \mathbb{R}^n \oplus \mathbb{R}^n\sqrt{-1}$ is called *obtuse* if with every point $u = x + y\sqrt{-1} \in U$, $x, y \in \mathbb{R}^n$, it contains the cone which has u as the vertex and the Euclidean ball $B_x(\rho) \in \mathbb{R}^n$, $\rho = 2\|y\|$, as the base. We denote by $W^u \subset U \cup \mathbb{R}^n \subset \mathbb{C}^n$ the n-dimensional manifold obtained by joining the complement $\mathbb{R}^n \setminus B_x(\rho)$ with the cone from $u \in U$ over the boundary sphere $\partial B(\rho)$. We restrict the standard holomorphic n-form $du_1 \wedge du_2 \wedge \cdots du_n$ to W^u and call it dw. If a function $f: \mathbb{R}^n \to \mathbb{R}$ extends to a (unique) function f^U on $\mathbb{R}^n \cup U$ which is holomorphic on U, then the integral $\int_{W^u} f^U(w)\,dw$ does not depend on the point u, and so it is equal to the Euclidean integral $\int_{\mathbb{R}^n} f(w)\,dw$.

For a function f as above we denote by $J^j(w)$ either the usual jet at $w = v \in \mathbb{R}^n$ or the "holomorphic" j-th orther jet of f^U at $w = u \in U$, and set $\|f\|_j^U = \sup_w J^j(w)$ for w running over the union $\mathbb{R}^n \cup U$. If there is no analytic continuation of f to U we assume $\|f\|_j^U = \infty$ for all $j = 0, 1, \ldots$.

Now, we consider a linear combination

$$S = \sum_{k=0}^{\bar{d}} a_k \exp(-\|\lambda_k v\|^2), \qquad v \in \mathbb{R}^n,$$

which is made (by an appropriate choice of λ_k and a_k) normalized and deep of depth $\bar{d}$. Every such function S extends to a holomorphic function $S(u)$, $u \in \mathbb{C}^n$, and $S(u - u_0)$ is uniformly rapidly decreasing on the manifold W^{u_0}. That is $\|J_S^j(u - u_0)\| \|u - u_0\|^j \leq \text{const}_j$, for all $u_0 \in \mathbb{C}^n$ and $u \in W^{u_0}$ and for all $j = 0, 1, \ldots,$ where the constants const_j depend on S but not on the point u_0.

Theorem (Whitney 1934; Nash 1966). *For all bounded C^0-functions $f\colon \mathbb{R}^n \to \mathbb{R}$ the "convolutions"*

$$S_\lambda * f = (S_\lambda * f)(v) = \int_{\mathbb{R}^n} S_\lambda(w) f(w + v)\, dw$$

are real analytic functions on $\mathbb{R}^n$ that extend to holomorphic functions on $\mathbb{C}^n$. Furthermore, the operators

$$S_i(f) = S_{\lambda_i} * f, \qquad \text{for } \lambda_i = (i+1)^2,$$

satisfy the estimates (1), (2) *and* (3) *for all integers α and β and for all norms $\| \ \|_\alpha = \| \ \|_\alpha^U$ with the constants C_α, $\alpha = 0, 1, \ldots,$ independent of the obtuse set $U \subset \mathbb{C}^n$.*

Proof. Write

$$(S_\lambda * f)(u) = \int_{\mathbb{R}^n} S_\lambda(w - u) f(w)\, dw, \qquad \text{for } u \in \mathbb{C}^n.$$

These integrals uniformly converge for $\|\text{Im}(u)\| \leq \text{const}$ and so $S_\lambda * f$ is holomorphic.

Now, we prove (1) (2) and (3) exactly as we did it before with only one new ingredient: when estimating our integrals, for example, $(\tilde{S}_\lambda * f)$ at $u \in U$, we employ W^u in place of $\mathbb{R}^n$,

$$(\tilde{S}_\lambda * f)(u) = \int_{W^u} \tilde{S}_\lambda(u - w) f(w)\, dw,$$

and we use Taylor's expansion of f at u. Since the function $\tilde{S}_\lambda$ rapidly decreases on W^u, all the calculations go through.

Exercise 2. Fill in the details for this argument.

Exercise 3. Prove Whiteney's theorem: *Every C^∞-function f on $\mathbb{R}^n$ admits a fine C^∞-approximation by real analytic functions.*

Hint. Take an exhaustion *of* $\mathbb{C}^n$ by relatively compact obtuse sets $\ldots U_k \supset U_{k-1} \ldots$ and, with a given number $j = 0, 1, \ldots,$ and with a function $\varepsilon = \varepsilon(v) > 0$ construct functions f_k such that

$$\|(J^j_{f_k} - J^j_f)(v)\| \leq \left(1 - \frac{1}{k}\right)\varepsilon(v), \qquad \text{for all } v \in U_k \cap \mathbb{R}^n,$$

$$\|f_k\|_j^{U_k} < \infty,$$

$$\|f_k - f_{k+1}\|_j^{U_k} \leq 2^{-k},$$

and take $\lim_{k\to\infty} f_k$ as an ε-approximation to f.

Exercise 4. Generalize Whitney's theorem to C^{an}-fibrations $X \to V$.

Hint. Use C^{an}-embeddings of X and V to $\mathbb{R}^N$.

2.3.5 The Existence and Convergence of Nash's Process

We work with a fixed local smoothing S_i in $\mathscr{G}$ of depth $\bar{d}$ and with an $\mathscr{A}$-stable smoothing $S_i = S_i^{x_0}$, $x_0 \in \mathscr{A}$, in $\mathscr{X}$, also of depth $\bar{d}$. We fix a compact set K in V and we write $\| \ \|_\alpha$ for $\| \ \|_\alpha(K)$.

For a section $x \in \mathscr{A}$ we denote by $R(x)$ the lower bound of those numbers $R > 0$ for which the inequality $\|y\|_d \leq R^{-1}$ implies $x + y \in \mathscr{A}$ for all sections y with support in K. Since the smoothing in $\mathscr{X}$ is $\mathscr{A}$-stable the upper bound $\sup_{0\leq i<\infty} R(S_i^{x_0}(x_0))$ is finite and so for some $R_0 < \infty$ the inequality $\|y\|_d \leq R_0^{-1}$ implies $\sup_{0\leq i<\infty} R(S_i^{x_0}(x_0 + y)) \leq R_0$. The lower bound of these numbers R_0 is denoted by

$$\bar{R}_0 = \bar{R}(x_0) = \bar{R}(x_0, S_i^{x_0}).$$

A section $g \in \mathscr{G}$ is called localized (in K) if the 1-neighborhood of its support is contained in K, that is $\operatorname{dist}(\operatorname{supp}(g), V\backslash K) \geq 1$. If $x(i)$, $i = 0, 1, \ldots, i_0$, is Nash's process directed by (x_0, g) and if the section g is localized, then the sections $y(i)$ have supports in K (see 2.3.3). Hence, the condition $\sum_{i=1}^{i_0} \|y(i)\|_d \leq \bar{R}_0^{-1}$is sufficient for the existence of $y(i_0 + 1) = M(\bar{x}(i_0), g'(i_0))$ (see 2.3.3) and of $x(i_0 + 1) = x(i_0) + y(i_0 + 1)$. Furthermore, since all differential operations (of M) in this expression for $y(i_0 + 1)$ are preceded by smoothing operators that are continuous maps $\mathscr{X}^0 \to \mathscr{X}^\infty$ and $\mathscr{G}^0 \to \mathscr{G}^\infty$, we have (by induction as in 2.3.3) *continuous* dependence of every $y(i) \in \mathscr{X}^\infty$ on the directing pair $(x_0, g) \in \mathscr{A}^d \times \mathscr{G}^0$. In particular, if $\|g\|_0 \to 0$, then $\|y(i_0)\|_\sigma \to 0$ for every $\sigma \in [0, \infty)$ (see 2.3.3) and we come to the following conclusion.

Initial Estimate. *For every* $i_0 = 1, \ldots,$ *there exists a positive number* $\varepsilon = \varepsilon(i_0, \bar{R}_0, \|x_0\|_d)$, *such that the inequality* $\|g\|_0 \leq \varepsilon$ *implies the existence of Nash's process* $x(i)$ *for* $i = 0, 1, \ldots, i_0$. *Furthermore,*

$$\sum_{i=1}^{i_0} \|y(i)\|_\sigma \leq \delta_\sigma(\varepsilon),$$

where $\delta_\sigma(\varepsilon) \underset{\varepsilon\to 0}{\to} 0$ *for every fixed* $\sigma \geq 0$.

Now, for a small g, we want to have an *infinite* Nash process and we proceed with an estimate for $i \to \infty$ as follows. First we recall the numbers d and s, the orders of $M(x, g)$ in x and g, and the order r of the operator L. We assume the depth $\bar{d}$ of the smoothing to be large compared to d, r and s, for example

$$\bar{d} \geq 10(\bar{s} + 1), \qquad \text{for } \bar{s} = \max(d, 2r + s).$$

Then we take four numbers α_0, σ_0, η_0 and v such that

$$(*) \qquad \begin{aligned} \bar{s} + 3v \leq \sigma_0 \leq \alpha_0 \leq \eta_0, \\ 0 < v < 1, \end{aligned}$$

and set

$$Q_0 = Q_0(i, \alpha) = Q_0(i, \alpha - \sigma_0, \bar{d}) = i^{-2\bar{d}} + i^{2(\alpha - \sigma_0) - 1}.$$

Next, with an arbitrary constant $C_0 > 0$ we consider all localized sections g and all $x_0 \in \mathscr{A}$ such that

$$(\mathrm{A}_0) \qquad \begin{aligned} \|x_0\|_{\eta_0 + r + s} \leq C_0 \\ \|g\|_{\sigma_0 + s} \leq C_0. \end{aligned}$$

Finally, we consider all Nash's processes $x(i) = x_0 + \sum_{j=1}^{i} y(i)$ directed by these pairs (x_0, g) and we ask ourselves under what condition the following two inequalities hold.

$$(\bar{\mathrm{B}}) \qquad \sum_{j=1}^{i} \|y(j)\|_d \leq \bar{R}_0^{-1} = \bar{R}^{-1}(x_0),$$

$$(\mathrm{C}_0) \qquad \|y(i)\|_\alpha \leq i^v C_0 Q_0(i, \alpha).$$

Lemma. *There exists a constant $C_0' = C_0'(C_0, \sigma_0, \eta_0, v)$ with the following property. If, for a given number i_0, the inequality $(\bar{\mathrm{B}})$ holds for $i = i_0$ and if (C_0) holds for all $\alpha \leq \alpha_0 + r + s$ and $i = 1, \ldots, i_0$, then the sections $y(i)$ for $i \leq i_0 + 1$ satisfy the inequality (C_0') below.*

$$(\mathrm{C}_0') \qquad \|y(i)\|_\alpha \leq C_0' Q_0(i, \alpha), \qquad \textit{for all } \alpha \leq \eta_0 + r + s.$$

Remark. The constant C_0' also depends on our fixed data, such as the "smoothing" constant C_α of 2.3.4 and on the operators L and M. But what is really important, C_0' does *not* depend on i_0.

Corollary. *There exists a positive number $\varepsilon = \varepsilon(\bar{R}_0, C_0)$, such that the conditions (A_0) together with the additional inequality $\|g\|_0 < \varepsilon$ imply the existence of Nash's process $x(i)$ for all $i = 0, 1, \ldots$, and for this process $x(i)$ the sections $y(i) = x(i) - x(i-1)$ satisfy the inequality (C_0') for all i.*

Proof. Since $v > 0$, for some i_0 we have $C_0' \leq i_0^v C_0$, and, if necessary, by choosing i_0 greater, we also make

$$(**) \qquad \sum_{i=i_0+1}^{\infty} i^v C_0 Q_0(i, \alpha = d) \leq \tfrac{1}{2} \bar{R}_0^{-1}.$$

Then, by the initial estimate, we have for $\|g\|_0 \leq \varepsilon$, with sufficiently small $\varepsilon =$

$\varepsilon(i_0) > 0$, the inequality

$$(\bar{\mathbf{B}}_0) \qquad \sum_{i=1}^{i_0} \|y(i)\|_d \leq \tfrac{1}{2}\bar{R}_0^{-1},$$

as well as the inequality

$$(\mathrm{C}_0\mathrm{i}_0) \qquad \|y(i)\|_\alpha \leq i^\nu C_0 Q_0(i,\alpha) \quad \text{for } i \leq i_0 \text{ and for } \alpha \leq \alpha_0 + r + s.$$

The inequalities (**) and $(\bar{\mathrm{B}}_0)$ show that $(\bar{\mathrm{B}})$ holds as long as (C_0) does, and since $C_0' \leq i^\nu C_0$ for $i \leq i_0$, the inequality (C_0') is stronger, for $i \geq i_0$, than (C_0). Therefore, by the lemma, "(C_0) for i" implies "(C_0) for $i + 1$" for all $i \geq i_0$ and with $(\mathrm{C}_0\mathrm{i}_0)$ we have (C_0) as well as (C_0') for all $i \times 0, 1, \dots,$

Proof of the Lemma. For two related positive numbers, or for two sequences of numbers, such as $\|y(i)\|_\alpha$ and $\|x(i)\|_\alpha$, we say that *the estimate*

$$\|y(i)\|_\alpha \ll Q(i,\alpha)|\alpha \leq \beta_0,$$

implies

$$\|x(i)\|_\alpha \ll Q'(i,\alpha)|\alpha \leq \beta_0',$$

if for every constant C there is another constant $C' = C'(C, Q, Q', \beta_0, \beta_0')$ such that the inequality

$$\|y(i)\|_\alpha \leq CQ(i,\alpha), \qquad \text{for all } \alpha \leq \beta_0,$$

implies the inequality

$$\|x(i)\|_\alpha \leq C'Q'(i,\alpha), \qquad \text{for all } \alpha \leq \beta_0',$$

for *all* sequences $y(i)$ and $x(i)$ in question, regardless of their length.

Example. We write

$$(0) \qquad \|x_0\|_{\eta_0+r+s} \ll 1,$$

$$(00) \qquad \|g\|_{\sigma_0+s} \ll 1,$$

$$(1) \qquad \|y(i)\|_\alpha \ll i^\nu Q_0(i,\alpha)|\alpha \leq \alpha_0 + r + s,$$

where $Q_0(i,\alpha) = i^{-2\bar{d}} + i^{2(\alpha-\sigma_0)-1}$, and

$$(1') \qquad \|y(i)\|_\alpha \ll Q_0(i,\alpha)|\alpha \leq \eta_0 + r + s,$$

and we now express the lemma by saying that (0), (00), (1) and the inequality $(\bar{\mathrm{B}})$ imply (1′).

We also write $\alpha \ll 1$ in place of $\alpha \leq \beta_0$ if a relevant estimate holds for all α in any given interval $[0, \beta_0]$.

Example. Since the constants C_α of the smoothing estimate (2) of 2.3.4 are bounded on finite intervals by $C_\alpha \ll 1$ for $\alpha \ll 1$, the estimate (00) implies the following estimate for $\tilde{g}(i) = (S_{i+1} - S_i)g$,

$(\tilde{0}\tilde{0})$ $$\|\tilde{g}(i)\|_\alpha \ll i^{-2\bar{d}-1} + i^{2(\alpha-\sigma_0-s)-1}|\alpha \ll 1.$$

Our next estimate concerns $x(i) = x_0 + \sum_{j=1}^{i} y(j)$. Since $v > 0$ and $d \geq 1$, the estimates (0) and (1) imply

(2) $$\|x(i)\|_\alpha \ll 1 + i^{2(\alpha-\sigma_0+v)}|\alpha \leq \alpha_0 + r + s$$

This estimate (2) implies [via (1) of 2.3.4] the following estimate for $\bar{x}(i) = S_i(x(i))$,

$(\bar{2})$ $$\|\bar{x}(i)\|_\alpha \ll 1 + i^{2(\alpha-\sigma_0+v)}|\alpha \leq \eta_0 + r + s + d,$$

(in fact, we could write $\alpha \ll 1$), and with (3) of 2.3.4 we have for $\hat{x}(i) = \bar{x}(i) - x(i)$,

$(\hat{2})$ $$\|\hat{x}(i)\|_\alpha \ll i^{-2\bar{d}} + i^{2(\alpha-\sigma_0+v)}|\alpha \leq \alpha_0 + r + s.$$

To proceed further we need some general facts concerning the differential operators L and M. First, for two functions in $\alpha \geq 0$, say for $Q(\alpha)$ and $Q'(\alpha)$, we write

$$Q(\alpha) * Q'(\alpha) = \sup_{0 \leq \beta \leq \alpha} Q(\beta)Q(\alpha - \beta).$$

Then we observe the following *Leibniz' inequality*

(3) $$\|JJ'\|_\alpha \ll \|J\|_\alpha * \|J'\|_\alpha,$$

where J and J' are two vector functions $V \to \mathbb{R}^N$, and JJ' may denote any given bilinear form in J and J'.

To prove (3) we write $\alpha = j + \theta$, for $j = 0, 1, \ldots$, and for $0 < \theta < 1$. Then the ordinary product of two real functions satisfies

$$\|JJ'\|_\theta \leq \|J\|_\theta \|J'\|_0 + \|J\|_0 \|J'\|_\theta.$$

Then we obtain (3) with the inequality

$$\|JJ'\|_\alpha \leq C_j \sup_{k+l \leq j} \|(\partial^k J)(\partial^l J')\|_\theta.$$

The next fact we need is the

Chain Rule Estimate. Consider a C^∞-function F on some domain $A \subset \mathbb{R}^N$, and a map $J\colon V \to A$. We denote by $R(J)$ the C^0-norm of the function $\delta^{-1} = \delta^{-1}(v)$ for $\delta(v) = \operatorname{dist}(J(v), \mathbb{R}^N \backslash A)$, that is $R(J) = \|\delta^{-1}\|_0$. Notice, that $R = 0$ for $A = \mathbb{R}^n$.

If $\|J\|_0 \ll 1$ *and* $R(J) \ll 1$, *then*

(4) $$\|F(J)\|_\alpha \ll 1 + \underbrace{\|J\|_\alpha * \cdots * \|J\|_\alpha}_{j+1},$$

where $j = \operatorname{ent}(\alpha)$.

Proof. Our assumptions imply that all partial derivatives $F_\mu(a) = (\partial^\mu F)(a)$ satisfy $\|F_\mu(a)\|_0 \ll 1$. Then we expand the partial derivatives $\partial^l(F(J))$ into sums of products like $\Pi_l = \partial^{l_1}(J) \ldots \partial^{l_k}(J) F_\mu(J)$, for $l \leq j$ and $l_1 + \cdots + l_k \leq j$. Since the norm $\|F(J)\|_\alpha$ for $\alpha = j + \theta$ is estimated by $\|\Pi_l\|_\theta$ for all products Π_l, $l \leq j$, we get (4) by using (3) and the following obvious inequality for $0 \leq \theta \leq 1$,

$$\|F_\mu(J)\|_\theta \leq \|J\|_\theta \|F_{\mu+1}(J)\|_0 + \|F_\mu(J)\|_0.$$

Example. *If*

$$\|J\|_\alpha \ll Q(i,\alpha) = 1 + i^{2\alpha-\gamma}|\alpha \leq \beta_0,$$

then for $\gamma \geq 0$ *we have the identical estimate for* $F(J)$,

(4*) $$\|F(J)\|_\alpha \ll Q(i,\alpha)|\alpha \leq \beta_0.$$

Indeed $Q(i,\alpha) * Q(i,\alpha) \ll 1 + i^{2\alpha-\gamma'}$, for $\gamma' = \min(\gamma, 2\gamma)$.

The Operator M. Finally we turn to differential operators [such as $M(x,g)$] linear in the second argument. With the jets $J = J_x^d$ and $J' = J_g^s$, we write $M(x,g) = F(J)J'$ for the usual local coordinate expression

$$M = \sum_k F^k(J)\partial^{l_k} g,$$

and since $\|J\|_\alpha = \|x\|_{\alpha+d}$ and $\|J'\|_\alpha = \|g\|_{\alpha+s}$, we have, under the assumptions used for (4) and (4*),

(5) $$\|M(x,g)\|_\alpha \ll (1 + \|x\|_{\alpha+d}) * \|g\|_{\alpha+s}.$$

The Operator L'. Let

$$L'(x,u,y) = \frac{d}{dt}L(x + tu, y)|t = 0.$$

Then

(–) $$L(x + u, y) - L(x,y) = \int_0^1 L'(x + tu, u, y)\,dt.$$

Observe, that L' is a differential operator of order r and it is bilinear in u and y, and so we write $L' = F'(J)\hat{J}\tilde{J}$, for $J = J_x^r$, $\hat{J} = J_u^r$ and $\tilde{J} = J_y^r$.

This operator L' is related to the operator L^* of 2.3.3 as follows. Put

$$L = L(x(i-1), y(i)),$$
$$L_t' = L'(x(i-1) + t\hat{x}(i-1), \hat{x}(i-1), y(i)),$$

and

$$L_{t,\tau}' = L'(x(i-1) + t\tau y(i), \tau y(i), y(i)).$$

Then with $\hat{x}(i-1) = \bar{x}(i-1) - x(i-1)$ we have, according to (–) above and to the formulae of 2.3.3,

$$e(i) = L^*(x(i-1), y(i)) = \int_0^1\int_0^1 (L_t' - L_{t\tau}')\,dt\,d\tau.$$

Let u denote $t\hat{x}(i-1)$ or $t\tau y(i)$ for some t and τ in the interval $[0,1]$. Then by (1) and $(\hat{2})$ we have (keeping in mind $\bar{d} \geq 1 \geq \nu$),

$$\|\hat{J} = J_u^r\|_\alpha \ll i^{-\bar{d}} + i^{2(\alpha-\sigma_0+r+\nu)}|\alpha \leq \alpha_0 + s,$$

and for $\tilde{J} = J_y^r$, $y = y(i)$ we have

$$\|\tilde{J}\|_\alpha \le i^{-\bar{d}} + i^{2(\alpha-\sigma_0+r)+v-1}|\alpha \le \alpha_0 + s.$$

These estimates imply

$$\|\hat{J}\tilde{J}\|_\alpha \ll \|\hat{J}\|_\alpha * \|\tilde{J}\|_\alpha \ll i^{-2\bar{d}} + i^{2\alpha-\gamma}|\alpha \le \alpha_0 + s,$$

where γ is the minimum of the following three numbers

$$\bar{d} + 2(\sigma_0 - r - v),$$

$$\bar{d} + 2(\sigma_0 - r) - v + 1$$

and

$$2(2\sigma_0 - 2r - v) - v + 1.$$

Next, with (2), (4*) and with the estimate for $\hat{J}$ above we come to the following estimates for $J = J^r_{x(i-1)} + \hat{J}$ and for $F'(J)$,

$$\|F'(J)\|_\alpha \ll \|J\|_\alpha \ll 1 + i^{2(\alpha-\sigma_0+r+v)}|\alpha \le \alpha_0 + s,$$

and since $\|e(i)\|_\alpha \ll \|F'(J)\hat{J}\tilde{J}\|_\alpha \ll \|J\|_\alpha * \|\hat{J}\tilde{J}\|_\alpha$, we have

$$\|e(i)\|_\alpha \ll i^{-2\bar{d}} + i^{2\alpha-\gamma'}|\alpha \le \alpha_0 + s,$$

where $\gamma' = \min(\gamma, \gamma + 2(\sigma_0 - r - v), 2(d + \sigma_0 - r - v))$.

Finally, we recall that $\sigma_0 \ge 2r + s + 3v$, and that $\bar{d} \ge 2(\bar{s} + v + 1)$ [see (*)], and get $\gamma' = \gamma > 2(\sigma_0 + s + v) + 1$; hence,

(6) $$\|e(i)\|_\alpha \ll i^{-2\bar{d}} + i^{2(\alpha-\sigma_0-s-v)-1}|\alpha \le \alpha_0 + s.$$

Estimates for $E^*(i)$ and for $g'(i)$. The estimate (6) implies

$$\|E(i-1) = \sum_1^{i-1} e(j)\|_\alpha \ll 1 + i^{2(\alpha-\sigma_0-s)-v}|\alpha \le \alpha_0 + s,$$

and with (1) and (2) of 2.3.4 we estimate $\bar{e}(i) = S_i(e(i))$ and $\tilde{E}(i-1) = (S_i - S_{i-1})E(i-1)$ as follows

$$\|\bar{e}(i)\|_\alpha \ll i^{-2d} + i^{2(\alpha-\sigma_0-s-v')-1}|\alpha \ll 1,$$

where $v' = \min(v, \alpha_0 - \sigma_0) \ge 0$, and

$$\|\tilde{E}(i-1)\|_\alpha \ll i^{-2\bar{d}-1} + i^{2(\alpha-\sigma_0-s)-v'-1}|\alpha \ll 1.$$

Finally we get

(7*) $$\|E^*(i) = \bar{e}(i) + \tilde{E}(i-1)\|_\alpha \ll i^{-2\bar{d}} + i^{2(\alpha-\sigma_0-s)-v'-1}|\alpha \ll 1,$$

where $v' = \min(v, \alpha_0 - \sigma_0) \ge 0$, and with $(\tilde{0}\tilde{0})$ we conclude

(7′) $$\|g'(i) = \tilde{g}(i) + E^*(i)\|_\alpha \ll i^{-2\bar{d}} + i^{2(\alpha-\sigma_0-s)-1}|\alpha \ll 1,$$

Estimate for $y(i+1) = M(\bar{x}(i), g'(i))$. We came full circle. The inequality $(\bar{B})$ allows us to use (4) and with (5) we derive from $(\bar{2})$ and (7′), the required estimate (1′),

$$\|y(i+1)\|_\alpha \ll \|\bar{x}(i)\|^*_{\alpha+d}\|g'(i)\|_{\alpha+s} \ll i^{-2\bar{d}} + i^{2\alpha-\sigma}|\alpha \le \eta_0 + r + s,$$

where $\sigma = \min(2\sigma_0 + 1, 4\sigma_0 - 2d - 2v + 1, 2(\bar{d} - d + \sigma_0 - v)) = 2\sigma_0 + 1$.

Proof of the Main Theoren of 2.3.2. Since Nash's process is local (see 2.3.3), its behavior at a given point $v \in V$ does not change if we localize g in the ball around v of radius $\rho \ge 2$ by taking Pg in place of g, where P is a C^∞-function with support in K and $P|B_1(v) \equiv 1$. Now, we can use the corollary of our lemma to obtain infinite Nash's processes $x(i)$ for all directions (x_0, g) provided $x_0 \in \mathscr{A}^{\sigma_0+r+s}$ and g is C^{σ_0+s}-small. This amounts to the neighborhood property for $\mathscr{D}^{-1}(x_0, g) = x(\infty)$, since the limit $x(\infty)$ does exist according to the estimate (1′). Furthermore, according to (6) the "total error" sequence $E(i)$ also converges, and so the inversion property holds as well (see 2.3.3).

Regularity and Continuity of $\mathscr{D}^{-1} = x(\infty)$. In addition, let us assume that x_0 is C^{η_1+r+s}-smooth and g is C^{σ_1+s}-smooth for $\sigma_0 \le \sigma_1 \le \eta_1$. Then, starting with the estimate (1′), namely,

$$\|y(i)\|_\alpha \ll Q_0(i, \bar{d}, \alpha - \sigma_0)|\alpha \le \eta_0 + r + s,$$

and by applying the lemma several times with σ_1 and η_1 instead of σ_0 and η_0, (i.e. with the estimates $\|x_0\|_{\eta_1+r+s} \ll 1$, and $\|g\|_{\sigma_1+s} \ll 1$ in place of (0) and (00) we come up with the following stronger estimate for $\|y(i)\|_\alpha$,

$$(1_1) \qquad \|y(i)\|_\alpha \ll Q_0(\alpha - \sigma_1) = i^{-2\bar{d}} + i^{2(\alpha-\sigma_1)-1}|\alpha \le \eta_1 + r + s.$$

With this (1_1) we get $\|x(i) - x(\infty)\|_\beta \to 0_{i\to\infty}$, for all $\beta < \sigma_1$, and for a non-integer σ_1 we also conclude that $x(\infty)$ is C^{σ_1}-smooth, (see Exercise 1 in 2.3.4). This settles the regularity of $\mathscr{D}^{-1}$ for these σ_1. Notice also, that the continuity of $\mathscr{D}^{-1}\colon \mathscr{B}^{\eta_1+r+s,\sigma_1+s} \to \mathscr{X}^\beta$, $\beta < \sigma_1$, is immediate from the continuity of every $x(i)$ in x_0 and g (see the initial continuity discussion at the beginning of this section), and from the uniformity of our estimates for $\|y(i)\|_\alpha$, and so for $\|x(i) - x(\infty)\|_\alpha$ as $i \to \infty$.

Now, let us assume σ_1 to be an integer and let us prove C^{σ_1}-convergence $x(i) \to x(\infty)$ for $\eta_1 > \sigma_1$. To do that, we rewrite $y(i+1) = x(i+1) - x(i)$ as

$$y(i+1) = M(\bar{x}(i), \tilde{g}(i) + E^*(i)) = M(\bar{x}(i), \tilde{g}(i)) + M(\bar{x}(i), E^*(i)),$$

and observe that (1_1) implies the following estimate (7_1^*) in place of (7^*) above,

$$(7_1^*) \qquad \|E^*(i)\|_\alpha \ll i^{-2\bar{d}} + i^{2(\alpha-\sigma_1-s)-1-v'}|\alpha \ll 1,$$

where v' is a *positive* number due to the inequality $\eta_1 > \sigma_1$. Furthermore, (1_1) implies

$$(2_1) \qquad \|x(i)\|_\alpha \ll 1 + i^{2(\alpha-\sigma_1+v)}|\alpha \le \eta_1 + r + s,$$

$$(\bar{2}_1) \qquad \|\bar{x}(i)\|_\alpha \ll 1 + i^{2(\alpha-\sigma_1+v)}|\alpha \le \eta_1 + r + s + d,$$

$$(\hat{2}_1) \qquad \|\hat{x}(i)\|_\alpha \ll i^{-2\bar{d}} + i^{2(\alpha-\sigma_1+v)}|\alpha \le \eta_1 + r + s,$$

for $\hat{x} = \bar{x} - x$, and from (1_1), and (2_1) we derive

$$\|\bar{x}(i) - x(\infty)\|_\alpha \ll i^{-2\bar{d}+1} + i^{2(\alpha-\sigma_1+\nu)} | \alpha \leq \eta_1 + r + s.$$

Observe, that these estimates hold for an arbitrary fixed $\nu > 0$.

Since $\|M(\bar{x}(i), E^*(i))\|_\alpha \ll \|\bar{x}(i)\|_{\alpha+d} * \|E^*(i)\|_{\alpha+s}$, we conclude as before that

$$\|M(\bar{x}(i), E^*(i))\|_\alpha \ll i^{-2\bar{d}} + i^{2(\alpha-\sigma_1)-1-\nu'} | \alpha \leq \eta_1 + r + s,$$

and with $\nu' > 0$ the sum $\sum_{i=1}^\infty M(\bar{x}(i), E^*(i))$ C^{σ_1}-converges. Thus, the C^{σ_1}-convergence $x(i) \underset{i=1}{\to} x(\infty)$ is reduced to C^{σ_1}-convergence of the sum $\sum_{i=1}^\infty M(\bar{x}(i), \tilde{g}(i))$, where $\tilde{g}(i) = (S_{i+1} - S_i)g$ and, since $\|g\|_{\sigma_1+s} \ll 1$, we have

$$(\sim) \qquad \|\tilde{g}(i)\|_\alpha \ll i^{-2\bar{d}-1} + i^{2(\alpha-\sigma_1-s)-1} | \alpha \ll 1.$$

Let us show that the estimates $(\bar{2}_1)$, $(\hat{2}_1)$ and $(\sim)$ imply the C^{σ_1}-convergence of the sum $\sum_{i=1}^\infty M(\bar{x}(i), \tilde{g}(i))$. We do this by proving C^0-convergence of the derivatives $\sum_i \partial^{\sigma_1} M$ as follows. We set $\bar{J}(i) = J^d_{\bar{x}(i)}$ and $\tilde{J}^l(i) = J^{s+l}_{\tilde{g}(i)}$. Then $M(\bar{x}(i), \tilde{g}(i)) = F(\bar{J}(i))\tilde{J}^0(i)$ and the derivatives $\partial^{\sigma_1} M$ expand into the sums of the following products $\Pi_{k,l}(i) = \partial^k F(\bar{J}(i))\tilde{J}^l(i)$, $k + l = \sigma_1$. Since $\bar{d}$ is large, we have

$$\|\Pi_{k,l}(i)\|_0 \ll \|\bar{x}(i)\|_{k+d} \|\tilde{g}(i)\|_{s+l}$$
$$\ll i^{-2\bar{d}-1} + i^{2(l-\sigma_1)-1} + i^{2(d+\nu-\bar{d})-1} + i^{2(d-\sigma_1+\nu)-1}$$

and with $\nu < 1$, we get the C^0-convergence of all sums $\sum_{i=1}^\infty \Pi_{k,l}(i)$ for $k \geq 1$ and $l < \sigma_1$. Let us concentrate on the remaining sum $\sum_{i=0}^\infty \Pi_{0,\sigma_1}(i) = \sum_{i=1}^\infty F(\bar{J}(i))\tilde{J}^{\sigma_1}(i)$. Set $J(\infty) = J^d_{x(\infty)}$ and $\hat{J}(i) = \bar{J}(i) - J(\infty)$. Then we have,

$$F(\bar{J}(i)) = F(J(\infty)) + \int_0^1 \frac{d}{dt} F(J(\infty) + t\hat{J}(i))\, dt.$$

The last derivative (compare with L') can be written as $F'(J(\infty) + t\hat{J}(i))\hat{J}(i)$, and since $\|x(\infty)\|_{\sigma_1-\nu} \ll 1$, we derive from $(\hat{2}_1)$ the following estimate for our integral

$$\int_0^1 = \int_0^1 \frac{d}{dt} F = \int_0^1 F'\hat{J}(i),$$

$$\left\| \int_0^1 \right\|_0 \ll i^{-2d+1} + i^{2(d-\sigma_1+\nu)},$$

and so the sum $\sum_{i=1}^\infty \|(\int_0^1)\tilde{J}^{\sigma_1}(i)\|_0$ converges.

Finally we are left with the sum

$$\sum_{i=0}^\infty F(J(\infty)) \cdot \tilde{J}^{\sigma_1}(i) = F(J(\infty)) \sum_{i=0}^\infty \tilde{J}^{\sigma_1}(i),$$

for $\tilde{J}^{\sigma_1}(i) = J^{\sigma_1+s}_{\tilde{g}(i)}$, $\tilde{g}(i) = (S_{i+1} - S_i)g$. Since g is C^{σ_1+s}-smooth the sums $\sum_{i=1}^k \tilde{g}(i) = S_{k+1}(g)$ C^{σ_1+s}-converge to g as $k \to \infty$ [see (7) of 2.3.4] and so the sums $\sum_{i=1}^k \tilde{J}^{\sigma_1}(i)$ converge in C^0-topology. Q.E.D.

With this proof of the regularity we automatically have the required continuity of the inversion

$$\mathscr{D}^{-1}\colon \mathscr{B}^{\eta_1+r+s,\sigma_1+s} \to \mathscr{X}^{\sigma_1},$$

for all $\sigma_1 = \bar{s} + 1, s + 2, \ldots$, and for all real $\eta_1 > \sigma_1$.

Exercise. Prove C^{σ_1}-continuity (but not the C^{σ_1}-convergence) of $\mathscr{D}^{-1}$ for all real $\sigma_1 < \eta_1$.

Hint. Apply the difference operators ∂_w of the form $(\partial_w f)(v) = f(v + w) - f(v)$, to the derivatives $\partial^j M$ for $j = \operatorname{ent}(\sigma_1)$.

2.3.6 The Modified Nash Process and Special Inversions of the Operator $\mathscr{D}$

We work in this section with a fixed smoothing in $\mathscr{X}$, rather than with an $\mathscr{A}$-stable family as in 2.3.5. We modify Nash's process of 2.3.3 by replacing $\bar{x}(i)$ in L^* and in M by $x_0 + \bar{z}(i)$, for $z(i) = \sum_{j=1}^{i} y(j)$ and $\bar{z}(i) = S_i(z(i))$. This modification does not effect the basic properties of $x(i) = x_0 + z(i)$ and the limit $x(\infty)$, still serves as $\mathscr{D}^{-1}(x_0, g)$. Furthermore, the condition $x_0 + \bar{z}(i) \in \mathscr{A}$ guarantying the existence of our new

$$y(i + 1) = M(x_0 + \bar{z}(i), g'(i)),$$

can be numerically expressed with $R(x)$ (see the beginning of 2.3.5) by the inequality $\|\bar{z}(i)\|_d < R^{-1}(x_0)$, and so the modified Nash process exists as long as this inequality holds. Moreover, now we are able to give an explicit bound for $\|g\|_{\sigma_1 + s}$ needed for the existence of the modified process $x(i)$. This estimate depends on the data of the following two types.

(1) *Fixed Data.* First, we fix a partition of unity on V and a compact $K \subset V$, so that we have specific norms $\| \;\; \|_\alpha$ in the spaces $\mathscr{X}$ and $\mathscr{G}$. Then, we have the smoothing operators S_i in $\mathscr{X}$ and in $\mathscr{G}$ of depth $\bar{d}$, where

$$\bar{d} \geq 10(\bar{s} + 1), \qquad \bar{s} = \max(d, 2r + s),$$

and we denote by $C(S)$ the upper bound of the smoothing constants C_α, for $0 \leq \alpha \leq \bar{d}$, of the inequalities (1), (2) and (3) of 2.3.4, that is

$$C(S) = \sup_{0 \leq \alpha \leq \bar{d}} C_\alpha.$$

Finally, we fix a number $\sigma_0 > \bar{s}$, we additionally assume $\sigma_0 \leq \bar{s} + 1$ and we take $v = \frac{1}{4}(\sigma_0 - \bar{s})$.

(2) *Variable Data.* We characterize sections $x \in \mathscr{A}$ by the following quantity

$$N(x) = \|x\|_{\beta_0} + 2R(x) + 1, \qquad \text{for } \beta_0 = \sigma_0 + r + s + d.$$

Then we need specific numerical bounds for the operators $L'_x = F'(J^r_x)\hat{J}J$ and $M_x = F(J^d_x)J'$ (see 2.3.5). In order to avoid cumbersome notations, we assume that $V = \mathbb{R}^n$ and that the bundles X and G are trivial. In this case our F' and F are actual Euclidean functions. Indeed, we have $X^{(r)} = \mathbb{R}^k$ for some k and F' maps this $\mathbb{R}^k = X^{(r)}$ into another Euclidean space, namely to the total space of the bundle $\operatorname{Hom}(X^{(r)} \oplus X^{(r)}, G) \to V$, and F maps the open set $A \subset X^{(d)} = \mathbb{R}^b$ to the space $\operatorname{Hom}(G^{(s)}, X) = \mathbb{R}^c$.

Now, the norms of the partial derivatives of F and F', in the estimates of 2.3.5, have specific values.

We denote by $\|F_j(x)\|_0$ the C^0-norm of the function $\|J^j_F(J^d_x(v))\|$, $v \in V$, for the j-th order jet J^j_F, and we define $\|F'_j(x)\|_0$ as the C^0-norm of $\|J^j_{F'}(J^r_x(v))\|$. Then with a given section $x \in \mathscr{A}$ we consider all sections $y \in \mathscr{X}^d$ such that $\|y\|_d \leq \min(1, \frac{1}{2}R^{-1}(x))$ and set $\|M\|_j(x) = \sup_y \|F_j(x+y)\|_0$. We also define $\|L'\|_j(x) = \sup_{\|y\|_r \leq 1} \|F_j'(x+y)\|_0$.

Let us give an explicit estimate of $y(i+1)$ in terms of $y(j)$ for $j \leq i$, and of the "norms" $N(x_0)$, $\|L'\|(x_0)$, $\|M\|(x_0)$ and $\|g\|_{\sigma+s}$. We agree to write "const" for positive constants that depend only on our fixed data.

First, we recall $Q_0 = i^{-2\bar{d}} + i^{2(\alpha-\sigma_0)-1}$, and we take a small positive number $\delta > 0$, such that

$$(*) \qquad 2C(S)\delta \sum_{i=1}^{\infty} Q_0(i, \alpha = d) \leq \min(1, \tfrac{1}{2}R^{-1}(x_0)).$$

This amounts to an inequality

$$\delta \leq \text{const}(R(x_0) + 1)^{-1}.$$

Then we consider the modified Nash process $x(i) = x_0 + z(i) = x_0 + \sum_{j=1}^{i} y(j)$ directed by (x_0, g) with a localized section g, and let

$$\|y(i)\|_\alpha \leq \delta Q_0(i, \alpha) | \alpha \leq \sigma_0 + r + s, \qquad \text{for } i = 1, 2, \ldots, i_0.$$

We estimate $z(i)$ and $\hat{z}(i) = \bar{z}(i) - z(i)$ as in 2.3.5 and *now* we denote by $u(i)$ either $t\hat{z}(i-1)$ or $t\tau y(i)$ for $0 \leq t, \tau \leq 1$. Since $r \leq d$, we have, according to $(*)$ above, $\|z(i-1)\|_r + \|u(i)\|_r \leq 1$ and so

$$\|F_\beta'(x(i-1) + u(i) = x_0 + z(i-1) + u(i))\|_0 \leq \|L'\|_\beta(x_0).$$

We also have [compare with $(\hat{2})$ of 2.3.5]

$$\|u(i)\|_\alpha \leq \text{const}\, \delta \hat{Q}(i, \alpha) | \alpha \leq \sigma_0 + r + s, \quad \text{for } \hat{Q} = i^{-2\bar{d}} + i^{2(\alpha-\sigma_0+\nu)}.$$

With the notation $\| \ \|_\alpha^{(l)} = \underbrace{\| \ \|_\alpha * \cdots * \| \ \|_\alpha}_{l}$ (see 2.3.5) and with $l = \bar{s} + r + 1$ we have for $\alpha \leq \sigma_0 + r$ the following relation

$$\|L'(x(i-1) + u(i), u(i), y(i)\|_\alpha$$
$$\leq \text{const}(\|L'\|_\alpha(x_0))\|x(i-1) + u(i) + u(i)\|_{\alpha+r}^{(l)} * \|u(i)\|_{\alpha+r} * \|y(i)\|_{\alpha+r},$$

and so for $l = \bar{s} + r + 1$ we obtain

$$\|e(i) = L'\|_\alpha \leq \text{const}\, \delta^2(\|L'\|_l(x_0))N^l(x_0)Q'(i, \alpha) | \alpha \leq \sigma_0 + s,$$

where $Q' = i^{-2\bar{d}} + i^{2(\alpha-\sigma_0-s-\nu)-1}$ [compare with (6) of 2.3.5]. Then, as in (7*) of 2.3.5, we get

$$\|E^*(i)\|_\alpha \leq \text{const}\, \delta^2 C_L(x_0)\tilde{Q}(i, \alpha) | \alpha \leq \bar{d},$$

where $C_L(x_0) = \|L'\|_l(x_0)N^l(x_0)$ for $l = \bar{s} + r + 1$, and $\tilde{Q} = i^{-2\bar{d}} + i^{2(\alpha-\sigma_0-s)-1}$.

Finally, under assumption $\|g\|_{\sigma_0+s} \leq \varepsilon$, we obtain

$$\|g'\|_\alpha \leq \text{const}(\varepsilon + \delta^2 C(x_0))\tilde{Q}(i, \alpha) | \alpha \leq \bar{d},$$

and for $y(i+1) = M(x_0 + \bar{z}(i), g'(i))$ we obtain the following estimate

$$(1') \qquad \|y(i+1)\|_\alpha \leq \text{const}\, C_M(x_0)(\varepsilon + \delta^2 C_L(x_0))Q_0(i, \alpha) | \alpha \leq \sigma_0 + r + s,$$

where Q_0 is the old expression $i^{-2\bar{d}} + i^{2(\alpha-\sigma_0)-1}$, and

$$C_M(x_0) = \|M\|_m(x_0)N^m(x_0) \qquad \text{for } m = \bar{s} + r + s + 1.$$

Observe, that here we do need the section x_0 to be of class $C^{\sigma_0+r+s+d}$, rather than only C^{σ_0+r+s}-smooth as in 2.3.5. Indeed the section x_0 is not acted upon by smoothing operators anymore, and so estimates for $\|y(i+1)\|_\alpha$ require $\|x_0\|_{\alpha+d}$. However, now we have an advantage of a better control over the relation $\|g\|_{\sigma_0+s} \leq \varepsilon(x_0)$ needed for the existence of Nash's process. Namely, *there is a positive number* $\varepsilon_0 > 0$, *that depends only on our fixed data, such that the inequalities*

$$\text{(0)} \qquad \|g\|_{\sigma_0+s} \leq \varepsilon = \varepsilon_0(1 + C_L(x_0))^{-2}(C_M(x_0))^{-2},$$

$$\text{(0*)} \qquad 2C(S)\varepsilon^{1/2} \sum_{i=1}^{\infty} Q_0(i,d) \leq \min(1, \tfrac{1}{2}R^{-1}(x_0)),$$

imply the existence of the modified Nash's process $x(i)$ *such that*

$$\text{(1)} \qquad \|y(i)\|_\alpha \leq \delta Q(i,\alpha) | \alpha \leq \sigma_0 + r + s,$$

for $\delta = \varepsilon^{1/2}$.

Indeed, under the condition (0), the inequality (1′) becomes $\|y(i+1)\|_\alpha \leq \varepsilon_0 \delta \operatorname{const} Q_0(i,\alpha)$, and for $\varepsilon_0 \leq \operatorname{const}^{-1}$ we get our assertion by induction, as (0*) implies (*) above.

The inequalities (0) and (0*) determine a neighborhood $\mathscr{B} = \mathscr{B}^{\beta_0,\sigma_0+s} \subset \mathscr{X}^{\beta_0} \times \mathscr{G}^{\sigma_0+s}$, for $\beta_0 = (\sigma_0 + r + s) + d$ and the new operator $\mathscr{D}^{-1}$ maps $\mathscr{B}$ to $\mathscr{A}^{\sigma_0}$. Furthermore, the argument of 2.3.5 applies to the new $\mathscr{D}^{-1}$ and shows this $\mathscr{D}^{-1}$ to be a continuous map

$$\mathscr{B}^{\beta_1,\sigma_1+s} \to \mathscr{A}^{\sigma_1} \qquad \text{for } \beta_1 - (r+s+d) > \sigma_1 \geq \sigma_0.$$

So, our new $\mathscr{D}^{-1}$ is somewhat less regular than the old one, but it behaves better in other respects and is easier to construct as well. Below we give three applications of the modified Nash process and of the new operator $\mathscr{D}^{-1}$.

(A) *Operators of Polynomial Growth.* We say that a C^∞-map $F\colon A \to \mathbb{R}^c$ for $A \subset \mathbb{R}^b$, has *polynomial* growth if all jets J_F^j, $j = 0, 1, \ldots$ satisfy the following inequalities

$$\|J_F^j(a)\| \leq C_j(\|a\| + (\operatorname{dist}(a, \mathbb{R}^b \backslash A))^{-1} + 1)^{k_j}, \quad a \in A,$$

for some positive numbers C_j and k_j. If $A = \mathbb{R}^b$, these inequalities become

$$\|J_F^j(a)\| \leq C_j(1 + \|a\|)^{k_j}.$$

Now, for a manifold V with a partition of unity $\{U_\mu, P_\mu\}$ as in 2.3.4, differential operators N between vector bundles are given over each $U_\mu = \mathbb{R}^n$, by maps of the corresponding jet spaces (who are Euclidean spaces) or of open subsets A in these spaces to Euclidean spaces. Then we put $N_\mu = P_\mu N$ and we say that N has *polynomial growth* if the maps corresponding to N_μ for all $\mu = 1, \ldots,$ have polynomial growth. Clearly, this definition does not depend on a particular choice of $\{U_\mu, P_\mu\}$ and on trivializations of vector bundles over U_μ.

Theorem. *If the operator $\mathscr{D}$ and its infinitesimal inversion have polynomial growth, then for every compact set $K \subset V$ there is a number $k = 1, 2, \ldots,$ such that the inequality*

$$\|g\|_{\sigma_0+s} \leq (2 + \|x_0\|_{\beta_0} + R(x_0))^{-k},$$

for $\beta_0 = \sigma_0 + r + s + d$ and for a given $\sigma_0 > \bar{s}$, implies the existence of the modified Nash process yielding the (new) operator $\mathscr{D}^{-1}(x_0, g)$ over K.

Proof. The polynomial gorwth of $\mathscr{D}$ implies such a growth for the operator L' and the proposition above applies.

Observe, that many interesting operators have polynomial growth. For example, the metric inducing operator on maps $x: V \to \mathbb{R}^q$, $\mathscr{D}(x) = f^*(h)$, for $h = \sum_1^q dx_i^2$, and its infinitesimal inversion (constructed in 2.3.2) have polynomial growth.

(B) *The Initial Value (Cauchy) Problem.* We call $\mathscr{D}^{-1}$ a *k-consistent inversion of $\mathscr{D}$ along a closed subset V_0 in V* if the equality $J_g^{k+s-1} | V_0 = 0$ implies $J_x^{k-1} | V_0 = J_{x_0}^{k-1} | V_0$, for $x = \mathscr{D}^{-1}(x_0, g)$.

Let $V_0 \subset V$ be a codimension one C^∞-submanifold without boundary that divides V, let $\sigma_0 > \bar{s}$ and let $k \geq 2r + s$ be an integer.

Theorem. *There exists a k-consistent inversion $\mathscr{D}^{-1}: \mathscr{B} \to \mathscr{A}$ for*

$$\mathscr{B} = \mathscr{B}^{\beta_0, \sigma_0+k+2s} \subset \mathscr{X}^{\beta_0} \times \mathscr{G}^{\sigma_0+k+2s}, \qquad \beta_0 = \sigma_0 + r + s + d,$$

that continuously maps every space

$$\mathscr{B}^{\beta_1, \sigma_1+k+2s} \text{ to } \mathscr{A}^{\sigma_1} \text{ for } \beta_1 - (r + s + d) > \sigma_1 \geq \sigma_0,$$

and this $\mathscr{D}^{-1}$ also enjoys all other properties [i.e. (2), (3) *and* (5)] *of $\mathscr{D}^{-1}$ of* 2.3.2.

Proof. Fix a C^∞-function φ on V which vanishes exactly on V_0 and whose differential $d\varphi$ does not vanish on V_0, and introduce, with the operators S_i, new smoothing operators, called S_i^φ, as follows,

$$S_i^\varphi(y) = \varphi^k S_i(\varphi^{-k} y) \qquad \text{for } y \in \mathscr{X};$$

$$S_i^\varphi(g) = \varphi^{k+s} S_i(\varphi^{-k-s} g) \qquad \text{for } g \in \mathscr{G}.$$

Next, we introduce new norms, $\|y\|_\alpha^\varphi = \|\varphi^{-k} y\|_\alpha$ in $\mathscr{X}$ and $\|g\|_\alpha^\varphi = \|\varphi^{-k-s} g\|_\alpha$ in $\mathscr{G}$ for $\| \ \|_\alpha = \| \ \|_\alpha(K)$, and we observe that relative to these new norms the operators S_i^φ satisfy the smoothing estimates (1), (2) and (3) of 2.3.4 as well as the locality and convergence properties (see 2.3.3). Furthermore, the operators M and L' satisfy, relative to the new norms, the same estimates as before. Indeed,

$$\|M(x, \varphi^{k+s} g)\|_\alpha^\varphi = \|M^\varphi(x, g)\|_\alpha$$

where $M^\varphi(x, g) = \varphi^{-k} M(x, \varphi^{k-s} g)$ is again a differential C^∞-*operator* of the same order as M, and for $k \geq 2r + s$ the corresponding operator $(L')^\varphi(x, u, y) = \varphi^{-k-s} L'(x, \varphi^k u, \varphi^k y)$ is also a differential C^∞-operator. Finally we denote by $\mathscr{G}^\varphi \subset \mathscr{G}$ the space of sections of the form $\varphi^{k+s} g$ for all $g \in \mathscr{G}$, and we observe that a section $h \in \mathscr{G}^{k+s}$ is contained in $\mathscr{G}^\varphi$ iff $J_h^{k+s-1} | V_0 = 0$. Since the operator $(\mathscr{G}^\varphi)^{\alpha+k+s} \to \mathscr{G}^\alpha$, given by $g \mapsto \varphi^{-k-s}g$, is continuous for all α, the estimate $\|g\|_{\sigma_0+k+2s} \ll 1$ for $g \in \mathscr{G}^\varphi$ implies $\|g\|^\varphi_{\sigma_0+s} \ll 1$. Now we can see that the modified Nash process constructed with the new operators S_i^φ in place of S_i and directed by (x_0, g) for $x_0 \in \mathscr{X}^{\beta_0}$ and $g \in (\mathscr{G}^\varphi)^{\sigma_0+k+2s}$, satisfies the same estimate relative to the norms $\| \ \|^\varphi_\alpha$ we had before for the norms $\| \ \|_\alpha$. Therefore for $x_0 \in \mathscr{X}^{\beta_1}$ and $g \in (\mathscr{G}^\varphi)^{\sigma_1+k+2s}$ we get $\|x_0 - x(\infty) = \sum_{i=1}^{\infty} y_i\|^\varphi_{\sigma_1} < \infty$, and in particular, $J^{k-1}_{x_0}|V_0 = J^{k-1}_{x(\infty)}|V_0$.

Remark. The non-modified Nash process does not work with the operators S_i^φ since the non-modified u contains x_0, which unlike $z(i)$ does not vanish on V_0 and so the corresponding (non-modified) operator L' is discontinuous in the norms $\| \ \|^\varphi$.

Corollary. *Let $x_0 \in \mathscr{G}^{\beta_0}$ and $g \in \mathscr{G}^{\beta_0 - r}$ be sections such that $J^l_{g_0}|V_0 = J^l_g|V_0$ for $g_0 = \mathscr{D}(x_0)$ and for some $l \leq \beta_0 - r$. If $\beta_0 > \bar{s} + r + s + \max(d, 2r + 2s)$ and if $l \geq \bar{s} + 2r + 3s$, then, for a non-integer β_0, there is a section $x_0' \in \mathscr{A}^{\sigma_0}$ for $\sigma_0 = \beta_0 - r - s - \max(d, 2r + 2s)$ such that $\mathscr{D}(x_0')|\mathscr{O}p(V_0) = g|\mathscr{O}p(V_0)$ and $J^{2r+s-1}_{x_0'}|V_0 = J^{2r+s-1}_{x_0}|V_0$. If β_0 is an integer, then such an x_0' exists in all spaces $\mathscr{A}^\sigma$ for $\sigma < \sigma_0$.*

Proof. The condition $J^l_{g_0}|V_0 = J^l_g|V$ implies that the section $g - g_0$ is C^l-small on $\mathscr{O}p(V_0) \subset V_0$ and under our assumption there is a k-consistent operator $\mathscr{D}^{-1}(x_0, g - g_0)$ for $k = 2r + s$.

Remark. If we drop the last requirement, $J^{2r+s-1}_{x_0'}|V_0 = J^{2r+s-1}_{x_0}|V_0$, then with our old non-modified $\mathscr{D}^{-1}$ we get x_0' under less restrictive conditions, namely for $l > \bar{s} + s$ and for $\beta_0 > \bar{s} + r + s$. This x_0' itself is of class $\beta_0 - r - s$, or anything less than that, if β_0 is an integer, [compare with (D) of 2.3.2].

Example [compare Jacobowitz (1974)]. For the isometric immersion problem we have $r = 1$, $s = 0$, $d = \bar{s} = 2$, so that the corollary applies to free maps x_0 of class C^{β_0}, $\beta_0 > 5$, which are *infinitesimally isometric* of order 4 along V_0. That is, the original metric g in V and the induced metric agree along V_0 with the derivatives of order ≤ 4. The corollary allows us to construct a new map, x_0' of class $\beta_0 - 3$, isometric on $\mathscr{O}p(V_0)$ such that $J^1_{x_0'}|V_0 = J^1_{x_0}|V_0$.

Question. Is it possible for metrices g of class C^∞ to extend x_0 from V_0 to an isometric immersion $\mathscr{O}p(V_0) \to \mathbb{R}^q$ without this unpleasant loss of three (and $3 + \varepsilon$ for an integer β_0) degrees of smoothness?

Exercise 1. Prove the theorem and the corollary for an arbitrary C^∞-submanifold $V_0 \subset V$ without boundary, and also analyze the case, when V_0 has a boundary, for example codim $V_0 = 0$.

(C) *Global Analytic Inversions.* Let us describe the "variable data", discussed earlier, in slightly different term. Let $\beta_0 = \bar{s} + r + s + d + 2$ and let us lift the set $A \subset X^{(d)}$ to $X^{(\beta_0)}$ by taking the pullback $A_0 = (p^{\beta_0}_d)^{-1}(A) \subset X^{(\beta_0)}$. The "functions" F and F' are also lifted to A_0 and denoted by F and F_0' respectively. (In fact F_0' is defined everywhere on $X^{(\beta_0)}$.) Now, for the same reason as before, we assume $V = \mathbb{R}^n$, so that

F_0 and F_0' are actual functions on the domain A_0 in the Euclidean space $\mathbb{R}^m = X^{(\beta_0)}$. We put $R^{-1}(a) = \operatorname{dist}(a, \mathbb{R}^m \backslash A_0)$, $a \in A_0$, and let J and J' denote the jets of order β_0 of the functions F_0 and F_0' respectively. Then, with a point $a \in A_0$, we consider all those points $b \in A_0$ for which $\operatorname{dist}(a, b) = \|a - b\| \leq \rho = \min(1, R^{-1}(a)/2)$ and set

$$\|M\|(a) = \sup_b \|J(b)\|, \qquad \|a - b\| \leq \rho,$$

$$\|L'\|(a) = \sup_b \|J'(b)\|, \qquad \|a - b\| \leq \rho.$$

Finally, we denote by $N(a)$ the sum $\|a\| + 2R(a) + 1$, and set

$$C_0(a) = (N(a) + 1)^{4\beta_0}(\|M\|(a) + 2)^2(\|L'\|(a) + 2)^2.$$

This huge "constant" is meant to dominate the constants $C_{L'}$ and C_M of the modified Nash process.

Now for the "fixed data" we take the *analytic* smoothing operators in $V = \mathbb{R}^n$ of 2.3.4 of the depth $\bar{d} \geq 10(\bar{s} + 1)$ and we take $\sigma_0 = \bar{s} + 1$ and $k = 2r + s + 2$.

Theorem. *Let φ: $A_0 \to \mathbb{R}$ be a positive real analytic function, such that $\varphi(a)C_0(a) \leq \varepsilon_0$ where ε_0 is a sufficiently small positive number which depends only on the fixed data above. Then there exists an inversion $\mathscr{D}^{-1}(x_0, g)$ defined for all those sections $x_0 \in \mathscr{A}^{\beta_0}$ and $g_0 \in \mathscr{G}^{\sigma_0 + s}$, for which*

$$\|J_g^{\sigma_0 + s}(v)\| \leq (\varphi(J_x^{\beta_0}(v)))^{k+s+1}, \qquad v \in V,$$

and the section $\mathscr{D}^{-1}(x_0, g)$ is C^{σ_0}-smooth. Furthermore, if the operators $\mathscr{D}$ and M are real analytic (that is the maps Δ: $\mathscr{X}^r \to G$ and F on $A \subset X^{(r)}$ are real analytic), then the section $\mathscr{D}^{-1}(x_0, g)$ is real analytic for real analytic pairs (x_0, g). Finally, this operator $\mathscr{D}^{-1}$ enjoys the normalization and inversion (properties) and also the following

Continuity. If

$$\|(x_0 - x_0')(\varphi(J_0))^{-\beta_0 - k}\|_{\beta_0} \leq \varepsilon,$$

and

$$\|(g - g')(\varphi(J_0))^{-k-s-1}\|_{\sigma_0 + s} \leq \varepsilon$$

for $J_0(v) = J_{x_0}^{\beta_0}(v)$, then

$$\|J_x^{\sigma_0}(v) - J_{x'}^{\sigma_0}(v)\| \leq \delta\varphi_0(J_0(v)) \qquad \textit{for all } v \in V,$$

where $x = \mathscr{D}^{-1}(x_0, g)$, $x' = \mathscr{D}^{-1}(x_0', g')$ and where $\delta \to 0$ as $\varepsilon \to 0$.

Proof. We introduce new smoothing operators with the function $\varphi_0(v) = \varphi(J_{x_0}^{\beta_0}(v))$ by putting

$$S_i^{\varphi}(y) = \varphi_0^k S_i(\varphi_0^{-k} y) \qquad \text{for } y \in \mathscr{X},$$

and

$$S_i^{\varphi}(g) = \varphi_0^{k+s+1} S_i(\varphi_0^{-k-s-1} g) \qquad \text{for } g \in \mathscr{G},$$

and we consider the modified Nash process with these smoothing operators. Then, with the norms $\|y\|_\alpha^\varphi = \|\varphi_0^{-k} y\|_\alpha$ and $\|g\|_\alpha^\varphi = \|\varphi_0^{-k-s-1} g\|_\alpha$, we get all our old estimates and thus the operator $\mathscr{D}^{-1}$ as well. Furthermore, for real analytic x_0 and g, we obtain these estimates in all small obtuse neighborhoods (see 2.3.4) $U \subset \mathbb{C}^n$ of $\mathbb{R}^n \subset \mathbb{C}^n$, since the real analytic operators M and L' satisfy in *small* neighborhoods U the same estimates as on V, and since the estimates for S_i extend to *obtuse* neighborhoods U (see 2.3.4). Therefore, the Nash process converges in the norm $\| \ \|_{\sigma_0}^U$ (see 2.3.4), and in particular, the limit $x(\infty) = \mathscr{D}^{-1}(x_0, g)$ is real analytic.

Exercise 2. Generalize this theorem to an arbitrary C^{an}-manifold V and show, that C^∞-solutions x of the equation $\mathscr{D}(x) = g$ can be approximated in the fine C^∞-topology by C^{an}-solutions, provided the operators $\mathscr{D}$ and M and the section g are real analytic.

Hint. Imbed $V \to \mathbb{R}^q$; first extend the functions from V to a normal neighborhood $N(V) \to V$, and then make them zero outside $N(V) \subset \mathbb{R}^q$. Now, analytically smooth the extended functions and then restrict them back to V. This gives a "good" analytic smoothing on V.

Exercise 3. Let V be compact and let x_0 and g be real analytic. Show that Nash's process with $S_i = \mathrm{Id}$ (that is Newton's process, see 2.3.3) converges, when directed by $(x_0, \varepsilon g)$ for small positive numbers ε, and it defines $\mathscr{D}^{-1}(x_0, \varepsilon g)$.

Hint. Pass to a complexification $\mathbb{C}V$ and use Cauchy's inequality. [Also see Gromov-Rochlin (1970), Green-Jacobowitz (1971).]

Final Remarks and References. Nash's theory has been developed and generalized in several directions (see Schwartz 1960; Moser 1961; Clarke 1970; Jacobowitz 1972; Sergeraert 1972; Gromov 1972; Hörmander 1976; Hamilton 1982), but unfortunately, there is no single general theorem that would cover all interesting possibilities.

2.3.7 Infinite Dimensional Representations of the Group Diff(V)

A vector bundle $G \to V$ is called *natural* if the action of the group of C^∞-diffeomorphisms, Diff(V) on V, lifts to fiberwise linear action on G.

Examples. Trivial bundles $V \times \mathbb{R}^n \to V$ with the obvious action of Diff(V) are natural. Invariant subbundles and factor bundles of natural bundles are natural. The bundle $\mathrm{Hom}(G_1, G_2)$ for two natural bundles G_1 and G_2 is natural. The jet bundles $G^{(r)} \to V$, $r = 1, 2, \ldots$, for natural G are natural.

All classical bundles of tensors on V, for example, the tangent and the cotangent bundles of V come from these constructions, but there are some "non-classical" examples as well.

Exercise 1. Construct a continuum of one dimensional natural bundles over $V = \mathbb{R}^1$.

Hint. Use one-dimensional linear representations of the structural group of the tangent bundle $T(\mathbb{R}^1)$.

The group Diff(V) acts by linear transformations on the spaces of sections of natural bundles, and such an action is called (algebraically) *irreducible* if there is no non-trivial invariant subspaces.

Exercise 2. Consider the action of Diff(V) on the space $C^\infty(V)$ of C^∞-functions $V \to \mathbb{R}$ and prove the following five facts

(a) If V is a compact connected manifold without boundary then the only non-trivial invariant subspace consists of constant functions.

(b) If $V = \mathbb{R}^n$, $n \geq 2$, then there are exactly six (non-trivial) invariant subspaces, and for $n = 1$ there are ten of them.

(c) If V has a non-empty boundary then there are uncountably many invariant subspaces. However, if V is compact, then there are only countably many spaces of C^{an}-functions invariant under Diff$^{an}(V)$.

(d) The number of invariant subspaces is finite *iff* V has no boundary and it has only finitely many ends.

(e) The action of Diff(V) on $C^\infty(V)$ has a cyclic vector [that is a vector in $C^\infty(V)$ whose orbit spans the whole space $C^\infty(V)$] *iff* there is no discrete infinite orbits of the action of Diff(V) on V and the number of compact orbits of this section on V is finite.

Now, look at the space $\mathscr{G}^\infty$ of quadratic differential C^∞-forms on V and at the (Diff)-invariant subspace $\mathscr{G}_0^\infty$ of the forms with compact supports. Denote by Diff$_0$ the subgroup in Diff(V) of those C^∞-diffeomorphisms, that are fixed outside compact sets $K \subset V$.

Lemma. *Let $g \in \mathscr{G}^\infty$ be a positively definite form (that is a Riemannian metric on V). Then the span Λ of the* Diff$_0$*-orbit of g contains the space $\mathscr{G}_0^\infty$.*

Proof. Take a form $g_0 \in \mathscr{G}_0^\infty$ and a small neighborhood U in V of its support $\operatorname{supp}(g_0) \subset V$. Then, for $m = 1, 2, \ldots$, we consider manifolds $(W, h) = \mathsf{X}_1^m(V, g)$ and let $x\colon V \to W$ be the diagonal imbedding. If we take a small perturbation x_0 of x which equals x outside U, then the induced form $x_0^*(h)$ on V is contained in Λ, since the projections $p_j \circ x_0\colon V \to V_m, j = 1, \ldots, m$, are diffeomorphisms in Diff$_0$ and $x_0^*(h) = \sum_{j=1}^m (p_j \circ x_0)^*(g)$. Now, for a large number m, generic maps x_0 are free on U (see 1.3.2) and by the implicit function theorem of 2.3.2 there is a small number $\lambda \neq 0$, such that the form $x_0^*(h) + \lambda g$ can be induced by a C^∞-map $x_1\colon V \to (W, h)$ that is C^1-close to x_0^* and that agrees with x_0 outside U. Therefore,

$$g_0 = \lambda^{-1} \sum_{j=1}^{m} ((p_j \circ x_0^*)(g) - (p_j \circ x_0^*)(g)) \in \Lambda.$$

Theorem. *If V is a compact connected manifold without boundary then the action of* Diff(V) *in the space of quadratic differential C^∞-forms on V is (algebraically) irreducible.* [See (D) of 2.3.8 for generalizations.]

Proof. Take an arbitrary form $g \in \mathcal{G}^\infty$ such that $g(v_0) \neq 0$ for some point $v_0 \in V$ and first consider the diffeomorphisms that keep this point fixed. Then we obtain an action on the tangent space $T_{v_0}(V) = \mathbb{R}^n$, that amounts to the standard action of the linear group GL_n on $\mathbb{R}^n$. Since the induced action of GL_n on the space of quadratic polynomials on $\mathbb{R}^n$ is irreducible, the span Λ of the (Diff)-orbit of g contains a form which is positive at v_0 and therefore, positive in a small neighborhood $V_0 \subset V$ of v_0. Then, by the lemma, all forms with supports in V_0 are contained in Λ. Finally, we cover V by diffeomorphic images of V_0 and with a partition of unity we put all forms into Λ.

Exercise 3. Let the group Diff(V) be transitive on a non-compact manifold V and let $\dim(V) \geq 2$. Show that the only non-trivial Diff(V)-invariant subspace in $\mathcal{G}^\infty$ consists of forms with compact supports. Analyze the case of $\dim(V) = 1$.

Warning. If a form $g_0 \in \mathcal{G}^\infty$ has *infinite* support, then, in the space $\mathcal{G}^\infty$ with the *fine* C^∞-topology, the convergence $\lambda \to 0$ does *not* imply $\lambda g_0 \to 0$.

Example. Consider the differential operator that sends C^∞-functions $f\colon V \to \mathbb{R}^1$ to the forms $(df)^2 \in \mathcal{G}^\infty$ induced by f from the standard quadratic form $(dx)^2$ on $\mathbb{R}^1$. This operator commutes with the action of Diff(V) and the span of its image consists exactly of those forms in $\mathcal{G}^\infty$ which can be induced by maps into pseudo-Euclidean spaces,

$$V \to \left(\mathbb{R}^{q_1, q_2}, h = \sum_1^{q_1} (dx_i)^2 - \sum_1^{q_2} (dx_j)^2 \right).$$

It follows, that every C^∞-form g can be induced from some pseudo-Euclidean space. This result also can be obtained by applying the implicit function theorem to *isotropic maps* $x\colon V \to \mathbb{R}^{q_1, q_2}$ for which $x^*(h) \equiv 0$.

(Green 1970). *If there is a free C^∞-smooth isotropic map x_0 of a compact manifold V into $\mathbb{R}^{q_1, q_2}$, then any given C^∞-form g on V can be induced by a C^∞-map $V \to \mathbb{R}^{q_1, q_2}$.*

Indeed, $\lambda g \to 0 = x_0^*(h)$ as $\lambda \to 0$, and for some $\lambda > 0$ one can induce λg by a map x that is close to x_0, and so g is realized by the map $\lambda^{-2} x$.

There is an easy way to get free isotropic maps $V \to \mathbb{R}^{q_1, q_2}$ for $q_1 = q_2 = q = \frac{1}{2} n(n+5)$, $n = \dim(V)$, by taking two copies of the same free map, one of V to $\mathbb{R}^{q,0}$ and the other to $\mathbb{R}^{0,q}$. But isometric imbeddings $V \to \mathbb{R}^{q,q}$ can be obtained for so large a q even without the implicit function theorem. Namely, for a form g on V and for an arbitrary free map $x_0\colon V \to \mathbb{R}^q$ we first solve the linearized equation (L) of 2.3.1

$$\text{(L)} \qquad L(x_0, y) = g,$$

and then we observe (see 2.3.1) that the form $(x_0 + y)^*$ induced by the map $(x_0 + y)\colon V \to \mathbb{R}^q$ satisfies the following identity,

$$(x_0 + y)^* = x_0^* + y^* + L(x_0, y) = g + x_0^* + y^*.$$

Now, to realize a given form g_0 by a map $V \to \mathbb{R}^{q,0} \oplus \mathbb{R}^{0,q}$, we solve L with $g = g_0 - x_0^*$ and get $(x_0 + y)^* = g_0 + y^*$, so that the map $(x_0 + y, y)$: $V \to \mathbb{R}^{q,0} \oplus \mathbb{R}^{0,q}$ induces the form $(x_0, y)^* - y^* = g_0$.

Notice that this algebraic construction works for C^2-forms g_0, while the implicit function theorem only applies to C^σ-forms for $\sigma > 2$. However, this theorem applies for smaller q_1 and q_2, where (known) algebraic constructions fail. Namely, free *isotropic maps* exist locally *iff* $q_1 \geq n$, $q_2 \geq n$ and $q_1 + q_2 \geq n(n+5)/2$, (see 3.3.4). Then the theorem of Hirsch yields free isotropic maps of all open parallelizable manifolds into these spaces $\mathbb{R}^{q_1,q_2}$; It follows that if such a V is compact, then *every C^σ-form, $\sigma > 2$, on V can be induced by a free C^σ-map $V \to \mathbb{R}^{q_1,q_2}$ for $q_1 + q_2 = n(n+5)/2$ and* $\min(q_1, q_2) \geq n$. In the general case with no parallelizability condition, the sheaf theory of 2.2 leads to a slightly less precise result, namely to the existence of immersions $V \to \mathbb{R}^{q_1,q_2}$ with $q_1 + q_2 = \frac{1}{2}(n^2 + 7n + 2)$ for $\min(q_1, q_2) \geq 2n - 1$. (See 3.3.)

Positive Forms. Consider in $\mathscr{G}^\infty$ the subspace of positive definite forms. These constitute a convex open cone, $\mathscr{G}_+^\infty \subset \mathscr{G}^\infty$, invariant under the action of Diff(V). We shall see later, in 3.1.3, that *there is no nontrivial* convex Diff(V)-*invariant subcone in* $\mathscr{G}_+^\infty$, and, in particular, all positive forms g on V are induced by maps x: $V \to \mathbb{R}^q$, for q large, that is, $g = \sum_{i=1}^q (dx_i)^2$, for $(x_1, \ldots, x_q) = x$. The reduction to the implicit function theorem is more complicated when it comes to maps to $\mathbb{R}^q$ (rather than maps $V \to \mathbb{R}^{q_1,q_2}$, $\min(q_1, q_2) \geq n$) and there is no purely algebraic construction of isometric immersions $V \to \mathbb{R}^q$ whatsoever.

Exercise 4. Show, that there is no algebraic immersion of the hyperbolic plane H^2 to $\mathbb{R}^q$ for any q.

Hint. The Poincaré metric is given by rational functions and the existence of an algebraic isometric immersion $H^2 \to \mathbb{R}^q$ would imply polynomial growth of (the area of) the balls $B(\rho) \subset H^2$ for $\rho \to \infty$.

2.3.8 Algebraic Solution of Differential Equations

We show in this section that many under-determined systems of linear P.D.E. can be solved by purely algebraic manipulations and we obtain in particular, the infinitesimal invertibility for generic "under-determined" non-linear differentia operators. (Compare Nash's theorem in 2.3.1.)

We denote by q and q' respectively the dimensions of our C^∞-smooth vector bundles X and G over V. We use the same symbol, say L, for linear differential operators, L: $\mathscr{X}^{r+\alpha} \to \mathscr{G}^\alpha$, and for the corresponding vector bundle homomorphisms, L: $X^{(r)} \to G$, so that $L(x) = L(J_x^r)$ for x: $V \to X$. We call the operator L and the system $L(x) = g$ *under-determined* if $q > q'$. An *inversion* of L is, by definition, a linear *differential* operator M: $\mathscr{G}^{s+\alpha} \to \mathscr{X}^\alpha$, such that $L \circ M = \mathrm{Id}$. Observe that the existence of such an M is a purely *local* problem. Indeed, if some operators M_μ invert L over neighborhoods $U_\mu \subset V$, then, with a partition of unity $\{P_\mu\}$, we set $M(g) = \sum_\mu M_\mu(P_\mu g)$ and obtain for all sections g: $V \to X$

$$(L \circ M)(g) = \sum_{\mu} (L \circ M_\mu)(P_\mu g) = \sum_{\mu} P_\mu g = g.$$

Now, we assume $V = \mathbb{R}^n$ and we start with the following example.

(A) *Operators with Constant Coefficients.* Let first $q' = 1$. Then the operators L send maps $x = (x_1, \ldots, x_q)\colon \mathbb{R}^n \to \mathbb{R}^q$ to functions $g\colon \mathbb{R}^n \to \mathbb{R}$, $g = L(x) = \sum_{j=1}^{q} L_j(x_j)$, and one represents such an $L = (L_1, \ldots, L_q)$ by q (characteristic) polynomials, $\hat{L}_1, \ldots, \hat{L}_q$, on $\mathbb{R}^n$. If an inversion $M = (M_1, \ldots, M_q)$, $M(g) = (M_1(g), \ldots, M_q(g))$, also has constant coefficients and is represented by polynomials $\hat{M}_1, \ldots, \hat{M}_q$, then the relation $L \circ M = \mathrm{Id}$ turns into the identity

$$\sum_{j=1}^{q} \hat{L}_j \hat{M}_j = 1.$$

This identity says that the ideal generated by the polynomials $\hat{L}_i$ equals the whole ring of the polynomials on $\mathbb{R}^n$, that is the (non-linear) algebraic system

$$(*) \qquad \begin{array}{c} \hat{L}_1(w) = 0 \\ \cdots\cdots\cdots \\ \hat{L}_q(w) = 0 \end{array}$$

has no solutions $w \in \mathbb{C}^n \supset \mathbb{R}^n$. Now, for *generic* polynomials $\hat{L}_j$, this system is solvable *iff* $q \leq n$ and so *generic operators L with constant coefficients are invertible for* $q \geq n + 1$.

Example. Let $n = 1$ and $q = 2$. Take $L_1(x) = a_1 x + b_1(dx/dt) + (d^2x/dt^2)$ and $L_2(x) = a_2 x + (dx/dt)$ for some real constants a_1, b_1, a_2 and for $t \in \mathbb{R}^1$. Now, the system $(*)$ reduces to the following two equations in the unknown $w \in \mathbb{C}^1$,

$$a_1 + b_1 w + w^2 = 0,$$

$$a_2 + w = 0.$$

These equations have a common solution *iff* $\Delta = a_1 - b_1 a_2 + a_2^2 = 0$ and so the genericity condition above amounts to $\Delta \neq 0$. Then for $M_1(g) = \Delta^{-1} g$ and for $M_2(g) = \Delta^{-1}(a_2 - b_1)g - \Delta^{-1}(dg/dt)$ we have $L_1 \circ M_1 + L_2 \circ M_2 = \mathrm{Id}$, that is $L_1(M_1(g)) + L_2(M_2(g)) = g$ for all g.

Now, in the general case of $q' \geq 1$, an operator L of order r is given by a matrix with polynomial entries, $(\hat{L}_{ij})$, $i = 1, \ldots, q'$, $j = 1, \ldots, q$ and this matrix is invertible by another such matrix *iff* rank $(\hat{L}_{ij}(w)) = q'$ for all $w \in \mathbb{C}^n$. It follows that *generic operators with constant coefficients are invertible iff* $q \geq q' + n$.

Exercise. Show that if an operator L with constant coefficient is invertible by an operator with *non-constant* coefficients, then L is also invertible by an operator with *constant* coefficients.

Remark 1. The genericity condition above can be specified as follows. Recall, that an *algebraic relation* between some given real numbers $l_1, \ldots, l_N$ is, by definition, a non-zero polynomial $Q = Q(z_1, \ldots, z_N)$ with integer coefficients such that $Q(l_1, \ldots, l_N) = 0$. The numbers $l_1, \ldots, l_N$ are called *k-independent*, if there is no relations Q whose degrees and the absolute values of the coefficients are bounded by k.

Now, each polynomial $\hat{L}_{ij}$ of degree r is determined by its $(n + r)!/n!r!$ coefficients and so we have $N = qq'(n + r)'/n!r!$ numbers $l_1, \ldots, l_N$ that determine the matrix $(\hat{L}_{ij})$, and the invertibility statement above holds if the genericity is understood as k-independence of these numbers for a sufficiently large k, for example for $k = \exp\exp N$.

Remark 2. If $q = q'$ then the invertibility of L amounts to the relation

$$\mathrm{Det}(\hat{L}_{ij}(w)) = \text{const} \neq 0,$$

and so we have an infinite dimensional group of polynomial maps $\hat{L}\colon \mathbb{R}^n \to SL_q$ that correspond to invertible operators.

(B) *Ordinary Differential Equations.* Let $L(x) = g$ be a first order system of ordinary differential equations, that is $L = A + B(d/dt)$, where $A = A(t)$ and $B = B(t)$ are matrix functions in $t \in \mathbb{R}$, $A, B\colon \mathbb{R} \to \mathrm{Hom}(\mathbb{R}^q, \mathbb{R}^{q'})$. Then, with $x = (x_1, \ldots, x_q)$ and $g = (g_1, \ldots, g_{q'})$, we express our system by

$$Ax + B\frac{dx}{dt} = g. \tag{1}$$

Next we introduce a new function $h = (h_1, \ldots, h_{q'})$, we consider the following *algebraic* system,

$$Bx = h, \tag{2}$$

and by differentiating (2) we obtain

$$B\frac{dx}{dt} = \frac{dh}{dt} - B'x \qquad \text{for } B' = \frac{dB(t)}{dt}. \tag{2'}$$

Now, the joint system (1) + (2) can be written as

$$\begin{aligned} Bx &= h \\ (A - B')x &= g - \frac{dh}{dt}. \end{aligned} \tag{3}$$

Example. Take the following single equation

$$\frac{dx_1}{dt} + b(t)\frac{dx_2}{dt} = g = g(t).$$

The system (3) in this case reads

$$\begin{aligned} x_1 + bx_2 &= h \\ -b'x_2 &= g - \frac{dh}{dt}, \qquad \text{for } b' = \frac{db(t)}{dt}. \end{aligned}$$

and so the functions

$$x_1 = b(b')^{-1}\left(g - \frac{dh}{dt}\right) + h$$

$$x_2 = (b')^{-1}\left(\frac{dh}{dt} - g\right)$$

satisfy our equation for all $h = h(t)$. This gives a regular solution $(x_1(t), x_2(t))$ as long as the derivative $b'(t)$ does not vanish. Furthermore, if $b'(t)$ has some zeros of finite multiplicity, then in order to have $x_1(t)$ and $x_2(t)$ regular, we choose the function $h(t)$ such that the difference $\frac{dh(t)}{dt} - g(t)$ has the same multiple zeros as $b'(t)$. It follows that the operator $L\colon (x_1(t); x_2(t)) \mapsto \frac{dx_1(t)}{dt} + b(t)\frac{dx_2(t)}{dt}$ *is invertible if the zeros of the derivative* $b'(t)$ *have finite multiplicities*. In particular, L is invertible if $b(t)$ is a nonconstant real analytic function or if it is a *generic* C^∞-function.

Now, we return to the system (3) and we suppose that the matrix $\begin{pmatrix} B \\ A - B' \end{pmatrix}(t)$ has constant rank $= \min(q, 2q')$. Then we express all linear relations between the rows of this matrix by another matrix $R = R(t)$ with $2q'$ columns and with at most $2q' - q$ rows. Thus we reduce the system (3) to the consistency condition expressed by the system

$$R\begin{pmatrix} h \\ g - \dfrac{dh}{dt} \end{pmatrix} = 0. \tag{3'}$$

In particular, for $q \geq 2q'$, the system (3′) contains no equations at all, and so this process delivers a zero order operator M for which $L \circ M = \mathrm{Id}$, for $L = A + B(d/dt)$, as in the example above. In the general under-determined case, for $q > q'$, the system (3′) contains less equations than the original system (1) and so by repeating this elimination process several times we again produce an operator

$$M = M^0 + M^1\frac{d}{dt} + \cdots + M^s\frac{d^s}{dt^s}, \qquad \text{for some } s \leq \frac{q}{2} - 1,$$

such that $L \circ M = \mathrm{Id}$. The coefficients $M^i = M^i(t)$ of this operator are $q \times q'$ matrices whose entries are rational functions in the entries a_{ij} and b_{ij} the matrices $A = (a_{ij}(t))$ and $B = (b_{ij}(t))$ and in the derivatives of these entries of order $\leq q/2$.

Exercise. Show that the common denominator of these rational functions, $\Delta = \Delta(a_{ij}, b_{ij}, a'_{ij}, \ldots)$, is *not* the identically zero polynomial. Prove, that for *generic* C^∞-functions $a_{ij}(t)$ and $b_{ij}(t)$ the function $\Delta(t) = \Delta(a_{ij}(t), \ldots)$ has zeros of finite multiplicity and show in the generic case that the under-determined operator L can be inverted by an operator M with C^∞-coefficients.

Hint. Use the same argument as in the example above. Also see section (E) where a more general case is studied.

(C) *Lie Equations*. Let $G \to V$ be a vector bundle with a flat connection. Then the covariant derivative $\nabla_L x$ equals the Lie derivative Lx for all vector fields L on V and all sections $x\colon V \to G$. We say that a system of vector fields, $(L_1, \ldots, L_p)$, is *large* if there are some C^∞-functions $r_1, \ldots, r_p$ on V for which $\sum_1^p r_j L_j = 0$ and such that the function $\sum_1^p L_j r_j$ does not vanish.

Example. Let $V = \mathbb{R}^n$, and let $L_1 = \dfrac{\partial}{\partial u_1}, \ldots, L_n = \dfrac{\partial}{\partial u_n}, L_{n+1} = \sum_1^n u_i \dfrac{\partial}{\partial u_i}$. Then the system $(L_1, \ldots, L_{n+1})$ is large.

Exercise. Show that generic systems $(L_1, \ldots, L_p)$ are large for $p \geq 2n + 1$.

Now, we consider the following P.D.E. system in unknowns $x_j\colon V \to G$, $j = 1, \ldots, p$,

$$(*) \qquad \sum_{j=1}^{p} L_j x_j = g,$$

and we observe that if $\sum_1^p r_j L_j = 0$, then the sections $x_j = r_j x$ for $x = (\sum_1^p L_j r_j)^{-1} g$ satisfy $(*)$. Therefore *the operator* $\bar{L} = \bigoplus_{j=1}^p L_j\colon (x_1, \ldots, x_p) \mapsto \sum_1^p L_j x_j$ *is invertible for large systems of vector fields* L_j.

Exercise. Generalize this to the operators ∇_{L_j} for an arbitrary (non-flat) connection ∇ in G.

Denote by $\mathscr{L}$ the Lie algebra generated by the fields L_j. Then, for any vector fields $L_\nu \in \mathscr{L}$, $\nu = 1, \ldots, N$, solutions of the equation

$$\sum_1^N L_\nu x_\nu = g$$

yield solutions of $(*)$. For example, if $[L_1, L_2] x = g$ then the tensors $x_1 = \frac{1}{2} L_2 x$ and $x_2 = -\frac{1}{2} L_1 x$ satisfy the equation $L_1 x_1 + L_2 x_2 = g$.

It follows, that if the Lie algebra $\mathscr{L}$ is large, i.e. if it contains a large system of vector fields, $(L_1, \ldots, L_\nu, \ldots, L_N)$, $L_\nu \in \mathscr{L}$, then the operator $\bar{L} = \bigoplus_1^p L_j$ is invertible.

Example. Let $V = \mathbb{R}^2$, and take $L_1 = \partial/\partial u_1$, and $L_2 = (u_1 + u_2)(\partial/\partial u_2)$. Then the system $(L_1, L_2, [L_1, L_2] = \partial/\partial u_2)$ is large, while the original system (L_1, L_2) is not large.

Exercise. Show that two *generic* fields L_1 and L_2 on an arbitrary manifold V generate a large Lie algebra, and so the system $L_1 x_1 + L_2 x_2 = g$ (containing $q' = \dim G$ equations and $2q'$ unknowns) is generically solvable.

(D) *Inversions of Zero Order*. Consider a differential operator $L\colon X^{(1)} \to G$, and express it in local coordinates $u_1, \ldots, u_n$ in V as

$$L(x) = Ax + \sum_{i=1}^{n} B_i \frac{\partial x}{\partial u_i},$$

where A and B_i are $q' \times q$ matrix functions in the coordinates $u_1, \ldots, u_n$ for $q' = \dim G$, $q = \dim X$, and where x is a vector function with q components. Consider all those sections $x: V \to X$ for which

$$(*) \qquad L(fx) = fL(x),$$

for all C^∞-functions $f: V \to \mathbb{R}$. The condition $(*)$ is expressed in the local coordinates by the following system of nq' algebraic equations in x,

$$(**) \qquad B_i x = 0, \qquad i = 1, \ldots n.$$

Assume the matrix $B = \begin{pmatrix} B_1 \\ \cdots \\ B_n \end{pmatrix}$ to have constant rank, call it r, and consider the $(q - r)$-dimensional subbundle in x, call it $X^* \subset X$, whose sections $x: V \to X^*$ are exactly the solutions of $(*)$.

Exercise. Show for $q \geq n(q' + 1)$ that generic matrices B have constant rank $r = nq'$ and then $\dim X^* = q - nq' \geq n$.

The condition $(*)$ implies that the operator L, restricted to sections $V \to X^*$, has order zero and so it is given by a homomorphism $L|X^* = L^*: X^* \to G$. Indeed, by differentiating the equations $B_i x = 0$ we get $B_i(\partial x/\partial u_i) = -B_i' x$ for $B_i' = \partial B_i/\partial u_i$ and so, on the solutions x of $(**)$, the operator L is given by the matrix $A^* = A - \sum_1^n B_i'$, that is

$$Lx = A^* x \qquad \text{for } x: V \to X^*.$$

We call the operator L *free* if the homomorphism L^* is surjective, that is if the matrix $\begin{pmatrix} B \\ A^* \end{pmatrix}$ has rank $r + q'$ everywhere. If L is free then it is invertible, since the system

$$Ax + \sum_1^n B_i \frac{\partial x}{\partial u_i} = g$$

reduces to the algebraic system

$$(***) \qquad \begin{aligned} Bx &= 0 \\ A^* x &= g. \end{aligned}$$

(Compare with Nash's theorem of 2.3.1.)

Exercise. Show that generic operators L are free for $q \geq nq' + n + q'$.

Families of Operators. Denote by $H \to V$ the bundle $\mathrm{Hom}(X^{(1)}, G)$ and let $\mathscr{H}$ be the space of C^∞-sections $V \to H$, that is the space of C^∞-operators $X^{(1)} \to G$. Consider a linear subspace $\mathscr{L} \subset \mathscr{H}$ and take some operators $L_\nu \in \mathscr{L}$, $\nu = 1, \ldots, N$,

$$L_\nu = A_\nu + \sum_{i=1}^n B_{i\nu} \frac{\partial}{\partial u_i}.$$

Let $\bar{L} = \bigoplus_1^N L_\nu : (\bigoplus_1^N X)^{(1)} \to G$, that is

$$L(x_1, \ldots, x_N) = \sum_{\nu=1}^{N} L_\nu(x_\nu), \qquad \text{for } x_\nu : V \to X.$$

Denote by $r(v)$, $v \in V$, the maximal value of rank $\bar{B}(v)$, for $\bar{B} = (B_{i\nu})$, over all choices of operators $L_1, \ldots, L_N$ in $\mathscr{L}$ and over all $N = 1, \ldots,$ and suppose that $r(v)$ does not depend on $v \in V$, that is $r(v) = r = r(\mathscr{L}) \leq nq'$.

Exercise. Show, that one needs at most $N = r + n \leq n(q' + 1)$ operators L_ν to achieve this maximal value r at every point $v \in V$.

Let us call the subspace $\mathscr{L}$ *large* if there are some operators $L_\nu \in \mathscr{L}$, $\nu = 1, \ldots, N$, such that $\bar{L} = \bigoplus_1^N L_\nu$ is a free operator.

Example. With a fixed operator $L: X^{(1)} \to G$ one gets a linear space of operators l acting on functions $f: V \to \mathbb{R}$, by taking $l(f) = L(fx)$ for all sections $x: V \to X$. This space is large iff L is a free operator.

Remark. Observe, that the "large" property depends only on the 1-jets of sections in $\mathscr{L}$, namely on the span of the one jets $J_L^1: V \to H^{(1)}$ for all $L \in \mathscr{L}$. In particular, if these jets span the whole bundle $H^{(1)} \to V$, then the space $\mathscr{L}$ is large.

Exercise. Show that a generic subspace $\mathscr{L} \subset \mathscr{H}$ of dimension $N \geq q'n + q' + n$ is large.

More on Natural Bundles. Let $G \to V$ be a tensor bundle and let $X \to V$ be the tangent bundle $T(V)$. For a fixed tensor $g \in G$, the Lie derivative $x(g) = \partial g / \partial x$, for tangent fields $x: V \to X$, defines a linear first order operator $L_g: X^{(1)} \to G$, namely $L_g(x) = x(g)$. These operators L_g for all $g: V \to G$ form a linear subspace $\mathscr{L}$ in $\mathscr{H}$. We call the tensor bundle G *large* if this space $\mathscr{L}$ is large.

Exercise. Show that the exterior powers of degrees $< \dim V$ and the symmetric powers of the cotangent bundle $T^*(V)$ are large bundles. (Probably, all tensor bundles are large with a few obvious exceptions.)

Now we take a C^∞-section $g: V \to G$ and ask ourselves under what condition the span $\mathscr{G}' = \mathscr{G}'(g) \subset \mathscr{G}^\infty$ of the Diff(V) orbit of g is equal to the whole space $\mathscr{G}^\infty$, that is the space of C^∞-sections $V \to G$. First we observe the following obvious necessary condition that is reminiscent of the h-principle.

If $\mathscr{G}' = \mathscr{G}^\infty$, then, for each $r = 0, 1, \ldots,$ *the jets* $J_{g'}^r: V \to G^{(r)}$, *for all* $g' \in \mathscr{G}'$, *span the bundle* $G^{(r)}$.

Theorem. *If V is a compact manifold without boundary, if the bundle $G \to V$ is large, and if the 2-jets $J_{g'}^2: V \to G^{(2)}$, for all $g' \in \mathscr{G}' = \mathscr{G}'(g)$, span the bundle $G^{(2)} \to V$, then $\mathscr{G}' = \mathscr{G}^\infty$.*

Proof. Let $\mathscr{L}'$ denote the linear subspace in $\mathscr{L}$ of the operators $L_{g'}: X^{(1)} \to G$ for all $g' \in \mathscr{G}'$. The correspondence $g \mapsto L_g$ is a first order operator, $G^{(1)} \to H$, and since the jets $J^2_{g'}: V \to G^{(2)}$ for $g' \in \mathscr{G}'$ span the bundle $G^{(2)}$ the 1-jets of sections $L': V \to H$ for all $L' \in \mathscr{L}'$ span the same subbundle in $H^{(1)}$ as the 1-jets of all sections $L \in \mathscr{L} \supset \mathscr{L}'$.

Now, as $\mathscr{L}$ is a large space, we conclude with the above remark that $\mathscr{L}$ is also large. It follows, that the operator $L = \bigoplus_1^N L_{g_\nu}$ is free for sufficiently many sections $g_1, \ldots, g_\nu, \ldots, g_N$ of the form $g_\nu = d_\nu(g)$, where $d_\nu \in \mathrm{Diff}(V)$ are *generic* diffeomorphisms. Therefore, by solving the system (***) above we invert all these operators L by some zero order operators M whose coefficients smoothly depend on diffeomorphisms d_ν and on their derivatives.

Finally, define a first order non-linear operator $\mathscr{D}$ by $\mathscr{D}(d_1, \ldots, d_n) = \sum_1^N g_\nu = \sum_1^N d_\nu(g)$. The linearization of $\mathscr{D}$ at $(d_1, \ldots, d_N)$, $d_\nu \in \mathrm{Diff}(V)$, is exactly the operator L, and so the inversions M for the linearizations near $(d_1, \ldots, d_N)$ yield an infinitesimal inversion of $\mathscr{D}$. Hence, the operator $\mathscr{D}$ has open range, and so $\mathscr{G}' = \mathscr{G}^\infty$, (compare with 2.3.7).

Exercises. Prove that the action of $\mathrm{Diff}(V)$ in the space of symmetric differential forms of degree $k \geq 2$ is (algebraically) irreducible.

Let $\mathscr{G}^\infty$ be the space of exterior C^∞-forms on V of degree $k \geq 0$. Show that every proper ($\neq \mathscr{G}^\infty$) $\mathrm{Diff}(V)$-invariant subspace in $\mathscr{G}^\infty$ consists of closed k-forms. Establish a one-to-one correspondence between invariant subspaces in $\mathscr{G}^\infty$ and invariant subspaces in the cohomology group $H^k(V, \mathbb{R})$.

Question. How does one classify invariant subspaces of sections of an arbitrary natural bundle $G \to V$?

(E) *Generic Under-determined Operators.* We now denote by H the bundle $\mathrm{Hom}(X^{(r)}, G)$ for some $r \geq 0$ and we consider its C^∞-sections that are differential operators $L: X^{(r)} \to G$. Then we take an open set (relation) $A \subset H^{(r+s)}$ for some $s \geq 0$, we denote by $\mathscr{L} = \mathscr{L}(A)$ the space of C^∞-solutions $L: V \to H$ of A and we consider differential C^∞-operators $M = M(L, g)$, $M: \mathscr{L} \times \mathscr{G}^\infty \to \mathscr{X}^\infty$, that have order $s + r$ in L and that are linear of order s in g. Such an operator M is called *a universal right inversion over A* if for all $L \in \mathscr{L}$ it satisfies

$$L \circ M(L, \ ) = \mathrm{Id},$$

that is $L(M(L, g)) = g$ for all C^∞-sections g. One also defines *universal left inversions* for operators $L: G^{(r)} \to X$ as differential operators $M(L, x)$, of order s in L and in x, such that $M(L, \ ) \circ L = \mathrm{Id}$ for all solutions L of a given open relation in the space of s-jets of the bundle $\mathrm{Hom}(G^{(r)}, X)$.

Observe that right inversions solve equations $L(x) = g$, while with left inversions one gets the uniqueness of solutions for the homogeneous equations $L(g) = 0$.

If one has two universal right inversions, M_1 over $A_1 \subset H^{(r+s)}$ and M_2 over $A_2 \subset H^{(r+s)}$, then with a partition of unity in $A = A_1 \cup A_2$ one constructs an inversion over A, and so there is a unique maximal open set $A = A(r + s) \subset H^{(r+s)}$ over which universal inversions exist. The same conclusion holds true for universal left inversions.

Example. If $r = 0$, then differential operators are homomorphisms $L: X \to G$ and they are universally right invertible over the set $A(0) \subset \mathrm{Hom}(X, G)$ of *surjective* homomorphisms. The complement to this $\Lambda(0)$, called $\Sigma^* = H \setminus \Lambda(0)$, is a stratified set of codimension $q - q' + 1$ for $q = \dim X$ and $q' = \dim G$, and so the set $\Lambda(0)$ is dense as well as open for $q \geq q'$. Futhermore, for $q \geq q' + n$ we have codim $\Sigma^* > n = \dim V$, and so *generic* homomorphisms $X \to G$ are right invertible.

Let now $r \geq 1$. Then the operator $L\colon X^{(r)} \to G$ is, definitely, not invertible, at the points where it is not surjective as a homomorphism, and so the complement $\Sigma^*(r + s) = H^{(r+s)} \setminus \Lambda(r + s)$ must have codimension $\leq p - q' + 1$, for $p = \dim X^{(r)} = q((n + r)!/n!r!)$.

Exercise 1. Show that determined operators, (i.e. $q = q'$) have no *universal* inversions at all. Moreover, generic determined operators of order $r \geq 1$ are not invertible. (Neither from the right nor from the left.)

Theorem. *If $q > q'$, then for all sufficiently large numbers $s \geq s_0(q, q', r, n)$, the set $\Lambda = \Lambda(r + s) \subset H^{(r+s)}$, over which universal right inversions exist, have the following properties.*

(i) *Λ is dense as well as open*;
(ii) *if $r \geq 1$, then the complement $\Sigma^* = \Sigma^*(r + s) = H^{(r+s)} \setminus \Lambda$ has codimension at least $n + 1$*;
(iii) *both statements*, (i) *and* (ii), *hold true for universal left inversions of operators $G^{(r)} \to X$.*

Remark. The statement (ii) sharpens (i) and we also shall see in the course of the proof that Σ^* is, in fact, a stratified set. The inequality codim $\Sigma^* \geq n + 1$ implies invertibility (from the right) of *generic* operators $L\colon X^{(r)} \to G$, since their jets, $J_L^{r+s}\colon V \to H^{(r+s)}$, do not intersect Σ^* (see 1.3.2).

Exercise 2. Let $H \to V$ be an arbitrary fibration and let H_0 be C^∞-submanifold in the jet space $H^{(r)}$. Show, that there is a *non-empty* open subset in H_0 consisting of those points $h \in H_0$, that can be represented by sections $V \to H$ whose r-jets $V \to H^{(r)}$ meet H_0 without tangency. For codim $H_0 \leq n$ this is equivalent to the existence of a non-empty open set of sections $V \to H$ whose r-jets intersect H_0. If codim $H_0 > n$ then, for every such $h \in H_0$, there exists a section $L\colon V \to H$, such that $J_L^r(v) = h$ for some $v \in V$, while the image of the differential $T_v(V) \to T_h(H^{(r)})$ of this jet has only zero intersection with $T_h(H_0) \subset T_h(H^{(r)})$.

Let us indicate three corollaries of the theorem. [Compare Ritt (1950).]

(1) Take some linear differential operators $L_1, \ldots, L_k$ acting on functions $V \to \mathbb{R}$ and consider the right ideal spanned by $L_1, \ldots, L_k$ in the ring of differential operators. *If $k \geq 2$, then for generic operators $L_1, \ldots, L_k$ this ideal is equal to the whole ring, and the same is true for the left ideal spanned by $L_1, \ldots, L_k$.*

Remark. A single operator of positive order never spans the whole ring.

(2) *A generic non-linear under-determined operator $\mathcal{D}\colon \mathcal{X}^{r+\alpha} \to \mathcal{G}^\alpha$ of order $r \geq 1$, admits an infinitesimal inversion over an open dense set $\mathcal{G} \subset \mathcal{X}^{2r+s}$ for some sufficiently large s.*

(3) Generic *over-determined* operators, $\mathscr{D}: \mathscr{G}^{r+\alpha} \to \mathscr{X}^{\alpha}$, enjoy the following *rigidity property*.

For every $x \in \mathscr{X}^{\infty}$ *the set of solutions of the equation* $\mathscr{D}(g) = x$ *is discrete, that is there is no smooth non-constant families of solutions,* $g_t \in \mathscr{G}^{\infty}$ *for* $t \in [0, 1]$.

Observe, that for an *arbitrary* over-determined C^{∞}-operator $\mathscr{D}$, the equation $\mathscr{D}(g) = x$ is C^{∞}-unsolvable for generic sections $x \in \mathscr{X}^{\infty}$.

The Proof of (i). We assume our fibrations to be trivial, $X = \mathbb{R}^q \times \mathbb{R}^n \to \mathbb{R}^n$ and $G = \mathbb{R}^{q'} \times \mathbb{R}^n \to \mathbb{R}^n$, and we write operators $L: X^{(r)} \to G$ as

$$L = \sum_{|I| \leq r} L^I \partial^I,$$

where $I = (i_1, \ldots, i_n)$, $\partial^I = \partial^I / \partial u_1^{i_1} \ldots \partial u_n^{i_n}$, $|I| = i_1 + \cdots + i_n$, and where $L^I = (L^I_{\alpha\beta})$ are $q' \times q$ matrix functions in the variables $u_1, \ldots, u_n$, called the *coefficients* of L.

The coefficients of the composition $P = L \circ M$ for $M = \sum_{|J| \leq s} M^J \partial^J$ are bilinear functions $P^K_{\mu\nu}$, for $\mu, \nu = 1, \ldots, q'$ and $|K| \leq r + s$, in the coefficients $L^I_{\alpha\beta}$ and in the derivatives $\partial^I \mathrm{M}^J_{\beta\alpha}$; therefore, the relation $L \circ M = \mathrm{Id}$ is expressed by a system of partial *differential* equations in the unknowns $M^J_{\beta\alpha}$. On the other hand, the left inversion equation, $M \circ L = \mathrm{Id}$ for a given operator $L: G^{(r)} \to X$, is expressed by a system of linear *algebraic* equations in the unknowns $M^J_{\alpha\beta}$, while the coefficients of these equations are linear combinations of the derivatives $\partial^J L^I_{\beta\alpha}$. This last algebraic system is much easier to handle and the following classical operation of *conjugation* shows that the solution of the P.D.E. system $L \circ M = \mathrm{Id}$ is equivalent to finding solutions $\bar{M}$ of the algebraic system $\bar{M} \circ \bar{L} = \mathrm{Id}$, where $\bar{M}$ and $\bar{L}$ are the (formal adjoint) operators conjugate to M and L. Namely, the operator $\bar{L} = \bar{L}(g) = \bar{L}(g_1, \ldots, g_{q'})$ for $L = \sum L^I \partial^I$, $L = L(x) = L(x_1, \ldots, x_q)$, is given by

$$\bar{L}(g) = \sum_{|I| \leq r} (-1)^{|I|} \partial^I ((L^I)^t g),$$

where $(\ )^t$ denotes the transposition of matrices, and $\bar{M}$ is defined in the same way. Observe that the functions $\bar{L}^I_{\beta\alpha}$, for $\bar{L} = \sum_{|I| \leq r} \bar{L}^I_{\beta\alpha} \partial^I$, are combinations of partial derivatives $\partial^{I'} L^I_{\alpha\beta}$ for $|I'| \leq |I|$, that is the operation $L \mapsto \bar{L}$ is, in fact, a linear partial differential operator, $H^{(r)} \to \bar{H} = \mathrm{Hom}(G^r, X)$. On the other hand, $\bar{\bar{L}} = L$ and $\overline{L \circ M} = \bar{M} \circ \bar{L}$, so that the equations $L \circ M = \mathrm{Id}$ and $\bar{M} \circ \bar{L} = \overline{\mathrm{Id}} = \mathrm{Id}$ are indeed equivalent.

Now we concentrate on the left inversion equation, $M \circ L = \mathrm{Id}$, and we write it as

$$(*) \quad \begin{cases} M^0 L^0 + \cdots + M^J \partial^J L^0 + \cdots + M^s \partial^s L^0 = \delta \\ M^0 L^1 + \cdots + M^J (\partial^J L^1 + \partial^{J-1} L^0) + \cdots + M^s (\partial^s L^1 + \partial^{s-1} L^0) = 0 \\ \cdots\cdots\cdots\cdots\cdots\cdots\cdots\cdots\cdots\cdots\cdots\cdots\cdots\cdots \\ M^s L^r = 0 \end{cases}$$

where $\partial^J L^I$, $|J| \leq s$, $|I| \leq r$, are given $q \times q'$ matrix functions on $\mathbb{R}^n$, where M^J are unknown $q' \times q$ matrix functions and where δ denotes the diagonal $q' \times q'$ matrix with the unit diagonal entries. Observe, that the first line in $(*)$ contains $(q')^2$ equations, the second line represents $n(q')^2$ equations and the last one, $M^s L^r = 0$, schematically expresses $(q')^2((n + r + s - 1)!/(n - 1)!(r + s)!)$ equations.

There are $(q')^2((n+r+s)!/n!(r+s)!)$ equations in $(*)$ altogether and the number of unknown functions $M^J_{\alpha\beta}$ is equal to $qq'((n+s)!/n!s!)$. The coefficients of these equations are linear combinations of the derivatives $\partial^J L^I_{\beta\alpha}$, and let us interpret these derivatives as independent variables, that parametrize the fibers of the fibration $\bar{H}^{(s)} = \mathbb{R}^N \times \mathbb{R}^n \to \mathbb{R}^n$ for $\bar{H} = \mathrm{Hom}(G^{(r)}, X)$, where $N = N(s) = qq'((n+r)!(n+s)!/(n!)^2 r!s!)$. Denote by $\mathcal{N}$ the field of rational functions in the variables $\partial^J L^I_{\beta\alpha}$.

Lemma 1. *If* $s \geq n((q/q')^{1/r} - 1)^{-1}$, *then the system* $(*)$ *admits a solution* $(M^J_{\alpha\beta})$, $\alpha = 1, \ldots, q'$, $\beta = 1, \ldots, q$, $|J| \leq s$, *in the field* $\mathcal{N}$.

Remark. Since the coefficients of $(*)$ are (linear) polynomials on $\mathbb{R}^N$ there is a uniquely defined *Zariski open* set $\tilde{\Lambda} \subset \mathbb{R}^N$, such that $(*)$ admits a solution by some C^∞-functions $M^J_{\alpha_\beta}$ on $\tilde{\Lambda}$, and there is no such solution near the points of the complement $\tilde{\Sigma} = \mathbb{R}^N \backslash \tilde{\Lambda}$. It follows that the set $\Sigma^* \subset H^{(r+s)}$ (of the statement of the theorem) is indeed a stratified set which is algebraic over each coordinate neighborhood in V. The lemma implies that $\tilde{\Sigma}$ is contained in the poles of the rational functions $M^J_{\alpha\beta}$ on $\mathbb{R}^N$ that satisfy $(*)$ and so the set Σ^*, as well as $\tilde{\Sigma}$, must have positive codimension.

Proof of the Lemma. For every $\alpha_0 = 1, \ldots, q'$ we take the row $M^J_{\alpha_0} = M_{\alpha_0\beta}$ in each of the matrices $M^J_{\alpha\beta}$ and then the system $(*)$ splits into q' independent systems

$$
(*\alpha_0) \qquad \begin{cases} M^0_{\alpha_0} L^0 + \cdots + M^s_{\alpha_0} \partial^s L^0 = \delta(\alpha, \alpha_0) \\ \cdots\cdots\cdots\cdots\cdots\cdots\cdots\cdots\cdots\cdots \\ \qquad\qquad\qquad M^s_{\alpha_0} L^r = 0, \end{cases}
$$

where $\delta(\alpha, \alpha_0)$ equals one for $\alpha = \alpha_0$ and is zero otherwise. There is only one non-homogeneous equation in this system, namely

$$
M^0_{\alpha_0} L^0_{\alpha_0} + \cdots + M^s_{\alpha_0} \partial^s L^0_{\alpha_0} = 1,
$$

where $L^0_{\alpha_0}$ denotes the column $L^0_{\beta\alpha_0}$, and to prove the existence of a solution $(M^J_{\alpha_0})$ of $(*\alpha_0)$ in the field $\mathcal{N}$ we must only show that the "non-homogeneous row,"

$$
(L^0_{\alpha_0}, \ldots, \partial^J L^0_{\alpha_0}, \ldots, \partial^s L^0_{\alpha_0}),
$$

is not a combination with coefficients in $\mathcal{N}$ of the rest of the rows, called "homogeneous rows," of the system $(*\alpha_0)$. Observe, that the entries of these "homogeneous rows" are linear combinations with integer coefficients of the variables $\partial^J L^0_{\beta\alpha}$ and $\partial^J L^I_{\beta\alpha}$ for $|I| > 0$. Furthermore, a variable $\partial^J L^0_{\beta\alpha}$, $|J| = 0, \ldots s$, appears with a non-zero coefficient in an entry of a "homogeneous row" only if this entry is contained in a column that *follows* the column containing this very $\partial^J L^0_{\beta\alpha}$ in the "non-homogeneous row". For example, the variable $L^0_{1\alpha_0}$ appears as the first entry in the "non-homogeneous row," but it does not appear anymore in the first column of the system $(*\alpha_0)$.

Now, suppose that the "non-homogeneous row" is represented as a combination with some coefficients $f_1, \ldots, f_k \in \mathcal{N}$, $k = q'[(n+s+r)!/n!(s+r)!] - 1$, of the "homogeneous rows". Then every entry $\partial^J L^0_{\beta\alpha_0}$ of the "non-homogeneous row" can be expressed as a polynomial in the variables $f_1, \ldots, f_k$ and $\partial^J L^I_{\beta\alpha}$ for $|I| > 0$. Indeed, we start with the first column in $(*\alpha_0)$ and get $L^0_{0\alpha_0}$ as a linear combination of some $L^I_{1\alpha}$, for $|I| > 0$, with the coefficients $f_1, \ldots, f_k$. Then we substitute $L^0_{1\alpha_0}$, as it appears in the following columns, by this combination and express the second entry in the "non-homogeneous row" as a polynomial quadratic in $f_1, \ldots, f_k$ and linear in some $\partial^J L^I_{\beta\alpha}$ for $|I| > 0$, and so on.

As the entries $\partial^J L^0_{\beta\alpha_0}$ are independent variables, in $\mathcal{N}$, their number, $q((n+s)!/n!s!)$, can not exceed the number k of the "homogeneous rows" and so we must have

$$q\frac{(n+s)!}{n!s!} < q'\frac{(n+s+r)!}{n!(s+r)!},$$

but this inequality is incompatible with our assumption,

$$s > n[(q/q')^{1/r} - 1]^{-1}. \text{ Q.E.D.}$$

The Proof of the Inequality codim $\Sigma^* > n$. Take an operator $L\colon X^{(r)} \to G$, for $r \geq 1$ on $V = \mathbb{R}^n$, namely $L = \sum_{|I| \leq r} L^I \partial^I$, and a C^∞-function $P\colon \mathbb{R}^n \to \mathbb{R}$. Let us expand $L(P^r x)$ according to Leibniz' formula and write

$$L(P^r x) = L_r(x) + PL'(x),$$

where L_r and L' are differential operators in x whose coefficients are polynomials in $L^I_{\alpha\beta}$ and $\partial^I P$. Observe, that L_r is, in fact, a zero order operator that is a homomorphism $L_r\colon X \to G$, whose coefficients are linear functions in L^I for $|I| = r$ and polylinear of degree r in the first derivatives of P. For example, if $P(u_1, \ldots, u_n) = u_1$, then $L_r = r!L^I$ for $I = (r, 0, \ldots, 0)$. Let $V_0 \subset V$ denote the zero set of P and let us observe that over each point $v \in V_0$ the homomorphism L_r, that is the linear map called $L_r(v)\colon X_v \to G_v$, depends only on the differential $dP(v)$ and on $L^I(v)$ for $|I| = r$. If $dP(v) = 0$, then $L_r(v) = 0$, and for $dP(v) \neq 0$ we call the operator L *transversal* to the tangent space $T_v(V_0) \subset T_v(V)$, or briefly, to V_0 at v, if the map $L_r(v)$ is surjective. We say, otherwise, that L is tangent to V_0 at v_0. This definition agrees with the usual notion of tangency of vectorfields, $L = \sum_{i=1}^n l_i(\partial/\partial u_i)$, to V_0.

A rational function is called regular at some point if its denominator is not zero at this point. If we have a rational function $F = F(\partial^I f_\mu)$ in the partial derivatives of some functions f_μ on $\mathbb{R}^n$, we call F *regular* at $v \in \mathbb{R}^n$, if it is regular at $(\partial^I f_\mu(v))$.

The following lemma is an algebraic extract from the Cauchy-Kovalevskaya theorem.

Lemma 2. *Suppose that L is transversal to V_0 at $v \in V_0$. Then, for every $\rho = 0, 1, \ldots$, there are two operators $\hat{M}\colon G^{(s)} \to X$ and $\tilde{M}\colon G^{(s)} \to G$, for some $s = s(r, \rho)$, whose coefficients are regular at v rational functions in the derivatives $\partial^J L_{\alpha\beta}$ and $\partial^J P$ for $|J| \leq s$, such that in a neighborhood $U \subset \mathbb{R}^n$ of v one has $L \circ \hat{M} + P^\rho \tilde{M} = \mathrm{Id}$, and this identity holds for small perturbations of L and P.*

Proof. Let us expand $L(P^{r+\rho}x)$ as above and get

$$L(P^{r+\rho}x) = P^\rho L_{r\rho}(x) + P^{\rho+1}L'_\rho(x).$$

Observe, that $L_{r\rho} = [(r + \rho)!/r!]L_r$ and so the homomorphism $L_{r\rho}: X \to G$ is invertible near v. Now, if we already have $\hat{M}$ and $\tilde{M}$ such that $L \circ \hat{M} + P^{\rho}\tilde{M} = \mathrm{Id}$, we take

$$\hat{M}' = \hat{M} + P^{r+\rho}L_{r\rho}^{-1} \circ \tilde{M},$$

$$\tilde{M}' = -L'_{\rho} \circ L_{r\rho}^{-1} \circ \tilde{M},$$

and obtain

$$L \circ \hat{M}' + P^{\rho+1}\tilde{M} = \mathrm{Id}.$$

The proof is concluded with an obvious induction, starting with $\hat{M} = 0$ and $\tilde{M} = \mathrm{Id}$ for $\rho = 0$.

Let now V_0 be a submanifold in $V = \mathbb{R}^n$ of codimension $k > 0$, and let us call V_0 *characteristic* for L if L is tangent to all those hyperplanes in $T_v(V)$ for all $v \in V_0$, which contain the tangent spaces $T_v(V_0) \subset T_v(V)$.

Lemma 3. *Generic under-determined operators L have no characteristic submanifolds of codimension* >0.

Proof. First let V_0 be given by the equations $u_1 = 0, \ldots, u_k = 0$ and consider the k coefficients L^{I_j}, $j = 1, \ldots, k$ at the derivatives $\partial^{I_j} = \partial^r/\partial u_j^r$ for $j = 1, \ldots, k$. If V_0 is characteristic, then these coefficients have rank $<q'$ for all points $v \in V_0$. The conditions rank $L^{I_j}(v) < q'$ for $j = 1, \ldots, k$ define a subset of codimension $k(q - q' + 1)$ in the fiber $H_v \subset H = \mathrm{Hom}(X^{(r)}, G)$, and by taking derivatives along V_0 we get for every $s = 0, 1, \ldots,$ a subset $\Sigma_s \in H_v^{(s)}$ of codimension $c(s, k) = k(q - q' + 1)\dfrac{(n - k + s)!}{(n - k)!s!}$, such that Σ_s contains the s-jets of all those operators $L: V \to H$ for which V_0 is a characteristic submanifold. Next we consider all submanifolds that pass through $v \in V$ and that are given by equations $u_j = f_j(u_{k+1}, \ldots, u_n), j = 1, \ldots k$, and we observe that the space of $(s + 1)$-jets of these submanifolds, that is the space of $(s + 1)$-jets of the corresponding maps $f = (f_1, \ldots, f_k): \mathbb{R}^{n-k} \to \mathbb{R}^k$, has dimension $d(s, k) = k\left(\dfrac{(n - k + s + 1)!}{(n - k)!(s + 1)!} - 1\right)$. Finally, for the set $\Sigma_k^s \subset H_v^{(s)}$ of the jets of those operators L who have characteristic submanifolds that pass through the point $v \in V$, we get

$$\operatorname{codim} \Sigma_k^s \geq c(s, k) - d(s, k),$$

and so for $q' < q$ and for large s we get codim $\Sigma_k^s > n = \dim V$. Q.E.D.

Exercise 3. Show for all $s \geq 0$ that if $q = 1$, then

$$\operatorname{codim} \Sigma_n^s = (q - q' + 1)\frac{(n + r - 1)!}{(n - 1)!r!} = c(r),$$

and prove that codim $\Sigma_n^s > c(r)$ for $q' \geq 2$ and $n \geq 2$. Prove that generic *determined* operators ($q = q'$) have no characteristic submanifold of codimension ≤ 2 for $q' \geq 2$.

Hint. Take $\tilde{\partial}^I = i_1(\partial/\partial u_1) + \cdots + (\partial/\partial u_n)$ for $I = (i_1, \ldots, i_n)$ and write operators L as

$$L = \sum_{m=0}^{r} \sum_{|I|=m} \tilde{L}^I(\tilde{\partial}^I)^m.$$

To prove our inequality codim $\Sigma^* > n$, we first observe that Lemmas 1, 2, and 3 extend to differential operators on $\mathbb{C}^n$ with holomorphic coefficients. Now we work in the complex analytic context. The derivatives $\partial^J L^I$ for $|J| \leq s$, $|I| \leq r$, are interpreted as the coordinates in $\mathbb{C}^N$, for $N = N(s) = qq' \dfrac{(n+r)!(n+s)!}{(n!)^2 r! s!}$, and let us denote by $\mathscr{P}(s)$ the ideal in the ring of polynomials on $\mathbb{C}^N$ that consists of those polynomials $P = P(\partial^J L^I)$ for which the equation $L \circ M = P$ Id has a solution M whose coefficients are polynomials in $\partial^J L^I$. Denote by $\Sigma(s) \subset \mathbb{C}^N$ the zero set of this ideal and observe that the natural projection $\Pi_s^{s'}: \mathbb{C}^{N(s')} \to \mathbb{C}^{N(s)}$ for $s' > s$, sends $\Sigma(s')$ to $\Sigma(s)$, and that according to Lemma 1 we have codim $\Sigma(s) > 0$ for a large. Let $\Sigma_0 \subset \Sigma(s)$ be an irreducible component of $\Sigma(s)$ such that codim $\Sigma_0 = k \leq n$.

Lemma 4. *For some* $s' > s$ *the pullback* $\Sigma_0' = \Sigma(s') \cap (\Pi_s^{s'})^{-1}(\Sigma_0) \subset \mathbb{C}^{N(s')}$ *satisfies the inequality* codim $\Sigma_0' \geq k + 1$.

Proof. Take some polynomials $P_1, \ldots, P_k$ that vanish on $\Sigma(s)$ and such that their differentials are linearly independent at some non-singular point $\sigma_0 \in \Sigma_0$. Then we take a generic operator L on $\mathbb{C}^n$ with holomorphic coefficients L^I such that for some point $v_0 \in \mathbb{C}^n$ the vector $(\partial^J L^I)(v_0 \in \mathbb{C}^{N(s)}$ is in Σ_0 close to σ_0, and such that the unduced functions

$$P_1^L = P_1^L(v) = P_1((\partial^J L_{\alpha\beta}^I)(v)), \ldots, P_k^L = P_k^L(v) = P_k((\partial^J L_{\alpha\beta}^I)(v)),$$

$v \in \mathbb{C}^n$, have independent differentials at the point v_0. Then the zero set of the function $P_1^L, \ldots, P_k^L$, called $\Sigma^L(s) \subset \mathbb{C}^n$, is a non-singular variety near $v_0 \in \Sigma^L(s)$ and by Lemma 3 it is not characteristic. Therefore, there is a linear combination $P^L = \lambda_1 P_1^L + \cdots + \lambda_k P_k^L$, such that the operator L is transversal to the zero set of P^L, called $V_0 \subset \mathbb{C}^n$, at some point $v_1 \in \Sigma^L(s) \subset V_0$ close to v_0. Next, by Hilbert's Nullstellensatz, for some number ρ the polynomial $P^\rho = (\lambda_1 P_1 + \cdots + \lambda_k P_k)^\rho$ is contained in $\mathscr{P}(s)$ and so we have an operator M_0 such that $L \circ M_0 = P^\rho$Id. Finally, according to Lemma 2, we have $L \circ \hat{M} + P^\rho \tilde{M} =$ Id, for some operators $\hat{M}$ and $\tilde{M}$, and with $P = P^L$, and so for $M = \hat{M} + M_0 \circ \tilde{M}$ we get $L \circ M =$ Id. The coefficients of this operator M, of some order $\bar{s} > s$, are rational functions in the derivatives $\partial^J L^I$ for $|J| \leq \bar{s}$ and these functions are regular at $v_1 \in \Sigma^L(s) \subset \mathbb{C}^n$. Since the point $(\partial^J L^I(v_1)) \in \mathbb{C}^{N(\bar{s})}$ projects to Σ_0, the common denominator of these functions, call it Q, is not identically zero on the pullback $\bar{\Sigma}_0 = (\Pi_s^{\bar{s}})^{-1}(\Sigma_0)$. On the other hand, the operator M' defined by $M'(g) = M(Q^{\bar{s}+1}(g))$ has for the coefficients some *polynomials* in $\partial^J L^I$ for $|J| = s' = 2\bar{s}$, and since $L \circ M' = Q^{\bar{s}+1}$Id we get $Q^{\bar{s}+1}$ in the ideal $\mathscr{P}(s')$ and thus we prove our lemma.

Now, with this lemma we get codim $\Sigma(s) > n$ for large s, so we have the same inequality for our set $\Sigma^* \subset H^{(s+r)}$ and the proof of the theorem is completed.

Exercise 4. Show that

$$\operatorname{codim} \Sigma^* \geq (q - q' + 1)\frac{(n + r - 1)!}{(n - 1)!r!}.$$

Hint. See Exercise 3.

(E′) *Non-free Isometric Immersions.* Let us return to the system (L) of 2.3.1 in the unknown map $y\colon V \to \mathbb{R}^q$,

$$\left\langle \frac{\partial x}{\partial u_i}, \frac{\partial y}{\partial u_j} \right\rangle + \left\langle \frac{\partial x}{\partial u_j} \frac{\partial y}{\partial u_i} \right\rangle = g_{ij}. \tag{L}$$

If the map $x\colon V \to \mathbb{R}^q$ is free, then this system is algebraically solvable and so for $q \geq [n(n + 1)/2] + n$ we have infinitesimal invertibility of the metric inducing operator $\mathscr{D}$ over a *non-empty* open set of maps $x\colon V \to \mathbb{R}^q$. This system (L) is still underdetermined for $[n(n + 1)/2] + n > q > n(n + 1)/2$, but unfortunately, it is not "sufficiently generic" to use our general theorem.

Problem. Is the operator $\mathscr{D}$ infinitesimally invertible over a dense (or at least non-empty) open set of maps $x\colon V \to \mathbb{R}^q$ for all $q > n(n + 1)/2$?

Let us indicate an approach to this problem for $[n(n + 1)/2] + n \geq q \geq [n(n + 1)/2] + n - m$, where $m \approx \sqrt{n/2}$. We modify the system (N) of 2.3.1 by introducing new unknown functions $h_i\colon V \to \mathbb{R}^q$, $i = 1, \ldots n$, and then we write

$$\left\langle \frac{\partial x}{\partial u_i}, y \right\rangle = h_i. \tag{Nh}$$

Next, we differentiate this system and reduce (L) to

$$\left\langle \frac{\partial^2 x}{\partial u_i \partial u_j}, y \right\rangle = \frac{1}{2}\left(\frac{\partial h_i}{\partial u_j} + \frac{\partial h_j}{\partial u_i} - g_{ij}\right), \tag{N*h}$$

as in 2.3.1.

The $[n(n + 1)/2] + n$ linear algebraic equations (Nh) + (N*h), are solvable in y, *iff* the right hand sides of these equation satisfy the consistency condition that is expressed for generic maps x by a system of P.D. equations in the unknowns h_i, [compare with (B)]

$$Ah + \sum_{i=1}^{n} B_i \frac{\partial h}{\partial u_i} = \tilde{g}, \tag{h}$$

where $h = (h_1, \ldots, h_n)$ and A and B_i are some $m \times n$ matrix functions on V for $m = q - [n(n + 1)/2] - n$, whose entries are some rational functions in the derivatives $\partial x/\partial u_i$ and $\partial^2 x/\partial u_i \partial u_j$. These functions are regular as long as the derivatives span the whole space $\mathbb{R}^q$, $q = [n(n + 1)/2] + n - m$. The system (h) contains m equations and so it is under-determined for $m < n$, but it fails to be generic since the $n(n + 1)m$ entries of the matrices A and B_i are expressed in the derivatives of q

functions, and there must be at least $n(n+1)m - q$ differential relations between these entries. Now, if we "truncate" the system (h) by letting $h_{m+2} = 0, \ldots, h_n = 0$, we get another system in the unknowns $(h_1, \ldots, h_{m+1}) = \tilde{h}$,

$$(\tilde{\text{h}}) \qquad \tilde{A}\tilde{h} + \sum_{i=1}^{n} \tilde{B}_i \tilde{h} = \tilde{g},$$

where the total number of the entries in $\tilde{A}$ and $\tilde{B}_i$ is $(n+1)m(m+1)$, and so for $q = [n(n+1)/2] + n - m \geq (n+1)m(m+1)$ the system $(\tilde{\text{h}})$ has a fair chance to be generic for generic maps $x: V \to \mathbb{R}^q$.

Exercise. Apply *the proof* of our "generic theorem" to the system (h), for $q \geq q_0 = [n(n+1)/2] + n - \sqrt{n/2}$, and prove $\mathscr{D}$ to be infinitesimally invertible over generic maps $x: V \to \mathbb{R}^q$ for $q \geq q_0$.

(F) *Completely Integrable Systems.* A first order system of P.D.E. in unknowns $(x_1, \ldots, x_q) = x: \mathbb{R}^n \to \mathbb{R}^q$ is called *complete* if it can be written as

$$(0) \qquad \Phi_\mu(x, u_1, \ldots, u_n) = 0, \qquad \mu = 1, \ldots, m,$$

$$(1) \qquad \Psi_i(x, u_1, \ldots, u_n) = \frac{\partial x}{\partial u_i}, \qquad i = 1, \ldots, n,$$

and so the 1-jets of x are uniquely determined by its 0-jets. The system (0) + (1) is called *integrable* if $(\partial \Phi_\mu / \partial x) \Psi_i + (\partial \Phi_\mu / \partial u_i) = 0$ and if the identities $\partial \Psi_i(x, \ldots,)/\partial u_j = \partial \Psi_j(x, \ldots,)/\partial u_i$, $i, j = 1, \ldots, n$, hold for all solutions x of the system (0). Frobenius' theorem says in the integrable case that *0-jets of solutions x of* (0) *uniquely extend to germs of solutions of the joint system* (0) + (1).

The integrability condition may also be expressed by saying that the 0-jets of solutions of (0) extend to 2-jets of infinitesimal solutions of (0) + (1). More generally, take a differential relation $\mathscr{R} \subset X^{(r)}$ and suppose $\mathscr{R}$ is a C^∞-submanifold in $X^{(r)}$. The completeness condition now says that the projection $\mathscr{R} \to X^{(r-1)}$ is an embedding, and we slightly relax this condition by allowing this projection to be an immersion. Then we consider the "lift" $\mathscr{R}^1 \subset X^{(r+1)}$ (compare 1.1.1) that consists of $(r+1)$-jets of germs of those sections $V \to X$, whose r-jets $V \to X^{(r)}$ are *tangent* to $\mathscr{R}$. To get a better picture, let us first call a tangent n-plane, $\tau \in T_x(X^{(r)})$ for $x \in X^{(r)}$ and $n = \dim V$, *holonomic* if there exists a holonomic (see 1.1.1) section $\varphi: V \to X^{(r)}$ tangent to τ, that is τ equals the image of the differential $d_v \varphi: T_v(V) \to T_x(X^{(r)})$ for $v = p^r(x) \in V$. Then the space $X^{(r+1)}$ is canonically isomorphic to the space of all holonomic n-planes in $X^{(r)}$ and $\mathscr{R}$ is equal to the space of holonomic planes in $T(\mathscr{R}) \subset T(X^{(r)})$.

If the projection $\mathscr{R} \to X^{(r-1)}$ is an immersion, then at each point $x \in \mathscr{R}$ there is at most one holonomic n-plane in $T_x(\mathscr{R})$, and we call such a relation $\mathscr{R}$ integrable if holonomic n-planes exist in all spaces $T_x(\mathscr{R})$, $x \in \mathscr{R}$, or equivalently, if the map $\mathscr{R}^1 \to \mathscr{R}$ is surjective as well as injective. In this case we get a smooth field of holonomic n-planes tangent to $\mathscr{R}$ and by Frobenius' theorem for every point $x \in \mathscr{R}$ there is a unique germ of a holonomic jet $\varphi: \mathcal{O}p(v) \to \mathscr{R}$ for $v = p^r(x) \in V$, such that $\varphi(v) = x$, and so we have a local solution of $\mathscr{R}$, $f: \mathcal{O}p(v) \to X$ for which $J_f^r = \varphi$. Therefore, the whole manifold $\mathscr{R}$ is, in fact, foliated by holonomic germs φ that may be interpreted as germs of leaves of this foliation; the global leaves project to V by immersions, and when such an immersion is a diffeomorphism we get a global solution of $\mathscr{R}$ over V.

There are two essential obstructions to the existence of global solutions: a continuation of a local solution goes to infinity in finite time, as for the equation $dx(t)/dt = x^2(t) + 1$, or solutions become multivalued over V for $\pi_1(V) \neq 0$, as for the equation $dx(t)/dt = 1$ for $t \in S^1 = \mathbb{R}/\mathbb{Z}$.

Let us indicate several examples where global solutions of integrable relations $\mathscr{R}$ do exist.

(1) The projection $p^r: \mathscr{R} \to V$ is a diffeomorphism. Then the section $(p^r)^{-1}: V \to \mathscr{R}$ is holonomic *iff* $\mathscr{R}$ is integrable, and so the integrability is sufficient for the existence of global solutions of $\mathscr{R}$. Of course, there is only one solution in this case.

(2) Let $X \to V$ be a vector bundle and let $\mathscr{R} \subset X^{(r)}$ be an affine subbundle of the bundle $X^{(r)} \to V$. Then, if V is simply connected, all local solutions of $\mathscr{R}$ extend to global ones.

(3) Let V and W be Riemannian manifolds and let $X = V \times W \to V$. Suppose that the projection $\mathscr{R} \to X$ is a proper map and let all local solutions of $\mathscr{R}$ be *short* maps (see 1.2.2(B) $\mathcal{Op}(v) \to W$, $v \in V$. If V is simply connected and W is complete, then all local solutions uniquely extend to global solutions $V \to W$.

Now, let $\mathscr{R} \subset X^{(r)}$ be a submanifold, that admits at every point $x \in \mathscr{R}$ a unique tangent holonomic n-plane, but the projection $\mathscr{R} \to X^{(r-1)}$ is not necessarily injective. Take the lift $\mathscr{R}^1 \subset X^{(r+1)}$ and observe that the projection $\mathscr{R}^1 \to \mathscr{R}$ is a diffeomorphism, but the induced field of n-planes on $\mathscr{R}^1$ may not be holonomic. Then we inductively define $\mathscr{R}^s = (\mathscr{R}^{s-1})^1 \subset X^{(r+s)}$, assuming that $\mathscr{R}^{s-1} \subset X^{(r+s-1)}$ is a submanifold, and we denote by $\mathscr{R}_s \subset \mathscr{R}$ the projection of $\mathscr{R}^s$ back to $\mathscr{R}$. If for some $s \geq 2$ we have $\mathscr{R}_s = \mathscr{R}_{s-1}$ then the relation $\mathscr{R}^{s-1}$ is integrable as well as complete, so $\mathscr{R}_{s-1}$ is also complete and integrable, and, if $\mathscr{R}_s$ is non-empty, we get local solutions of the relation $\mathscr{R}_{s-1}$.

Example. Consider two Riemannian manifolds, (V, g) and (W, h) of dimension n, call a germ $x_0: \mathcal{Op}(v) \to W$, for $v \in V$, *an infinitesimal isometry of order* s if the induced metric $g_0 = x_0^*(h)$ satisfies $J^s_{g_0}(v) = J^s_g(v)$, and denote by $\mathscr{R}(s) \subset X^{(s+1)}$, for $X = V \times W \to V$, the set of the $(s+1)$-jets of such germs for all $v \in V$. Observe that $\mathscr{R}(0)$ coincides with the isometric immersion condition, $\mathscr{R}(0) = \mathscr{R} \subset X^{(1)}$, that is the principle $O(n)$-fibration, $\mathscr{R} \to X$, whose fiber $\mathscr{R}_x$, for $x = (v, w)$, consists of the isometric linear maps $T_v(V) \to T_w(W)$. Then we also have $\mathscr{R}(1) = \mathscr{R}^1$ and $\mathscr{R}_1 = \mathscr{R}$, that is each isometric map $l: T_v(V) \to T_w(W)$ extends to an infinitesimal isometry of order one. Futhermore $\mathscr{R}(2) = \mathscr{R}^2$, but $\mathscr{R}_2 \subset \mathscr{R}$ may be not equal to $\mathscr{R}$. In fact, the map l extends to a second order infinitesimal isometry *iff* it establishes an isomorphism between the respective curvature tensors, of V in $T_v(V)$ and of W in $T_w(W)$. For example, if $W = \mathbb{R}^n$, then the fibration $\mathscr{R} \to V$ is a principal $\mathrm{Is}(\mathbb{R}^n)$-fibration and the field of holonomic n-planes tangent to $\mathscr{R}$ is called the *canonical affine conection* of V. The identity $\mathscr{R}_2 = \mathscr{R}$ here amounts to the vanishing of the curvature of V and so, with Frobenius' theorem, we come to the following classical fact, *V is locally isometric to $\mathbb{R}^n$, iff its curvature is everywhere zero.*

A Riemannian manifold V is called *infinitesimally homogeneous of order* s_0, if for any two points v_1 and v_2 in V there is an infinitesimal isometry $\mathcal{O}p(v_1) \to \mathcal{O}p(v_2)$ of order s_0, sending v_1 to v_2. Observe that all manifolds have such homogeneity of order one, and that homogeneous manifolds are infinitesimally homogeneous for all s_0. If V is infinitesimally homogeneous of order s_0, then for all $s \leq s_0$ we have $\mathscr{R}(s) = \mathscr{R}^s$ and all subsets $\mathscr{R}_s \subset \mathscr{R}$ are subfibrations of the fibration $\mathscr{R} \to V$. Futhermore, if the projection $\mathscr{R}(s_0) \to X = V \times W$ is onto, then each subset $\mathscr{R}(s)$, for $s \leq s_0$, is also a subfibration of the fibration $\mathscr{R} \to X$ whose fiber $\mathscr{R}_x(s) \subset \mathscr{R}_x = O(n)$, $x \in X$, is a closed subgroup in $O(n)$, called $G(s) \subset O(n)$.

Exercise. Show that $\mathscr{R}(s)$ always contains $\mathscr{R}^s$ but in general, $\mathscr{R}(s) \not\subset \mathscr{R}^s$ for $s \geq 3$.

Denote by $\delta(n)$ the maximal length of descending chains of connected subgroups in $O(n)$.

Exercise. Show that $\delta(2) = 2$, $\delta(3) = 3$, $\delta(4) = 5$ and, in general, $\delta(n) < \frac{3}{2}n$.

Theorem. *If for every two points, $v \in V$ and $w \in W$, there is an infinitesimal isometry $\mathcal{O}p(v) \to \mathcal{O}p(w)$ of order $s_0 > \delta(n)$, which sends v to w, then*
(a) *there is a local isometry that sends v to w.*
(b) *If V is simply connected and if W is complete, then there is a global isometric immersion $V \to W$ that sends a given point $v \in V$ to a given point $w \in W$.*

Proof. Under our assumptions, the manifold V is infinitesimally homogeneous of order s_0 and since $s_0 > \delta(n)$ we have, for some $s < s_0$, $\dim G(s+1) = \dim G(s)$ and so $\dim \mathscr{R}_{s+1} = \dim \mathscr{R}_s$. Therefore, the connected component of $\mathscr{R}_s$ that contains some component of $\mathscr{R}_{s+1}$ is integrable. We get (a) by applying Frobenius' theorem and for (b) we use (3) above.

Question. What is the minimal s_0 for which the theorem holds? It is known that the theorem fails for $s_0 = 2$ if n is large. (Ferus-Karcher-Munzner 1981.)

Exercise. Generalize (a) to pseudo-Riemannian manifolds and also generalize (b) under the additional assumption of W being compact and simply connected.

Corollary (Singer 1960). *Complete simply connected Riemannian manifolds are homogeneous iff they are infinitesimally homogeneous of a sufficiently high order.*

A pseudo-Riemannian manifold V is called *weakly locally homogeneous* if for any two points v_1 in v_2 in V there are arbitrary small *isometric* neighborhoods, $\mathcal{O}p(v_1)$ and $\mathcal{O}p(v_2)$. One calls *V locally homogeneous* if there is an isometry between $\mathcal{O}p(v_1)$ and $\mathcal{O}p(v_2)$ which sends v_1 to v_2.

Exercise. Show, that if a pseudo-Riemannian C^∞-manifold V is weakly locally homogeneous then it is infinitesimally homogeneous (of any given order) and so it is locally homogeneous.

Question. Does the last conclusion hold for C^r-manifolds for some $r \leq \delta(n)$?

Exercises. Show that compact simply connected locally homogeneous pseudo-Riemannian C^r-manifolds are homogeneous for $r \geq 4$.

Hint. Show, with (2) above, that local Killing fields extend to global ones.

Show that complete simply connected locally homogeneous Riemannian C^0-manifolds are homogeneous.

Hint. See the book of Montgomery-Zippin (1955).

Question. What is the regularity condition under which the (weak) local homogeneity of a geometric structure on a compact simply connected manifold implies homogeneity?

(G) *Generic Over-determined Systems.* Consider a (non-linear) C^∞-operator $\mathscr{D}$: $\mathscr{X}^{\alpha+r} \to \mathscr{G}^\alpha$ and let Δ^s: $X^{(r+s)} \to G^{(s)}$ denote the associated maps of jets. These maps are generically surjective in the (under) determined case, for $\dim X = q \geq q' = \dim G$, but now let $q < q'$. Then, for large s, we have $\dim X^{(r+s)} = q((n+r+s)!/n!(r+s)!) < q'((n+s)!/n!s!) = \dim G^{(s)}$ and so the map Δ^s is far from being surjective. Therefore, there is a non-vacuous *consistency condition* for solvability of the equation $\mathscr{D}(x) = g$. Namely, the jet J_g^s: $V \to G^{(s)}$ must send V *into* the image of the map Δ^s for all s. If we denote by $\mathscr{R}(g,s) \subset X^{(r+s)}$ the pullback of the image $J_g^s(V) \subset G^{(r)}$ under the map Δ^s, that is $\mathscr{R}(g,s) = (\Delta^s)^{-1}(J_g^s(V))$, then this consistency condition can be equally expressed by saying that the projection $\mathscr{R}(g,s) \to V$ is onto. Observe, that for every $s' > s$, the projection $p_s^{s'}$: $X^{(r+s')} \to X^{(r+s)}$ sends $\mathscr{R}(g,s')$ into $\mathscr{R}(g,s)$. Let us stabilize the conditions $\mathscr{R}(g,s)$ by intersecting the images $p_s^{s'}(\mathscr{R}(g,s')) \subset \mathscr{R}(g,s)$ and put

$$\mathscr{R}[g,s] = \bigcap_{s' \geq s} p_s^{s'}(\mathscr{R}(g,s')) \subset \mathscr{R}(g,s) \subset X^{(r+s)}.$$

Now, for a *generic* operator $\mathscr{D}$, we have left inversions of its linearization $L_{\mathscr{D}}$, and it follows [as in the "rigidity corollary" to the theorem of (E)] that the projections p^s: $\mathscr{R}[g,s] \to V$ are *zero-dimensional* maps, that is every "fiber" $(p^s)^{-1}(v) \subset \mathscr{R}[g,s]$, for $v \in V$, is a zero-dimensional set.

Exercise. Let Δ: $\mathbb{R}^1 \to \mathbb{R}^2$ be an immersion with a non-empty set of transversal double points and let $\mathscr{D}(x) = \Delta \circ x$, for functions x: $\mathbb{R} \to \mathbb{R}^1$. This operator $\mathscr{D}$ has order zero and it sends functions x to maps g: $\mathbb{R} \to \mathbb{R}^2$. Analyse the consistency condition for the equation $\mathscr{D}(x) = g$ and study the sets $\mathscr{R}[g,s]$. Show, that the consistency condition is *not* sufficient for solvability of this equation.

Let us sharpen the consistency condition by requiring the existence of a C^∞-section $V \to \mathscr{R}[g,s_0]$ for some $s_0 \geq 0$.

Exercise. Show that if $\mathscr{D}$ is a generic operator and s_0 is sufficiently large, for example $s_0 \geq \exp\exp(n+r+q')$, then the sharpened consistency condition is sufficient for the solvability of the equation $\mathscr{D}(x) = g$, for any given C^∞-section g: $V \to G$. Show that the equations $\mathscr{D}(x) = g$ abides by the h-principle.

There are many interesting "not quite generic" cases when the projections $\mathscr{R}[g,s] \to V$ are not zero dimensional maps but the following two properties hold.

Stability. For all sufficiently large s_0 the projections $\mathscr{R}[g,s] \to \mathscr{R}[g,s_0]$ are injective for all $s > s_0$.

Regularity. For all $s > s_0$ the subsets $\mathscr{R}[g,s] \subset X^{(r+s)}$ are smooth submanifolds, the projections $\mathscr{R}[g,s] \to V$ are smooth fibrations and $(\mathscr{R}[g,s])^{[1]} = \mathscr{R}[g,s+1]$.

For such stable and regular relations $\mathscr{R}[g,s]$ one may apply Frobenius' theorem above and thus solve the equations $\mathscr{D}(x) = g$.

Example. Let $\mathscr{D}$ be a linear operator, such that for $g = 0$ the conditions $\mathscr{R}[g = 0, s]$ are stable and regular for large s, i.e. each $\mathscr{R}[g = 0, s] \subset X^{(r,s)}$ is a smooth subbundle of the bundle $X^{(r+s)} \to V$ and the dimension $\dim \mathscr{R}[g = 0, s]$ does not depend on s for $s \geq s_0$. Then, under the consistency condition for an equation $\mathscr{D}(x) = g$, the sets $\mathscr{R}[g,s] \subset X^{(r+s)}$ are also stable and regular for $s > s_0$, and by Frobenius' theorem, consistent equations $\mathscr{D}(x) = g$ are globally solvable if V is a simply connected manifold.

Exercise. Let $\mathscr{D} = (L_1, \ldots, L_{q'})$ for operators L_i with constant coefficients who act on functions $\mathbb{R}^n \to \mathbb{R}$. Show that the corresponding sets $\mathscr{R}[g = 0, s]$ are stable *iff* the characteristic equations $\hat{L}_1(w) = 0, \ldots, \hat{L}_{q'}(w) = 0$ have only finitely many common solutions $w \in \mathbb{C}^n$, and prove in this case the solvability of the system $\mathscr{D}(x) = g$, that is the system $L_i(x) = g_i$, $i = 1, \ldots, q'$, under the consistency condition.

Further Examples and Exercises. Let (V, g) be a Riemannian manifold of dimension $n \geq 3$. Then the isometric immersion equation, $\mathscr{D}(x) = g$, for maps $x\colon V \to \mathbb{R}^{n+1}$ is over-determined.

Show that there is no stability if g is a *flat* metric.

Show, for metrics g of *positive* sectional curvature, that under the consistency condition the equation $\mathscr{D}(x) = g$ is locally solvable, and if $\pi_1(V) = 0$, it is also globally solvable.

Give an example of a C^∞-metric g of *non-negative* curvature on $V = S^3$, such that the equation $\mathscr{D}(x) = g$ is consistent, moreover it is solvable near each point $v \in S^3$, but it is not globally solvable. Show that there are no C^{an}-metrics with such properties. Prove, that if a C^{an}-metric on a compact simply connected manifold V of dimension $n \geq 3$ without boundary admits a local isometric C^∞-immersion into $\mathbb{R}^{n+1}$ at every $v \in V$, then there is a global isometric C^{an}-immersion $x\colon V \to \mathbb{R}^{n+1}$. Show, this is in general not true for complete *non-compact* manifolds.

The stability condition for a general relation $\mathscr{R} \subset X^{(r)}$ says, in effect, that $\dim \mathscr{R}^s$ does not depend on s for s large. In some cases, the condition $\mathscr{R}$ is acted upon by a large symmetry group that makes this stabilization impossible, but Frobenius' theorem still works, when applied to an appropriate space of orbits. For example, consider the operator $\mathscr{D}$ that relates to a Riemannian metric g its curvature tensor $R = \mathscr{D}(g)$. For $R = 0$ this equation is consistent and it is also solvable, though the stabilization does not occur.

Questions. Is the consistency condition for the equation $\mathscr{D}(g) = R$ always sufficient for solvability?

As another interesting example consider the differential relation $\mathscr{R}$ expressing the property of infinitesimal homogeneity of order 2 for Riemannian metrics on V.

Does the dimension of the space of orbits, $\mathscr{R}^s/\mathrm{Diff}(V)$, stabilizes for large s? Does Frobenius' theorem apply?

Let V_0 be a non-complete manifold that is infinitesimally homogeneous of order 2. Under what condition there is a *complete* manifold V which is infinitesimally, of order 2, isometric to V_0?

Exercise. Give an example of a non-complete *locally* homogeneous Riemannian manifold V_0, such that no *complete* manifold V is locally isometric to V_0.

2.4 Convex Integration

The global techniques of the previous sections only apply to a rather narrow class of differential relations $\mathscr{R}$. The continuous sheaves theory (see 2.2) depends upon the invariance of $\mathscr{R}$ under some diffeomorphisms of the underlying manifold V, while the removal of singularities of maps $V \to \mathbb{R}^q$ (see 2.1) exploits the symmetry of $\mathscr{R}$ under certain transformations of $\mathbb{R}^q$.

In this section we approach more general relations $\mathscr{R}$ and we establish h-principle under mild geometric assumptions on $\mathscr{R}$.

2.4.1 Integrals and Convex Hulls

The key idea of the *method of convex integration* can be seen in the following

(A) **Example.** Let $f: S^1 \to \mathbb{R}^q$ be a C^1-map whose derivative $\varphi = df/ds: S^1 \to \mathbb{R}^q$ sends the circle S^1 into a given subset $A \subset \mathbb{R}^q$. Then the convex hull Conv A contains the origin $0 \in \mathbb{R}^q$.

In fact, the path connected component A_0 in A which receives S^1 satsifies Conv $A_0 \ni 0$. Indeed, $\int_{S^1} \varphi(s)\,ds = 0$ and this integral is the limit of Riemann sums that are positive linear combinations of some vectors $\varphi(s) \in \varphi(S^1) \subset A_0$. Q.E.D.

The converse is not true. For instance, the (connected!) ark $A = \{x_1^2 + x_2^2 = 1, x_1 \geq 1\}$ in the plane $\mathbb{R}^2$ admits no continuous maps $\varphi: S^1 \to A$ for which $\int_{S^1} \varphi(s)\,ds = 0$, although $0 \in \mathrm{Conv}\,A$.

However, if the convex hull of some path connected subset $A_0 \subset \mathbb{R}^q$ contains a small neighborhood of the origin, Conv $A_0 \supset \mathcal{O}_{\!\not p}0$, then there exists a map $f: S^1 \to \mathbb{R}^q$ whose derivative sends S^1 into A_0.

Proof. Take some points $a_1, \ldots, a_k$ in A_0 whose convex hull contains $\mathcal{O}_{\!\not p}0$. Since A_0 is path connected there exists a continuous map $\psi: S^1 \to A_0$ such that $\psi(s_i) = a_i$ for some $s_i \in S^1$, $i = 1, \ldots, k$. Let a *continuous* measure $d\mu_i$ on S^1 (i.e. the density function $d\mu_i/ds$ is strictly positive and continuous) approximate in the weak topology the Dirac measure $\delta(s - s_i)\,ds$. Then the integral $b_i = \int_{S^1} \psi(s)\,d\mu_i \in \mathbb{R}^q$ is close to $a_i = \psi(s_i) = \int_{S^1} \psi(s)\delta(s - s_i)\,ds$ for $i = 1, \ldots, k$. Since the origin $0 \in \mathbb{R}^q$ lies in the *interior* of the convex hull $\operatorname{Conv}\{a_i\}$, $i = 1, \ldots, k$, and since small perturbations of vectors only slightly move their convex hull, we have $0 \in \operatorname{Conv}\{b_i\}$, $i = 1, \ldots, k$, as the vectors b_i are close to a_i. Now write $0 = \sum_{i=1}^k p_i b_i$ for some $p_i \geq 0$, $\sum_{i=1}^k p_i = 1$, and observe that the *continuous* measure $d\mu = \sum_{i=1}^k p_i\,d\mu_i$ satisfies

$$\int_{S^1} \psi(s)\,d\mu = \sum_{i=1}^k p_i \int_{S^1} \psi(s)\,d\mu_i = \sum_{i=1}^k p_i b_i = 0.$$

Finally, we integrate the measure $d\mu$ to a C^1-diffeomorphism $\mu_*\colon S^1 \to S^1$ which pushes forward the Lebesgue measure ds to $d\mu$. Namely, we assume both measures to have total mass one, we put $\mu(s) = \int_{s_0}^s d\mu$ and we take the inverse of the diffeomorphism $s \mapsto \mu(s)$ for μ_*. Then the compositions of maps, $\varphi = \psi \circ \mu_*$, has $\int_{S^1} \varphi\,ds = \int_{S^1} \psi\,d\mu = 0$ and the map $\varphi\colon S^1 \to A_0$ integrates to the required map $f\colon S^1 \to \mathbb{R}^q$, whose derivative $df/ds = \varphi$ sends S^1 to A_0.

We need a generalization of this construction to families of maps.

Definition. We say that a map $\psi\colon V \times S \to \mathbb{R}^q$ *strictly surrounds* a given map $f_0\colon V \to \mathbb{R}^q$ if the vector $f_0(v) \in \mathbb{R}^q$ is contained in the interior of the convex hull of the ψ-image of the fiber $S_v = v \times S \subset V \times S$, that is

$$\mathcal{O}\!\mathit{p} f_0(v) \subset \operatorname{Conv}\psi(S_v) \qquad \text{for all } v \in V.$$

(B) **The Reparametrization Lemma.** *Let V be a smooth manifold and $S = [0, 1]$. If a C^r-map $\psi\colon V \times S \to \mathbb{R}^q$ strictly surrounds a C^r-map $f_0\colon V \to \mathbb{R}^q$ then there exists a fiber preserving C^r-diffeomorphism $\mu_*\colon V \times S \to V \times S$ ["fiber preserving" means $\mu_*(S_v) = S_v$ for all $v \in V$], such that the composition $\varphi = \psi \circ \mu_*\colon V \times S \to \mathbb{R}^q$ satisfies,*

$$\int_0^1 \varphi(v, s)\,ds = f_0(v) \qquad \textit{for all } v \in V.$$

Proof. As $\mathcal{O}\!\mathit{p} f_0(v) \subset \operatorname{Conv}\psi(S_v)$, there exist some measures on the fiber S_v, say $d\mu_i = d\mu_i(v)$, $i = 1, \ldots, k$, on S_v for all $v \in V$, such that the density functions $d\mu_i/ds$ are positive and C^∞-smooth in the variables s and v and such that the integrals $I_i(v) = \int_{S_v} \psi(v, s)\,d\mu_i \in \mathbb{R}^q$ strictly surrounds the vector $f_0(v) \in \mathbb{R}^q$. Namely,

$$\operatorname{Conv}\{I_i(v)\}_{i=1,\ldots,k} \supset \mathcal{O}\!\mathit{p} f_0(v) \qquad \text{for all } v \in V.$$

Then, there is a C^r-smooth partition of unity $p_i\colon V \to [0, 1]$, $\sum_{i=1}^k p_i \equiv 1$, such that $\sum_{i=1}^k p_i(v) I_i(v) = f_0(v)$ for all $v \in V$, and so the measure $d\mu(v) = \sum_{i=1}^k p_i(v)\,d\mu_i(v)$ on S_v has $\int_{S_v} \psi(v, s)\,d\mu(v) = f_0(v)$ for all $v \in V$. Finally, there exists a fiber preserving C^r-diffeomorphism μ_* of $V \times S$ which pushes forward Lebesgue measure $ds = ds(v)$ on S_v to $d\mu(v)$ for all $v \in V$. (This μ_* is unique if we require $\mu | V \times 0 = \mathrm{Id}$.) Clearly, $\int_0^1 \varphi(v, s)\,ds = f_0(v)$. Q.E.D.

The following proposition (which is not used in the sequel) explains the geometric significance of the lemma.

Convex Decomposition. Let A be a *connected* C^l-submanifold in $\mathbb{R}^q$ for some $l \geq 1$ and let f_0 be a C^r-map, $r \leq l$, of a *compact* manifold V into $\operatorname{Int} \operatorname{Conv} A \subset \mathbb{R}^q$. Then f_0 is a convex combination of some C^r-maps $\varphi_\nu\colon V \to A$, that is $f_0 = \sum_\nu \lambda_\nu \varphi_\nu$ for some *constants* $\lambda_\nu \geq 0$, $\sum_\nu \lambda_\nu = 1$.

Proof. As A is path connected there obviously exists a C^r-map $\psi\colon V \times S \to A$ which strictly surrounds f_0 and so there is a C^r-map $\varphi\colon V \times S \to A$ whose integral $I = \int_S \varphi(\cdot, s)\, ds\colon V \to \mathbb{R}^q$ equals f_0. Since V is compact, this integral is the uniform limit of Riemann sums $\Sigma = N^{-1} \sum_{\nu=1}^N \varphi(\cdot, s_\nu)$ and so each C^r-map $f_0\colon V \to \operatorname{Int} \operatorname{Conv} A$ admits a C^r-approximation by the sums Σ that are convex combinations of some C^r-maps $\varphi(\cdot, s_\nu)\colon V \to A$.

Since the (connected!) manifold A has $\operatorname{Int} \operatorname{Conv} A \neq \varnothing$, some tangent spaces $T_{a_i}(A) \subset T_{a_i}(\mathbb{R}^q) = \mathbb{R}^q, a_i \in A, i = 1, \ldots, q$, linearly span $\mathbb{R}^q$. It follows with the implicit function theorem that every C^r-map $\delta\colon V \to \mathbb{R}^q$ which is C^0-close to the constant map $x_0\colon V \to \frac{1}{q} \sum_{i=1}^q a_i$ is a convex combination of C^r-maps $V \to A$ which are C^0-small perturbations of the constant maps $V \to a_i$, $i = 1, \ldots, q$.

Now, for small $\alpha > 0$, the map f_0 is the convex combination, $f_0 = \alpha x_0 + (1 - \alpha) f_0'$, where the map f_0' sends V into $\operatorname{Int} \operatorname{Conv} A$. Hence, $f_0 = \alpha x_0 + (1 - \alpha)(\Sigma' + \delta')$ where Σ' is a convex combination of maps $V \to A$ and where δ' is an arbitrarily small C^r-map. Then the map $\delta = x_0 + [(1 - \alpha)/\alpha]\delta'$ is also a convex combination of maps $V \to A$ and so $f_0 = \alpha\delta + (1 - \alpha)\Sigma'$ is such a combination as well.

Exercises. Generalize the convex decomposition to those subsets $A \subset \mathbb{R}^q$ which are images of C^l-maps of connected manifolds into $\mathbb{R}^q$.

Let A be a *disconnected* compact submanifold in $\mathbb{R}^q$, such that no component A_0 of A has $\operatorname{Conv} A_0 = \operatorname{Conv} A$. Prove the existence of a C^r-map $f_0\colon V \to \operatorname{Int} \operatorname{Conv} A$ which is not a convex combination of continuous maps $V \to A$.

Questions. How does one estimate (from above and from below) the minimal number N for which some convex decomposition $f_0 = \sum_{\nu=1}^N \lambda_\nu \varphi_\nu$ exists? Is the convex decomposition possible for non-compact manifolds V? What happens to path-connected subsets $A \subset \mathbb{R}^q$ which are not C^1-submanifolds?

$C^\perp$-Approximation over Split Manifolds. Let a smooth manifold V split into the product, $V = V' \times [0, 1]$. If we fix local coordinates $u_1, \ldots, u_{n-1}$ in V', then the rth order jet of a C^r-map $f\colon V \to \mathbb{R}^q$ is given by the totality of the partial derivatives in the variable $u_1, \ldots, u_{n-1}, t$. Namely, $J_f^r = \{\partial_U^K \partial_t^l f\}$, where $K = (k_1, \ldots, k_{n-1})$, such that $|K| = k_1 + , \ldots, + k_{n-1} \leq r - l$, and where $\partial_U^K \partial_t^l$ stands for the partial derivative $(\partial^{k_1 + \cdots + k_{n-1} + l})/(\partial u_1^{k_1} \cdots \partial u_{n-1}^{k_{n-1}} \partial t^l)$. We assemble in the notation $J^\perp$ those derivatives $\partial_U^K \partial_t^l$ for which $l \leq r - 1$ and thus we split the jet J_f^r into $J_f^r = J_f^\perp \oplus \partial_t^r f$. This splitting $J^r = J^\perp \oplus \partial_t^r$ only depends on the splitting $V = V' \times [0, 1]$, but it does not depend on the local coordinates $u_1, \ldots, u_{n-1}$ in V'. We introduce *$C^\perp$-convergence* of maps as the C^0-convergence of the respective jets $J^\perp$.

(C) **The $C^\perp$-Approximation Lemma.** *Let V be a compact split C^∞-manifold and let $S = [0, 1]$. Let $f_0\colon V \to \mathbb{R}^q$ be an arbitrary C^{2r}-map and let $\psi\colon V \times S \to \mathbb{R}^q$ be a*

C^r-map which strictly surrounds the derivative $\partial_t^r f_0 = \dfrac{\partial^r f_0}{\partial t^r}: V \to \mathbb{R}^q$. Then the map f_0 admits an arbitrary close $C^\perp$-approximation by those maps f, whose derivatives $\partial_t^r f: V \to \mathbb{R}^q$ lift to C^r-sections $V \to V \times S$. That is $\partial_t^r f(v) = \psi(v, \eta(v))$ for some C^r-function $\eta: V \to S$.

Proof. We may assume with the reparametrization lemma that $\int_0^1 \psi(v,s)\,ds = \partial_t^r f_0(v)$. Then we also have for $|K| \le r$,

$$(*)\qquad \int_0^1 \partial_U^K \psi(v,s)\,ds = \partial_U^K \partial_t^r f_0(v).$$

Next, for every positive $\varepsilon < \frac{1}{2}$, we choose some disjoint ε-subintervals in $[0,1]$ of total length $\ge 1-\varepsilon$, and let $h_\varepsilon: [0,1] \to S = [0,1]$ be a C^∞-function which linearly maps every chosen subinterval onto $[\varepsilon, 1-\varepsilon] \subset [0,1] = S$. This h_ε is given on each subinterval, say on $[t_0, t_0+\varepsilon]$, by one of the two formulas:

$$(1)\qquad h(t) = at + b, \quad a = \varepsilon^{-1} - 2, \quad b = \varepsilon - at_0;$$

$$(2)\qquad h(t) = -at + b', \quad b' = 1 - \varepsilon + at_0.$$

Set $\eta_\varepsilon(v', t) = h_\varepsilon(t)$, where $(v', t) = v \in V$, take $g_\varepsilon(v', t) = \psi(v', t, \eta_\varepsilon(v', t))$ and show that the derivatives $\partial_U^K g_\varepsilon$, $|K| \le r$, satisfy

$$(+)\qquad \varepsilon^{-1} \int_{t_0}^{t_0+\varepsilon} \partial_U^K g_\varepsilon(v', t)\,dt \underset{\varepsilon \to 0}{\to} \partial_U^K \partial_t^r f_0(v', t_0).$$

Indeed, in case (1) (the second case is left to the reader) the substitution $t = h^{-1}(s) = (s-b)/a$ gives the identity

$$\varepsilon^{-1} \int_{t_0}^{t_0+\varepsilon} \partial_U^K \psi(v', t, at+b)\,dt = \frac{1}{1-2\varepsilon} \int_\varepsilon^{1-\varepsilon} \partial_U^K \psi(v', t_0 + \delta_\varepsilon, s)\,ds,$$

where $\delta_\varepsilon = \dfrac{(s-\varepsilon)}{1-2\varepsilon} \to 0$. Therefore,

$$\varepsilon^{-1} \int_{t_0}^{t_0+\varepsilon} \to \int_0^1 \partial_U^K \psi(v', t_0, s)\,ds, \qquad \varepsilon \to 0,$$

and then $(+)$ follows from $(*)$.

The map $f_0(v', t)$ is $C^\perp$-approximated by the r-fold integral in t of $g_\varepsilon(v', t)$ for $\varepsilon \to 0$. This integral, called $f_\varepsilon(v', t)$, is defined by the conditions $\partial_t^r f_\varepsilon = g_\varepsilon$ and $J_{f_\varepsilon}^\perp | V' \times 0 = J_{f_0}^\perp | V' \times 0$, which implies for $i = r, \ldots, 1$,

$$\partial_t^{i-1} f_\varepsilon(v', t) = \partial_t^{i-1} f_0(v', 0) + \int_0^t \partial_t^i f_\varepsilon(v', t)\,dt.$$

The uniform in ε bounds $\|\partial_U^K g_\varepsilon\| \le \text{const}$ and the relation $(+)$, which applies to $\partial_t^r f_\varepsilon = g_\varepsilon$ over the ε-subintervals of the total length $\ge 1-\varepsilon$, show that for $|K| \le r$,

$$\|\partial_U^K \partial_t^{r-1} f_\varepsilon - \partial_U^K \partial_t^{r-1} f_0\| \to 0, \qquad \varepsilon \to 0.$$

Then we obtain with successive integration in t the uniform convergences $\partial_U^K \partial_t^i f_\varepsilon \to$

$\partial_U^K \partial_t^i f_0$, $\varepsilon \to 0$, for all $i = r-1, \ldots, 0$, and $|K| \le r$. This yields the required $C^\perp$-convergence $f_\varepsilon \to f_0$.

Finally, the derivative $\partial_t^r f_\varepsilon = g_\varepsilon$ lifts to a section $V \to V \times S$ by the very definition of g_ε and so the proof is concluded.

The Jet Space $X^\perp$. Let $p: X \to V$ be an arbitrary smooth fibration with q-dimensional fibers and let τ be a continuous tangent hyperplane field on V that is a codimension one subbundle of the tangent bundle $T(V)$. Then, there are for each $r = 1, 2, \ldots,$ a unique manifold $X^\perp$ and continuous maps $p^r: X(r) \to X^\perp$ and $p_{r-1}^\perp: X^\perp \to X^{(r-1)}$, such that $p_{r-1}^\perp \circ p_\perp^r = p_{r-1}^r$ for the natural projection $p_{r-1}^r: X^{(r)} \to X^{(r-1)}$ of the jet spaces, and such that the $J^\perp$-*jet* $J^\perp \overset{\text{def}}{=} p_\perp^r \circ J^r$ satisfies the following conditions for all pairs of (germs of) sections f_1 and $f_2: V \to X$,

$$J_{f_1}^\perp = J_{f_2}^\perp \Leftrightarrow DJ_{f_1}^{r-1}|\tau = DJ_{f_2}^{r-1}|\tau,$$

for the differentials

$$DJ_{f_1}^{r-1} \text{ and } DJ_{f_r}^{r-1}: T(V) \to T(X^{(r-1)}).$$

The maps $X^{(r)} \to X^\perp$ and $X^\perp \to X^{(r-1)}$ carry natural structures of affine bundles which are compatible with the affine bundle $X^{(r)} \to X^{(r-1)}$. The fibers of the bundles $X^{(r)} \to X^\perp$ for all fields τ on V form a certain distinguished class of affine q-dimensinal subspaces in the affine fibers $X_y^{(r)} \subset X^{(r)}$, $y \in X^{(r-1)}$ which are called *principal subspaces in* $X^{(r)}$.

The Convex Hull $\operatorname{Conv}_\tau(\mathscr{R})$. Consider a subset $\mathscr{R} \subset X^{(r)}$, take the convex hull of the intersection $\mathscr{R}_x = \mathscr{R} \cap X_x^{(r)} \subset X_x^{(r)}$ for $X_x^{(r)} = (p_\perp^r)^{-1}(x)$, $x \in X^\perp$, and denote by $\operatorname{Conv}_\tau(\mathscr{R})$ the union of these convex hulls over all $x \in X^\perp$.

(D) The $C^\perp$-Dense h-Principle. *A differential relation $\mathscr{R} \subset X^{(r)}$ satisfies the C^i-dense h-principle, for $i = \perp$, if the following four conditions are satisfied.*

(1) *The field τ is integrable;*
(2) *The subset $\mathbb{R} \subset X^{(r)}$ is open;*
(3) $\operatorname{Conv}_\tau(\mathscr{R}) = X^{(r)}$;
(4) *The intersection $\mathscr{R}_x = \mathscr{R} \cap X_x^{(r)}$ is connected for all $x \in X^\perp$.*

In particular, if there is a section $X^\perp \to \mathscr{R}$ then every C^r-section $V \to X$ admits a fine $C^\perp$-approximation by solutions of the relation $\mathscr{R}$, where the $C^\perp$-topology by definition is induced from the C^0-topology in the space of the $J^\perp$-jets $V \to X^\perp$.

Proof. We may assume the fibration $X \to V$ to be a vector bundle as a small neighborhood of (the image of) each section $V \to X$ admits a vector bundle structure.

Since the field τ is integrable, every point $v \in V$ lies inside a small *split* submanifold $U = U' \times [0,1] \subset V$, for which the slices $U' \times t$, $t \in [0,1]$ are tangent to the field τ, and over which the fibration is trivial, $X|U = U \times \mathbb{R}^q \to U$.

Let us prove the h-principle over V by solving the extension problem over U for all small split submanifolds $U \subset V$ (compare 1.4.3) as follows.

Take a C^r-section $f_0: U \to X$ and a C^0-section $\varphi_0: U \to \mathcal{R}$, such that $J^{\perp}_{f_0} = p^r_{\perp} \circ \varphi_0$ and such that $J^r_{f_0}|U_0 = \varphi_0|U_0$ for a given closed subset $U_0 \subset U$. To prove the $C^{\perp}$-dense h-principle for extensions (of solutions of $\mathcal{R}$ from U_0 to U) we must homotope φ_0 to a holonomic section $J^r_{f_1}: U \to \mathcal{R}$, such that f_1 is $C^{\perp}$-close to f_0. Moreover, the homotopy in question must be constant near U_0 while the projection of this homotopy to $X^{\perp}$ must be nearly constant over all of U (compare 1.2.2).

The conditions (2)–(4) obviously yield a C^{∞}-homotopy of (non-holonomic) sections, $\varphi_s: U \to \mathcal{R}$, $s \in S = [0, 1]$, with the following three properties.

(a) The homotopy $\varphi: U \times S \to \mathcal{R} \subset X^{(r)}$ lies *over* $J^{\perp}_{f_0}$, that is $p^r_{\perp} \circ \varphi_S = J^{\perp}_{f_0}$ for all $s \in S$.

(b) The homotopy φ *strictly* surrounds the section $J^r_{f_0}$. Namely, the vector $J^r_{f_0}(v) \in X^{(r)}_x$, for $x = J^{\perp}_{f_0}(v) \in X^{\perp}$, lies in the interior of the convex hull of the path $s \mapsto \varphi_s(v) \in X^{(r)}_x$, $s \in S$, for all $v \in U$.

(c) The sections $\varphi_s|\mathcal{O}p\,U_0$ for all $s \in S$ are as close to $\varphi_0|\mathcal{O}p\,U_0$ as we wish. (One even may have $\varphi_s|\mathcal{O}p\,U_0 = \varphi_0|\mathcal{O}p\,U_0$ if "strictly surrounds" in (b) is relaxed to "surrounds").

Now, the splitting of jets, $J^r = J^{\perp} \oplus \partial^r_t$, induces a splitting of the jet bundle, $X^{(r)} = X^{\perp} \times \mathbb{R}^q$ over U. The $\mathbb{R}^q$-component of the homotopy $\varphi: U \times S \to X^{(r)}$ is a C^{∞}-map, say $\psi: U \times S \to \mathbb{R}^q$, to which the $C^{\perp}$-approximation lemma applies. Thus we obtain a section $f: U \to X$ whose jet J^r_f nearly factors through a section $U \to U \times S$, that is $J^r_f(u)$ is C^0-close to $\varphi(u, \eta(u)$ for some function $\eta: U \to S$. This implies the existence of a homotopy of sections $U \to \mathcal{R}$ between φ_0 and J^r_f. Moreover, as the sections $\varphi_s|\mathcal{O}p\,U_0$, $s \in S$, are close to $\varphi_0|\mathcal{O}p\,U_0$, the jet $J^r_f|\mathcal{O}p\,U_0$ is also close to φ_0, and so the above homotopy can be chosen almost constant on $\mathcal{O}p\,U_0$. Hence, a small C^r-perturbation f of f_1 equals to f_0 on $\mathcal{O}p\,U_0$ and the jet $J^r_{f_1}: U \to \mathcal{R}$ admits the required homotopy to φ_0. Q.E.D.

(D′) **Remark.** The results in 2.4.4 yield the C^i-dense h-principle for all $i = 0, 1, \ldots,$ under the above assumptions on $\mathcal{R}$.

(E) **Corollary.** *Let V be an arbitrary manifold and let $\Sigma \subset X^{(r)}$ be a closed stratified subset of codimension ≥ 2. Furthermore, assume that the intersection of Σ with every principal subspace $R \approx \mathbb{R}^q$ in $X^{(r)}$ has codimension ≥ 2 in R. Then Σ-nonsingular sections $V \to X$ satisfy all forms of the h-principle* (see 1.2.1, 1.4.3, 1.5.2).

Proof. First let V admit an integrable field τ. Then the complement $\mathcal{R} = X^{(r)} \backslash \Sigma$ satisfy the assumptions of (D) and we get with (D′) all non-parametric h-principles for Σ-nonsingular sections that, by definition, are solutions of $\mathcal{R}$. To get the parametric h-principle, we interprete families of sections $\varphi_p: V \to X$, $p \in P$, as sections of the fibration $X \times P \to V \times P$ and then apply (D′) to the respective problem over $V \times P$ for all manifolds P.

Now, if V is open, it admits a function h without critical points (see 1.4.4) and the above applies to the kernel τ of the differential dh. If V is closed, then we have our field τ on V minus a point. Since the h-principle is trivially true at every single point $v_0 \in V$, the extension h-principle over $V \backslash v_0$ yields the h-principle over all of V. Q.E.D.

Remark. The condition $\operatorname{codim}(R \cap \Sigma) \geq 2$ is *generic* for singularities Σ of codimension ≥ 2. But this condition is not satisfied in most geometrically interesting cases, such as immersions, free maps etc. These classes of maps are covered by the techniques developed in the following sections.

2.4.2 Principal Extensions of Differential Relations

A differential relation *over* the jet space $X^{(r)}$, for a fibration $X \to V$ is by definition an arbitrary topological space $\mathscr{R}$ with a given map $\mathscr{R} \to X^{(r)}$. This generalizes our notion of a relation *in* $X^{(r)}$, where the structure map is the inclusion $\mathscr{R} \subsetneq X^{(r)}$. A relation $\mathscr{R}$ over $X^{(r)}$ also lies over every space Y *under* $X^{(r)}$, where "under" means a fixed map $X^{(r)} \to Y$ and $\mathscr{R}$ goes to Y by the composition of maps. For example, $\mathscr{R}$ lies over $X^{(s)}$, $s \leq r$, and over V.

Sections $V \to \mathscr{R}$ are those continuous maps for which the composed map $\overrightarrow{V \to \mathscr{R} \to V}$ is the identity Id: $V \to V$. A section $V \to \mathscr{R}$ is called *holonomic* if the composition $\overrightarrow{V \to \mathscr{R} \to X^{(r)}}$ is a holonomic section $V \to X^{(r)}$. A relation $\mathscr{R}$ over $X^{(r)}$ is said *to satisfy the h-principle* if every section $V \to \mathscr{R}$ admits a homotopy of sections to a holonomic section $V \to \mathscr{R}$. We define in a similar way the dense h-principle, the parametric one etc (compare 1.2).

Extensions. We call *an extension* of a differential relation $\rho: \mathscr{R} \to X^{(r)}$ another relation $\tilde{\rho}: \tilde{\mathscr{R}} \to X^{(r)}$ which comes with an embedding $E: \mathscr{R} \to \tilde{\mathscr{R}}$ and with a retraction (or projection) $\Pi: \tilde{\mathscr{R}} \to \mathscr{R}$, such that the diagrams

$$\begin{array}{ccc} E: \mathscr{R} \to \tilde{\mathscr{R}} & \text{and} & \Pi: \tilde{\mathscr{R}} \to \mathscr{R} \\ \searrow \swarrow & & \searrow \swarrow \\ X^{(r)} & & X^{(r-1)} \end{array}$$

commute.

(A) **Example.** The relation $\tilde{\mathscr{R}}_r = \{a \in \mathscr{R}, x \in X^{(r)} | \rho(a) = p^r_{r-1}(x)\}$, $\tilde{\rho}: (a, x) \mapsto x$, extends $\mathscr{R}$ for $E: a \mapsto (a, \rho(a))$ and $\Pi: (a, x) \mapsto a$. This $\tilde{\mathscr{R}}_r$ *trivially* satisfies the C^{r-1}-dense h-principle. Indeed, sections of $\tilde{\mathscr{R}}_r$ are pairs (φ, ψ), where $\varphi: V \to \mathscr{R}$ is an arbitrary continuous map and $\psi: V \to X^{(r)}$ is a section. Hence, any homotopy of ψ to a holonomic section $V \to X^{(r)}$ also makes the section $(\varphi, \psi): V \to \tilde{\mathscr{R}}_r$ holonomic.

An extension $\tilde{\mathscr{R}} \supset \mathscr{R}$ is called *h-stable* (h for holonomy) if every holonomic section $\varphi_0: V \to \tilde{\mathscr{R}}$ admits a homotopy of sections $\varphi_t: V \to \tilde{\mathscr{R}}$, $t \in [0, 1]$, such that φ_1 is a holonomic section $V \to \mathscr{R} \subset \tilde{\mathscr{R}}$ and such that $\varphi_t(v) = \varphi_0(v)$ for all those points $v \in V$ for which $\varphi_0(v) \in \mathscr{R}$ and for all $t \in [0, 1]$. We do not require the sections φ_t for $0 < t < 1$ to be holonomic. However, one can construct a homotopy of holonomic sections φ_t, $t \in [0, 1]$, with little extra work for all h-stable extensions which appear in this chapter (see "convex extensions" below and in the following section).

The important (and obvious) property of h-stable extensions $\tilde{\mathscr{R}} \supset \mathscr{R}$ (which does not depend upon the holonomy of φ_t for $0 < t < 1$) is expressed by the following implication,

$$[\text{the } C^i\text{-dense } h\text{-principle for } \tilde{\mathcal{R}}] \Rightarrow [\text{the } C^i\text{-dense } h\text{-principle for } \mathcal{R}],$$
$$\text{for all } i = 0, \ldots, r-1.$$

Convex Hull Extensions. Let τ be a tangent hyperplane field on an open subset $U \subset V$ and let $p_\perp^r \colon X_U^{(r)} \to X_U^\perp$ be the associated affine bundle whose fibers are principal subspaces $R \approx \mathbb{R}^q$ in $X^{(r)}$. Take a point in the pullback $\rho^{-1}(R) \subset \mathcal{R}$ for $\rho \colon \mathcal{R} \to X^{(r)}$, say $a \in \rho^{-1}(R)$, and denote by $\mathrm{Conv}_R(\mathcal{R}, a)$ the convex hull in R of the ρ-image of the path connected component in $\rho^{-1}(R)$ of the point $a \in \rho^{-1}(R)$. In other words, the convex subset $\mathrm{Conv}_R(\mathcal{R}, a) \subset R$ consists of those points in R which are surrounded by homotopies of the point a over R.

Now we define the *τ-convex hull* $\tilde{\mathcal{R}} = \mathrm{Conv}_\tau(\mathcal{R})$ as the set of those pairs $(a, x) \in \mathcal{R} \times X^{(r)}$ for which either $x = \rho(a)$, or $x \in \mathrm{Conv}_R(\mathcal{R}, a)$ for some principal subspace $R \subset X^{(r)}$ associated to τ. This $\tilde{\mathcal{R}}$ lies over $X^{(r)}$ for $\tilde{\rho} \colon (a, x) \to x$, and the relation $\tilde{\rho} \colon \tilde{\mathcal{R}} \to X^{(r)}$ extends $\rho \colon \mathcal{R} \to X^{(r)}$ for $E \colon a \mapsto (a, \rho(a))$ and $\Pi \colon (a, x) \mapsto a$.

Remark. This elaborate convex hull is needed to distinguish the convex hulls of different path connected components of $\rho^{-1}(R)$. The impossibility to do this *inside* $X^{(r)}$ is the main reason for the "over" generalization.

Open Relations over $X^{(r)}$. A relation $\mathcal{R}$ over $X^{(r)}$ is called *open* if the implied map $\rho \colon \mathcal{R} \to X^{(r)}$ is a microfibration (see 1.4.2). For example, if ρ is a submersion, then $\mathcal{R}$ is open over $X^{(r)}$. Observe that the *openness of the map* ρ is strictly weaker than *the openness of the relation* $\mathcal{R}$ over $X^{(r)}$.

(B) The *h*-Stability Theorem. *If $\mathcal{R} \to X^{(r)}$ is an open relation, then the extension $\tilde{\mathcal{R}} = \mathrm{Conv}_\tau(\mathcal{R}) \supset \mathcal{R}$ is h-stable for every continuous hyperplane field τ on an open subset $U \subset V$. In particular if $U = V$ and if $\mathrm{Conv}_R(\mathcal{R}, a) = R$ for all principal subspaces R associated to τ and for all $a \in \rho^{-1}(R)$ then the relation $\mathcal{R}$ satisfies the $C^\perp$-dense h-principle.* [Compare (D) in 2.4.1.]

Proof. Sections of open relations $\mathcal{R}$ over $X^{(r)}$ may be dealt with as if $\mathcal{R}$ were an open subset in $X^{(r)}$ because of the following

(B′) *Properties of Open Relations $\rho \colon \mathcal{R} \to X^{(r)}$.* (a) If a section $\varphi \colon V \to X^{(r)}$ lifts to $\mathcal{R}$ (i.e. φ lifts to a section $\tilde{\varphi} \colon V \to \mathcal{R}$, $\rho \circ \tilde{\varphi} = \varphi$), then also small (in the fine C^0-topology) perturbations of φ lift to $\mathcal{R}$.

(a′) If $\mathcal{R}$ is homeomorphic to a finite polyhedron, then there exists a neighborhood of $\mathcal{R}$ in the extension $\mathcal{R} \times X^{(r)} \supset \mathcal{R}$ [see (A)], say $\mathcal{O}p\mathcal{R} \subset \mathcal{R}X^{(r)}$, such that the map $\mathcal{O}p\mathcal{R} \to X^{(r)}$, $(a, x) \mapsto x$, lifts to a map $\tilde{\alpha} \colon \mathcal{O}p\mathcal{R} \to \mathcal{R}$ such that $\rho \circ \tilde{\alpha}(a, x) = x$ for all $(a, x) \in \mathcal{O}p\mathcal{R}$ and $\tilde{\alpha}(a, \rho(a)) = a$ for all $a \in \mathcal{R}$.

(a″) Let $\rho \colon \mathcal{R} \to X^{(r)}$ be an arbitrary relation. Then $\mathcal{R}$ is open if and only if (a′) holds true for all maps of polyhedra into $\mathcal{R}$, say for $\beta \colon P \to \mathcal{R}$. Namely, there is a neighborhood $\mathcal{O}pP \subset P \times X^{(r)}$ of $P = \{p, \rho \circ \beta(p)\} \subset P \times X^{(r)}$, such that the map

$\mathcal{O}p P \to X^{(r)}$, $(p,x) \mapsto x$, lifts to a map $\tilde{\alpha}: \mathcal{O}p P \to \mathcal{R}$ for which $\rho \circ \tilde{\alpha}(p,x) = x$ and $\tilde{\alpha}(p, \rho \circ \beta(p)) = p$ for all $(p,x) \in \mathcal{O}p P$.

(b) The sheaf of holonomic sections $V \to \mathcal{R}$, called $\Phi_h(\mathcal{R})$, is microflexible (compare 1.4.2, 2.2). Recall that the sheaf of all continuous sections, $\Phi_c(\mathcal{R})$, is flexible for all (not necessarily open) relations $\mathcal{R} \to X^{(r)}$.

(b′) The map $\rho: \mathcal{R} \to X^{(r)}$ induces *microflexible* homomorphisms (see 2.2.4) of sheaves, $\Phi_c(\mathcal{R}) \to \Phi_c(X^{(r)})$ and $\Phi_h(\mathcal{R}) \to \Phi_h(X^{(r)})$. Namely, the maps $\eta = \eta(A, B)$ defined in 2.2.4 are microfibrations for these homomorphisms. Furthermore, the images of these homomorphisms are the subsheaves $\Phi_c(\rho(\mathcal{R})) \subset \Phi_c(X^{(r)})$ and $\Phi_h(\rho(\mathcal{R})) \subset \Phi_h(X^{(r)})$ respectively for the image $\rho(\mathcal{R}) \to X^{(r)}$, and the corresponding homomorphisms $\Phi_c(\mathcal{R}) \to \Phi_c(\rho(\mathcal{R}))$ and $\Phi_h(\mathcal{R}) \to \Phi_h(\rho(\mathcal{R}))$ are microextension (see 2.2.4).

(c) The extension $\tilde{\mathcal{R}} = \mathrm{Conv}_\tau(\mathcal{R})$ is an *open* relation over $X^{(r)}$ for all open relations $\mathcal{R} \to X^{(r)}$ and for all hyperplane fields τ on a given open subset $U \subset V$.

The proof of (a″) is immediate with the definition of a microfibration (see 1.4.2), and (a″) obviously implies (a), (a′), (b) and (b′). The proof of (c) is straightforward.

Now, we obtain the h-stability of the extension $\tilde{\mathcal{R}} = \mathrm{Conv}_\tau(\mathcal{R}) \supset \mathcal{R}$ by generalizing the proof of the $C^\perp$-dense h-principle in (D) of 2.4.1 as follows. We are given a holonomic section $V \to \tilde{\mathcal{R}} \subset \mathcal{R} \times X^{(r)}$, that is a pair of sections $\varphi_0: V \to \mathcal{R}$ and $f_0: V \to X^{(r)}$, where the section f_0 is holonomic and such that

(i) the section f_0 equals $\rho \circ \varphi_0$ outside the open subset $U \subset V$ on which the field τ is defined,
(ii) $p_\perp^r \circ f_0 = p_\perp^r \circ \rho \circ \varphi_0$ for the affine bundle $p_\perp^r: X_U^{(r)} \to X_U^\perp$,
(iii) $f_0(u) \in \mathrm{Conv}_R(\mathcal{R}, \varphi_0(u))$ for all $u \in U$ and for the fiber $R = (p_\perp^r)^{-1}(x)$ for $x = p_\perp^r \circ \rho \circ \varphi_0(u)$.

We must construct a homotopy of sections $(\varphi_t, f_t): V \to \tilde{\mathcal{R}} \to \mathcal{R}$, $t \in [0,1]$, which is fixed outside U and such that $f_1 = \rho \circ \varphi_1: V \to X^{(r)}$ is a holonomic section. In fact, we only need the homotopy $\varphi_t: V \to \mathcal{R}$ of φ_0 to a holonomic section $\varphi_1: V \to \mathcal{R}$. The homotopy of f_t then obviously follows with the linear homotopy $(\varphi_0, (1-\theta)f_0 + \theta(\rho \circ \varphi_0)): V \to \tilde{\mathcal{R}}$, $\theta \in [0,1]$ which brings the section (φ_0, f_0) to $\mathcal{R} \subset \tilde{\mathcal{R}}$ for $\theta = 1$.

Obviously, there exists a homotopy of (non holonomic!) sections $\varphi_s: U \to \mathcal{R}$, $s \in S = [0,1]$, whose projection to $X^{(r)}$ surrounds the section f_0 in the fibers of the fibration $p_\perp^r: X_U^{(r)} \to X_U^\perp$, and then (an obvious modification of) the proof of (D) in 2.4.1 works in the present case provided τ is a *split integrable field*. This means the existence of a split submanifold $V_0 = V_0' \times [0,1] \subset V$ which contains U, such that V_0 is a disjoint union of countably many compact submanifolds in V and such that the field τ is tangent to the submanifold $V_0' \times t \subset V_0$, at all points $(v', t) \in U$.

Next, we reduce the general case of the h-stability theorem to the split integrable case by first localizing the problem to a small neighborhood of each point $u \in U$ and then by approximating the field τ near u by an integrable split field.

Localization. Let open subsets $U_i \subset U$, $i = 1, \ldots, k$, cover U and let τ_i be a hyperplane field on U_i. Put $\mathcal{R}_0 = \mathcal{R}$ and $\mathcal{R}_i = \mathrm{Conv}_{\tau_i}(\mathcal{R}_{i-1})$ for $i = 1, \ldots, k$. Suppose that the fields τ_i are the restrictions $\tau | U_i$, $i = 1, \ldots, k$, and let us "lift" a given section $\tilde{\varphi} = (\varphi_0, f_0)$: $V \to \tilde{\mathscr{R}}$ to a section φ_k: $V \to \mathscr{R}$ as follows. Fix a partition of unity $b_1, \ldots, b_k$ on U inscribed in the cover $\{U_i\}$ and assume the fibration $X \to V$ to be a vector bundle [compare the proof of (D) in 2.4.1]. Define a section f_1: $U \to X_U^{(r)}$ by $f_1 = \rho \circ \varphi_0 + b_1(f_0 - \rho \circ \varphi_0)$ and then by induction take $f_i = f_{i-1} + b_i(f_0 - \rho \circ \varphi_0)$ for $i = 2, \ldots, k$. Observe that the sections f_i continuously extend to $V \supset U$ by $f_i | V \backslash U = f_0 | V \backslash U$ and that the (extended) pairs $\varphi_1 = (\varphi_0, f_1)$, $\varphi_2 = (\varphi_1, f_2), \ldots, \varphi_k = (\varphi_{k-1}, f_k)$ are sections of the relations $\mathscr{R}_1, \mathscr{R}_2, \ldots, \mathscr{R}_k$ respectively. Furthermore,

$$f_k = \rho \circ \varphi_0 + \sum_{i=1}^{k} b_i(f_0 - \rho \circ \varphi_0) = f_0.$$

This amounts to the equality $\tilde{\rho} \circ \tilde{\varphi} = \rho_k \circ \varphi_k$ for ρ_k: $\mathscr{R}_k \to X^{(r)}$, which by definition expresses the "lift" property of our construction $\tilde{\varphi} \rightsquigarrow \varphi_k$.

If the extensions $\mathscr{R}_i \supset \mathscr{R}_{i-1}$, $i = 1, \ldots, k$, are h-stable, then the extension $\mathscr{R}_k \supset \mathscr{R}_0 = \mathscr{R}$ also is stable and so φ_0 admits the required deformation, provided $f_0 = f_k$ is a holonomic section $V \to X^{(r)}$. Now, let τ'_i, $i = 1, \ldots, k$, be sufficiently small (in the fine C^0-topology in the space of the hyperplane fields on U) perturbations of the field τ and let $\bar{\tau}_i = \tau'_i | U_i$. Then the perturbed relation $\bar{\mathscr{R}}_k$ still admits a section $\bar{\varphi}_k$ such that $\bar{\rho}_k \circ \bar{\varphi}_k = \rho_k \circ \varphi_k = \tilde{\rho} \circ \tilde{\varphi} = f_0$ and such that $\bar{\varphi}_k$ is close to φ_k in the ambient relation $\mathscr{R} \times \underbrace{X^{(r)} \times \cdots \times X^{(r)}}_{k}$ which contains $\bar{\mathscr{R}}_k$ as well as $\mathscr{R}_k$. The existence of $\bar{\varphi}_k$ for *small* perturbations τ'_i of τ is immediate from the openness of the relations $\mathscr{R}_i$, $i = 0, \ldots, k$. Moreover, the required "smallness" of these perturbations depends only on the original section $\tilde{\varphi} = (\varphi_0, f_0)$: $V \to \tilde{\mathscr{R}}$ but not on specific open subsets U_i and (or) the partition of unity b_i. Indeed, let φ'_i, $i = 1, \ldots, k$, be small perturbations of φ_0 which are equal to φ_0 outside U and such that φ'_i agrees with f_0 under the projection $p'_i \overset{\text{def}}{=} p^r_{\perp}$: $X_U^{(r)} \to X_U^{\perp}$, which corresponds to the field τ'_i as follows,

$$p'_i \circ \rho \circ \varphi'_i = p'_i \circ f_0 \qquad \text{for } i = 1, \ldots, k.$$

Then, the sum $\sum_{i=1}^{k} b_i \rho \circ \varphi'_i$ is close to $\sum_{i=1}^{k} b_i \rho \circ \varphi_0 = \rho \circ \varphi_0$ and so there is a small perturbation $\bar{\varphi}_0$ of φ_0 such that $\rho \circ \bar{\varphi}_0 = \sum_{i=1}^{k} b_i \rho \circ \varphi'_i$.

Take $\bar{f}_1 = \rho \circ \bar{\varphi}_0 + b_1(f_0 - \rho \circ \varphi'_1)$ and $\bar{f}_i = \bar{f}_{i-1} + b_i(f_0 - \rho \circ \varphi'_i)$ for $i = 2, \ldots, k$. The pairs $\bar{\varphi}_1 = (\bar{\varphi}_0, \bar{f}_1), \ldots, \bar{\varphi}_k = (\varphi_{k-1}, f_k)$ are sections of the perturbations $\bar{\mathscr{R}}_1, \ldots, \bar{\mathscr{R}}_k$ of $\mathscr{R}_1, \ldots, \mathscr{R}_k$ and the section $\bar{\varphi}_k$ is a lift of f_0 as

$$\bar{f}_k = \rho \circ \bar{\varphi}_0 + \sum_{i=1}^{k} b_i(f_0 - \rho \circ \varphi'_i) = f_0.$$

Now, we take for τ'_i (arbitrarily) small perturbations of τ for $i = 1, \ldots, k = \dim V + 1$, for which there is a cover $\{U_i\}$ of U such that each field $\bar{\tau}_i = \tau'_i | U_i$, $i = 1, \ldots, k$, is an integrable split field (i.e. $\bar{\tau}_i$ is tangent to the slices $V'_i \times t$ of some split manifolds $V_i = V'_i \times [0, 1] \supset U_i$). The section f_0: $V \to X^{(r)}$ lifts to a section $V \to \bar{\mathscr{R}}_k$ and so the h-stability theorem reduces to the split case which was considered earlier. Q.E.D.

Principal Convex Extensions. We define the (first) *principal extension* $Pr_1 \mathscr{R}$ of a relation ρ: $\mathscr{R} \to X^{(r)}$ as the subrelation in the extension $\tilde{\mathscr{R}}_r$ of (A), such that $(a, x) \in$

$Pr_1\mathscr{R}$ if and only if there is a principal subspace R in $X^{(r)}$ which contains the points $\rho(a)$ and $x \in X^{(r)}$. This R is uniquely determined by a and x unless $\rho(a) = x$. *The principal convex extension* $\mathrm{Conv}_1\mathscr{R}$ of $\mathscr{R}$ is the subset of those $(a, x) \in Pr_1\mathscr{R}$ for which either $x \in \mathrm{Conv}_R(\mathscr{R}, a)$ or $\rho(a) = x$. Then we define by induction the principal extensions $Pr_N\mathscr{R} = Pr_1 Pr_{N-1}\mathscr{R}$ and $\mathrm{Conv}_N\mathscr{R} = \mathrm{Conv}_1 \mathrm{Conv}_{N-1}\mathscr{R}$ for $N = 2, 3, \ldots$.

Next, we define $Pr_N\mathscr{R}$ for $N = \infty$ as the space of pairs $(a, y = y(t))$, where $a \in \mathscr{R}$ and where $y\colon [0,1] \to X^{(r)}$ is a *piecewise principal* path which issues from $\rho(a) \in X^{(r)}$. That is

(i) $y(0) = \rho(a)$,
(ii) there is a *principal subdivision* of the path $y = y(t)$ by finitely many points, say by $0 = t_0 \le t_1 \le \cdots \le t_N = 1$, such that each segment $[t_{i-1}, t_i]$, $i = 1, \ldots, N$, is sent by y into some principal subspace $R = R(y[t_{i-1}, t_i]) \subset X^{(r)}$ and $y(t)$ linearly interpolates between $y(t_{i-1})$ and $y(t_i)$ for $t \in [t_{i-1}, t_i]$, $y(t) = \frac{t_i - t}{t_i - t_{i-1}} y(t_{i-1}) + \frac{t - t_{i-1}}{t_i - t_{i-1}} y(t_i)$ for $i = 1, \ldots, N$.

There is a unique *minimal* principal subdivision of y for which the number N is the least possible. This N is called the *principal length* of y and of (a, y). Every principal subdivision of y refines the minimal one.

The relation $Pr_\infty\mathscr{R}$ lies over $X^{(r)}$ for the map $\rho_\infty\colon (a, y(t)) \to y(1)$ and it extends $\mathscr{R}$ for the embedding $E_\infty\colon a \mapsto (a, y(t) \equiv \rho(a))$ and for the projection $\Pi_\infty\colon (a, y) \mapsto a$. Furthermore, every principal subdivision of $(a, y) \in Pr_\infty\mathscr{R}$ defines a point in the union $\bigcup_{N<\infty} Pr_N\mathscr{R}$ as follows. If $0 = t_0 \le t_1 \le \cdots \le t_N = 1$ are the division points, then

$$(a, y) \mapsto (a, y)_N = (a, y(t_1), \ldots, y(t_N)) \in Pr_N\mathscr{R}.$$

Now, we define the extension $\mathrm{Conv}_\infty\mathscr{R} \subset Pr_\infty\mathscr{R}$ of $\mathscr{R}$ as the set of those pairs $(a, y) \in Pr_\infty\mathscr{R}$, for which there is a principal subdivision into N segments, for some $N = 1, 2, \ldots$, such that the corresponding sequence $(a, y)_N \in Pr_N\mathscr{R}$ is contained in $\mathrm{Conv}_N\mathscr{R} \subset Pr_N\mathscr{R}$. Observe that this definition is stable under refinements of subdivisions. Namely, if $(a, y)_{N_1} \in \mathrm{Conv}_{N_1}\mathscr{R}$ for some principal subdivision into N_1 segments, then also $(a, y)_{N_2} \in \mathrm{Conv}_{N_2}\mathscr{R}$ for every refined subdivision into $N_2 > N_1$ segments.

The space of piecewise principal paths $y\colon [0,1] \to X^{(r)}$ comes with the C^0-topology and thus we have natural topologies in the spaces $Pr_\infty\mathscr{R}$ and $\mathrm{Conv}_\infty\mathscr{R}$. Next, a continuous map f of a locally compact space $P \to Pr_\infty\mathscr{R}$ is called *locally finite* if the principal length of $f(p) \in Pr_\infty\mathscr{R}$ is bounded from above by some positive function on P which is *continuous* in $p \in P$. All maps $P \to Pr_\infty\mathscr{R}$ as well as maps $P \to \mathrm{Conv}_\infty\mathscr{R} \subset Pr_\infty\mathscr{R}$ are assumed from now on to be *continuous and locally finite*. In particular, *sections $V \to Pr_\infty\mathscr{R}$ and homotopies of these are always assumed locally finite as well as continuous.*

If $\mathscr{R}$ is open, then the relations $Pr_N\mathscr{R}$ and $\mathrm{Conv}_N\mathscr{R} \to X^{(r)}$, $N < \infty$, are open by (c) in (B′) and also Pr_∞ and Conv_∞ are open relative to the quasitopology defined by locally finite maps (compare 1.4.2).

(C) Principal Stability Theorem. *The principal convex extensions* $\mathrm{Conv}_N \mathscr{R} \supset \mathscr{R}$, $N = 1, 2, \ldots, \infty$, *are h-stable for all open relations* $\mathscr{R} \to X^{(r)}$.

Proof. Every section (φ, f): $V \to Pr_1 \mathscr{R}$ for φ: $V \to \mathscr{R}$ and f: $V \to X^{(r)}$ defines a unique hyperplane field τ on the open subset $U \subset V$ of those points $v \in V$ for which $\rho \circ \varphi(v) \neq f(v)$. This is the field which corresponds to the field of principal subspaces R in $X^{(r)}$ through the points $\rho \circ \varphi(v)$ and $f(v)$ in $X^{(r)}$. The section (φ, f) is contained in $\mathrm{Conv}_1 \mathscr{R} \subset Pr_1(\mathscr{R})$ if and only if it is a section of $\mathrm{Conv}_\tau(\mathscr{R})$ and so the h-stability theorem implies the h-stability of the extension $\mathrm{Conv}_1 \mathscr{R}$ for open relations $\mathscr{R}$. Then, we conclude with an obvious induction to the h-stability of $\mathrm{Conv}_N \mathscr{R}$ for $N < \infty$. If the manifold V is compact, then the h-stability of $\mathrm{Conv}_N \mathscr{R}$ for $N < \infty$ implies that for $N = \infty$, as the sections $V \to \mathrm{Conv}_\infty \mathscr{R}$ in question have bounded principal length and so they lie in some $\mathrm{Conv}_N \mathscr{R}$ for $N < \infty$. If V is non-compact, then the above applies to compact parts of V and the h-stability of $\mathrm{Conv}_\infty \mathscr{R}$ is obtained with a compact exhaustion of V. Q.E.D.

Applications of the Theorem (C) depend upon the structure of *piecewise principal* homotopies of sections f_t: $V \to X^{(r)}$ which by definition are (continuous and locally finite!) sections

$$V \to (\text{the space of piecewise principal paths } [0, 1] \to X^{(r)}),$$

where $v \mapsto f_t(v)$, $t \in [0, 1]$.

(D) Lemma. *If two sections* f_0 *and* f_1: $V \to X^{(r)}$ *have equal projections to* $X^{(r-1)}$, *that is if* $p^r_{r-1} \circ f_0 = p^r_{r-1} \circ f_1$, *then* f_0 *and* f_1 *can be joined by a piecewise principal homotopy of sections* f_t: $V \to X^{(r)}$, $t \in [0, 1]$. *Moreover, every continuous homotopy, whose projection to* $X^{(r-1)}$ *is constant in* t, *admits a piecewise principal approximation.*

Proof. We must show that there are "sufficiently many" piecewise principal paths in the fiber $X^{(r)}_x$ of the affine bundle $X^{(r)} \to X^{(r-1)}$. We need for this "sufficiently many" principal subspaces in $X^{(r)}_x$. The required sufficiency of principal subspaces is immediate from the following

(D′) Sublemma. *Let* $(X^{(r)}_x, 0)$ *be the vector space with some point* $0 \in X^{(r)}_x$ *taken for the origin in the affine space* $X^{(r)}_x$. *Then the principal subspaces in* $X^{(r)}_x$ *which pass through* 0 *linearly span the space* $(X^{(r)}_x, 0)$.

Proof. The space $(X^{(r)}_x, 0)$ is naturally isomorphic to the space of maps Q: $\mathbb{R}^n \to \mathbb{R}^q$ (where $n = \dim V$ and q is the dimension of the fiber of the underlying fibration $X \to V$) whose components Q_i, $i = 1, \ldots, q$, are homogeneous polynomials of degree r. Such a $Q \in (X^{(r)}_x, 0)$ lies in a principal subspace $R \subset X^{(r)}_x$ associated to a hyperplane $\tau \subset \mathbb{R}^n = T_v(V)$ if and only if $Q_i = c_i l^r$, $i = 1, \ldots, q$, where l: $\mathbb{R}^n \to \mathbb{R}$ is a linear form for which $\operatorname{Ker} l = \tau$ and where c_i are arbitrary constants. Now, the sublemma follows from the existence of a decomposition of every homogeneous polynomial $\mathbb{R}^n \to \mathbb{R}$ of degree r into a linear combination $\sum_\mu c_\mu l^r_\mu$ for some linear forms l_μ on $\mathbb{R}^n$.

(D″) **Corollary.** *The relation $Pr_\infty \mathscr{R} \to X^{(r)}$ trivially* [compare (A)] *satisfies the C^{r-1}-dense h-principle for all relations $\mathscr{R} \to X^{(r)}$.*

Proof. Let (φ, f_t) be an arbitrary section of $Pr_\infty \mathscr{R}$ for $\varphi\colon V \to \mathscr{R}$ and $f_t\colon V \to X^{(r)}$, $t \in [0, 1]$, and let $f_2\colon V \to X^{(r)}$ be a holonomic section whose projection to $X^{(r-1)}$ equals to that of f_1. We lift the section f_2 to $Pr_\infty \mathscr{R}$ by joining f_1 and f_2 by a piecewise principal homotopy of sections $V \to X^{(r)}$ and thus we obtain the required holonomic section $V \to Pr_\infty \mathscr{R}$ which is homotopic to the original section (φ, f_t). Q.E.D.

2.4.3 Ample Differential Relations

A differential relation $\mathscr{R} \to X^{(r)}$ is called *ample over* $X^{(r)}$ if $\mathrm{Conv}_\infty \mathscr{R} = Pr_\infty \mathbb{R}$. As $\mathscr{R}$ lies over the jet spaces $X^{(s)}$ for $s \le r$, one may also speak of the ampleness over $X^{(s)}$. We say that $\mathscr{R}$ is *ample* if it is ample over $X^{(s)}$ for all $s = 1, \ldots, r$. We shall frequently use the following

Sufficient Condition for the Ampleness. If $\mathrm{Conv}_1 \mathscr{R} = Pr_1 \mathscr{R}$ (over $X^{(r)}$) then $\mathscr{R}$ is ample over $X^{(r)}$.

Indeed, $Pr_1 \,\mathrm{Conv}_1 \mathscr{R} \subset \mathrm{Conv}_1 Pr_1 \mathscr{R}$ for all $\mathscr{R}$, and so the equality $\mathrm{Conv}_1 \mathscr{R} = Pr_1 \mathscr{R}$ by induction implies $\mathrm{Conv}_N \mathscr{R} = Pr_N \mathscr{R}$ for $N = 2, 3, \ldots, \infty$.

(A) **Theorem.** *If an open relation $\mathscr{R} \to X^{(r)}$ is ample over $X^{(r)}$ then it satisfies the C^{r-1}-dense principle for extensions. Furthermore, if $\mathscr{R}$ is ample over all $X^{(s)}$ for $s \le r$, then it satisfies the h-principle. Moreover the h-principle is C^{s-1}-dense for $s = 1, \ldots, r$ and the parametric h-principle holds true as well. In fact, $\mathscr{R}$ satisfies all the h-principles discussed in Chapter* 1.

Proof. The C^{r-1}-density immediately follows from the principal stability theorem (C) in 2.4.2 and the trivial h-principle for $Pr_\infty \mathscr{R} = \mathrm{Conv}_\infty \mathscr{R}$. [See (D″) in 2.4.2. The latter h-principle obviously holds for extensions as the proof of (D″) shows.]

Next we observe that the C^{i-1}-dense principle for $\mathscr{R} \to X^{(s)}$, where $r \ge s > i > 0$, formally follows from the C^{i-1}-dense h-principle for $\mathscr{R} \to X^{(s-1)}$ and the C^{s-1}-dense h-principle for $\mathscr{R} \to X^{(s)}$. Thus by induction we obtain the C^{s-1}-dense h-principle for $\mathscr{R} \to X^{(r)}$ which implies for $s - 1 = 0$ the ordinary h-principle.

In order to prove the parametric h-principle we apply the above to families of sections $f_p\colon V \to X$, $p \in P$, which are viewed as sections $V \times P \to X \times P$ for all manifolds P (compare 1.2.1). Q.E.D.

Examples of Ample Relations. We start with the immersion condition $\mathscr{R} \subset X^{(1)}$ for $X = V \times W \to V$. The fiber $X_x^{(1)}$, $x = (v, w)$, of the fibration $X^{(1)} \to X$ is the space of linear maps $\mathbb{R}^n = T_v(V) \to T_w(W) = \mathbb{R}^q$. A linear map $L_0 \in X_x^{(1)}$ belongs to $\mathscr{R}$ if and only if rank $L_0 = n = \dim V$. The principal subspace $R = R(L_0, \tau) \subset X_x^{(1)}$ through L_0 for some hyperplane $\tau \subset \mathbb{R}^n = T_v(V)$ consists of those linear maps $L\colon \mathbb{R}^n \to \mathbb{R}^q$ for which $L|\tau = L_0|\tau$. If $L_0 \in \mathscr{R}$, then the intersection $R \cap \mathscr{R}$ equals R minus a subspace $R' \subset R$ of dimension $n - 1 = \dim \tau$. If $\dim R = q > n$, then codim $R' \ge 2$

and so the intersection $R \cap \mathscr{R}$ is path connected as well as dense in R and so $\mathrm{Conv}_1 \mathscr{R} = Pr_1 \mathscr{R}$. Hence, $\mathscr{R}$ is ample for $q > n$ ($\mathscr{R}$ obviously is not ample for $n = q$) and so we conclude to *the Hirsch immersion theorem in the extra dimension case.*

Exercises. Derive Hirsch's theorem for equidimensional immersions of open manifold from the Theorem (A).

Hint. Restrict the immersion condition to the $(n-1)$-skeleton of an appropriate triangulation of V (compare 1.4.1).

Prove *Feit's k-immersion theorem* by showing $\mathrm{Conv}_1 \mathscr{R}_k = Pr_1 \mathscr{R}_k$, $k \neq \dim W$, for the differential relation $\mathscr{R}_k \subset X^{(1)}$ which governs maps $V \to W$ of rank $\geq k$.

Show that the map $\rho_1: Pr_1 \mathscr{R}_k \to X^{(1)}$ sends $Pr_1 \mathscr{R}_k$ onto $\mathscr{R}_{k-1} \subset X^{(1)}$ and that $Pr_N \mathscr{R}_k$ goes onto $\mathscr{R}_{k-N}$ for all $N = 1, 2, \ldots$.

Let $A \to V$ and $B \to W$ be vector bundles and let $\alpha: A \to T(V)$ and $\beta: T(W) \to \mathrm{B}$ be homomorphisms such that rank $\beta \geq k_0$ for some integer k_0.

Prove the h-principle for those C^1-maps $f: V \to W$ for which the composed homomorphism $\beta \circ D_f \circ \alpha: A \to B$ has rank $\leq k$ for a given integer $k < k_0$.

Free Maps. Let W be a Riemannian manifold of dimension q. Then the freedom condition for C^2-maps $V \to W$ (see 1.1.4) is ample in the extra dimension case, that is for $q \geq [n(n+1)/2] + n + 1$. This also is true (and equally obvious) for kth-order free maps, $k = 2, 3, \ldots$, in the extra dimension case. Thus we obtain the h-principle for free maps (of any order) in the extra dimension case.

Systems of Sections of D-rank $\geq k$. Let Y and Z be vector bundles over V and let $D: \Gamma^r(Y) \to \Gamma^0(Z)$ be a differential operator of order r which is given by a continuous vector bundle homomorphism $\Delta: Y^{(r)} \to Z$ (compare 2.1.2). *The D-rank* of a system of C^r-sections $f_j: V \to Y$, $j = 1, \ldots, q$, is the infimum of the dimensions of the subspaces $\mathrm{span}\{Df_j(v)\} \subset Z_v$ over all fibers $Z_v \subset Z$, $v \in V$. We have indicated in 2.1.2 the h-principle for systems of D-rank $\geq k < q$ for C^∞-smooth homomorphisms Δ. This h-principle can be recovered (for all *continuous* Δ) with the convex integration as the corresponding differential relation $\mathscr{R} \subset X^{(r)}$, for $X = \underbrace{Y \oplus \cdots \oplus Y}_{q}$, is ample for $k < q$. This is an exercise for the reader. Moreover, the convex integration allows the case $k = q$ for the homomorphisms Δ of *principal* rank ≥ 2, which means that the rank of Δ on each principal subspace in $Y^{(r)}$ is at least 2. Indeed, the relation $\mathscr{R} \subset X^{(r)}$ is ample in this case as the following consideration shows. Every principal subspace $R \subset X_v^{(r)}$, $v \in V$, is associated with some hyperplane $\tau \in T_v(V)$ and R is the Cartesian product of q principal subspaces $R_j \subset Y_v^{(r)}$, $j = 1, \ldots, q$, which are also associated to τ. The homomorphism Δ sends the subspaces $R_j \subset Y_v^{(r)}$, $j = 1, \ldots, q$, which are also associated to τ. The homomorphism Δ sends the subspaces R_j on mutually parallel subspaces in Z_v, say onto $R'_j \subset Z_v$ and $\dim R'_j \geq 2$, $j = 1, \ldots, q$, provided principal rank $(\Delta) \geq 2$.

Lemma. *Let R' denote the Cartesian product of the subspaces $R'_j \subset Z_v$, $j = 1, \ldots, q$, and let $\Sigma' \subset R'$ denote the subset of the q-tuples of those vectors $z_j \in R'_j \subset Z_v$, $j =$*

$1, \ldots, q$, which are linearly dependent in the ambient space Z_v. If $d = \dim R'_j \geq 2$, then the convex hull of every connected component of the complement $R' \backslash \Sigma'$ equals R' in so far as $\Sigma' \neq R'$.

Proof. Let first $d = 2$ and $\dim Z_v = q$. Choose some vectors a, b and c_j in Z_v, $j = 1, \ldots, q$, such that every vector $z_j \in R'_j$ is a combination, $z_j = x_j a + y_j b + c_j$ for some real coefficients x_j and y_j, and such that the vectors $a, b, c_2, c_3, \ldots, c_q$ are linearly independent in Z_v. The exterior product $z_1 \wedge \cdots \wedge z_q$ is a (non-homogeneous) quadratic polynomial $Q(x_j, y_j) = Ax_1 y_2 + Bx_2 y_1 + \cdots$, where $A, B \neq 0$, and the subset Σ' is given by the equation $Q = 0$. The complement to the hypersurface $Ax_1 y_2 + Bx_2 y_1 = \text{const}$ in $\mathbb{R}^4$ has two connected components whose convex hulls equal $\mathbb{R}^4$, as $A, B \neq 0$. Hence, the complement to Σ' in R' has the same property.

Now we still let $d = 2$ but let $q' = \dim Z_v > q$. For $q = 1$ the lemma is obvious. If $q \geq 2$, then we take a linear map $Z_v \to \mathbb{R}^q$ which sends the planes R'_j to parallel *planes*, say onto $\bar{R}_j \subset \mathbb{R}^q$. The above case of $q = q'$ applies to the corresponding singularity $\bar{\Sigma}' \subset \bar{R}' = R'$ and then the lemma follows for Σ' as $\Sigma' \subset \bar{\Sigma}'$.

Finally, let $d > 2$. If $z = (z_1, \ldots, z_q)$ is an arbitrary point in $R' \backslash \Sigma'$ and if $R''_j \subset R'_j$ are parallel planes through the points $z_j \in R'_j$, then, by the above, the convex hull C_z of the component of $z \in R' \backslash \Sigma'$ contains the Cartesian product $R'' \subset R'$ of these planes. Take a linear projection of $Z_v \supset R'_j$ onto a hyperplane $H \subset Z_v$ such that the subspaces R'_j go to $(d-1)$-dimensional subspaces, say onto $\bar{R}'_j = R'_j \cap H \subset H$, and the planes R''_j go to lines $\bar{R}''_j = R''_j \cap H \subset H$. The projections $\bar{z}_j \in H$ of the vectors z_j span a subspace in H of dimension $\geq q - 1$, and so we may assume that the vectors $\bar{z}_1, \ldots, \bar{z}_{q-1}$ are linearly independent. We apply, by induction, the lemma to the subspaces $\bar{R}'_j \subset H$, $j = 1, \ldots, q - 1$, and we conclude that the convex hull $C_z \subset R'$ goes *onto* the product of the subspaces $\bar{R}'_j \subset H$, $j = 1, \ldots, q$, under the projection $Z_v \to H$. Therefore, the inclusion $R'' \subset C_z$ implies $C_z = R'$. Q.E.D.

The lemma shows that $\text{Conv}_1 \mathcal{R} = Pr_1 \mathcal{R}$ and so $\mathcal{R}$ is ample.

Examples. A differential operator D by definition is *elliptic* if the principal rank equals $\dim Z_v$ and so the relation $\mathcal{R}$ is ample for elliptic operators unless $\dim Z_v = 1$.

Let Y be an exterior power of the cotangent bundle, $Y = \Lambda^m(V)$, let $Z = \Lambda^{m+1}(V)$ and let D be the exterior differential d. It is clear, that the principal rank of d is at least $n - 1$, unless $m = 0$ or $m + 1 = n = \dim V$, and so we have by the h-principle the following

Theorem. *Let ω_j, $j = 1, \ldots, q$, be differential $(m+1)$-forms on V which are linearly independently at every point $v \in V$. If $1 \leq m \leq n - 1$, then there exist exact independent forms $\tilde{\omega}_j$, $j = 1, \ldots, q$ on V, such that the systems of forms $\{\omega_j\}$ and $\{\tilde{\omega}_j\}$ can be joined by a homotopy of linearly independent forms. In particular, if the manifold V is parallelizable, then it supports q exact $(m+1)$-forms $(1 \leq m \leq n-1)$ for $q = n!/(m+1)!(n-m-1)!$ which are linearly independent at every point $v \in V$.*

Corollary (Divergence Free Vector Fields). *Let Ω be a nonvanishing n-form on V and let $L_1, \ldots, L_q$ be linearly independent vector fields on V. If* $\dim V \geq 3$, *then there exist*

q independent fields $\tilde{L}_j$ on V whose flows preserve Ω and such that there is a homotopy of independent fields between $\{\tilde{L}_j\}$ and $\{L_j\}$. In particular, every parallelizable manifold supports $n = \dim V$ independent divergence free vector fields.

Proof. The flow of a field L on V preserves Ω if and only if the $(n-1)$-form $\omega = L\Lambda\Omega$ is closed. Recall the definition:

$$\omega(t_1, \ldots, t_{n-1}) = \Omega(L, t_1, \ldots, t_{n-1}).$$

As the correspondence $L \mapsto \omega$ is induced by an *isomorphism* $T(V) \to \Lambda^{n-1}(V)$, exact $(n-1)$-forms give us the required divergence free (i.e. preserving Ω) fields on V. Q.E.D.

Exercises. Generalize the above h-principle to divergence free fields on non-orientable manifolds V.

Show that the h-principle for non-vanishing divergence free fields fails for the 2-torus as every such field on T^2 is homotopic to a standard linear field.

Let us sketch another construction of non-vanishing divergence free fields on split manifolds $V = V_1 \times V_2$ for $\dim V_1 \geq 2$ and $\dim V_2 \geq 2$. We may assume (see Moser 1966) that the volume form Ω also splits, $\Omega = \Omega_1 \oplus \Omega_2$ and then we deform a given non-vanishing field L on V to a non-vanishing divergence free field $\tilde{L}$ as follows. The splitting $V = V_1 \times V_2$ induces the splitting of the tangent spaces,

$$T_v(V) = T_v(V_1 \times v_2) \oplus T_v(v_1 \times V_2) \qquad \text{for } v = (v_1, v_2) \in V$$

and thus the splitting of fields, $L = L_1 + L_2$. Since the form Ω splits, this splitting of fields respects the div $= 0$ condition. We assume the original field $L = L_1 + L_2$ to be generic, such that the zero set $\Sigma(L_1) \subset V$ of the field L_1 meets every submanifold $v_1 \times V_2 \subset V$, $v_1 \in V_1$, at (at most) finitely many points. Then we deform L to a split field $L' = L_1 + L_2'$, where L_2' is a divergence free field, which does not vanish on $\Sigma(L_1)$ and such that the intersection $\Sigma(L_2') \cap (V_1 \times v_2)$ is finite for all $v_2 \in V$. The existence of L_2' is an easy "zero dimensional" (compare 1.4.4) problem. Then we pass from L' to the required field $\tilde{L} = \tilde{L}_1 + L_2'$ where $\tilde{L}_1$ is a divergence free field which does not vanish on $\Sigma(L_2')$. Q.E.D.

An advantage of this new proof is a possibility of applications to complex analytic and algebraic manifolds (compare 2.1.5).

Yet, another construction of non-vanishing divergence free fields is due to Asimov (1976). Amisov uses his round handle decomposition of V and he produces fields with controlled dynamic properties.

Exercise. Fill in the details in the construction of the fields L_2' and $\tilde{L}_1$ and generalize the construction of $\tilde{L}$ to all (non-split) manifolds V of dimension ≥ 4.

2.4.4 Fiber Connected Relations and Directed Immersions

A continuous map between two topological spaces, $f\colon A \to B$, is called a *0-fibration* if the homotopy lifting property holds for maps of zero-dimensional polyhedra. That is every path $g: [0,1] \to B$ admits a lift to a path $G: [0,1] \to A$ which issues from a given point $a \in A$ over $g(0) \in B$. We say that f is *fiber connected* if the pullback $f^{-1}(b) \subset A$ is path connected for all $b \in B$. The following two properties of *fiber connected microfibrations* $f: A \to B$ are obvious.

(i) If the intersection of the image $f(A) \subset B$ with an open subset $U \subset B$ is path connected then the pullback $f^{-1}(U) \subset A$ is path connected.
(ii) The map of A onto its image, $f: A \to f(A) \subset B$ is a 0-fibration.

Consider two relations $\rho: \mathscr{R} \to X^{(r)}$ and $\tilde{\rho}: \tilde{\mathscr{R}} \to X^{(r)}$ and let $\sigma: \tilde{\mathscr{R}} \to \mathscr{R}$ be a continuous map such that $\tilde{\rho} = \rho \circ \sigma$. Let $\sigma_N: Pr_N \tilde{\mathscr{R}} \to Pr_N \mathscr{R}$, $N = 1, 2, \ldots, \infty$, denote the induced maps of the principal extensions.

(A) **Lemma.** *Let the map σ be a 0-fibration and let $(\tilde{a}, x) \in Pr_1 \tilde{\mathscr{R}}$ and $(a, x) \in Pr_1 \mathscr{R}$ be points, such that $\sigma(\tilde{a}) = a$. Then the point $(\tilde{a}, x)$ is contained in* $\mathrm{Conv}_1 \tilde{\mathscr{R}} \subset Pr_1 \tilde{\mathscr{R}}$ *if and only if $(a, x) \in \mathrm{Conv}_1 \mathscr{R}$.*

Proof. The inclusion $\sigma_1(\mathrm{Conv}_1 \tilde{\mathscr{R}}) \subset \mathrm{Conv}_1 \mathscr{R}$ is obvious. Now, if $(a, x) \in \mathrm{Conv}_1 \mathscr{R}$, then by definition there is a path $\alpha: [0,1] \to \mathscr{R}$, $\alpha(0) = a$ which lies over the principal subspace $R \subset X^{(r)}$ through the points $\rho(a)$ and x in $X^{(r)}$, such that the point $x \in X^{(r)}$ is surrounded by the path $\rho \circ \alpha: [0,1] \to R$. This α lifts to a path $\tilde{\alpha}$ in $\tilde{\mathscr{R}}$ with the same image $\tilde{\rho} \circ \tilde{\alpha} = \rho \circ \alpha$ in R. Q.E.D.

(A′) **Corollary.** *Let σ be a 0-fibration. Then the map σ_N is a 0-fibration of* $\mathrm{Conv}_N \tilde{\mathscr{R}}$ *over* $\mathrm{Conv}_N \mathscr{R}$ *for $N = 1, \ldots, \infty$. Furthermore, if $\mathscr{R}$ is ample then $\tilde{\mathscr{R}}$ also is ample. In particular, if $\tilde{\rho}: \tilde{\mathscr{R}} \to X^{(r)}$ is a 0-fibration then $\tilde{\mathscr{R}}$ is ample.*

Proof. First, the map $\sigma_1: \mathrm{Conv}_1 \tilde{\mathscr{R}} \to \mathrm{Conv}_1 \mathscr{R}$ is a 0-fibration. Indeed, if $(a(t), x(t))$ is a path $[0,1] \to \mathrm{Conv}_1 \mathscr{R}$, then there is a lift $\tilde{a}(t)$ of $a(t)$ to $\tilde{\mathscr{R}}$ for every given point $\tilde{a}(0) \in \tilde{\mathscr{R}}$ over $a(0)$ and by the previous lemma the path $(\tilde{a}(t), x(t))$ lies in $\mathrm{Conv}_1 \tilde{\mathscr{R}}$. Then by induction we obtain the same result for all $N = 2, 3, \ldots$, and for $N = \infty$. Also by this argument a point $(\tilde{a}, x(t)) \in Pr_\infty \tilde{\mathscr{R}}$ lies in $\mathrm{Conv}_\infty \mathscr{R}$ if and only if $(a, x(t)) \in \mathrm{Conv}_\infty \mathscr{R}$ for $a = \sigma(a)$. Q.E.D.

(B) **Lemma.** *Let $a: [0,1] \to \mathscr{R}$ be a continuous path whose projection $y = \rho \circ a$ to $X^{(r)}$ is piecewise principal. Let $b = (a(1), x = x(t))$ and $b^* = (a(0), z = z(t))$ be two points in $Pr_\infty \mathscr{R}$, where the path z is the composition of paths, $z = y * x$. Then* $[b \in \mathrm{Conv}_\infty \mathscr{R}] \Leftrightarrow [b^* \in \mathrm{Conv}_\infty \mathscr{R}]$.

Proof. The implication $\Rightarrow$ is immediate from the definition of Conv_∞ and the implication $\Leftarrow$ follows from $\Rightarrow$. Q.E.D.

Two points a_0 and a_1 in $\mathscr{R}$ by definition *lie in the same principal component* of $\mathscr{R}$ if they can be joined by a path whose projection to $X^{(r)}$ is piecewise principal. Each principal component in $\mathscr{R}$ is sent by the map $\rho: \mathscr{R} \to X^{(r)}$ to some fiber $X_x^{(r)}$ of the fibration $X^{(r)} \to X^{(r-1)}$, $x \in X^{(r-1)}$. If the relation $\mathscr{R}$ is *open*, then the principal components of $\mathscr{R}$ are exactly and only the path connected components of the pullbacks $\rho^{-1}(X_x^{(r)} \subset \mathscr{R}$ for all $x \in X^{(r-1)}$, as paths $[0,1] \to \rho^{-1}(X_x^{(r)}$ may be perturbed to those which have piecewise principal projections. Lemma (B) now implies the following

(B′) **Corollary.** *Let the image of the map $\sigma\colon \tilde{\mathscr{R}} \to \mathscr{R}$ meet every principal component of $\mathscr{R}$. If $\tilde{\mathscr{R}}$ is ample over $X^{(r)}$ then $\mathscr{R}$ also is ample over $X^{(r)}$.*

(C) *Recipes for Checking the Ampleness.* A relation $\mathscr{R} \to X^{(r)}$ is ample over $X^{(r)}$ if and only if each path connected component of the intersection $\rho^{-1}(X_x^{(r)})$ is ample for all $x \in X_x^{(r-1)}$. (Observe that empty relations are ample). Thus the verification of the ampleness reduces to those path connected relations $\mathscr{R}$ which lie over a single fiber, $\mathscr{R} \to X_x^{(r)} \subset X^{(r)}$, $x \in X^{(r-1)}$. The fiber $X_x^{(r)}$ is naturally isomorphic to the space $H_n = H_n^r(q)$ of homogeneous polynomial maps $\mathbb{R}^n \to \mathbb{R}^q$ of degree r. The ampleness of $\mathscr{R} \to H_n$ depends on how $\mathscr{R}$ interacts with principal subspaces $R \approx \mathbb{R}^q$ in H_n, where "ampleness" is understood (here and below) as the ampleness over $X^{(r)}$. We obtain with (A′) and (B′) the following list of implications which helps us to establish the ampleness of relations $\rho\colon \mathscr{R} \to H_n$.

(a) $\mathscr{R}$ is ample $\Leftrightarrow$ each path connected component of $\mathscr{R}$ is ample.
(a′) $\mathscr{R}$ is ample $\Leftrightarrow$ each principal component of $\mathscr{R}$ is ample.
[If $\mathscr{R}$ is open then (a) is equivalent to (a′).]
(b) $\mathscr{R}$ is ample $\Leftrightarrow$ $\mathrm{Conv}_1 \mathscr{R}$ is ample.
(b′) $\mathscr{R}$ is ample $\Leftrightarrow$ $\mathrm{Conv}_\infty \mathscr{R}$ is ample.
(c) Let $\mathscr{R}$ be a 0-fibration over its image $\mathscr{R}' \subset H_n$. Then
$\mathscr{R}$ is ample $\Leftrightarrow$ $\mathscr{R}'$ is ample.
(d) Let $\mathscr{R}$ consists of a single principal component and let $\mathscr{R}' \subset \mathscr{R} \to H_n$. Then
$\mathscr{R}'$ is ample $\Rightarrow$ $\mathscr{R}$ is ample, unless $\mathscr{R}'$ is an empty set.
Finally, the empty relation and $\mathscr{R} = H_n$ are ample.

One can generate with (a)–(d) a great deal of ample relations over (as well as in) H_n and so over $X^{(r)}$.

Example. *Triangular Relations.* Let $r = 1$. [See (C) in 2.4.4 for $r \geq 2$.] Then every linear embedding $\mathbb{R}^{n-1} \to \mathbb{R}^n$ induces an obvious linear map $h_n\colon H_n \to H_{n-1}$ whose fibers are principal subspaces in H_n. By induction we define a class of relation $\mathscr{R} \subset H_i$, $i = 0, 1, \ldots, n$, which are called *triangular*, as follows. If $i = 0$ then all relations in $H_0 = 0$ (i.e. $\mathscr{R} = \varnothing$ and $\mathscr{R} = 0$) are triangular. A relation $\mathscr{R} \subset H_i$ is triangular if and only if there is an embedding $\mathbb{R}^{i-1} \subset \mathbb{R}^i$ such that the corresponding projection $h_i\colon H_i \to H_{i-1}$ satisfies three conditions,

(i) the image $h_i(\mathscr{R}) \subset H_{i-1}$ is triangular;
(ii) the map h_i is a 0-fibration of $\mathscr{R}$ over $h_i(\mathscr{R})$. [For example, the map $h_i\colon \mathscr{R} \to h_i(\mathscr{R})$ is a fiber connected submersion.];
(iii) the convex hull of each path connected component of $\mathscr{R} \cap (h_i^{-1})(x)$ equals $h_i^{-1}(x)$ for all $x \in h_i(\mathscr{R})$.

Triangular conditions are ample [use (b) and (c)] and so we have the following

Theorem. *Let $\mathscr{R} \subset X^{(1)}$ be an open relation such that each connected component of the intersection $\mathscr{R} \cap X_x^{(1)}$ contains a non-empty triangular relation for all $x \in X$. Then $\mathscr{R}$ satisfies the dense h-principle.*

Let us apply this theorem to immersions of an oriented n-dimensional manifold, $f: V \to \mathbb{R}^{n+1}$, which are *directed* (compare 1.4.4), by a given open connected subset $A \subset Gr_n(\mathbb{R}^{n+1}) = S^n$, that is $G_f(V) \subset A$ for the oriented normal map $G_f: V \to S^n$. The corresponding differential relation $\mathscr{R} = \mathscr{R}(A) \subset H_n = \mathrm{Hom}(\mathbb{R}^n \to \mathbb{R}^{n+1})$ by definition consists of those injective linear maps $L: \mathbb{R}^n \to \mathbb{R}^{n+1}$ for which $L(\mathbb{R}^n) \in A$. Take a small open ε-ball B_0 in S^n, fix a hyperplane $\mathbb{R}_0^{n-1} \subset \mathbb{R}^n$, and denote by $\mathscr{R}_0 \subset H_n$ the set of those maps $L: \mathbb{R}^n \to \mathbb{R}^{n+1}$ which send $\mathbb{R}_0^{n-1}$ into some hyperplane $b \in B_0 \subset Gr_n(\mathbb{R}^{n+1}) = S^n$. A straightforward check up shows,

If every great circle S^1 in S^n, which meets B_0, intersects the subset A over an arc of length $> \pi$, then the relation $\mathscr{R} \cap \mathscr{R}_0 \subset H_n$ is triangular.

Now fix a point $b_0 \in S^n$ and consider the pencil of great circles $S^1 \subset S^n$ through b_0. Let A satisfy the following condition.

(D) The intersection $A \cap S^1$ contains an arc of length $> \pi$ for all S^1 in the pencil.

This (D) obviously implies the existence of a triangular subset $\mathscr{R}_0 \subset \mathscr{R} = \mathscr{R}(A)$ and so $\mathscr{R}$ is ample. Hence we have the following

(D′) **Theorem.** *If A satisfies* (D) *then A-directed immersions abide the C^0-dense h-principle. In particular every continuous map of a parallelizable manifold V into $\mathbb{R}^{n+1}$ admits a fine approximation by A-directed immersions $V \to \mathbb{R}^{n+1}$ (notice that V is necessarily parallelizable if there exists an A-directed immersion $V \to \mathbb{R}^{n+1}$ for $A \neq S^n$).*

The condition (D) is satisfied, for instance, in the following two cases.
(1) The complement $\Sigma = S^n \backslash A$ is a finite subset with no pairs of opposite points.
(2) The subset A contains a closed hemisphere.

Exercise. Give a direct geometric construction of an immersion of the n-torus $T^n \to \mathbb{R}^{n+1}$ whose oriented spherical image is contained in a small neighborhood of the hemisphere $S_+^n \subset S^n$.

Question. Is there a "simple" immersion $T^2 \to \mathbb{R}^3$ whose spherical image misses the four vertices of a regular tetrahedron in S^2?

(E) *Necessary Conditions for the Existence of A-Directed Immersions $V \to \mathbb{R}^q$.* There is a wide gap between known suffficient and necessary conditions for the validity of the h-principle for A-directed immersions. For example the only known necessary condition for the existence of an A-directed immersion $f: V \to \mathbb{R}^{n+1}$ for closed parallelizable manifolds V is as follows. The double cover map $S^n \to P^n$ sends every path connected component of A *onto* the projective space P^n. Indeed, every linear function $\mathbb{R}^n \to \mathbb{R}$ has critical points on $f(V) \subset \mathbb{R}^n$.

Exercise. Show that there is no A-directed *embeddings* of a closed manifold $V \to \mathbb{R}^{n+1}$ unless $A = S^n$.

A similar necessary condition holds true for immersions f of *closed* manifolds $V \to \mathbb{R}^q$ for all $q > n = \dim V$.

(i) If the non-oriented tangential map $G_f: V \to Gr_n(\mathbb{R}^q)$ sends V into a given subset $A \subset Gr_n(\mathbb{R}^q)$ then A intersects the Grassmann manifold $Gr_n(H) \subset Gr_n(\mathbb{R}^q)$ for all hyperplanes $H \subset \mathbb{R}^q$.

There is an additional condition on A (for $q \geq n + 2$) which is due to the obvious fact that every exact n-form on V necessarily vanishes. Namely, let ω denote an n-form in $\mathbb{R}^q$ with constant coefficients and let $\Sigma(\omega) \subset Gr_n(\mathbb{R}^q)$ be the subset of those n-dimensional subspaces in $\mathbb{R}^q$ on which ω vanishes.

(ii) The set A intersects $\Sigma(\omega)$ for all n-forms ω.

Questions. Let an open connected subset $A \subset Gr_n(\mathbb{R}^q)$ satisfy (i) and (ii). Does the h-principle hold for A-directed immersions? Does there exist an A-directed immersion of at least one closed manifold $V_0 \to \mathbb{R}^q$? Does the existence of a single A-directed immersion imply the h-principle for A-directed immersions of all n-dimensional manifolds $V \to \mathbb{R}^q$?

The latter question is motivated by the following

Example. A C^1-map $E: V_0 \to V_0$ is called *expanding* if some iterate E^N of E strictly enlarges the length of all non-zero tangent vectors in V_0 relative to a fixed Riemannian metric on V_0. (If V_0 is compact then the expanding property does not depend on the choice of the metric). For instance, the map $x \mapsto 2x$ of the n-torus $T^n = \mathbb{R}^n/\mathbb{Z}^n$ onto itself is expanding. [See Shub (1969) for additional examples.]

Let a compact manifold V_0 admit an expanding map $E: V_0 \to V_0$. If there exists an A-directed immersion $f_0: V_0 \to \mathbb{R}^q$ for some open subset $A \subset Gr_n(\mathbb{R}^q)$, then an arbitrary continuous map $f: V \to \mathbb{R}^q$ can be uniformely approximated by A-directed immersions.

Proof. We may assume the map f to be C^1. Then the maps $f + N^{-1} f_N$ for $f_N = f \circ E^N$ are A-directed immersions for large N and the sequence $f + N^{-1} f_N$ uniformely converges to f as $N \to \infty$.

Exercise. Show that every compact manifold V_0 which admits an expanding map has $\mathbb{R}^n$ for the universal covering.

Remark. Farrel and Jones (1981) introduced a notion of a compact *branched* manifold V_0 and produced many examples of expanding maps $V_0 \to V_0$ such that the above discussion applies. Their results indicate the existence of many compact n-dimensional subpolyhedra $P \subset Gr_n(\mathbb{R}^q)$ such that immersions $V \to \mathbb{R}^q$ which are directed by arbitrarily small neighborhoods $\mathcal{O}p\,P \subset Gr_n(\mathbb{R}^q)$ satisfy the h-principle.

(E′) *Directed Immersions of Open Manifolds.* If V is open, then the h-principle for A-directed immersions $f: V \to \mathbb{R}^q$ holds true for all *open* subsets $A \subset Gr_n(\mathbb{R}^q)$ (see 2.2.2). However, if we additionally require the immersions f to be *complete*, which means the completeness of the induced Riemannian metric g on V, then we can partially recover the necessary conditions (i) and (ii).

Examples. (a) Let V have no boundary and let $f: V \to \mathbb{R}^{n+1}$, $n = \dim V$, be a complete immersion whose non-oriented normal map $G_f: V \to P^n$ sends V into a proper *closed* subset $\bar{A} \subset P^n$. Then the map f is unbounded and the manifold V topologically splits, $V = V' \times \mathbb{R}$. Indeed, the normal projection to a line $l \in P^n \setminus \bar{A}$ gives us an infinitesimally enlarging (in the metric εg on V for some $\varepsilon > 0$) map $V \to l$ which is a fibration (see 1.2.3).

(a′) If $G_f(V \setminus V_0) \subset \bar{A}$ for some compact subset $V_0 \subset V$, then the map $V \to l$ is infinitesimally enlarging (in εg) at infinity. Hence, the map f is unbounded, the manifold V has at most finitely many ends and the intersection map on homology, $H_*(V) \otimes H_*(V) \to H_*(V)$, has finite rank, provided V is connected. This implies for $\dim V = 2$ the finiteness of the topological type of V.

Subexample. Let the map $G_f: V \to P^n$ be an open map outside a compact subset in V. (For instance, G_f is an immersion or a branched covering at infinity). If the pullback $G_f^{-1}(p) \subset V$ is compact for all $p \in P^n$, then the image $G_f(V \setminus V_0) \subset P^n$ is not dense for some compact subset $V_0 \subset V$ and so (a′) applies.

(a″) If the closure of the image $G_f(V \setminus V_0) \subset P^n$ misses a hyperplane in P^n, then the projection of V to this "missing" hyperplane $H \subset \mathbb{R}^{n+1}$ is infinitesimally enlarging (for εg) outside V_0 and so the projection $V \to H$ is a covering map at infinity. Hence, each end of V is diffeomorphic to $S^{n-1} \times \mathbb{R}$.

Question. For which open subsets $A \subset S^n$ is there a complete immersion of some oriented manifold, $f: V \to \mathbb{R}^{n+1}$, whose oriented Gauss map is a diffeomorphism of V onto A? [See Verner (1970) and Burago (1968) for partial results.]

Here is a related result by Burago:

(b) Let V be a connected open surface without boundary and let $f: V \to \mathbb{R}^3$ be a complete immersion of infinite total area. If the Gauss map $G_f: V \to P^2$ has finite total area (i.e. the Jacobian J of G_f has $\int_V |J| < \infty$) then the map f is unbounded and the Euler characteristic of V is finite.

Proof. Let $\lambda_1(v)$ and $\lambda_2(v)$ denote the principal curvatures of the immersion at the points $v \in V$. Then the (intrinsic) Gauss curvature $K(v) = \lambda_1(v)\lambda_2(v)$ has $\int_V K(v) > -\infty$, and so by the Gauss-Bonnet theorem (see Huber 1957) the Euler characteristic of V is finite.

Now, suppose the image $f(V) \subset \mathbb{R}^3$ is contained in a ball; say in $B \subset \mathbb{R}^3$. Take a *projective* map P of B into the unite sphere S^3. Then the principal curvatures λ_1'

and λ_2' of the immersion $P \circ f\colon V \to S^3$ satisfy $\lambda_1'(v)\lambda_2'(v) \geq \text{const}\, \lambda_1(v)\lambda_2(v)$ for all $v \in V$ and for some const > 0 (compare 3.2.3).

The Riemannian metric induced by $P \circ f$ is complete and its total area is infinite. Since the Gauss curvature is $K'(v) = 1 + \lambda_1'(v)\lambda_2'(v)$, we conclude

$$\int_V K'(v)\,dv = \text{Area}\, V - \text{const} \int_V K(v)\,dv = +\infty.$$

This contradicts a theorem of Cohn-Vossen (1933).

(c) Let $f\colon V \to \mathbb{R}^q$, $q \geq n + 1$, be a complete bounded immersion, such that the closure of the image of the non-oriented tangential map $G_f\colon V \to Gr_n(\mathbb{R}^q)$ misses the subset $\Sigma(\omega) \subset Gr_n(\mathbb{R}^q)$ for some n-form ω on $\mathbb{R}^q$ with constant coefficients. Then V has the exponential growth relative to the induced Riemannian metric,

$$(*) \qquad \liminf_{R\to\infty} R^{-1} \log \text{Vol}\, B_v(R) > 0,$$

for the concentric balls $B_v(R)$ around each point $v \in V$.

Proof. If $(*)$ fails, then

$$\liminf_{R\to\infty} \text{Vol}\, \partial B_v(R)/\text{Vol}\, B_v(R) = 0.$$

Since the tangential image $G_f(V) \subset Gr_n(\mathbb{R}^q)$ lies away from $\Sigma(\omega)$,

$$\left| \int_{B_v(R)} f_*(\omega) \right| \geq \delta\, \text{Vol}\, B_v(R),$$

for all $R > 0$ and for some $\delta > 0$.

Let $\omega = d\lambda$ for some $(n-1)$-form λ on $\mathbb{R}^q$. Since f is bounded,

$$\left| \int_{\partial B_v(R)} f_*(\lambda) \right| \leq \text{const Vol}\, \partial B_v(R),$$

and Stokes formula, $\int_B f_*(\omega) = \int_{\partial B} f_*(\lambda)$, yields the contradiction.

Subexample. Look at immersions of surfaces $V \to \mathbb{R}^4 = \mathbb{C}^2$ and let $\omega = dx_1 \wedge dy_1 + dx_2 \wedge dy_2$ for the coordinates $x_1 + iy_1$ and $x_2 + iy_2$ in $\mathbb{C}^2$. The form ω does not vanish on *complex* lines in $\mathbb{C}^2$ and so the set $\Sigma(\omega) \subset Gr_2(\mathbb{R}^4)$ is disjoint from the complex projective line $\mathbb{C}P^1 \subset Gr_2(\mathbb{R}^4)$. The immersions $V \to \mathbb{C}^2$ directed by $\mathbb{C}P^1$ are *holomorphic curves* in $\mathbb{C}^2$, and we see that all complete bounded holomorphic curves in $\mathbb{C}^2$, as well as *nearly holomorphic* curves whose tangential image is close to $\mathbb{C}P^1 \subset Gr_2(\mathbb{R}^4)$, have exponential growth.

2.4.5 Directed Embeddings and the Relative *h*-Principle

Consider relations $\mathcal{R}_0 \subset X^{(r)}$ and $\rho\colon \mathcal{R} \to \mathcal{R}_0 \subset X^{(r)}$. Denote by $Pr_1(\mathcal{R}|\mathcal{R}_0) \subset Pr_1\mathcal{R}$ the subset of those pairs $(a, x) \in Pr_1\mathcal{R}$ for which the segment $(1-t)\rho(a) + tx$, $t \in [0, 1]$, is contained in $\mathcal{R}_0$. Then by induction we define

$$Pr_N(\mathscr{R}|\mathscr{R}_0) = Pr_1(Pr_{N-1}(\mathscr{R}|\mathscr{R}_0)|\mathscr{R}_0),$$

for $N = 2, 3, \ldots$, and we stabilize to $Pr_\infty(\mathscr{R}|\mathscr{R}_0) \subset Pr_\infty \mathscr{R}$ that is the space of those pairs $(a, y(t)) \in Pr_\infty \mathscr{R}$ for which the piecewise principal path y sends $[0, 1]$ into $\mathscr{R}_0$.

The lemma (D) in 2.4.2 immediately implies the following trivial

h-Principle for $\mathscr{R}$ Relative to (the h-Principle for) $\mathscr{R}_0$. Let $\mathscr{R}_0 \subset X^{(r)}$ and $p^r_{r-1} \circ \rho\colon \mathscr{R} \to X^{(r-1)}$ be open relations. If the extension $Pr_\infty(\mathscr{R}|\mathscr{R}_0) \supset \mathscr{R}$ is h-stable then

$$[\text{the } C^{r-1}\text{-dense } h\text{-principle for } \mathscr{R}_0] \Rightarrow [\text{the } C^{r-1}\text{-dense } h\text{-principle for } \mathscr{R}].$$

Next we define

$$\mathrm{Conv}_N(\mathscr{R}|\mathscr{R}_0) = \mathrm{Conv}_N(\mathscr{R}) \cap Pr_N(\mathscr{R}|\mathscr{R}_0),$$

for $N = 1, 2, \ldots, \infty$, and we call the relation $\mathscr{R}$ *ample over* $\mathscr{R}_0$ if $\mathrm{Conv}_\infty(\mathscr{R}|\mathscr{R}_0) = Pr_\infty(\mathscr{R}|\mathscr{R}_0)$. The principal stability theorem [see (B) and (C) in 2.4.2] implies *the h-stability of the extension* $\mathrm{Conv}_\infty(\mathscr{R}|\mathscr{R}_0) \supset \mathscr{R}$ *for open relations* $\mathscr{R} \to \mathscr{R}_0 \subset X^{(r)}$, *and so we have the relative h-principle for ample relations over* $\mathscr{R}_0$.

Now, the (obvious modification of) implications (a)–(d) in (C) of 2.4.4 hold true over $\mathscr{R}_0$ and so we have a list of sufficient conditions for the ampleness of $\mathscr{R}$ over $\mathscr{R}_0$. Unfortunately, the inclusion $Pr_1 \mathrm{Conv}_1 \subset \mathrm{Conv}_1 Pr_1$ may fail over $\mathscr{R}_0 \neq X^{(r)}$ and so we lose the sufficient condition $\mathrm{Conv}_1 = Pr_1$ (compare 2.4.3). However, this condition still works if modified as follows. We restrict to an individual fiber $X^{(r)}_x = H_n$ [compare (C) in 2.4.4] and we fix a Euclidean metric "dist" in H_n. Next we introduce another metric, $\mathrm{dist}_P(x, y)$ (P for principal), as the lower bound of the lengths of piecewise principal paths between the points x and y in H_n.

Observe that

$$\mathrm{dist} \le \mathrm{dist}_P \le \mathrm{const}\,\mathrm{dist},$$

for $\mathrm{const} = \mathrm{const}(\dim H_n)$. We impose on relations $\mathscr{R}_0 \subset H_n$ and $\rho\colon \mathscr{R} \to \mathscr{R}_0$ the following

λ-Condition. If $\delta = \mathrm{dist}_P(\rho(a), x) \le \lambda^{-1} \mathrm{dist}_P(x, \partial\mathscr{R}_0)$, for the boundary $\partial\mathscr{R}_0$ of $\mathscr{R}_0 \subset X^{(r)}$, then

$$(a, x) \in \mathrm{Conv}_1(\rho^{-1}(B_x(\lambda\delta))|\mathscr{R}_0)$$

for all $(a, x) \in Pr_1(\mathscr{R}|\mathscr{R}_0)$, where $B_x(\lambda\delta)$ denotes the dist_P-ball around $x \in \mathscr{R}_0$ of radius $\lambda\delta$, and where $\lambda \ge 1$ is a given number.

Denote by $Pr^\lambda \subset Pr_\infty(\mathscr{R}|\mathscr{R}_0)$ the subset of those pairs $(a, x(t)) \in Pr_\infty(\mathscr{R}|\mathscr{R}_0)$ for which the path $x(t)$, $t \in [0, 1]$, has length $< (\lambda + 1)^{-1} \mathrm{dist}_P(x(1), \partial\mathscr{R}_0)$.

It directly follows from these definitions that

$$[\lambda\text{-condition}] \Rightarrow [Pr^\lambda \subset \mathrm{Conv}_\infty(\mathscr{R}|\mathscr{R}_0)].$$

(A) **Corollary.** *If open relations $\mathscr{R}_0 \subset H_n$ and $\rho\colon \mathscr{R} \to \mathscr{R}_0$ satisfy the λ-condition for some $\lambda \ge 1$ and if the map $Pr^\lambda \to \mathscr{R}_0$ is a 0-fibration then the relation $\mathscr{R}$ is ample over $\mathscr{R}_0$. In particular, if* $\mathrm{Conv}_1(\rho^{-1}(U)|U) = Pr_1(\rho^{-1}(U)|U)$ *for all open subsets $U \subset \mathscr{R}_0$, then $\mathscr{R}$ is ample over $\mathscr{R}_0$.*

(B) Examples. (a) Let $\Sigma \subset X^{(r)}$ be a closed stratified subset such that the intersection $R \cap \Sigma$ is either equal to R or $\operatorname{codim}(R \cap \Sigma \subset R) \geq 2$ for all principal subspaces $R \subset X^{(r)}$. Then the difference $\mathscr{R} = \mathscr{R}_0 \backslash \Sigma \subsetneq \mathscr{R}_0$ is an ample relation over (in fact, in) $\mathscr{R}_0$ for all open subsets $\mathscr{R}_0 \subset X^{(r)}$, as the equality $\operatorname{Conv}_1(U \backslash \Sigma | U) = Pr_1(U \backslash \Sigma | U)$ is satisfied for all open subsets $U \subset X^{(r)}$. Hence, we get the h-principle for $\mathscr{R}$ relative to $\mathscr{R}_0$. This applies, for instance, to the relations $\mathscr{R}_k \subset \mathscr{R}_0 \subset X^{(1)}$, $X = V \times W$, which define *strictly short maps* $V \to W$ of rank $\geq k$ between Riemannian manifolds V and W, provided $k < \dim W$. Therefore,

Every strictly short map $V \to W$, which is homotopic to a C^1-map of rank $\geq k$ admits a fine C^0-approximation by strictly short C^1-maps $V \to W$ of rank $\geq k$, provided $k <$ dim W. (Compare 1.1.5.)

(b) Let $\tau \subset T(W)$ be a codimension n subbundle and let the relation $\mathscr{R}_0 \subset X^{(1)}$, $X = V \times W$, correspond to C^1-maps $V \to W$ which are transversal to τ. The convex integration completely fails in proving the h-principle for such an $\mathscr{R}_0$ as the "convex hulls" $\operatorname{Conv}_N \mathscr{R}_0 \to X^{(1)}$, $N = 1, \ldots, \infty$, are not greater than $\mathscr{R}_0$ (in fact, their images in $X^{(1)}$ equal $\mathscr{R}_0$). However, the h-principle does hold for $\mathscr{R}_0$ if the subbundle τ is "sufficiently non-integrable" (see 2.2.3). Now, we take $\mathscr{R} = \mathscr{R}_0 \backslash \Sigma$ for the above Σ and we derive the h-principle for $\mathscr{R}$ from that for $\mathscr{R}_0$.

The proof of the h-principle for $\mathscr{R}_0$ depends upon Nash's implicit function techniques as well as on the formalism of continuous sheaves (see 2.2, 2.3). One expects a simpler argument which would incorporate the non-integrability of τ into the convex integration scheme thus giving a direct geometric proof of the h-principle for $\mathscr{R}_0$ and for $\mathscr{R}$.

(C) *The Relative h-Principle over Embeddings.* Consider a relation $\mathscr{R}_0 \subset X^{(r)}$ and let Φ_0 be a subset in the space of holonomic sections $V \to \mathscr{R}_0$. Take a relation $\mathscr{R} \to \mathscr{R}_0$ and denote by $\widetilde{\mathscr{R} \cap \Phi_0}$ the pullback of Φ_0 under the map between the spaces of sections, $\Gamma(\mathscr{R}) \to \Gamma(\mathscr{R}_0) \supset \Phi_0$. An extension $\mathscr{R}_1 \supset \mathscr{R}$ is called *stable over* Φ_0 (or Φ_0*-stable*) if every section $\tilde{a}_0 \in \widetilde{\mathscr{R}_1 \cap \Phi_0}$ admits a homotopy $\tilde{a}_t \in \widetilde{\mathscr{R}_1 \cap \Phi_0}$, $t \in [0, 1]$, such that $\tilde{a}_1 \in \widetilde{\mathscr{R} \cap \Phi_0}$ and such that $\tilde{a}_t(v) = a_0(v)$, $t \in [0, 1]$, for those $v \in V$, for which $\tilde{a}_0(v) \in \mathscr{R} \subset \mathscr{R}_1$ (compare 2.4.2).

We fix a metric in the space $X^{(r)}$ and then we may speak of ε-neighborhoods $U_\varepsilon(\varphi) \subset \Gamma(\mathscr{R}_0)$ of sections $\varphi \colon V \to \mathscr{R}_0$, where $\varepsilon = \varepsilon(x)$ stands for a continuous positive function on $\mathscr{R}_0$. We also associate to such an $\varepsilon \colon \mathscr{R}_0 \to \mathbb{R}$ the following

ε-*Condition.* Let $\varphi \in \Phi_0$ be an arbitrary section. Then there exists a fine C^0-neighborhood of the section $p^r_{r-1} \circ \varphi \colon V \to X^{(r-1)}$, say $U' \subset \Gamma^0(X^{(r-1)})$, such that every *holonomic* section $\psi \subset U_\varepsilon(\varphi)$, whose projection to $\Gamma^0(X^{(r-1)})$ lies in U', is contained in Φ_0 and there is a homotopy in Φ_0 between φ and ψ which is constant at those $v \in V$ where $\varphi(v) = \psi(v)$.

Example. Let $X = V \times W$, let $\mathscr{R}_0 \subset X^{(1)}$ be the immersion relation and let Φ_0 be the space of the 1-jets of C^1-*embeddings* $V \to W$. Then Φ_0 satisfies the ε-condition for a fixed metric on $X^{(1)}$ and for all sufficiently small ε.

Indeed, if an immersion $\psi: V \to W$ is C^0-close to a given embedding $\varphi: V \to W$ and if the angle between the tangent spaces $T_v(\varphi(V))$ and $T_v(\psi(V))$ in $T(W)$ is at most $\frac{\pi}{2} - \delta$ for all $v \in V$ and for a fixed $\delta > 0$, then the immersion ψ clearly is an embedding which is isotopic to φ.

We assume the metric on $X^{(r)}$ to be Euclidean on the fibers of the fibration $X^{(r)} \to X^{(r-1)}$ and we take a function ε on $\mathscr{R}_0$ for which $\operatorname{dist}(x, \partial\mathscr{R}_0) > \varepsilon(x)$ for all $x \in \mathscr{R}_0$. Let $\operatorname{Conv}_1(\mathscr{R}, \varepsilon) \subset \operatorname{Conv}_1(\mathscr{R}|\mathscr{R}_0)$, for $\rho: \mathscr{R} \to \mathscr{R}_0$, denote the union over $x \in \mathscr{R}_0$ of the subsets $\operatorname{Conv}_1(\rho^{-1}(B_x(\varepsilon(x)))|B_x(\varepsilon(x))) \subset \operatorname{Conv}_1(\mathscr{R}|\mathscr{R}_0)$ for the ε-balls $B_x(\varepsilon(x))$. Then by induction we define $\operatorname{Conv}_N(\mathscr{R}, \varepsilon) \subset \operatorname{Conv}_N(\mathscr{R}|\mathscr{R}_0)$ for $N = 2, 3, \ldots$, and as earlier we define $\operatorname{Conv}_\infty(\mathscr{R}, \varepsilon) \subset \operatorname{Conv}_\infty(\mathscr{R}|\mathscr{R}_0)$.

The principal stability theorem [see (C) in 2.4.2] implies:

If Φ_0 satisfies the ε-condition, then the extension $\operatorname{Conv}_\infty(\mathscr{R}, \varepsilon) \supset \mathscr{R}$ *is Φ_0-stable.*

Next, the proof of (A) implies the following result for open relations $\mathscr{R}_0 \subset X^{(r)}$ and $\rho: \mathscr{R} \to \mathscr{R}_0$.

If $\operatorname{Conv}_1(\rho^{-1}(U)|U) = Pr_1(\rho^{-1}(U)|U)$ *for all open subsets $U \subset \mathscr{R}_0$, then*

$$\operatorname{Conv}_\infty(\mathscr{R}, \varepsilon) = \operatorname{Conv}_\infty(\mathscr{R}|\mathscr{R}_0) = Pr_\infty(\mathscr{R}|\mathscr{R}_0),$$

for all positive continuous functions $\varepsilon: \mathscr{R}_0 \to \mathbb{R}$, and so the extension $Pr_\infty(\mathscr{R}|\mathscr{R}_0)$ is Φ_0-stable for all Φ_0 which satisfy the ε-condition for some continuous function $\varepsilon > 0$.

Recall that a pair $(a, x(t)) \in Pr_\infty(\mathscr{R}|\mathscr{R}_0)$ by definition is contained in $\operatorname{Conv}_\infty(\mathscr{R}, \varepsilon)$ if there is a subdivision of *some* $N = 1, 2, \ldots$, such that $(a, x(t)) \in \operatorname{Conv}_N(\mathscr{R}, \varepsilon) \subset Pr_N(\mathscr{R}|\mathscr{R}_0)$ relative to this subdivision (compare 2.4.2).

Totally Real Embeddings. Let W be a smooth manifold with a complex (or quasi-complex) structure. A smooth map $f: V \to W$ is called *totally real* if the complexified differential $D_f: \mathbb{C}T(V) \to T(W)$ has $\operatorname{rank}_{\mathbb{C}} D_f \geq \min(\dim V, \dim_{\mathbb{C}} W)$.

Every real analytic map $f: V \to W$ extends to a holomorphic map $\mathbb{C}f: \mathbb{C}V \to W$ for some (small) complexification $\mathbb{C}V \supset V$. Totally real maps extend to holomorphic immersions $\mathbb{C}V \to W$ for $n = \dim V \leq \dim_{\mathbb{C}} W$ and to holomorphic submersions for $n \geq \dim_{\mathbb{C}} W$.

The total reality condition $\mathscr{R} \subset X^{(1)}$, $X = V \times W$, is of the form $\mathscr{R} = X^{(1)} \backslash \Sigma$, where Σ is a stratified subset such that $\operatorname{codim}(R \cap \Sigma \subset R) \geq 2$ for all principal subspaces $R \subset X^{(1)}$ which are not completely contained in Σ. This implies the h-principle for totally real maps $V \to W$.

Now, let $\mathscr{R}_0 \subset X^{(1)}$ be the immersion condition, let Φ_0 be the space of the 1-jets of embeddings $V \to W$ and let $\mathscr{R} = \mathscr{R}_0 \backslash \Sigma$ for the above Σ. The Φ_0-stability of the extension $Pr_\infty(\mathscr{R}|\mathscr{R}_0)$ leads to the following

Theorem. *Let $f_0: V \to W$ be a smooth embedding whose differential $D_{f_0}: T(V) \to T(W)$ is homotopic (via fiberwise injective homomorphisms) $T(V) \to T(W)$!) to a totally real homo-φ_1: $T(V) \to T(W)$* [i.e. $\operatorname{rank}_{\mathbb{C}} \mathbb{C}\varphi_1 \geq \min(n, \dim_{\mathbb{C}} W)$]. *Then f_0 is isotopic to a totally real embedding $f_1: V \to W$.*

Example. *A totally real embedding $S^n \to \mathbb{C}^n$ exists if and only if $n = 1, 3$.*

Proof. The multiplication by $i = \sqrt{-1}$ establishes an isomorphism between the tangent and normal bundles of a totally real immersion $V^n \to \mathbb{C}^n$. As embedded spheres $S^n \to \mathbb{R}^{2n}$ have trivial normal bundles, the existence of a totally real embedding $S^n \to \mathbb{C}^n$ implies the parallelizability of S^n. Hence $n = 1, 3$ or 7. [We refer to Bott (1969) for the basic algebraic topology which is used here and below.]

The case $n = 1$ is trivial as all immersions $S^1 \to \mathbb{C}^1$ are totally real.

Let St_n denote the Stiefel manifolds of n-frames in $\mathbb{R}^{2n} = \mathbb{C}^n$ and let $\alpha\colon U(n) \to \mathrm{SO}(2n)$ and $\beta\colon \mathrm{SO}(2n) \to \mathrm{St}_n$ be the obvious maps.

The above theorem shows that an embedding $S^n \to \mathbb{C}^n$ with a normal frame $v\colon S^n \to \mathrm{St}_n$ is isotopic to a totally real embedding if and only if there exists a homotopy class $\mu \in \pi_n(U(n))$ for which $(\beta \circ \alpha)_*(\mu) = v_*[S^n]$.

If $n = 3$, then $\alpha_*\colon \pi_n(U(n)) \to \pi_n(\mathrm{SO}(2n))$ is an isomorphism while the homomorphism $\beta_*\colon \pi_n(\mathrm{SO}(2n)) \to \pi_n(\mathrm{St}_n)$ is onto. Hence, the standard embedding $S^3 \to \mathbb{C}^3$ is isotopic to the totally real one.

If $n = 7$ then the image of the homomorphism β_* is divisible by 2 [in $\pi_7(\mathrm{SO}(14)) \approx \mathbb{Z}$], while $\pi_7(\mathrm{St}_7) = \mathbb{Z}_2$. Hence $(\beta \circ \alpha)_* = 0$. On the other hand, every embedding $f\colon S^7 \to \mathbb{R}^{14}$ carries a normal frame $v\colon S^7 \to \mathrm{St}_7$ which is not homotopic to zero and so f is not isotopic to a totally real map $S^7 \to \mathbb{R}^{14}$. Q.E.D.

Corollary. *There exists a domain of holomorphy in $\mathbb{C}^3$ which is diffeomorphic to $S^3 \times \mathbb{R}^3$.*

Proof. Take a real analytic totally real embedding $f\colon S^3 \to \mathbb{C}^3$ and then analytically continue f to a small complexification $\mathbb{C}S^3 \approx S^3 \times \mathbb{R}^3$ with a pseudo-convex boundary.

Remark. There are other interesting classes of maps besides embeddings to which the above techniques may apply. For example, one may try maps $V \to \mathbb{R}^q$ which meet every k-dimensional subspace at no more than m points for given k and m [compare Szücs (1982)].

Exercise. The total reality condition for immersions $V \to \mathbb{C}^n$ can be expressed by the non-vanishing of the angle $\measuredangle(t, i\tau) \neq 0$ for all non-zero tangent vector t and $\tau \in T_v(V)$ and for all $v \in V$. A stronger condition, say $\mathscr{R}_\alpha \in X^{(1)}$, is $\measuredangle(t, i\tau) > \alpha$ for given α in the interval $0 < \alpha < \frac{\pi}{2}$. Prove the h principle for $\mathscr{R}_\alpha$ as well as the relative h-principle over embeddings. Show that every embedding $f_0\colon V \to \mathbb{C}^n$ which satisfies $\mathscr{R}_\alpha$ admits an isotopy of embeddings f_t, $t \in [0, 1]$, which also satisfy $\mathscr{R}_\alpha$ and such that f_1 satisfies $\mathscr{R}_\beta$ for a given β in the interval $\alpha \leq \beta < \frac{\pi}{2}$.

(C′) *Directed Embeddings of Open Manifolds.* Take an open subset in the Grassmann bundle of n-planes in W, say $A \subset Gr_n W$, and let $\mathscr{R} = \mathscr{R}(A) \subset \mathscr{R}_0$ be the subrelation in the immersion relation $\mathscr{R}_0$ which distinguishes immersions $f\colon V \to W$ whose tangential lift $G_f\colon V \to Gr_n W$ sends V into A.

Theorem. *If V is an open manifold, then $\mathscr{R}$ satisfies the relative h-principle over embeddings $V \to W$. Namely, every embedding $f_0: V \to W$ whose tangential lift $G_{f_0}: V \to Gr_n W$ is homotopic to a map $G_1: V \to A \subset Gr_n(W)$ can be isotoped to an embedding f_1 for which $G_{f_1}(V) \to A$.*

Proof. First, let $V = V' \times \mathbb{R}^i$, $i \geq 1$, and let $\mathscr{R}' = p_1^{-1}(V' \times 0)$ for the projection $p_1: X^{(1)} \to V$, where $X = V \times W \to V$. This $\mathscr{R}'$ naturally lies *over* $X_0^{(1)}$ for $X_0 = (V' \times 0) \times W \to V' \times 0$ and the earlier consideration yields the h-principle for $\mathscr{R}'$ over embeddings $V' \to W$; this implies the theorem for $V = V' \times \mathbb{R}^i$. Now, any open manifold decomposes into handles $D^{n-i} \times \mathbb{R}^i$ for $i = 1, \ldots, n$, to which the above yields the h-principle for extensions from $\partial D^{n-i} \times \mathbb{R}^i$ to $D^{n-i} \times \mathbb{R}^i$. This implies the theorem for all open manifolds V.

2.4.6 Convex Integration of Partial Differential Equations

We want to reduce the h-principle for a *non-open* relation $\rho: \mathscr{R} \to X^{(r)}$ to that for an auxiliary *open* relation $\rho^*: \mathscr{R}^* \to X^{(r)}$. We assume that $\mathscr{R}$ admits a metric, say dist on $\mathscr{R}$, such that $(\mathscr{R}, \text{dist})$ is a complete metric space. Denote by $P\mathscr{R}$ the space of pairs $(a, x(t))$, where $a \in \mathscr{R}$ and where $x(t)$, $t \in [0, 1]$, is a path in $X^{(r)}$ such that $x(0) = a$ and such that the path $x(t)$ is piecewise principal on the semiopen interval $(0, 1]$. We allow this path to have infinitely many principal segments in $(0, 1]$ which may accumulate to $0 \in [0, 1]$. The space $P\mathscr{R}$ lies over $X^{(r)}$ for $(a, x(t)) \mapsto x(1)$, and $P\mathscr{R}$ extends $\mathscr{R}$ for obvious maps $E: \mathscr{R} \subset P\mathscr{R}$ and $\Pi: P\mathscr{R} \to \mathscr{R}$. The principal extension $Pr_\infty \mathscr{R}$ clearly is a subextension of $P\mathscr{R}$. Furthermore, there is a natural isomorphism $Pr_\infty(P\mathscr{R}) \simeq P\mathscr{R}$, where $(a, x(t), y(t) \mapsto (a, z(t))$ for

$$z(t) = \begin{cases} x(2t), & 0 \leq t \leq \frac{1}{2}, \\ y(2t - 1), & \frac{1}{2} \leq t \leq 1. \end{cases}$$

Consider a subset $\mathscr{R}' \subset P\mathscr{R}$, take $\varepsilon > 0$ and denote by $\mathscr{R}'_a(\varepsilon) \subset \mathscr{R}'$, $a \in \mathscr{R}$, the intersection $\mathscr{R}' \cap \Pi^{-1}(B_a(\varepsilon))$ for the ε-ball $B_a(\varepsilon) \subset \mathscr{R}$ around $a \in \mathscr{R}$. Consider the union $\bigcup_{a \in \mathscr{R}} \text{Conv}_\infty \mathscr{R}'_a(\varepsilon) \subset Pr_\infty(P\mathscr{R})$ and let $C_\infty^\varepsilon \mathscr{R}' \subset P\mathscr{R}$ denote the image of this union under the isomorphism $Pr_\infty(P\mathscr{R}) \simeq P\mathscr{R}$.

In-extensions. A subextension $\mathscr{R}^+ \subset P\mathscr{R}$ of $\mathscr{R}$ is called an *in-extension* (of $\mathscr{R}$) if the following three conditions are satisfied

(1) If $\mathscr{R}' \subset \mathscr{R}^+$ is an arbitrary neighborhood of $\mathscr{R} \subset \mathscr{R}^+$, then the subrelation $\mathscr{R}^* = \mathscr{R}^+ \setminus \mathscr{R} \subset P\mathscr{R}$ is contained in the image of $\text{Conv}_\infty(\mathscr{R}' \cap \mathscr{R}^*)$ under the isomorphism $Pr_\infty(P\mathscr{R}) \simeq P\mathscr{R}$.
(2) There exists, for an arbitrary $\varepsilon > 0$, a neighborhood $\mathscr{R}^\varepsilon \subset \mathscr{R}^+$ of $\mathscr{R}$ such that $\mathscr{R}^\varepsilon \cap \mathscr{R}^* \subset C_\infty^\varepsilon(\mathscr{R}' \cap \mathscr{R}^*)$ for *all* neighborhoods $\mathscr{R}' \subset \mathscr{R}^+$ of $\mathscr{R} \subset \mathscr{R}^+$.
(3) The subrelation $R^+ \subset P\mathscr{R}$ is invariant under those piecewise linear homeomorphisms $h: [0, 1] \to [0, \delta]$, $0 < \delta \leq 1$, for which $h(0) = 0$; that is

$$(a, x(t)) \in \mathscr{R}^+ \Rightarrow (a, x(h(t))) \in \mathscr{R}^+.$$

(A) **Theorem.** *Let $\mathcal{R}^+ \subset \mathcal{R}$ be an in-extension for which the relation $\mathcal{R}^* = \mathcal{R}^+ \backslash \mathcal{R}$ is open over $X^{(r)}$. Then the extension $\mathcal{R}^+ \supset \mathcal{R}$ is h-stable.*

Proof. Take a sequence $\varepsilon_i > 0, i = 1, \ldots$, such that $\sum_{i=1}^{\infty} \varepsilon_i < \infty$ and let $\mathcal{R}_i = \mathcal{R}^{\varepsilon_i}$ [see (2) above]. Let $\varphi_0 \colon V \to \mathcal{R}^+$ be a holonomic section which is to be deformed into $\mathcal{R} = \bigcap_{i=1}^{\infty} \mathcal{R}_i \subset \mathcal{R}^+$. Suppose for the moment that the image $\varphi(V) \subset \mathcal{R}^+$ does not meet $\mathcal{R} \subset \mathcal{R}^+$. Then the principal stability theorem (see 2.4.2) implies with (1) above that φ admits a homotopy to a holonomic section $\varphi_1 \colon V \to \mathcal{R}_1 \cap \mathcal{R}^*$. Property (2) allows one to homotope φ_1 to a holonomic section $\varphi_2 \colon \mathcal{R}_2 \cap \mathcal{R}^*$, then to $\varphi_3 \colon \mathcal{R}_3 \cap \mathcal{R}^*$ and so on. Moreover, the holonomic section $\varphi_i \colon V \to \mathcal{R}_i \cap \mathcal{R}^*$ can be deformed to $\varphi_{i+1} \colon \mathcal{R}_{i+1} \cap \mathcal{R}^*$, such that $\operatorname{dist}(\Pi \circ \varphi_{i+1}(v), \Pi \circ \varphi_i(v)) \leq \varepsilon_i$ for the projection $\Pi \colon \mathcal{R}^+ \to \mathcal{R}$ and for all $v \in V$. This is done by considering the ε_i-neighborhood $U_i = U_{\varepsilon_i}(\Pi \circ \varphi_i(V)) \subset \mathcal{R}$ and by applying the principal stability theorem to the extension $\operatorname{Conv}_{\infty}(\mathcal{R}_{i+1} \cap \mathcal{R}^* \cap \Pi^{-1}(U_i))$. Since the space $(\mathcal{R}, \operatorname{dist})$ is complete, there exists a common limit of the sequences of sections $\Pi \circ \varphi_i \colon V \to \mathcal{R}$ and $\varphi_i \colon V \to \mathcal{R}^* \subset \mathcal{R}^+$ which is the desired holonomic section $\lim_{i \to \infty} \varphi_i = \varphi \colon V \to \mathcal{R}$.

Now, if the section $\varphi_0 \colon V \to \mathcal{R}^+$ meets $\mathcal{R}$, we apply the above argument over the complement $V \backslash \varphi_0^{-1}(\mathcal{R})$ thus concluding the proof of the theorem.

(A′) *In-deformations.* We have reduced the h-principle for $\mathcal{R}$ to that for $\mathcal{R}^+$. But the relation $\mathcal{R}^+ \to X^{(r)}$ may not be open and so we need an additional deformation of (sections of) $\mathcal{R}$ to $\mathcal{R}^*$. Namely, we need an *in-deformation* of $\mathcal{R}$ to $\mathcal{R}^*$, that by definition is a continuous map $\sigma \colon \mathcal{R} \to \mathcal{R}^*$ such that $\Pi \circ \sigma = \operatorname{Id}$. Such a map σ is given by a continuous family of paths $x_a(t)$, $a \in \mathcal{R}$, such that $x_a(0) = a$ and $(a, x_a(t)) \in \mathcal{R}^*$ for all $a \in \mathcal{R}$.

Now (A) implies the following

Corollary. *If $\mathcal{R}$ admits an in-deformation to the open relation $\mathcal{R}^* = \mathcal{R}^+ \backslash \mathcal{R}$ for an in-extension $\mathcal{R}^+ \supset \mathcal{R}$ then the h-principle for $\mathcal{R}$ reduces to that for $\mathcal{R}^*$.*

2.4.7 Underdetermined Evolution Equations

Let τ be a continuous tangent hyperplane field on V and let $p_{\perp} \colon X^{(r)} \to X^{\perp}$ be the corresponding affine fibration whose fibers $X_x^{(r)} \approx \mathbb{R}^q, x \in X^{\perp}$ are principal subspaces in $X^{(r)}$ (see 2.4.1). We study relations $\mathcal{R} \subset X^{(r)}$ which are locally closed subsets in $X^{(r)}$ (i.e. $\mathcal{R}$ is closed in some open subset $U \subset \mathcal{R}$). Such an $\mathcal{R}$ always carries some complete metric. If $\mathcal{R}$ meets each fiber $X_x^{(r)}$, $x \in X^{\perp}$, at a unique point $a = \mathcal{R} \cap X_x^{(r)}$ then the relation $\mathcal{R}$ can be expressed by a *determined* evolution system of P.D.E. of Cauchy-Kovalevskaja form, $\partial_t^r f = \psi(J_f^{\perp})$, where $f \colon V \to X$ is the unknown section, where $\psi(x) = a = \mathcal{R} \cap X_x^{(r)}$, and where ∂_t^r denotes the rth order derivative in the direction of some fixed vector field transversal to τ. (Compare 1.1.1) Here we are concerned with relations $\mathcal{R}$ whose intersections $\mathcal{R} \cap X_x^{(r)}$ have positive dimension and which correspond to *underdetermined* P.D.E. systems. [Compare Gromov (1973); Spring (1983, 1984).]

Nowhere Flat Sets and Maps. A continuous map $\alpha: A \to \mathbb{R}^q$ is called *nowhere flat* if the pullback of every affine hyperplane, $\alpha^{-1}(H) \subset A$, is a nowhere dense subset in A. In particular, a subset $A \subset \mathbb{R}^q$ is nowhere flat if the intersection $A \cap H$ is nowhere dense in A for all hyperplanes $H \subset \mathbb{R}^q$. The empty set is nowhere flat according to this definition. Generic maps of manifolds A of $\dim A \geq 1$ are nowhere flat.

A path $x: [0,1] \to \mathbb{R}^q$ is called an *in-path* at $a \in A$ if $x(0) = \alpha(a) \in \mathbb{R}^q$ and if $x(t) \in \operatorname{Int}\operatorname{Conv}\alpha(U)$ for $0 < t < \varepsilon$, where $U \subset A$ is an arbitrary neighborhood of $a \in A$, and $\varepsilon = \varepsilon(U) > 0$. For example, in-paths of a convex hypersurface $A \subset \mathbb{R}^n$ are those which approach A from inside.

Let A be a metric space and let the map $\alpha: A \to \mathbb{R}^q$ be nowhere flat. Then the convex hull of the α-image of the δ-ball around $a \in A$ has non-empty interior, $\operatorname{Int}\operatorname{Conv}\alpha(B_a(\delta)) \neq \varnothing$ for all $a \in A$ and $\delta > 0$, and so every value $\alpha(a) \in \mathbb{R}^q$ can be approached by an in-path $x_a(t)$. In fact, let $y_a(\delta)$ be the barycenter of the set $\operatorname{Conv}\alpha(B_a(\delta)) \subset \mathbb{R}^q$. Then $x_a(t) = t^{-1}\int_0^t y_a(\delta)\,d\delta$ is the required in-path which is continuous in the variables $(t, a) \in [0,1] \times A$.

We return to the fibration $p_\perp: X^{(r)} \to X^\perp$ and we consider a relation $\rho: \mathscr{R} \to X^{(r)}$ which satisfies the following four conditions.

(1) $\mathscr{R}$ carries a complete metric.
(2) $\mathscr{R}$ is open over $X^\perp$. That is the map $p_\perp \circ \rho: \mathscr{R} \to X^\perp$ is a microfibration. For example, A is a submanifold in $X^{(r)}$ which is transversal to the fibers of the fibration $p_\perp: X^{(r)} \to X^\perp$.
(3) The subset $\mathscr{R}_x = \rho^{-1}(X_x^{(r)}) = (p_\perp \circ \rho)^{-1}(x) \subset \mathscr{R}$ is locally path connected for all $x \in X^\perp$.
(4) The map $\rho: \mathscr{R}_x \to X_x^{(r)} \approx \mathbb{R}^q$ is nowhere flat for all $x \in X^\perp$.

Take a point $a \in \mathscr{R}_x$, consider the path connected component of a in $\mathscr{R}_x$ and let $C_a \subset X_x^{(r)}$ be the interior of the convex hull (in $X_x^{(r)} \approx \mathbb{R}^q$) of the ρ-image of this component. Let $\mathscr{R}^* \subset P\mathscr{R}$ be the subset of those $(a, z(t)) \in P\mathscr{R}$, for which the path $z(t)$ lies in C_a and which is an in-path at $a \in \mathscr{R}_x$ for the map $\rho: \mathscr{R}_x \to X_x^{(r)}$. The conditions (2) and (4) show the relation $\mathscr{R}^* \to X^{(r)}$ to be open. The condition (3) implies that $\mathscr{R}^+ = \mathscr{R}^* \cup \mathscr{R}$ is an in-extension of $\mathscr{R}$. Thus, we obtain with (A)

The Local Solvability of $\mathscr{R}$. *If $U \subset V$ is a small neighborhood of a point $v \in V$ which lies in the image of the map $\mathscr{R} \to V$, then there exists a holonomic section $U \to \mathscr{R}$.*

Example. Let $\mathscr{R}$ be a submanifold in $X^{(r)}$ of codimension s which locally represents a system of s P.D.E. in the unknown map $V \to \mathbb{R}^q$. If $s \leq q - 1$ and if $\mathscr{R}$ is a *generic* submanifold, then there is a principal subspace $R \approx \mathbb{R}^q$ in $X^{(r)}$ which is transversal to $\mathscr{R}$ and such that the intersection $\mathscr{R} \cap R$ is a nowhere flat submanifold in R of positive dimension. A small neighborhood in $\mathscr{R}$ of each point $a \in \mathscr{R} \cap R$ satisfies the conditions (1)–(4) for some (locally defined) field τ and so $\mathscr{R}$ admits a C^r-solution $U \to X$ over some neighborhood $U \subset X$.

This local solvability differs from the similar C^∞-result (see 2.3.8) in several respects.

(a) The genericity condition in 2.3.8 assumes $\mathscr{R} \subset X^{(r)}$ to be a C^∞-manifold (or at least a C^k-manifold for large k), while $\mathscr{R}$ is only C^1-smooth here.
(b) The "nowhere flat" condition excludes linear and quasilinear systems of P.D.E. for which $\mathscr{R}_x = \mathscr{R} \cap X_x^{(r)}$ is an affine subspace in the fiber $X_x^{(r)}$ of the fibration $p_{r-1}^r \colon X^{(r)} \to X^{(r-1)}$ for all $x \in X^{(r-1)}$, while the assumptions in 2.3.8 allow many linear and quasilinear systems.
(c) Local solutions delivered in 2.3.8 are C^∞-smooth, provided $\mathscr{R} \subset X^{(r)}$ is a C^∞-smooth submanifold. But no (known) regularity assumption on $\mathscr{R}$ helps the convex integration to produce a C^k-solution for $k \geq r + 1$.
(d) The space of local solutions $U \to X$ under the above assumptions (1)–(4) is "large": The C^{r-1}-closure of this space has a non-empty interior in the space of C^{r-1}-sections $U \to X$. (In fact, the $C^\perp$-closure has a non-empty interior.) A similar result fails to be true in the context of 2.3.8. Indeed, linear P.D.E.-systems are functionally closed (see 1.2.3) and so their solutions are nowhere C^i-dense for all $i = 0, 1, \ldots$.

One has with these remarks (a)–(d) several open

Questions. Let $\mathscr{R} \subset X^{(r)}$ be a generic C^k-submanifold of codimension $s \leq q - 1$. For which $l = l(k, r, \dim X)$ does there exist local C^l-solutions $U \to X$ of $\mathscr{R}$? Under what assumptions are the C^{r+1}-solutions $U \to X$ C^0-dense in some open subset in the space of continuous sections $U \to X$?

Now, we turn to the

Global Solution of $\mathscr{R}$. *Let $\rho \colon \mathscr{R} \to X^{(r)}$ satisfy (1)–(4) and let the convex subset $C_a \subset X_x^{(r)}$ equal $X_x^{(r)}$ for all $a \in \mathscr{R}_x = \rho^{-1}(X_x^r)$ and for all $x \in X^\perp$. Then the relation $\mathscr{R}$ abides the C^i-dense h-principle for $i = 0, \ldots, r - 1, \perp$.*

Proof. There obviously exists a continuous family $z_a(t)$ of in-paths in the fibers $X_x^{(r)}$, which provides an in-deformation of $\mathscr{R}$ to $\mathscr{R}^*$. Then the h-principle for $\mathscr{R}$ is reduced with the above (A) and (A′) to that for the *open* relation $\mathscr{R}^*$ to which the results of 2.4.2 apply.

Examples. Consider a relation $\mathscr{R} \subset X^{(r)}$ and let $X^\perp \times P \to X^\perp$ be a trivial fibration for some connected manifold P. Let $\rho \colon X^\perp \times P \to X^{(r)}$ be a fiberwise map with the following two properties

(i) the image of ρ is contained in $\mathscr{R}$;
(ii) the map of each fiber, $\rho \colon x \times P \to X_x^{(r)}$, $x \in X^\perp$, is nowhere flat and $\operatorname{Conv} \rho(x \times P) = X_x^{(r)}$ for all $x \in X^\perp$.

Then every C^{r-1}-section $V \to X$ admits a fine C^{r-1}-approximation by C^r-solutions $V \to X$ of the relation $\mathscr{R}$. Moreover every C^r-section admits a fine $C^\perp$-approximation by C^r-solutions $V \to X$ of $\mathscr{R}$.

Proof. Apply the above h-principle to the relation $\rho \colon X^\perp \times P \to X^{(r)}$.

Let us describe a class of equations in two unknown functions f_1 and f_2: $\mathbb{R}^2 \to \mathbb{R}$ to which the general theory applies. Let Φ_1: $\mathbb{R}^2 \to \mathbb{R}$ be a real analytic function without critical points such that the level $\{\Phi_1(x, y) = c\}$ is a connected curve in $\mathbb{R}^2$ whose convex hull equals $\mathbb{R}^2$ for all c. For example, $\Phi_1(x, y) = x + y^3$. Consider the following single equation

$$(*) \qquad \Phi_1\left(\frac{\partial^r f_1}{\partial u_1^r}, \frac{\partial^r f_2}{\partial u_1^r}\right) = \Phi_2$$

where Φ_2 is an arbitrary continuous function with entries u_1, u_2, $\frac{\partial^s f}{\partial u_1^{s_1} \partial u_2^{s_2}}$, where $f = (f_1, f_2)$, $s = 0, 1, \ldots, r$, and $s_1 = 0, \ldots, r - 1$.

The above h-principle shows that C^r-solutions f: $\mathbb{R}^2 \to \mathbb{R}^2$ of $(*)$ are C^{r-1}-dense in the space of C^{r-1}-maps $\mathbb{R}^2 \to \mathbb{R}^2$.

2.4.8 Triangular Systems of P.D.E.

Let ∂_i, $i = 1, \ldots, k$, be continuous linearly independent vector fields on V. For example, $V = \mathbb{R}^n$ and $\partial_i = \partial/\partial u_i$, $i = 1, \ldots, n$. Let Φ_i be smooth vector valued functions, such that Φ_i takes values in $\mathbb{R}^{s_i}$ and Φ_i has entries $v, f, \partial_1 f, \partial_2 f, \ldots, \partial_i f$, where f is the unknown map $V \to \mathbb{R}^q$. In other words Φ_i: $V \times \mathbb{R}^{q(i+1)} \to \mathbb{R}^{s_i}$. Consider the following (triangular) systems of $s = \sum_{i=1}^k s_i$ P.D.E.

$$(**) \qquad \begin{aligned} &\Phi_1(v, f, \partial_1 f) = 0 \\ &\Phi_2(v, f, \partial_1 f, \partial_2 f) = 0 \\ &\cdots\cdots\cdots\cdots\cdots\cdots \\ &\Phi_k(v, f, \partial_1 f, \ldots, \partial_k f) = 0. \end{aligned}$$

(A) **Local Solvability.** *If $s_i \leq q - 1$ for all $i = 1, \ldots, k$ and if the functions Φ_i, $i = 1, \ldots, k$, are generic, then the system $(**)$ admits a local C^1-solution f: $U \to \mathbb{R}^q$, for some open subset $U \subset V$. Moreover, C^1-solutions $U \to \mathbb{R}^q$ are C^0-dense in some open subset in the space of C^0-maps $U \to \mathbb{R}^q$.*

This result is an immediate corollary of the Theorem (C) below.

Observe that the system $(**)$ may have $n(q - 1)$ equations (for $k = n$ and $s_i = q - 1$), and then it is highly overdetermined for $n \geq 2$ and for large q. However, the above result does not claim the local solvability of generic over-determined systems. In fact, the equation $\Phi_i = 0$ is not generic for $i < n$, since Φ_i is a generic function in the variables $v, f, \partial_1 f, \ldots, \partial_i f$ rather than in $v, f, \partial_1 f, \ldots, \partial_n f$, as true genericity requires.

Question. Does there exist a generic map Φ: $V \times \mathbb{R}^{q(n+1)} \to \mathbb{R}^q$ (or rather an open subset in the space of C^∞-maps Φ: $V \times \mathbb{R}^{q(n+1)} \to \mathbb{R}^q$) for which local C^1-solutions of the (determined) system $\Phi(v, f, \partial_1 f, \ldots, \partial_n f) = 0$ are C^0-dense in an open subset in the space of C^0-maps $U \to \mathbb{R}^q$ for some neighborhood $U \subset V$?

(B) *Triangular Differential Relations.* Let τ_i, $i = 1, \ldots, k$, be continuous hyperplane fields on V which are *r-independent* in the following sense. If l_i: $T(V) \to \mathbb{R}$ are linear forms on V for which $\operatorname{Ker} l_i = \tau_i$, $i = 1, \ldots, k$, then the (symmetric differential of degree r) forms l_i^r are linearly independent, that is the (homogeneous of degree r) polynomials l_i^r, $i = 1, \ldots, k$, on $T_v(V)$ are linearly independent for all $v \in V$. For example, [1-independence] = [independence]. Furthermore, if the forms l_i are independent then the forms $l_{ij} = l_i + l_j$, $1 \le i \le j \le k$, are 2-independent, the forms $l_i + l_j + l_m$, $1 \le i \le j \le m \le k$, are 3-independent and so on.

Take a point $0 \in X_x^{(r)} \subset X^{(r)}$ for the origin in the (affine) fiber $X_x^{(r)}$ of the fibration $X^{(r)} \to X^{(r-1)}$ for some $x \in X^{(r-1)}$ and let $R_i \subset X_x^{(r)}$ be the principal subspaces through $0 \in X_x^{(r)}$ which correspond to the fields τ_i. The (linear) subspaces $R_i \subset (X_x^{(r)}, 0)$, $i = 1, \ldots, k$, clearly are linearly independent for r-independent fields τ_i. Put $X_x^k = X_x^{(r)}$ and $X_x^{k-i} = X_x^k/\operatorname{Span}(R_1, \ldots, R_i)$. Thus we obtain affine bundles $X^{k-i} \to X^{(r-1)}$, $i = 0, \ldots, k$, with the fibers X_x^{k-i}, and affine homomorphisms, say P_{k-i}: $X^{k-i} \to X^{k-i-1}$. The fibers of the affine fibration (homomorphism) P_{k-i} are (naturally identified with) principal subspaces (now in X^{k-i} rather than in $X^{(r)}$) associated to τ_{i+1}.

Consider a locally closed relation $\mathscr{R} = \mathscr{R}_k \subset X^k = X^{(r)}$ and denote by $\mathscr{R}_{k-i} \subset X^{k-i}$ the image of $\mathscr{R}$ under the (composed) map $X^k \to X^{k-i}$. Let us impose on $\mathscr{R}$ the following

Triangular Condition. The subset $\mathscr{R}_0 \subset X^0$ is open and the map P_{k-i}: $\mathscr{R}_{k-i} \to \mathscr{R}_{k-i-1}$ is a microfibration for all $i = 1, \ldots, k$.

Example. Start with an arbitrary open subset $\mathscr{R}_0 \subset X^0$. Next, take a submanifold $\mathscr{R}_1 \subset P_1^{-1}(\mathscr{R}_0)$ of codimension $s_1 \le q$ which is *transversal* to the fibers $P_1^{-1}(x)$, $x \in \mathscr{R}_0$, and is mapped by P_1 onto $\mathscr{R}_0$. As the map $P_1: \mathscr{R}_1 \to \mathscr{R}_0$ is a submersion it also is a microfibration. We proceed with taking a similar submanifold $\mathscr{R}_2 \subset P_2^{-1}(\mathscr{R}_1)$ of codimension s_2 and we continue up to a submanifold $\mathscr{R} = \mathscr{R}_k \subset P_k^{-1}(\mathscr{R}_{k-1})$ of codimension s_k.

Notice that generic systems (**) locally (i.e. in some open subset in the jet space) satisfy the above transversality assumption and so they locally fit this example.

Next, we add the

Nowhere Flat Condition. The intersection of $\mathscr{R}_{k-i} \subset X^{k-i}$ with every fiber $R \approx \mathbb{R}^q$ of the fibration $X^{k-i} \to X^{k-i-1}$ is a locally path connected nowhere flat subset in R for all $i = 0, \ldots, k-1$.

This property is satisfied for generic submanifolds $\mathscr{R}$ in the above example, provided $s_i \le q - 1$ for $i = 0, \ldots, k-1$.

In-paths. Consider a path z: $[0, 1] \to X^{(r)}$ which consists of at most countably many principal segments (with the only admissible accumulation point $t = 0$), such that each segment lies in a principal subspace associated to some of the fields τ_i, $i =$

$1, \ldots, k$. This path is necessarily contained in $\mathrm{Span}\{R_i\}_{i=1,\ldots,k}$, where $R_i \subset X^{(r)}$ is the principal subspace through $z(0) \in X^{(r)}$ which is associated to τ_i. Then we use the canonical projection $\mathrm{Span}\{R_i\} \to R_i$ and we invoke the projection of R_i into X^{k-i+1} which isomorphically maps R_i onto some fiber, say $\bar{R}_i \subset X^{k-i+1}$, of the fibration $X^{k-i+1} \to X^{k-i}$. Thus, we send the path z to a path in $\bar{R}_i$, say $z_i\colon [0,1] \to \bar{R}_i$.

A path $z(t)$ with the above properties is called an *in-path* at $a \in \mathscr{R}$ if $z(0) = a$ and if the path $z_i(t)$ in $\bar{R}_i$ is an in-path at $z_i(0) \in \mathscr{R}_{k-i+1} \cap \bar{R}_i$ for the subset $\mathscr{R}_{k-i+1} \cap \bar{R}_i \subset \bar{R}_i \approx \mathbb{R}^q$ for all $i = 1, \ldots, k$.

(C) **Theorem.** *Let $\mathscr{R} \subset X^{(r)}$ satisfy the triangular and the nowhere flat conditions. Then $\mathscr{R}$ admits a local C^r-solution $U \to X$ for some open subset $U \subset V$, unless $\mathscr{R}$ is empty. Furthermore, let $\mathscr{R}_0 = X^0$, let the maps $P_{k-i}\colon \mathscr{R}_{k-i} \to \mathscr{R}_{k-i-1}$ be 0-fibrations and let each path connected component of the intersection $\mathscr{R}_{k-i} \cap P_{k-i}^{-1}(x) \subset P_{k-i}^{-1}(x) \approx \mathbb{R}^q$ have convex hull $= P_{k-i}^{-1}(x)$ for all $i = 0, \ldots, k-1$, and for all $x \in \mathscr{R}_{k-i-1}$. Then the relation $\mathscr{R}$ satisfies the C^l-dense h-principle for $l = 0, \ldots, r-1$.*

Proof. Let $\mathscr{R}^* \subset P\mathscr{R}$ be the subset of those pairs $(a, z(t)) \in P\mathscr{R}$ for which $z(t)$ is an in-path at $a = z(0) \in \mathscr{R}$ and let $\mathscr{R}^+ = \mathscr{R}^* \cup \mathscr{R}$. Then $\mathscr{R}^*$ is open over $X^{(r)}$ and $\mathscr{R}^+ \supset \mathscr{R}$ is an in-extension. Hence, the local solvability follows from (A) in 2.4.6 (compare 2.4.7).

Now, the additional assumptions on $\mathscr{R}^*$ show the relation $\mathscr{R}$ to be ample. Then an obvious in-deformation of $\mathscr{R}$ into $\mathscr{R}^*$ (compare 2.4.7) allows us to apply (A′) of 2.4.6 and to obtain the h-principle for $\mathscr{R}$. Q.E.D.

(D) **Examples and Exercises.** (1) Take the differential operators $\Delta_{11} = \frac{\partial}{\partial u_1} \times \left(\frac{\partial}{\partial u_1} + \frac{\partial}{\partial u_2}\right)$, $\Delta_{12} = \frac{\partial^2}{\partial u_1 \partial u_2}$ and $\Delta_{22} = \frac{\partial}{\partial u_2}\left(\frac{\partial}{\partial u_1} + \frac{\partial}{\partial u_2}\right)$ on $\mathbb{R}^2$ and consider the following system of three second order equations in the unknown map $f = (f_1, f_2)\colon \mathbb{R}^2 \to \mathbb{R}^2$,

$$(+) \qquad \begin{cases} \Phi_1(\Delta_{11} f_1, \Delta_{11} f_2) = \psi_1(J_f^1) \\ \Phi_2(\Delta_{12} f_1, \Delta_{12} f_2) = \psi_2(J_f^1, \Delta_{11} f) \\ \Phi_3(\Delta_{22} f_1, \Delta_{22} f_2) = \psi_3(J_f^1, \Delta_{11} f, \Delta_{12} f), \end{cases}$$

where $J_f^1 = \left(u_1, u_2, f, \frac{\partial f}{\partial u_1}, \frac{\partial f}{\partial u_2}\right)$, where ψ_i, $i = 1, 2, 3$, are arbitrary continuous functions, and where $\Phi_i = \Phi_i(x, y)$, $i = 1, 2, 3$, are real analytic functions without critical points whose levels $\{\Phi_i(x, y) = c\}$ are connected and $\mathrm{Conv}\{\Phi_i(x, y) = c\} = \mathbb{R}^2$ for $i = 1, 2, 3$ and for all $c \in \mathbb{R}$. The differential relation expressed by $(+)$ is triangular for the line fields $\tau_1 = \mathrm{Ker}\, du_2$, $\tau_2 = \mathrm{Ker}(du_1 - du_2)$ and $\tau_3 = \mathrm{Ker}\, du_1$. The Theorem (C) shows C^2-solutions $f\colon \mathbb{R}^2 \to \mathbb{R}^2$ to be C^1-dense in the space of C^1-maps $\mathbb{R}^2 \to \mathbb{R}^2$.

(2) Let ∂_i, $i = 1, \ldots, k$, be linearly independent C^1-smooth vector fields on V. Set

$$\Delta_{ij} = \begin{cases} \partial_i \partial_j & \text{for } 1 \le i \le j \le k \\ \partial_i \sum_{j=1}^{k} \partial_j & \text{for } i = j = 1, \ldots, k. \end{cases}$$

Let $A_{ij} \subset \mathbb{R}^2$, $1 \le i \le j \le k$, be path connected, locally path connected nowhere flat locally closed subsets. Consider C^2-maps $f\colon V \to \mathbb{R}^2$ whose derivative $\Delta_{ij} f\colon V \to \mathbb{R}^2$ sends V into $A_{ij} \subset \mathbb{R}^2$ for all $1 \le i \le j \le k$.

(2′) *If* Conv $A_{ij} = \mathbb{R}^q$ *for* $1 \le i \le j \le k$, *then every* C^1-*map* $V \to \mathbb{R}^q$ *admits a fine* C^1-*approximation by the above maps* f.

Proof. Take linear differential forms l_i on V such that $l_i \partial_j = \delta_{ij}$, $1 \le i, j \le k$. Then the pertinent differential relation is triangular for the hyperplane fields

$$\tau_{ij} = \begin{cases} \operatorname{Ker}(l_i - l_j) & \text{for } 1 \le i \le j \le k \\ \operatorname{Ker} l_i & \text{for } i = j = 1, \ldots, k. \end{cases}$$

Observe that the above maps f are C^2-solutions of a system of $s = \sum_{1 \le i \le j \le k} s_{ij}$ differential equations of second order for $s_{ij} = \operatorname{codim} A_{ij}$. The most overdetermined case to which (2′) applies is $k = n = \dim V$ and $s_{ij} = q - 1$ for all $1 \le i \le j \le n$. This gives a system of $s = n(n+1)(q-1)/2$ P.D.E. in q unknown functions.

(3) Show the conclusion of (2′) to hold true under the following weaker assumptions on A_{ij}. Each A_{ij} is an arbitrary subset in $\mathbb{R}^q$ which receives a continuous map $\alpha_{ij}\colon P \to A_{ij}$ of a connected manifold P, such that

(i) the map α_{ij} is nowhere flat for all $1 \le i \le j \le k$;
(ii) Conv $\alpha_{ij}(P) = \mathbb{R}^q$ for $1 \le i \le j \le k$.

Hint. Prove (C) for triangular relations *over* (rather than in) $X^{(r)}$.

(3′) Let the maps α_{ij} satisfy (i) and let each map α_{ij} strictly surround the origin in $\mathbb{R}^q$, i.e. $0 \in \operatorname{Int} \operatorname{Conv} \alpha_{ij}(P)$. Prove the existence of at least one C^2-map f for which $(\Delta_{ij} f)(V) \subset A_{ij}$ for all $1 \le i \le j \le k$.

Hint. Prove a relative version of (C) (compare 2.4.5).

(4) Let $\bar{\Delta}_{ij} = \partial_i \partial_j$ for $1 \le i \le j \le k$. Prove (3) for these $\bar{\Delta}_{ij}$ in place of Δ_{ij} under the additional assumption: the subset $A_{ij} \subset \mathbb{R}^q$ is open for $1 \le i < j \le k$. (It is unclear if this assumption is necessary.)

(4′) Show (3′) to be false, in general, for $\bar{\Delta}_{ij}$ in place of Δ_{ij}.

Hint. Consider the torus T^2 with the standard fields $\partial_1 = \partial/\partial u_1$ and $\partial_2 = \partial/\partial u_2$. Take the unit disk $\{x^2 + y^2 \le 1\} \subset \mathbb{R}^2$ for A_{11} and for A_{22} and take the exterior $\{x^2 + y^2 \ge 10\}$ for A_{12}.

(5) Generalize (1)–(4) to pertinent derivations of an arbitrary order r.

2.4.9 Isometric C^1-Immersions

Let V be an n-dimensional Riemannian manifold and let ∂_i, $i = 1, \ldots, n$ be a frame of independent vector fields on V. The isometric immersion system for maps

$f: V \to \mathbb{R}^q$ (compare 1.1.5) is triangular in such a frame,

(1) $$\{\langle \partial_1 f, \partial_1 f\rangle = \langle \partial_1, \partial_1\rangle$$

(2) $$\begin{cases}\langle \partial_1 f, \partial_2 f\rangle = \langle \partial_1, \partial_2\rangle \\ \langle \partial_2 f, \partial_2 f\rangle = \langle \partial_2, \partial_2\rangle\end{cases}$$

⋮

(n) $$\begin{cases}\langle \partial_i f, \partial_n f\rangle = \langle \partial_i, \partial_n\rangle, \\ i = 1, \dots, n,\end{cases}$$

where the scalar products on the left hand side are in $\mathbb{R}^q$ and the right hand side ones are taken in the tangent bundle $T(V)$. Unfortunately, the "nowhere flat" condition is not satisfied by the equations (k) for $k = 2, \dots, n$. Indeed solutions $y_k \in \mathbb{R}^q$ of the algebriac system

(k)′ $$\langle y_i, y_k\rangle = g_{ik},$$

for $i = 1, \dots, k$ and for fixed vectors $y_1, \dots, y_{k-1}$ in $\mathbb{R}^q$ form a round $(q-k)$-dimensional sphere in $\mathbb{R}^q$ which is everywhere flat in $\mathbb{R}^q$. However, successive convex hull extensions sufficiently enlarge these spheres for $q - k \geq 1$ and so the theorems (A) and (A′) of 2.4.6 do apply to isometric immersions of V into Riemannian manifolds W, $q = \dim W > n = \dim V$, as follows.

The relevant jet space $X^{(1)}$ consists of the linear maps $T_v(V) = \mathbb{R}^n \to \mathbb{R}^q = T_w(W)$ and the isometric immersion relation is, in each fiber $X_x^{(1)} \approx \mathrm{Hom}(\mathbb{R}^n \to \mathbb{R}^q)$, $x = (v, w) \in X$, the subset of *isometric* homomorphisms $\mathbb{R}^n = (\mathbb{R}^n, g) \to \mathbb{R}^q$, called $Is = Is_g \subset H_n^q \overset{\text{def}}{=} \mathrm{Hom}(\mathbb{R}^n \to \mathbb{R}^q)$, where g denotes the given Euclidean metric in $\mathbb{R}^n$.

Let $R \subset H_n^q$ be a principal subspace associated to some hyperplane $\tau \subset \mathbb{R}^n$. The intersection $Is \cap R$ is either empty or a round sphere $S^{q-n} \subset R = \mathbb{R}^q$ obtained by "rotating" an isometry $h: \mathbb{R}^n \to \mathbb{R}^q$, $h \in Is \cap R$, around the "axis" $h|\tau$ in $\mathbb{R}^q$ (compare 2.4.3). The ball $B^{q-n+1} \subset R$, bounded by this S^{q-n}, consists of those *short* homomorphisms $h': \mathbb{R}^n \to \mathbb{R}^q$, for which $h'|\tau = h|\tau$, where "short" means the positive semidefiniteness of the quadratic form $g - g'$ for the form g' on $\mathbb{R}^n$ induced by h' (from the given form on $\mathbb{R}^q$). The form $g - g'$ has rank $= 1$ unless $h' \in S^{q-n} = \partial B^{q-n+1}$ (and $g = g'$), and so $g = g' + l^2$ for a unique (up to a sign) linear form l on $\mathbb{R}^n$ which satisfies $\operatorname{Ker} l = \tau$. This gives us a geometric description of the convex extension $\mathrm{Conv}_1(Is) \supset Is$ as follows. If $q = n$, then $\mathrm{Conv}_1(Is) = Is$, as the sphere $S^0 \subset R$ consists of two single points and so the "component-wise convex hull" of S^0 equals S^0. However, if $q > n$, then $\mathrm{Conv}_1 Is$ consists of the above pair (h, h'), where $h \in Is$ and where $h' \subset B^{q-n+1} \subset R$ for some principal subspace $R \subset H_n^q$ through h.

Another useful description of $\mathrm{Conv}_1 Is$ *starts* with a homomorphism $h' \in H_n^q$ such that the induced form g' equals $g - l^2$ for some linear form l. Then, one takes a hyperplane $\tau' \subset \operatorname{Ker} l$ (if $g' \neq g$, then $\tau = \operatorname{Ker} l$ and so τ' is unique) and the associated principal subspace R' in H_n^q through h'. Finally, one takes an arbitrary point in $Is \cap R' \approx S^{q-n}$ for h.

Denote by $\mathrm{Conv}_N^+ Is \subset Pr_N Is$ the subset of those $(N+1)$-triples $(h_0, h_1, \dots, h_N)$ for which $(h_{i-1}, h_i) \in \mathrm{Conv}_1 Is_{g_{i-1}}$, for all $i = 1, \dots, N$. Here g_{i-1} is the quadratic form induced by $h_{i-1}: \mathbb{R}^n \to \mathbb{R}^q$ and $h_0 \in Is$, that is $g_0 = g$. The above description of $\mathrm{Conv}_1\, Is$ immediately implies several useful facts.

(1) $\mathrm{Conv}_N^+\, Is \subset \mathrm{Conv}_N\, Is$ for $N = 1, 2, \ldots, \infty$.
(2) The image of the map $\mathrm{Conv}_N^+ \to H_n^q$, $(h_0, \ldots, h_N) \mapsto h_N$ consists of those short homomorphisms $h': \mathbb{R}^n \to \mathbb{R}^q$, for which the induced form g' satisfies $\operatorname{rank}(g - g') \leq N$. In particular, the image of the map $\mathrm{Conv}_N\, Is \to H_n^q$ consists, for $N \geq n$, of all short homomorphisms $\mathbb{R}^n \to \mathbb{R}^q$.
(3) Let $\mathscr{R} \subset X^{(1)}$ be the isometric immersion relation, that is $\mathscr{R} \cap X_x^{(1)} = Is$ for all $x \in X$, and let a section $\varphi': V \to X^{(1)}$ correspond to a *strictly short* homomorphism $T(V) \to T(W)$. Then the quadratic form $g - g'$ by definition is positive definite on $T(V)$ (where g is the original form on $T(V)$ and g' is induced by φ') and so $g - g' = \sum_{i=1}^{2n} l_i^2$ for some linear forms l_i on $T(V)$. Hence, there is a piecewise principal path $\psi: V \to \mathrm{Conv}_{2n}^+ \mathscr{R} \subset \mathrm{Conv}_{2n} \mathscr{R} \subset Pr_{2n} \mathscr{R}$ which terminates in φ', that is $\psi = (\varphi_0, \varphi_1, \ldots, \varphi_{2n} = \varphi')$ for some section $\varphi_0: V \to \mathscr{R}$, and the pair $(\varphi_{i-1}, \varphi_i)$ is contained in $\mathrm{Conv}_1^+ \mathscr{R}_{g_{i-1}}$ for all $i = 1, \ldots, 2n$.

Now, we define a piecewise principal path $z: [0, 1] \to H_n^q$ to be an *in-path* at $a = z(0) \in Is$, if the homomorphism $z(t) \in H_n^q$ is strictly short for all $t > 0$ and if each principal segment, say $[z(t), z(t')]$, for $0 < t < t' \leq 1$, satisfies $(z(t), z(t')) \in \mathrm{Conv}_1^+\, Is'$, where $Is' = Is_g$, for the quadratic form induced by the homomorphism $z(t') \in H_n^q$. Here the path z may have countably many principal segments which accumulate at $t = 0$.

We take the space of these in-paths in the fibers $X_x^{(1)}$, $x \in X$, for the relation $\mathscr{R}^* \subset P\mathscr{R}$. It is obvious that $\mathscr{R}^*$ is open over $X^{(1)}$ and that there is an in-deformation of $\mathscr{R}$ to $\mathscr{R}^*$. The above discussion shows that relation $\mathscr{R}^+ = \mathscr{R}^* \cup \mathscr{R}$ to be an in-extension and we see with (3) above $\mathscr{R}^*$ is ample over strictly short homomorphisms $T(V) \to T(W)$. Hence, the Theorem (A′) implies

(A) **The Theorem of Nash-Kuiper** (compare 1.1.5). *The h-principle for isometric C^1-immersions $V \to W$ is C^0-dense in the space of strictly short maps $V \to W$, provided* $\dim W > \dim V$.

(B) **Exercises and Generalizations.** (1) Show that every strictly short *embedding* $f_0: V \to W$, $\dim W > \dim V$, admits an isotopy to an *isometric C^1-embedding*. Moreover, let $f_0: V \to W$ be an *immersion with normal crossings*. Construct a C^1-diffeomorphism $g: W \to W$, such that the composed map $g \circ f_0: V \to W$ is isometric.

(1′) Let ∂_i, $i = 1, \ldots, k$, be linearly independently vector fields on V. Construct a C^1-map $f: V \to \mathbb{R}^{k+1}$, such that $\langle \partial_i f, \partial_j f \rangle = \delta_{ij}$, $1 \leq i \leq j \leq k$, where $\delta_{ii} = 1$ and $\delta_{ij} = 0$ for $i < j$. Construct for $k = 1$ a *C^∞-map* $f: V \to \mathbb{R}^2$ for which $\langle \partial_1 f, \partial_1 f \rangle = 1$, provided the field ∂_1 is C^∞-smooth.

Hint. Prove the h-principle for C^1-maps $V \to \mathbb{R}^{k+1}$ which are isometric on a given k-dimensional subbundle $\tau \subset T(V)$.

(2) *Indefinite Forms*. Let $\Phi = \sum_{1 \leq i \leq j \leq q} \Phi_{ij}(w)\, dx_i\, dx_j$ be a quadratic differential form on W and let $(q_+(w), q_-(w))$ denote the type (signature) of $\Phi | T_w(W)$, $w \in W$, so that $q_+(w) + q_-(w) = \operatorname{rank} \Phi | T_w(W) \leq q = \dim W$. The form Φ is called *non-*

singular if $q_+(w) + q(w) = q$ for all $w \in W$. A linear map $h: \mathbb{R}^n \to T_w(W)$ is called *Φ-regular* if it misses $\mathrm{Ker} = \mathrm{Ker}(\Phi | T_w(W)) \subset T_w(W)$ which is a linear subspace in $T_w(W)$ of codimension $q_+(w) + q_-(w)$. This means the identity $h^{-1}(\mathrm{Ker}) = 0$. A C^1-map $f: V \to W$ is called *Φ-regular* if the differential $D_f: T(V) \to T(W)$ is Φ-regular on each tangent space $T_v(V)$, $v \in V$. If the form Φ is non-singular, then Φ-regular maps are exactly immersions $V \to W$.

Let g be a quadratic form on V of type $(n_+(v), n_-(v))$, $n_+ + n_- \leq n = \dim V$, $v \in V$, and let $n - n_-(v) + 1 \leq q_+(w)$ and $n - n_+(v) + 1 \leq q_-(w)$ for all $v \in V$ and $w \in W$. Prove the C^0-dense h-principle for Φ-regular isometric C^1-maps $f: V \to W$ ("isometric" means $f^*(\Phi) = g$) and derive the following

(2′) **Corollary.** *An arbitrary continuous map $V \to W$ admits a fine C^0-approximation by isometric C^1-immersions $f: V \to W$ in the following three cases.*

(i) $q_+(w) \geq 2n - n_-(v)$ *and* $q_-(w) \geq 2n - n_+(v)$ *for all* $(v, w) \in V \times W$.
(ii) *The manifold V is topologically contractible, the forms g and Φ are nonsingular and* $q_+ \geq n_+ + 1$, $q_- \geq n_- + 1$. (Since the forms are non-singular their types are constant.)
(iii) *The manifold V is parallelizable, the manifold W is contractible* $q_+(w) \equiv \mathrm{const}_+ \geq n + 1$, $q_-(w) \equiv \mathrm{const}_- \geq n + 1$ *for all* $w \in W$ *and* $g \equiv 0$.

(3) *Forms of Degree* $d \geq 2$.Let Φ be a symmetric form of degree $d \geq 2$ on $\mathbb{R}^q$ (that is a homogeneous polynomial on $\mathbb{R}^q$). Then each homomorphism $h \in \mathrm{Hom}(\mathbb{R}^n \to \mathbb{R}^q) = H_n^q$ induces a form, called $g = h^*(\Phi)$ on $\mathbb{R}^n$. Let $S_d = S_d(\mathbb{R}^n)$ be the linear space of forms of degree d on $\mathbb{R}^n$ and denote by $\Delta = \Delta_\Phi: H_n^q \to S_d$ the map $h \mapsto h^*(\Phi)$. A homomorphism $h \in H_n^q$ is called *Φ-regular* if the differential $D_\Delta: T(H_n^q) \to T(S_d)$ has $\mathrm{rank}\, D_\Delta | T_h(H_n^q) = s = s(n, d) = \dim S_d = (n + d)!/n!d!$. We call h *isometric* for a given form g on $\mathbb{R}^n$ if $h^*(\Phi) = g$ and we denote by $Is = Is_g \subset H_n^q$ the subset of Φ-regular isometric homomorphisms. A straightforward calculation shows that $\mathrm{rank}\, D_\Delta | R = s(n, d - 1)$ for all principal subspaces $R \subset H_n^q = T_h(H_n^q)$ if and only if the homomorphism h is Φ-regular. It follows that $Is \cap R$ is a *non-singular* subvariety of codimension $s(n, d - 1)$ in R for all $R \subset H_n^q$ in so far as $Is \cap R$ is non-empty.

Example. Let $d = 3$ and let $g = g_{ijk} = g(e_i \otimes e_j \otimes e_k)$ for a fixed basis $e_1, \ldots, e_n$ in $\mathbb{R}^n$. Then the relation $h^*(\Phi) = g$ is expressed by $s(n, 3) = n(n + 1)(n + 2)/6$ equations in $y_i = h(e_i) \in \mathbb{R}^q$,

$$\Phi(y_i \otimes y_j \otimes y_k) = g_{ijk}, \qquad 1 \leq i \leq j \leq k \leq n.$$

If we restrict these equations to some principal subspace $R = \mathbb{R}^q \subset H_n^q$ associated to the hyperplane $\mathrm{Span}(e_1, \ldots, e_{n-1}) \subset \mathbb{R}^n$, then we obtain $s(n, 2) = n(n + 1)/2$ equations in y_n for fixed vectors $y_1, \ldots, y_{n-1}$,

$$\Phi(y_i \otimes y_j \otimes y_n) = g_{ijn}, \qquad 1 \leq i \leq j \leq n.$$

Denote by $\Phi_{ij}: \mathbb{R}^q \to \mathbb{R}$ the linear forms $\Phi_{ij}: z \mapsto (y_i \otimes y_j \otimes z)$ for $1 \leq i \leq j \leq n$. Then the Φ-regularity of h is equivalent to linear independence of these forms Φ_{ij} for $y_i = h(e_i)$.

We call a homomorphism $h: \mathbb{R}^n \to \mathbb{R}^q$ *hyper regular* if h extends to a Φ-regular homomorphism $\tilde{h}: \mathbb{R}^{n+1} \to \mathbb{R}^q$ for $\mathbb{R}^{n+1} \supset \mathbb{R}^n$, such that $\tilde{h}^*(\Phi) = p^*(g)$ for the normal projection $p: \mathbb{R}^{n+1} \to \mathbb{R}^n$.

Example. Let g and Φ be quadratic forms of types (n_+, n_-) and (q_+, q_-) respectively and let $h: \mathbb{R}^n \to \mathbb{R}^q$ be a Φ-regular isometric homomorphism. Then h is hyper regular if and only if $q_+ \geq n - n_- + 1$ and $q_- \geq n - n_+ + 1$.

Now, let g and Φ by symmetric differential forms of degree d on manifolds V and W respectively. A C^1-map $f: V \to W$ is called (*hyper regular*) *isometric* if the differential $D_f: T(V) \to T(W)$ is (hyper regular) isometric on each tangent space $T_v(V)$, $v \in V$.

(3') **Exercises.** Prove the C^0-dense h-principle for hyper regular isometric C^1-maps $f: V \to W$.

Let the form Φ on W be *diagonal*, $\Phi = \sum_{i=1}^{q_+} (L_i)^d - \sum_{j=1}^{q_-} (L_j)^d$ for some linearly independent linear differential forms L_i and L_j on W. Show for $q_+ \geq s(n+1, d-1) + n$ and $q_- \geq s(n+1, d-1) + n$ every continuous map $V \to W$ to admit a fine C^0-approximation by hyper regular isometric C^1-maps $f: V \to W$ for all V, $\dim V = n$, and for all forms g on V of degree d.

Questions. How can one improve the above lower bound on q_+ and q_-? What is, for example, the minimal number $q = q(n)$, such that every cubic form g on V is a sum of q cubes of exact linear C^0-forms, $g = \sum_{i=1}^{q} (df_i)^3$? [The functions f_i define an *isometric* map $f = (f_1, \ldots, f_q): V \to \mathbb{R}^q$ for the diagonal form $\Phi = \sum_{i=1}^{q} (dx_i)^3$ on $\mathbb{R}^q$.]

(4) *Definite Forms of Even Degree.* Let $C \subset S_d = S_d(\mathbb{R}^n)$ be a nonempty convex cone which is invariant under the natural action of the linear group $L_n(\mathbb{R})$ on the space S_d. If some form $g(x) = g(x_1, \ldots, x_n) = ax_1^d + \ldots$ is contained in C then the form $(\operatorname{sign} a)x_1^d$ is contained in the topological closure of C as $\lambda^{-d} g(\lambda x_1, x_2, \ldots, x_n) \to (\operatorname{sign} a)x^d$ for $\lambda \to \infty$. It follows that there are, for $d = 2m$, exactly two minimal closed convex invariant cones in S_d, called $S_d^+ \subset S_d$ and $S_d^- = -S_d$, where S_d^+ is the convex hull of the $L_n(\mathbb{R})$-orbit of the form x_1^d. Denote by $S_d^* \subset S_d^+$ the interior of S_d^+ and call a form g *positive definite* (*semidefinite*) if it is contained in S_d^* (in S_d^+). For example, the form $\sum_{i=1}^{n} x_i^d$ is positive semidefinite but it is not definite for $d = 2m \geq 4$, $n \geq 2$. In fact, if a form g is positive definite, then every *even* monomial, $(x_{i_1} x_{i_2}, \ldots, x_{i_m})^2$, enters g with a positive coefficient.

There is a unique (up to a scalar) form $g_0 \in S_d$, which is invariant under the action of the orthogonal group $O_n \subset L_n(\mathbb{R})$, namely $g_0 = (\sum_{i=1}^{n} x_i^2)^m$. This g_0 is positive definite. Indeed, the ρ-transforms $g_\rho \in S_d^*$, $\rho \in O_n$, of an arbitrary form $g_\rho \in S_d^*$ average to g_0,

$$g_0 = c \int_O g_\rho d_\rho, \qquad \text{for } c = c(g) > 0,$$

and for the Haar measure $d\rho$ on O_n. It follows, in particular, that

$$(*)\qquad \left(\sum_{i=1}^{n} x_i^2\right)^m = \sum_{j=1}^{q} b_j l_j^d, \qquad b_j > 0,\ d = 2m,$$

for some linear forms $l_j = \sum_{i=1}^n a_{ij} x_i$ and for $q = s(n,d) - 1$.

Remark (Hurwitz-Hilbert). Since the rational points are dense in the *open* cone $S_d^* \subset S_d$ the identity $(*)$ holds true for some *rational* coefficients $b_j > 0$ and a_{ij}. This reduces Waring's problem of the representation of positive integers as sums of dth powers to that for mth powers, $d = 2m$, as every positive integer is $\sum_{i=1}^4 x_i^2$, $x_i \in \mathbb{Z}$, by Lagrange's theorem.

Let us concentrate on representations $g = \sum_{j=1}^q l_j^d$ that correspond to isometric maps $(\mathbb{R}^n, g) \to (\mathbb{R}^q, \Phi)$ for the diagonal form $\Phi = \sum_{j=1}^q x_j^d$ on $\mathbb{R}^q$.

Exercises. Show that Φ-regular homomorphisms $\mathbb{R}^n \to \mathbb{R}^q$ induce positive definite forms g on $\mathbb{R}^n$.

Let a homomorphism $h\colon \mathbb{R}^n \to \mathbb{R}^q$ be given by q linear forms h_j on $\mathbb{R}^n$. Show that h is Φ-regular for $\Phi = \sum_{j=1}^q x_j^d$ if and only if the forms $h_j^{d-1} \subset S_{d-1}$ span the space S_{d-1}. Prove generic homomorphisms $h\colon \mathbb{R}^n \to \mathbb{R}^q$ to be Φ-regular for $q \geq s(n, d-1)$.

Show that a generic positive definite form g on $\mathbb{R}^n$ admits no isometric homomorphisms $(\mathbb{R}^n, g) \to (\mathbb{R}^q, \Phi)$ for $nq < s(n,d)$.

Prove that the boundary of the cone $S_d^+ \subset S_d$ contains no affine simplices of dimension $s(n,d) - n + 1$. Using this, show the existence of an isometric homomorphism $(\mathbb{R}^n, g) \to (\mathbb{R}^q, \Phi = \sum_{j=1}^q x_j^d)$ for all positive semidefinite forms g on $\mathbb{R}^n$ and for $q = s(n,d) - n$. Prove the existence of a *Φ-regular* isometric homomorphism $\mathbb{R}^n \to \mathbb{R}^q$ for all positive *definite* forms g on $\mathbb{R}^n$ and for $q = s(n,d) + s(n,d-1) - n$.

Now, let g be a differential C^0-form on V of degree $d = 2m$ which is positive definite on [the tangent spaces $T_v(V)$ of V], and let Φ be the diagonal differential form $\sum_{j=1}^q (dx_j)^{2m}$ on $\mathbb{R}^q$.

Questions. Do Φ-regular isometric maps $f\colon V \to \mathbb{R}^q$ satisfy the C^0-dense h-principle for large $q \geq q_0(n,d)$? What is the minimal q for which an isometric C^1-map $f\colon V \to \mathbb{R}^q$ exists for all g on V?

Exercises. Prove for $q = s(n,d) + s(n,d-1) - n$ the existence of a Φ-regular isometric C^1-map $f\colon U \to \mathbb{R}^q$ for a small neighborhood $U \subset V$ of a given point $v \in V$.

Show that every positive definite form g on V admits a decomposition $g = \sum_{j=1}^q l_j^d$ for some linear differential forms l_j on V and for $q = (n+1)(s(n,d) - n)$.

Prove generic C^∞-maps $V \to \mathbb{R}^q$ to be Φ-regular for $q \geq s(n,d-1) + n$.

Construct Φ-regular isometric C^1-maps $V \to \mathbb{R}^q$, $q = (n+1)(s(n,d) - n) + s(n,d-1) + n$, for all positive definite forms g on V.

Hint. Consult 3.1.4 and Gromov (1972) for a similar study of isometric C^r-immersions of forms of degree $d \geq 2$.

(5) *Nonsymmetric Forms and Further Generalizations.* Let Φ and g be forms of degree d on W and on V respectively having the same type of symmetry. The "isometric" map equation $f^*(\Phi) = g$ for $f: V \to W$ is of the triangular form for all Φ and g, but the convex integration may fail to prove the h-principle in certain cases. Namely, if the form Φ is *antisymmetric* (i.e. exterior), then the corresponding relation $\mathscr{R} \subset X^{(1)}$ is linear on every principle subspace $R \subset X^{(1)}$: the intersection $\mathscr{R} \cap R$ is an affine subspace in R for all $R \subset X^{(1)}$. For example, if $d = 2$, then the intersection $\mathscr{R} \cap R \subset R = \mathbb{R}^q$ is given by the following equations which are linear in the unknown vector $y_n \in R = \mathbb{R}^q$,

$$(y_i, y_n) = g_{in},$$

where y_i, $i = 1, \ldots, n-1$ are fixed vectors in R. Hence, the convex extensions do not enlarge the relation $\mathscr{R}$. In fact, the h-principle fails for the equation $f^*(\Phi) = g$ unless the Stokes-De Rham theorem is taken into account (see 2.2.7 and 3.4.1).

If the form Φ of degree $d \geq 2$ is not antisymmetric, then the relation $f^*(\Phi) = g$ looks sufficiently ample and one expects the h-principle under appropriate regularity assumptions on f.

Here are two further examples of differential equations to which the convex integration may apply.

(a) $\langle \partial_i \partial_j f, \partial_k \partial_l f \rangle = g_{ijkl}$, where $f: V \to \mathbb{R}^q$ is the unknown C^2-map, where ∂_i, $i = 1, \ldots, n$, are given C^1-smooth vector fields on V and where g_{ijkl} are continuous functions on V for $i \leq j$, $k \leq l$ and $i \leq k$.

(b) Let Λ^m denote the mth exterior power of the cotangent bundle of V and let $(\Lambda^m)^2$ denote the *symmetric* square of Λ^m. Our equation is

$$\sum_{j=1}^{q} (df_j)^2 = g,$$

where $f_j: V \to \Lambda^{m-1}$ are the unknown $(m-1)$-forms and where $g: V \to (\Lambda^m)^2$ is a given form of degree $2m$ on V.

2.4.10 Isometric Maps with Singularities

A *continuous* map between Riemannian manifolds $f: V \to W$, is called *isometric* if it preserves the length of all rectifiable curves C in V.

Examples. (a) Let V be a closed n-dimensional manifold, such that V minus a point, $V \backslash v_0$, is a parallelizable manifold. (If $n = 4$, then this is equivalent to the vanishing of the first two Stiefel-Whitney classes, $w_1 = 0$, $w_2 = 0$.) Then there exists an isometric C^1-immersion $V \backslash v_0 \to \mathbb{R}^{n+1}$ which obviously extends to a continuous isometric map $f: V \to \mathbb{R}^{n+1}$. This map may be extremely irregular near the point $v_0 \in V$, and if V admits no smooth immersions into $\mathbb{R}^{n+1}$ (for instance, $n = 4$ and the first Pontryagin class $p_1 \neq 0$), then this singularity cannot be removed. However, the map f can be modified in order to have the following (a') *Conical Singularity at* v_0. Fix a diffeomorphism of a small neighborhood $U \subset V$ of v_0 into the tangent space, say $e_0: U \to T_{v_0}(V)$, such that the differential $D_{e_0}: T_{v_0}(V) \to T_{v_0}(V)$ is the identity map (in particular $e_0(v_0) = 0$) and such that h_0 is strictly short on $U \backslash v_0$. Next we take an isometric map $f_0': \mathbb{R}^n = T_{v_0}(V) \to \mathbb{R}^{n+1}$ which is the cone from the origin over some isometric C^1-map between the unit spheres, say over $g_0': S^{n-1} \to S^n$, such that the immersion $e_0 \circ f_0': U \backslash v_0 \to \mathbb{R}^{n+1}$ is regularly homotopic to $f | U \backslash v_0$. The existence of such a g_0' is obvious with the h-principle for isometric C^1-immersions $S^{n-1} \to S^n$. Then the map $e_0 \circ f_0'$ extends to a map $f_0: V \to \mathbb{R}^{n+1}$ which is a strictly short immersion on $V \backslash v_0$. Finally, we take an isometric C^1-map $V \backslash v_0 \to \mathbb{R}^{n+1}$ which finely C^0-approximates f_0 on $V \backslash v_0$ and thus we get an isometric map $f_1: V \to \mathbb{R}^{n+1}$ which is C^1-smooth on $V \backslash v_0$ and which has the isometric map f_0' for the tangent cone at the point v_0 [compare (A') below].

(b) Let $C_0 \subset V$ be a nowhere rectifiable curve, that is $\dim_H(C_0 \cap C) < 1$ for all rectifiable curves C in V where $\dim_H$ stands for the Hausdorff dimension. Take a strictly short map $f_0: V \to W$ which collapses C_0 to a single point in W and then finely C^0-approximate the map f_0 on $V \backslash C_0$ by an isometric C^1-immersion $f: V \backslash C_0 \to W$ (this is possible, for example, if $\dim W \geq 2 \dim V$). Then the map f extends to an isometric map $V \to W$ which sends the curve $C_0 \subset V$ to a single point. This collapse can be avoided with the following

(b') *Strongly Isometric Maps.* Take two points x and y in V and consider chains of points, $x = x_0, x_1, \ldots, x_k = y$ in V, $k = 1, 2, \ldots$, such that $\operatorname{dist}_V(x_i, x_{i+1}) \leq \varepsilon$ for $i = 0, 1, \ldots, k-1$, and for a given $\varepsilon > 0$. Then we define $\operatorname{dist}_\varepsilon(x, y)$ to be the lower bound of the sums $\sum_{i=0}^{k-1} \operatorname{dist}_W(f(x_i), f(x_{i+1}))$ over all such chains of point $(x_0, \ldots, x_k)$ between x and y. Finally, we call a map f *strongly isometric* if $\lim_{\varepsilon \to 0} \operatorname{dist}_\varepsilon(x, y) = \operatorname{dist}_V(x, y)$ for all $x, y \in V$.

Exercise. Show every strongly isometric map to be isometric. Show that a strongly isometric map collapses no connected subset in V to a point.

(A) *Stratumwise Smooth Maps.* Let $\{\Sigma^i\}$, $i \in I$, be a C^1-stratification of V (see 1.3.2). Then a continuous map $f: V \to W$ is called *stratumwise smooth* on $\{\Sigma^i\}$ if the restriction $f | \Sigma^i: \Sigma^i \to W$ is C^1-smooth for all strata Σ^i. For example, if the stratification corresponds to a triangulation of V, then f is C^1-smooth on the *open* simplices of all dimensions. Such an f may fail the *piecewise* smooth condition, that requires the C^1-smoothness of f on all *closed* simplices.

Theorem. *Let $f_0: V \to W$ be a strictly short map, such that the restriction $f_0 | \Sigma^i: \Sigma^i \to W$ is homotopic to an immersion $\Sigma^i \to W$ for all strata Σ^i. If $\dim W > \dim V$ then f_0 admits a fine C^0-approximation by isometric stratumwise smooth maps $f: V \to W$.*

Proof. The convex integration does not directly apply here because the map f in question is not C^1-smooth. However, an obvious induction by strata (compare 1.4.4, 2.1) reduces the problem to the following lemma to which the convex integration does apply (compare 2.4.9).

Lemma. *Let U be an open subset in V and let $f_0\colon U \to W$ be a strictly short map whose restriction to a given submanifold $\Sigma \subset U$ is homotopic to an immersion $\Sigma \to W$. Then f_0 admits a fine C^0-approximation by C^1-maps $U \to W$ which are isometric on Σ and strictly short on $U \backslash \Sigma$.*

Exercises. Fill in the details of the above proof.

Show with an appropriate stratification of V that every strictly short map $V \to W$, $\dim V < \dim W$, admits an approximation by continuous isometric maps $V \to W$.

(A′) *Isometric Maps with Tangent Cones.* Let $f\colon \mathbb{R}^n \to \mathbb{R}^q$, $f(0) = 0$, be a continuous map for which the maps $f_\lambda\colon \mathbb{R}^n \to \mathbb{R}^q$, $f_\lambda(x) = \lambda^{-1} f(\lambda x)$, uniformly converge on compact subsets in $\mathbb{R}^n$ for $\lambda \to 0$. Then the limit $\lim_{\lambda \to 0} f_\lambda$ is called the *tangent cone* $T_0 f\colon \mathbb{R}^n \to \mathbb{R}^q$ of f at $0 \in \mathbb{R}^n$. (If f is smooth, this is the ordinary differential of f.) This definition, being purely local, generalizes to maps between smooth manifolds, $f\colon V \to W$, and so one may speak of (the existence of) tangent cones $T_v f\colon T_v(V) \to T_w(W)$, $w = f(v)$, at points $v \in V$.

A map between Riemannian manifolds, $f\colon V \to W$ is called T^1-*isometric* (T for tangent) if the tangent cone $T_v f\colon T_v(V) \to T_w(W)$ exists for all $v \in V$ and if the map $T_v f$ maps the unit sphere $S_v^{n-1} \subset T_v(V)$, $n = \dim V$, into the unit sphere $S_w^{q-1} \subset T_w(W)$, $q = \dim W$ for all $v \in V$.

Exercise. Let $f\colon V \to W$ be a T^1-isometric map. Show that f is strongly isometric and that the pull-back $f^{-1}(w)$ is a *discrete* subset in V for all $w \in W$.

Next, by induction we define a T^1-isometric map f to be T^i-*isometric*, $i = 2, 3, \ldots, n$, if the map $T_v f | S_v^{n-1}\colon S_v^{n-1} \to S_w^{q-1}$ is T^{i-1}-isometric for all $v \in V$ and $w = f(v)$. Finally, T-*isometric* maps, by definition are T^n-isometric.

Exercise. Show the map constructed in the above Example (a′) to be T-isometric.

Let us fix a smooth triangulation in V. Then the tangent spaces to the simplices at every point $v \in V$ partition the tangent space $T_v(V)$ into finitely many convex (polyhedral) cones.

Exercise. Show that a T-isometric map $f\colon V \to W$ is piecewise smooth (for a given triangulation) if and only if the tangent cone map $T_v f\colon T_v(V) \to T_w(W)$ is piecewise linear (i.e. linear on every cone of the above partition) for all $v \in V$.

Take a point in an open k-dimensional simplex of our triangulation, $v \in \Delta^k \subset V$. The above partition of $T_v(V)$ induces a triangulation of the unit sphere $S_v^{n-k-1} \subset T_v^\perp(\Delta^k) \subset T_v(V)$, where $T_v^\perp(\Delta^k)$ denote the normal space to Δ^k at v. Then by induction in $n = \dim V$ we define *regular* maps f. All maps are assumed regular for $n = 0$ and we call f regular for $n > 0$ if it is T-isometric, stratumwise smooth (i.e. C^1-smooth on every open simplex Δ^k, $k = 0, \ldots, n$) and the map $T_v f | S_v^{n-k-1}$ is regular for the above triangulation of S_v^{n-k-1} for all Δ^k and $v \in \Delta^k$.

Exercise. Combine the argument of the Example (a′) with an induction by skeletons and prove the following

Theorem. *If* $\dim V < \dim W$ *then every strictly short map* $V \to W$ *admits a fine* C^0*-approximation by* T*-isometric regular maps* $V \to W$.

(B) *Piecewise Linear Isometric Map.* Let K be an n-dimensional simplicial polyhedron with k^0 vertices, $K^0 = \{v_1, v_2, \ldots, v_{k^0}\}$, and with k^1 edges, $K^1 = \{e_1, e_2, \ldots, e_{k^1}\}$. An arbitrary *simplicial* map $f: K \to \mathbb{R}^q$ by definition is linear on every simplex in K and so these maps are given by vectors $(f(v_1), \ldots, f(v_{k^0})) \in \mathbb{R}^{qk^0}$. Every simplicial map f induces a flat metric on every simplex in K and so a (singular Riemannian) metric on the space K itself. Such a metric g is uniquely determined by the vector $(g_1, \ldots, g_{k^1}) \in \mathbb{R}^{k^1}$, where g_i denotes the (length)^2 of the edge e_i. We study the map $\mathscr{G}: \mathbb{R}^{qk^0} \to \mathbb{R}^{k^1}$ which assigns the induced metric $g = \mathscr{G}(f)$ to each simplicial map $f: K \to \mathbb{R}^q$. Since the map $\mathscr{G}$ is polynomial, the image $\mathscr{G}(\mathbb{R}^{qk^0})$ is a semialgebraic subset in $\mathbb{R}^{k^1}$. If $q \geq q_0 = k^0 - 1$, then this subset $\mathscr{G}(\mathbb{R}^{qk^0}) \subset \mathbb{R}^{k^1}$ obviously is a convex cone in $\mathbb{R}^{k^1}$ which does not depend on $q \geq q_0$ and which has $\dim \mathscr{G}(\mathbb{R}^{qk^0}) = k^1$. But if $q < k^0 - 1$ then only few facts are known about (singularities of) the map $\mathscr{G}$ and the dimension of its image.

Examples. Let K be homeomorphic to the sphere S^2. Then, obviously, $k^1 = 3k^0 - 6$. A theorem of Steinitz claims the existence of a simplicial map f of K onto a convex (with all dihedral angles $<\pi$) hypersurface in $\mathbb{R}^3$. The *rigidity* theorem of Dehn estimates the rank of the map $\mathscr{G}: \mathbb{R}^{3k^0} \to \mathbb{R}^{k^1}$ at such "convex" maps $f \in \mathbb{R}^{3k^0}$,

$$(*) \qquad \operatorname{rank}_f \mathscr{G} = 3k^0 - 6 = k^1.$$

Hence, convex polyhedra $f: K \hookrightarrow \mathbb{R}^3$ are rigid: every small deformation of f which does not change $g = \mathscr{G}(f)$ is a rigid motion. Since the map $\mathscr{G}$ is polynomial, the identity $(*)$ holds on a (Zariski) open subset in $\mathbb{R}^{3k^0}$. Hence, $\dim \mathscr{G}(K) = k^1$ and so *generic* maps $f: K \to \mathbb{R}^3$ are rigid (Gluck 1975). However, there are many (nongeneric) nonrigid maps. In fact, there exist, for certain $K \approx S^2$, submanifolds $F \subset \mathbb{R}^{3k^0}$ of positive dimension, such that the maps $f \in F$ are *embeddings* $K \to \mathbb{R}^3$ which are not rigid motions of a fixed map, but for which the induced metric is constant for all $f \in F$ (see Connelly 1979).

The definition of $\mathscr{G}$ generalizes to *spherical simplicial* maps $f: K \to S^q$ which send every simplex in K onto a convex spherical simplex in S^q. We write $f = \{f(v_i)\}, f(v_i) \in S^q$, $i = 1, \ldots, k^0$, and we assign to each edge $e_l \subset K$ the (length)^2 of the geodesic segment $f(e_l) \subset S^q$ that is $(\operatorname{dist}(f(v_i), f(v_j))^2$ for the ends v_i and v_j of e_l. Thus, we get $\mathscr{G}: (S^q)^{k^0} \to \mathbb{R}^{k^1}$ for the Cartesian product $(S^q)^{k^0} = S^q \times \cdots \times S^q$. This map $\mathscr{G}$ is real analytic on those f, for which length $f(e_l) < \pi$ for all $l = 1, \ldots, k^1$.

Since the metric in S^q locally is a small analytic perturbation of the flat Euclidean metric, the identity $(*)$ extends to an open dense subset of maps $f: K \to S^3$ for triangulations K of S^2. In other words, generic polyhedra $f: K \hookrightarrow S^3$ are rigid as well as polyhedra in $\mathbb{R}^3$ for $K \approx S^2$.

Now, let K be a triangulation of a 3-dimensional manifold without boundary and let $f: K \to \mathbb{R}^4$ be a simplicial map. Then the tangent cone $T_{v_i} f$ of f at every vertex $v_i \in K$ is a (spherical) simplicial map of the *link* $L(v_i) \subset K$ (this is a triangulated 2-sphere which by definition consists of all simplices in K opposite to v_i) into the unit sphere $S^3 \subset \mathbb{R}^4$ around $f(v_i) \in \mathbb{R}^4$. The rigidity of generic simplicial maps $L(v_i) \to S^3$ for all $v_i \in K$ implies (this is an easy exercise) the rigidity of generic simplicial maps of *connected* three manifolds into $\mathbb{R}^4$. This argument equally applies to maps $f: K \to S^n$, $\dim K = n - 1$, for $n \geq 4$ and then by induction on n we obtain the rigidity of generic maps $f: K \to S^n$ for all $n \geq 4$. Finally we conclude to the *rigidity of generic simplicial maps* $g: K \to \mathbb{R}^{n+1}$ *for all connected triangulated manifolds* K, $\dim K = n \geq 3$. Since the isometry group $Is(\mathbb{R}^{n+1})$ has dimension $(n+1)(n+2)/2$, this rigidity claims the following property of the map $\mathscr{G}: \mathbb{R}^{(n+1)k^0} \to \mathbb{R}^{k^1}$

$$(**) \qquad \dim \mathscr{G}(\mathbb{R}^{(n+1)k^0}) = \operatorname{rank}_f \mathscr{G} = (n+1)k^0 - \frac{(n+1)(n+2)}{2},$$

for all f in some (Zariski) open dense subset $F \subset \mathbb{R}^{(n+1)k^0}$.

The dimension of the image $\mathscr{G}(\mathbb{R}^{(n+1)k^0}) \subset \mathbb{R}^{k^1}$ does not exceed $k^1 = \dim \mathbb{R}^{k^1}$. Thus we arrive at the following *lower bound* [due to Barnette (1973)] *on the number of edges in* K,

$$(+) \qquad k^1 \geq (n+1)k^0 - \frac{(n+1)(n+2)}{2}$$

for all connected triangulated manifolds K of dimension $n \geq 3$. This $(+)$ is false for $n = 1$ but it does hold true for $n = 2$ as

$$k^1 = 3k^0 - 3\chi \geq 3k^0 - 6,$$

for all closed surfaces K with the Euler characteristic $\chi \leq 2$.

The above geometric proof of $(+)$ can be reduced to a purely combinatorial argument with the following definition. Let K be an arbitrary simplicial complex and let $L \subset K$ be a subcomplex. The pair (K, L) is called *q-rigid* for some $q > 0$ if every subset of m_0 vertices in $K \backslash L$ has at least qm_0 edges in K which meet (the union of) these vertices. [Compare Kalai (1986).]

Exercise. Let $f: K \to \mathbb{R}^q$, $q \geq n = \dim K$, be a generic simplicial map, such that no isometric deformation of f remains fixed on the subcomplex L. Show the pair (K, L) to be q-rigid.

Next, we call the polyhedron K itself q rigid if the pair (K, Δ^n) is q-rigid for all top dimensional simplices $\Delta^n \subset K$.

The following three properties of this rigidity are immediate from the definition.
(a) Let K be the simplicial cone over some $(n-1)$-dimensional complex L. If L is $(q-1)$-rigid then K is q-rigid.
(b) Let K be divided into a union, $K = K_1 \cup K_2$, for some subcomplexes K_1 and K_2 in K such that $\dim K_1 \cup K_2 = n = \dim K$ (i.e. $K_1 \cap K_2$ contains some top dimensional simplex). If K_1 and K_2 are q-rigid complexes, then K also is q-rigid;
(c) Let K be a *connected* complex of *pure* dimension n (i.e. each simplex in K is a

face of some top dimensional simplex $\Delta^n \subset K$). If the link $L(v_i)$ is $(q-1)$-rigid for all vertices $v_i \in K$, $i = 1, \ldots, k^0$, then the complex K is q-rigid.

In order to apply (c) to manifolds K of dimension $n \geq 3$ we need the following elementary fact.

(d) If K is (homeomorphic to) a connected closed surface, then K is 3-rigid.

Proof. Let $L \subset K$ be the union of the (closed) 2-simplices in K which meet the given m_0 vertices in $K \setminus \Delta^2$ for some triangle $\Delta^2 \subset K$. The *boundary* ∂L of L by definition is the union of those closed edges in L which do not meet the above m_0 vertices. The first Betti number of ∂L is related to the number m_1' of edges in ∂L by the obvious inequality

$$b_1(\partial L) \leq \tfrac{1}{3} m_1'.$$

Let m_1 denote the number of those edges in L which meet the given m_0 vertices. Then the number l_2 of the triangles in L satisfy $3l_2 = 2m_1 + m_1'$. The Euler characteristic of the pair $(L, \partial L)$ is

$$\chi(L, \partial L) = m_0 - m_1 + l_2 = m_0 - \tfrac{1}{3} m_1 + \tfrac{1}{3} m_1'.$$

On the other hand,

$$\chi(L, \partial L) \leq b_2(L, \partial L) \leq b_1(\partial L) \leq \tfrac{1}{3} m_1',$$

and so $m_0 - \frac{1}{3} m_1 \leq 0$. Q.E.D.

The propositions (c) and (d) by induction imply the $(n+1)$-rigidity of closed connected manifolds K of dimension $n \geq 2$ and then the inequality (+) obviously follows.

Let us return to the geometric rigidity problem and to the map $\mathscr{G}: \mathbb{R}^{qk^0} \to \mathbb{R}^{k^1}$ The algebra-geometric behaviour of $\mathscr{G}$ depends on the combinatorial structure of the graph (K^0, K^1) that is the 1-skeleton of the polyhedron K.

Question. Is there an efficient combinatorial description in terms of the graph (K^0, K^1) of basic invariants of the map G? For instance, let $\Sigma_m = \{x \in \mathbb{R}^{k^1} | \dim \mathscr{G}^{-1}(x) \geq m\}$. What is the dimension of this subvariety $\Sigma_m \subset \mathbb{R}^{k^1}$?

Example. Let K_d denote the 1-skeleton of the d-dimensional cube. If $d \geq 3$, then the generic maps $f: K_d \to \mathbb{R}^2$ are obviously rigid and so $\operatorname{codim} \Sigma_m > 0$ for $m > 3 = \dim(\operatorname{Isom} \mathbb{R}^2)$. Furthermore, the subvariety $\mathscr{G}^{-1}(x) \subset \mathbb{R}^{2k^0}$, $k^0 = 2^d$, clearly has $\dim \mathscr{G}^{-1}(x) \leq 2 + d$ for $d = 1, 2, \ldots$, and for all $x \in \mathbb{R}^{k^1}$, $k^1 = d 2^{d-1}$. The subvariety $\Sigma_{2+d} \subset \mathbb{R}^{k^1}$ is the principal diagonal in $\mathbb{R}^{k^1}$ and so the corresponding (most flexible) simplicial maps $f: K_d \to \mathbb{R}^2$ are those which give equal length to all edges in K_d. To see these maps we use the standard (unit) cube $I^d \subset \mathbb{R}^d$ which is spanned by a fixed orthonormal basis $\{e_1, \ldots, e_d\}$. Every *linear* map $f: \mathbb{R}^d \to \mathbb{R}^2$ for which $\| f(e_i) \| = 1$, $i = 1, \ldots, d$, gives a simplicial map of $K_d \subset I^d \subset \mathbb{R}^d$ to $\mathbb{R}^2$ with edges of unit length. Thus we obtain the top dimensional component in $\mathscr{G}^{-1}(x)$, $x \in \Sigma_{2+d}$, that is the product $\mathbb{R}^2 \times \underbrace{S^1 \times S^1 \times \cdots \times S^1}_{d}$.

Every linear map $f\colon \mathbb{R}^d \to \mathbb{R}^2$ for which $\|f(e_i)\| = 1, i = 1, \ldots, d$, gives rise to pretty configurations of unit circles in $\mathbb{R}^2$ as follows. Let $\mathbb{Z}^d$ be the integral lattice in $\mathbb{R}^d$ spanned by the vectors $e_i \in \mathbb{R}^d$, $i = 1, \ldots, d$. Call a point $\bar{z} = (z_1, \ldots, z_d) \in \mathbb{Z}^d$ *even* if the sum $z_1 + \cdots + z_d$ is even, and call $\bar{z}$ *odd* if this sum is odd. Let $K \subset \mathbb{R}^d$ be a 1-dimensional polyhedron (graph) with vertices in $\mathbb{Z}^d$ whose edges are straight unit segments in $\mathbb{R}^d$ parallel to the vectors e_i. The above graph $K_d \subset \mathbb{R}^d$ is an example. Let $K^0_{ev} \subset K$ be the set of even vertices in K and let $K^0_{odd} \subset K$ be odd vertices. Then the unit circles in $\mathbb{R}^2$ around the points $f(\bar{z}) \in \mathbb{R}^2, \bar{z} \in K^0_{ev}$, meet at the images of some odd vertices in K^0. For example, if $K = K_3$ is the 1-skeleton of the 3-dimensional cube, then we get (a well known 2-parametric family of) four unit circles, such that every three of them have a point in common.

Exercises. (a) Show, that one cannot find the center of a unit circle in $\mathbb{R}^2$ if the only operation one is allowed consists in drawing unit circles through pairs of points in $\mathbb{R}^2$.

Hint. Every configuration of unit circles in $\mathbb{R}^2$ comes from the above map $f\colon \mathbb{Z}^d \to \mathbb{R}^2$. Then, there is a perturbation f_1 of f, for which $\|f_1(e_i)\| = \|f(e_i)\| = 1$, $i = 1, \ldots, d$, such that the images $f_1(\mathbb{Z}^d_{ev})$ and $f_1(\mathbb{Z}^d_{odd})$ in $\mathbb{R}^2$ do not intersect.

(b) Show generic simplicial maps $f\colon K_d \to \mathbb{R}^q$ to be rigid for $d \geq 2q$.

Isometric Maps of Subdivided Polyhedra. The isometric map problem for piecewise linear metrics on K (for which every simplex in K is isometric to an affine simplex in $\mathbb{R}^n$) approaches its Riemannian counterpart if one asks for *piecewise linear* isometric maps $f\colon K \to \mathbb{R}^q$ which by definition are linear isometric on the simplices of some simplicial subdivisions of K. Two facts are known about these maps.
(1) If $\dim K = n \leq 4$, then K admits an [equidimensional ! Compare (B) below] piecewise linear isometric map into $\mathbb{R}^n$ (Zalgaller 1958) for an arbitrary piecewise linear metric on K.
(2) Every compact oriented surface with a piecewise linear metric can be piecewise linearly isometrically *embedded* into $\mathbb{R}^3$. (Burago-Zalgaller 1960.)

Exercise. Show that the torus T^n with an arbitrary flat Riemannian metric admits a piecewise linear isometric map $T^n \to \mathbb{R}^n$.

Hint. Use maps with normal foldings along flat subtori in T^n [compare (B) below].

Question. Does every n-dimensional Riemannian manifold V admit a piecewise smooth (for some triangulation of V) isometric map $f\colon V \to \mathbb{R}^{n+1}$? (The solution requires a preliminary study of the tangent cones $T_v f\colon T_v(V) \to \mathbb{R}^{n+1}$ which are *piecewise linear* isometric maps.)

Generalizations. Many partial differential relations admit interesting piecewise linear (and piecewise smooth) versions. Here are several sample questions.

(i) Take a simplicial polyhedron K and let us assign to each simplex Δ in K a subset $\mathscr{R}_\Delta$ in the space of *projective* maps $\Delta \to \mathbb{R}^q$ (one could take an arbitrary projectively flat manifold, for example S^q, instead of $\mathbb{R}^q$). Under what condition on

$\mathscr{R}_\Delta$ can one find a continuous map $f: K \to \mathbb{R}^q$, such that the restriction $f|\Delta: \Delta \to \mathbb{R}^q$ is a projective map and $f|\Delta \in \mathscr{R}_\Delta$ for all $\Delta \in K$?

(ii) If K' is a subdivision of K, then one extends the "relations" $\mathscr{R}_\Delta$ to the simplices Δ' in K' as follows. The subset $\mathscr{R}_{\Delta'}$ consists of those projective maps $f': \Delta' \to \mathbb{R}^q$ such that $f' = f|\Delta$ for $\Delta \supset \Delta'$ and for some $f \in \mathscr{R}_\Delta$. Now, one takes a continuous map $f_0: K \to \mathbb{R}^q$ and one asks for a C^0-approximation of f_0 by those piecewise projective maps $f: K' \to \mathbb{R}^q$ of subdivisions K' of K for which $f|\Delta' \in \mathscr{R}_{\Delta'}$ for all $\Delta' \in K'$.

Example. A piecewise linear (in some triangulation) map f of a smooth oriented n-dimensional manifold $V \to \mathbb{R}^q$ is called *A-directed*, for a given subset A in the Grassmann manifold $Gr_n(\mathbb{R}^q)$ of oriented n-dimensional subspaces in $\mathbb{R}^q$, if the differential $D_f: T_v(V) \to \mathbb{R}^q$ sends the tangent space $T_v(V)$ onto some subspace $a \in A$ for all regular points $v \in V$ of f (these are the interior points of the top dimensional simplices). For which A does an arbitrary continuous map $f_0: V \to \mathbb{R}^q$ admit a C^0-approximation by piecewise linear A-directed maps $f: V \to \mathbb{R}^q$? If $q = n + 1$ then a necessary condition on A is the existence of an affine simplex $\Delta^{n+1} \subset \mathbb{R}^{n+1}$ whose boundary $\partial\Delta^n \approx S^n \subset \mathbb{R}^{n+1}$ is directed by A (i.e. each oriented n-face of Δ^{n+1} is parallel to some hyperplane $a \in A$). Is this condition also sufficient for the approximation of continuous maps $f_0: V \to \mathbb{R}^{n+1}$ by A-directed maps?

(iii) A continuous map $f: V \to \mathbb{R}^q$ is called *locally K-flat* if each point $v \in V$ admits a neighborhood $U \subset V$ whose image $f(U) \subset \mathbb{R}^q$ lies in some k-dimensional affine subspace in $\mathbb{R}^q$. For which k, q and $n = \dim V$ does every continuous map $f_0: V \to \mathbb{R}^q$ admit a C^0-approximation by piecewise linear locally k-flat maps $f: V \to \mathbb{R}^q$? Notice, that the k-flat condition cannot be expressed with the above "relatives" $\mathscr{R}_\Delta$.

2.4.11 Equidimensional Isometric Maps

Let V and W be n-dimensional Riemannian manifolds. A continuous map $f: V \to W$ is called normally *folded* along a hypersurface $V_0 \subset V$ (compare 1.3.6) if the following condition is satisfied for every small open subset $U \subset V$ which is divided by V_0 into two submanifolds with boundaries, say into U' and U'', such that $U' \cap U'' = \partial U' = \partial U'' = U \cap V_0$.

The Folding Condition. The restricted maps $f' = f|U': U' \to W$ and $f'' = f|U''$ are C^1-smooth immersion (including the boundaries), whose differentials are related on the bundle $T(V)|U \cap V_0$ by the orthogonal reflection in the subbundle $T(V_0)|U \cap V_0$, where "orthogonal" applies to the Riemannian metric in W. Namely, if $v \in T_w(W)$, for $w = f(v)$, $v \in V_0$, is a normal vector to the hyperplane $Df'(T_v(V_0)) = Df''(T_v(V_0)) \subset T_w(W)$, then $(Df')^{-1}(v) = -(Df'')^{-1}(v)$ for all $v \in V_0$. (The Riemannian metric on V plays no role in this definition.)

If f is a normally folded map, then the Riemannian metrics induced by f' and f'' agree on the bundle $T(V)|V_0$, and so the map f induces a *continuous* Riemannian metric, say $f^*(\Phi)$ on V, where Φ denotes the Riemannian metric on W.

Suppose that the manifold V splits, $V = V' \times [0,1]$ for a closed manifold V', $\dim V' = n-1$, and let $f_0: V \to W$ be a C^1-map which is an immersion near the boundary $\partial V \subset V, \partial V = (V' \times 0) \cup (V' \times 1)$, and which also is an immersion on the submanifolds $V' \times t \subset V$, $t \in [0,1]$. Consider the (positive semidefinite) quadratic forms $g_0 = f_0^*(\Phi)$ and $g = g_0 + a(dt)^2$ on V for a continuous non negative function $a: V \to \mathbb{R}_+$ which vanishes on the boundary $\partial V \subset V$.

Stretching Lemma. *The map f_0 admits a uniform approximation by normally folded maps $f_\varepsilon: V \to W$, $f_\varepsilon \to f_0$ for $\varepsilon \to 0$ such that the induced forms $g_\varepsilon = f_\varepsilon^*(\Phi)$ uniformly converge to g for $\varepsilon \to 0$, and such that each map f_ε equals f_0 on a small neighborhood $U_\varepsilon \subset V$ of the boundary $\partial V \subset V$.*

The proof is given at the end of this section.

Now let V be an arbitrary n-dimensional manifold without boundary. A quadratic form g' on V is called *elementary* if there exists a split submanifold $U = U' \times [0,1] \subset V$, where U' is a finite or countable union of closed $(n-1)$-dimensional manifolds, $n = \dim V$, such that g' is zero outside U and $g'|U = a(dt)^2$ for some continuous function $a: U \to \mathbb{R}_+$, which vanishes on $\partial U \subset U$.

Decomposition Lemma. *An arbitrary Riemannian metric on V is a finite sum of elementary quadratic forms.*

This is a trivial corollary of the Nash-Kuiper theorem (see 2.4.9). An alternative proof is suggested in the following

Exercise. Let G be a nonempty set of continuous positive semidefinite quadratic forms on V with the following three properties:

(i) If $g \in G$ then also $ag \in G$ for all continuous functions $a: V \to \mathbb{R}_+$;
(ii) If $d: V \to V$ is C^1-diffeomorphism, then $d^*(g) \in G$ for all $g \in G$;
(iii) If the forms $g_i \in G$, $i = 1, 2, \ldots$, have mutually disjoint supports, then $\sum_i g_i \in G$.

Show that every positive definite form is contained in G and thus prove the decomposition lemma.

Proposition. *Let g and g_0 be Riemannian metrics on V, such that the form $g - g_0$ is positive definite. Then the identity map $V \to V$ admits a fine C^0-approximation by isometric maps $f: (V, g) \to (V, g_0)$.*

Proof. Write $g - g_0 = \sum_{j=1}^{N} e_j$ for some elementary forms e_j. The stretching lemma implies the existence of Riemannian metrics g_j on V and of normally folded strictly short maps $f_j: (V, g_j) \to (V, g_{j-1})$, $j = 1, \ldots, N$, which are C^0-close to the identity, such that the forms $g_j - f_j^*(g_{j-1})$ and $g_j - g_{j-1} - e_j$ are C^0-small for $j = 1, \ldots, N$. Thus we obtain a continuous map $F_1: V \to V$, namely the composed map $F_1 = f_1 \circ \cdots \circ f_N$, with the following three properties,

(i) The map F_1 is (arbitrary!) C^0-close to the identity map;
(ii) The induced form $g_0^1 = F_1^*(g_0)$ is continuous and positive definite;
(iii) The form $g - g_0^1$ is positive definite and it can be made (by some choice of F_1) as small as one wishes.

Then we take g_0^1 in place of g_0 and construct a map F_2 which also satisfies (i) and for which the metric $g_0^2 = F_2^*(g_0^1)$ satisfies (ii) and (iii). We go on and obtain continuous positive forms g_0^i on V, $i = 1, 2, \ldots$, and continuous isometric maps $F_{i+1}: (V, g_0^{i+1}) \to (V, g_0^i)$ (each F_i is a finite composition of some folded maps), such that the forms g_0^i uniformly converge to g and such that the composed map $F_1 \circ F_2 \circ \cdots \circ F_i$ uniformly converges for $i \to \infty$ to the desired isometric map $f: (V, g) \to (V, g_0)$.

Corollary. *Let V be an n-dimensional stably parallelizable manifold. Then V admits an isometric map $V \to \mathbb{R}^n$.*

Proof. Poenaru's folding theorem (see 2.1.3) yields a strictly short normally folded map $f_0: V \to \mathbb{R}^n$ (Compare the proof of the theorem below). Since the induced form g_0 on V is continuous and since the form $g - g_0$ is positive for the given Riemannian form g on V, there is an isometric map $f: (V, g) \to (V, g_0)$. Then the composed map $f_0 \circ f: V \to \mathbb{R}^q$ is isometric.

Theorem. *Let $f_0: V \to W$ be a strictly short map between Riemannian manifolds of dimension n. Then f_0 admits a fine C^0-approximation by isometric maps $f: V \to W$. In particular, there exists an isometric map $f: V \to \mathbb{R}^n$ for all Riemannian manifolds V of dimension n.*

Proof. We assume [compare (A) in 2.4.10] the map f_0 to admit a stratification $\{\Sigma^i\}$, $i = 0, \ldots, k$, such that the map $f_0|\Sigma^i$ is an immersion $\Sigma^i \to W$ for all $i = 1, \ldots, k$. Then by induction we make the map f_0 isometric on the stratum Σ^j, $j = 1, \ldots, k$, by applying to $U = V \setminus \bigcup_{i=0}^{j-1} \Sigma^i$ the following

Sublemma [compare (A) in 2.4.10]. *Every strictly short map $U \to W$ homotopic to $f_0|U: U \to W$ admits a fine C^0-approximation by continuous maps $U \to W$ which are isometric on $\Sigma^j \subset U$ and strictly short on $U \setminus \Sigma^j$.*

Proof. First let $\dim \Sigma^j < n$. Then the given map $U \to W$ admits a C^0-approximation by another strictly short map, say by $\tilde{f}_0: U \to W$ which is a C^1-*immersion* on Σ^j. This follows from the relative h-principle (see 2.4.5). Then the above proposition yields an isometric map, say $\tilde{f}: (\Sigma^j, g) \to (\Sigma^j, \tilde{f}_0^*(\Phi))$. This map $\tilde{f}$ is obtained as a limit of strictly short maps $\Sigma^j \to \Sigma^j$ which can be chosen arbitrarily close to the identity, and then the composition $\tilde{f}_0 \circ \tilde{f}: \Sigma^j \to W$ admits a strictly short extension to $U \setminus \Sigma^j$.

Now, let $\dim \Sigma^j = n$, which amounts to $\Sigma^j = U$ for our stratification. Then there is a strictly short *normally folded* map $\tilde{f}_0: U \to W$ which approximates the given map $U \to W$. To obtain $\tilde{f}_0$ we take an arbitrary triangulation of U and we first construct a strictly short map $\tilde{f}_0'$: $U \to W$ which is a C^1-immersion near the $(n-1)$-skeleton. Here again we use 2.4.5 and the obvious induction by skeletons. Next, we perturb $\tilde{f}_0'$ inside every open n-dimensional simplex Δ as follows. Let S_t^{n-1}, $t \in (0,1)$, be smooth concentric spheres in Δ which approach the boundary $\partial\Delta$ for $t \to 1$ and which shrink to the barycenter $s_0 \in \Delta$ as $t \to 0$. We require the perturbation $\tilde{f}_0''$: $U \to W$ to be a strictly short C^1-map which is an immersion on the sphere $S_t^{n-1} \subset \Delta$ for all Δ and $t \in (0,1)$, and which is also an immersion near the center $s_0 \in \Delta$ for all Δ. Recall that $\tilde{f}_0''$ equals $\tilde{f}_0'$ near the $(n-1)$-skeleton, and so $\tilde{f}_0''$ is an immersion outside the cylinders $S^{n-1} \times [\varepsilon, 1-\varepsilon] \subset \Delta$, for all $\Delta \subset U$ and for some $\varepsilon > 0$. Finally, we apply the stretching lemma to the map $f_0''|S^{n-1} \times [\varepsilon, 1-\varepsilon]$, for all cylinders $S^{n-1} \times [\varepsilon, 1-\varepsilon]$, thus obtaining a strictly short normally folded map $\tilde{f}_0$: $U \to W$. As the induced metric $\tilde{f}_0^*(\Phi)$ is continuous, the above proposition yields an isometric map $\tilde{f}: (U, g) \to (U, \tilde{f}_0^*(\Phi))$ and then the composition $\tilde{f}_0 \circ \tilde{f}: U \to W$ also is isometric. Q.E.D.

Proof of the Stretching Lemma. We may assume with an obvious approximation that the map f_0: $V \to W$ is C^∞-smooth and that the image $f_0(V) \subset W$ lies in the interior of W. We also assume the metric Φ on W to be C^∞-smooth and the function $a(v', t)$ to be C^∞-smooth and strictly positive for $0 < t < 1$. This allows us to use normal geodesic coordinates in (V, g) and in (W, Φ) near a given submanifold $V_0' = V' \times t_0 \subset V$, $0 < t_0 < 1$. Indeed, each point $v \in V$ which is close to V_0' is uniquely expressed by a pair (v', t'), where $v' \in V_0'$ is the g-normal geodesic projection of v to V', where $t' \in \mathbb{R}$ has $|t'| = \mathrm{dist}_g(v, V_0')$ and where sign t' is determined with a chosen orientation in the normal bundle of the submanifold $V_0' \subset V$. Similarly, we use Φ-normal coordinates in W for the immersed manifold f_0: $V_0' \to W$ and express the points $w = f(v) \in W$, for maps f which are close to $f_0|V_0'$, by pairs, $w = (w', s)$, $w' \in V_0'$, $s \in \mathbb{R}$.

Next, we choose a small $\varepsilon > 0$, such that $[t_0, t_0 + \varepsilon] \subset (0,1)$, and we stretch the map f_0 on the submanifold $V_\varepsilon = V' \times [t_0, t_0 + \varepsilon]$ as follows. Denote by $\varepsilon' = \varepsilon'(v)$ the g-distance, $\mathrm{dist}_g(v', v'')$, where $v'' \in V' \times (t_0 + \varepsilon) \subset V_\varepsilon$ is a (unique!) point whose normal projection to V_0' equals v' which also is the normal projection of v to V_0', and denote by $\delta' = \delta'(v)$ the s-coordinate of $f_0(v'') \in W$. Now we consider the (stretched) map f_ε: $V_\varepsilon \to W$, whose Φ-normal projection to the manifold V_0' (immersed by f_0 to W) equals that of f_0 and whose s-coordinate $s(v) = s(v', t')$ is uniquely determined by the conditions $s(v', 0) = 0$ and

$$\frac{ds}{dt'}(v', t') = \begin{cases} -1 & \text{for } t' \in [\alpha, \varepsilon' - \alpha],\ \alpha = \frac{1}{4}(\varepsilon' + \delta'); \\ +1 & \text{for } t' \in [0, \alpha) \cup (\varepsilon' - \alpha, \varepsilon']. \end{cases}$$

The map f_ε equals f_0 on the boundary $\partial V_\varepsilon = (V' \times t_0) \cup (V' \times (t_0 + \varepsilon))$ and it folds along two hypersurfaces which are C^1-close to $V_0' = V' \times t_0 \subset V$. This folding is nearly normal for small $\varepsilon \to 0$, and a small perturbation of f_ε near the two folds satisfies on V_ε the requirements of the stretching lemma.

We obtain the stretching on all of V by first subdividing the interval $[\varepsilon_0, 1 - \varepsilon_0] \subset [0,1]$ into ε-subintervals, then by applying the above stretching on each ε-subinterval $[t_0, t_0 + \varepsilon]$ of this subdivision and finally by letting $\varepsilon \to 0$ and $\varepsilon_0 \to 0$. Q.E.D.

Exercises and Generalizations. (a) Prove that Lipschitz' maps are almost everywhere differentiable and then show that no map $V^n \to W^{n-1}$ is isometric.
(a') Let V be an n-dimensional manifold with a C^2-smooth non flat Riemannian metric and let $f\colon V \to \mathbb{R}^n$ be an arbitrary isometric map. Find a point $w \in \mathbb{R}^n$ whose pullback $f^{-1}(w) \subset V$ contains a non empty perfect subset and thus show that f is never T_1-isometric [see (A') in 2.4.10].

(b) Let V and W be pseudo-Riemannian manifolds of the same type (n_+, n_-) for $n_+ + n_- = n = \dim V = \dim W$, such that $\min(n_+, n_-) \geq 1$. Show that every continuous map $V \to W$ admits a fine C^0-approximation by isometric maps, which by definition are Lipschitz'maps $f\colon V \to W$ preserving the pseudo-Riemannian length of all rectifiable curves in V.

Hint. Define and use maps with normal folds and cusps (compare 1.3.1).

P-Convexity and $C^{r-1,1}$-Solutions of Partial Differential Relations. A relation $\mathscr{R} \subset X^{(r)}$ is called *P-convex* (P for principal) if the intersection of $\mathscr{R}$ with every principal subspace $R \approx \mathbb{R}^q$ in $X^{(r)}$ is a convex subset in R. Denote by $\operatorname{Conv}_P \mathscr{R} \subset X^{(r)}$, now for an arbitrary $\mathscr{R} \subset X^{(r)}$, the intersection of all P-convex subsets in $X^{(r)}$ which contain $\mathscr{R}$.

Let a relation $\mathscr{R} \subset X^{(r)}$ admit an (in)approximation by some open relations $\mathscr{R}_i \subset X^{(r)}$, $i = 0, 1, \ldots$, such that:

(i) $\mathscr{R}_i \subset \operatorname{Conv}_P \mathscr{R}_{i+1}$ for all $i = 0, 1, \ldots$.
(ii) The relations $\mathscr{R}_i$ are uniformly bounded in the fibers of the fibration $p^r_{r-1}\colon X^{(r)} \to X^{(r-1)}$. That is the union $\bigcup_{i=0}^{\infty} \mathscr{R}_i \cap (p^r_{r-1})^{-1}(Y)$ is a relatively compact subset in $X^{(r)}$ for all compact subsets $Y \subset X^{(r-1)}$.
(iii) If a sequence $a_i \in \mathscr{R}_i$ converges as $i \to \infty$ to some point $x \in X^{(r)}$, then $x \in \mathscr{R}$.

Theorem. *Let $f_0\colon V \to X$ be a C^r-section for which $J^r_{f_0}(V) \subset \mathscr{R}_0$. Then f_0 admits a fine C^{r-1}-approximation by C^{r-1}-maps $f\colon V \to X$ which almost everywhere satisfy the relation $\mathscr{R}$ in the following sense. The jet $J^{r-1}_f\colon V \to X^{(r-1)}$ is an almost everywhere differentiable Lipschitz map and $J^r_f(v) \in \mathscr{R}$ for almost all $v \in V$.*

Idea of the Proof. The solutions f are obtained as limits of piecewise C^r-smooth solutions of the relations $\mathscr{R}_i$. The basic ingredients of the proof are similar to (in fact, much simplier than) those in 2.4.6. The key step is a piecewise smooth version of the $C^\perp$-approximation lemma (see 2.4.1). The actual proof is left to the reader.

Example. If $\mathscr{R}$ is the isometric immersion relation then one takes the strict shortness relation for $\mathscr{R}_0$ and let $\mathscr{R}_i$, $i = 1, \ldots$, correspond to those strictly short maps $V \to W$ for which the induced Riemannian metric on V is ε_i-close, $\varepsilon_i \to 0$ for $i \to \infty$ to the given metric on V. Then the theorem delivers a Lipschitz map $f\colon V \to W$, for $\dim W \geq \dim V$, whose differential $D_f\colon T_v(V) \to T_w(W)$, $w = f(v)$ is isometric for almost all $v \in V$. Such a map f, however, may fail to be isometric. In fact, f may collapse an arbitrary submanifold of positive codimension in V to a single point in W. Probably, there is a refinement of the above general theorem which produces "stronger" solutions, like actual isometric maps for $\mathscr{R} = \mathscr{R}_{\text{isom}}$.

2.4.12 The Regularity Problem and Related Questions in the Convex Integration

If $\mathscr{R} \subset X^{(r)}$ is a generic C^s-smooth submanifold whose intersections with the principal subspaces $R \subset X^{(r)}$, $R \approx \mathbb{R}^q$, have positive dimension (this corresponds to underdetermined systems of P.D.E. with C^s-smooth coefficients), then one expects this $\mathscr{R}$ to admit many C^s-smooth holonomic sections $V \to \mathscr{R}$ that correspond to C^{r+1}-solution $V \to X$ of $\mathscr{R}$. The convex integration only delivers C^r-solutions and no regularity assumption on $\mathscr{R}$ helps to smooth these solutions. One might try to apply the convex integration to the lift $\mathscr{R}^s \subset X^{(r+s)}$ that corresponds to the differentiation s times the corresponding P.D.E. system (see 1.1.1). Unfortunately, the differentiated system is linear in the top derivatives (of order $r + s$) and so the convex hull extensions $\text{Conv}_N \mathscr{R}^s$ do not enlarge $\mathscr{R}^s$ for $s \geq 1$. Thus the convex integration fails to produce holonomic sections $V \to \mathscr{R}^s$ that would correspond to C^{r+1}-solutions $V \to X$ of $\mathscr{R}$. Observe that quasilinear relations like $\mathscr{R}^s$ lie on the very boundary of the convex integration domain as small perturbation make them ample. One hopes that "nonintegrability" of generic quasilinear relations may substitute "nonlinearity" in the convex integration scheme [compare 2.3.8 and (B) in 2.4.5], but one does not know how to make this idea work.

A more realistic goal is that of obtaining solutions $V \to X$ of $\mathscr{R} \subset X^{(r)}$ in Hölder classes $C^{r+\alpha}$ for some $0 < \alpha < 1$. Some results in this direction are known for the isometric immersion problem. Borisov (1965) announced the existence of isometric $C^{1+\alpha}$-immersions $V^n \to \mathbb{R}^{n+1}$ for all $\alpha < (n^2 + n + 1)^{-1}$, and he also indicated some obstruction to isometric $C^{1+\alpha}$-immersions for $\alpha > \frac{2}{3}$. Furthermore, Källen (1978) proved that every n-dimensional manifold V with a C^β-smooth metric, $0 < \beta < 2$, admits an isometric $C^{1+\alpha}$-embedding $V \to \mathbb{R}^q$, $q = 3(n + 1)(n^2 + n + 2) + 2n$, for every $\alpha < \beta/2$.

Convex Integration with Elliptic Operators. The success of the convex integration depends on our complete understanding of the (elliptic) operator d^r/dt^r on maps $\mathbb{R} \to \mathbb{R}^q$ (see the $C^\perp$-approximation lemma in 2.4.1). The question is whether there exists a convex integration type theory which relies upon another elliptic operator, such as $\bar{\partial}$ or the Dirac operator.

Integrodifferential Inequalities. The convex integration works in those cases when an inequality like $\| f(b) - f(a) \| \leq (b - a) \int_a^b \| f'(t) \| \, dt$, for $f\colon [a, b] \to \mathbb{R}^q$, fails. For example, if a map $V \to \mathbb{R}^q$ is short, then it can be approximated by isometric maps. One asks himself if a similar role may be plaid by more interesting inequalities, such as *the isoperimetric inequality*,

$$(*) \qquad \left(\int_{\mathbb{R}^n} |f|^{n/(n-1)} \right)^{(n-1)/n} \leq \text{const}_n \int_{\mathbb{R}^n} \| \text{grad} \, f \|$$

for C^1-functions $f\colon \mathbb{R}^n \to \mathbb{R}$ with compact support. Unfortunately, one does not know *all* inequalities like (*) that are linear inequalities between the integrals $\int_{\mathbb{R}^n} \| f \|^i \| \text{grad} \, f \|^j$ and their powers. The problem amounts to understanding the class of measures on $\mathbb{R}^2$ which are the push-forwards of the Lebesgue's measure on $\mathbb{R}^n$ under the maps $x \mapsto (f(x), \| \text{grad} \, f \| (x))$ for all C^1-functions $f\colon \mathbb{R}^n \to \mathbb{R}$ with compact supports. Observe that measures on $\mathbb{R}^q$ which are push-forwards of smooth measures under smooth maps $F\colon V \to \mathbb{R}^q$ are rather special even for those F which are free of any additional (holonomy) condition. For example, let V be a closed connected manifold with a positive C^∞-measure $d\mu$ on V, such that $\int_V d\mu = 1$ (such a measure is unique up to a C^∞-diffeomorphism of V). Functions $f_i\colon V \to \mathbb{R}$, $i = 1, \ldots, q$, are called *independent* if they are independent as random variables on the probability space $(V, d\mu)$. That is the push-forward measure $F_*(d\mu)$ on $\mathbb{R}^q$ for $F = (f_1, \ldots, f_q)$ equals the product of the measures $(f_i)_*(d\mu)$ on $\mathbb{R}$, $i = 1, \ldots, q$. One has with this definition the following unsolved

Problem of V.Eidlin. Find a topological characterization of those manifolds V which admits nonconstant independent *real analytic* functions $f_i\colon V \to \mathbb{R}$, $i = 1, \ldots, q$.

Exercises. (a) Show that V admits independent nonconstant C^∞-functions $f_1, \ldots, f_q$ if and only if $q \leq \dim V$.

(b) Show that Cartesian products $V = V_1 \times \cdots \times V_q$, $\dim V_i > 0$, $i = 1, \ldots, q$, admit q nonconstant independent C^{an}-functions $V \to \mathbb{R}$.

(c) Let $F = (f_1, \ldots, f_q)\colon V \to \mathbb{R}^q$ be a C^∞-map whose components f_i, $i = 1, \ldots, q$, are independent. Show that a point $x = (x_1, \ldots, x_q) \in \mathbb{R}^q$ is a noncritical value of F if and only if $x_i \in \mathbb{R}$ is a noncritical value of f_i for $i = 1, \ldots, q$.

(d) Denote by $M(f) \subset V$ for $f\colon V \to \mathbb{R}$ the union of those points $v \in V$ at which the function f assumes local maximum or local minimum. Let $f_1, \ldots, f_q$ be independent C^∞-functions, such that the function f_i, $i = 2, \ldots, q$, is nonconstant in an arbitrarily small neighborhood $\mathcal{O}p(v) \subset V$ of a given point $v \in M(f_1) \subset V$. Show that the intersection $M(f_1) \cap \mathcal{O}p(v)$ goes under the map $(f_2, \ldots, f_q)\colon V \to \mathbb{R}^{q-1}$ onto a subset in $\mathbb{R}^{q-1}$ of dimension $q - 1$ (i.e. the interior is nonempty).

Hint. Apply the above (c).

(e) Let $f\colon V \to \mathbb{R}$ be a non-constant real analytic function such that every connected component of the subset $M(f) \subset V$ has dimension $n - 1$ for $n = \dim V$. Show that the fundamental group $\pi_1(v)$ admits a surjective homomorphism either on $\mathbb{Z}$ or on $\mathbb{Z}_2 * \mathbb{Z}_2$.

Hint. The map f admits a (unique) factorization,

$$V \xrightarrow{f_1} X \xrightarrow{f_2} \mathbb{R}, \quad f = f_2 \circ f_1$$

where X is a one-dimensional polyhedron and such that the pullback $f_1^{-1}(x) \subset V$ is connected for all $x \in X$. Study the map f_1 and use the fact that every analytic subset $M \subset V$ (of dimension $m = n - 1$) is a $\mathbb{Z}_2$-cycle (of dimension m).

(e′) Prove the converse to (e): If $\pi_1(V)$ surjects onto $\mathbb{Z}$ or $\mathbb{Z}_2 * \mathbb{Z}_2$ then there is a C^{an}-function $f\colon V \to \mathbb{R}$ for which the subset $M(f) \subset V$ is a C^{an}-hypersurface in V.

Hint. Use harmonic maps $\tilde{V} \to S^1$ for a double covering $\tilde{V} \to V$.

(f) Let V admit n nonconstant independent C^∞-functions for $n = \dim V$. Prove with (e) and (d) the existence of a surjective homomorphism of $\pi_1(V)$ either onto $\mathbb{Z}$ or onto $\mathbb{Z}_2 * \mathbb{Z}_2$.

Part 3. Isometric C^∞-Immersions

The theory of topological sheaves (see 2.2) and the implicit function theorem (see 2.3) provide a general framework for the isometric immersion problem which consists of the construction and classification of isometric C^∞-maps $f:(V,g)\to(W,h)$, for given forms g on V and h on W (compare 2.4.9). However, the direct application of 2.2 and 2.3 does not lead to geometrically significant results unless specific geometrical features of the forms in question are taken into account.

3.1 Isometric Immersions of Riemannian Manifolds

Let $V=(V,g)$ be a C^∞-smooth Riemannian manifold of dimension n. We want to give an upper bound to the minimal dimension q for which there exists an isometric C^∞-immersion $V\to\mathbb{R}^q$. A similar question is studied for an arbitrary Riemannian manifold $W=(W,h)$, $\dim W=q$, in place of $\mathbb{R}^q$.

3.1.1 Nash's Twist and Approximate Immersions; Isometric Imbeddings into $\mathbb{R}^q$

Let $S_\varepsilon^{q-1}\subset\mathbb{R}^q$ denote the ε-sphere in $\mathbb{R}^q$, let $f_0\colon V\to S_\varepsilon^{q-1}$ be a C^1-map and let $g_0=f_0^*(h)$ be the induced quadratic differential form on V for $h=\sum_{i=1}^q dx_i^2$ on $\mathbb{R}^q$. If $\varphi\colon V\to\mathbb{R}$ is any C^1-function, then the form $g_1=f_1^*(h)$ for $f_1=\varphi f_0\colon V\to\mathbb{R}^q$, satisfies the following

Fundamental Formula.

$$(*)\qquad g_1=\varphi^2 g_0+\varepsilon^2(d\varphi)^2.$$

Proof. The equation $\langle f_0,f_0\rangle=\varepsilon^2$ implies $\langle f_0,\partial f_0\rangle=0$ for all vector fields ∂ on V. Therefore,

$$\begin{aligned} g_1(\partial,\partial)&=\langle\partial f_1,\partial f_1\rangle=\langle\partial(\varphi f_0),\partial(\varphi f_0)\rangle\\ &=\varphi^2\langle\partial f_0,\partial f_0\rangle+2\varphi\partial\varphi\langle f_0,\partial f_0\rangle+(\partial\varphi)^2\langle f_0,f_0\rangle\\ &=\varphi^2 g_0(\partial,\partial)+\varepsilon^2(\partial\varphi)^2. \end{aligned}$$

Let us apply $(*)$ to the map f_0 given by $(y_\varepsilon,z_\varepsilon)\colon V\to S_\varepsilon^1\subset\mathbb{R}^2$, for $y_\varepsilon=\varepsilon\sin(\varepsilon^{-1}x)$,

$z_\varepsilon = \varepsilon\cos(\varepsilon^{-1}x)$, where $x\colon V \to \mathbb{R}$ is an arbitrary C^1-function. Then we obtain, with $g_0 = (dy_\varepsilon)^2 + (dz_\varepsilon)^2 = (dx)^2$,

Nash's Formula.

$$(**) \qquad \varphi^2(dx)^2 = (dy_\varepsilon)^2 + (dz_\varepsilon)^2 - \varepsilon^2(d\varphi)^2.$$

Exercise. Let the manifold (V,g) admit an isometric immersion $f\colon (V,g) \to S^{q-1} \subset \mathbb{R}^q$. Construct with $(*)$ an isometric immersion

$$(V, \varphi^2 g) \to \left(\mathbb{R}^{q,1}, -dx_{q+1}^2 + \sum_{i=1}^{q} dx_i^2\right).$$

for all $\varphi\colon V \to \mathbb{R}$.

Prove the existence of an isometric C^∞-immersion $(S^2, g) \to \mathbb{R}^{3,1}$ for all C^∞-metrics g on S^2 by using a conformal diffeomorphism $(S^2, g) \to S^2 \subset \mathbb{R}^3$.

A Decomposition $g = \sum \varphi_i^2\, dx_i^2$. Let g be a continuous Riemannian metric on V. Clearly there exist C^∞-functions $x_i\colon V \to \mathbb{R}$, $i = 1, \ldots, l$, for some l, such that the form $g \mid T_v(V)$ is contained in the interior of the convex hull of the forms $(dx_i)^2 \mid T_v(V)$ for each tangent space $T_v(V)$, $v \in V$.

Exercise. Prove the existence of such x_i, $i = 1, \ldots, l$, for $l = [n(n+1)/2] + n$.

Next, if the form g is C^α-smooth for some $\alpha \leq \infty$, then there are (use a partition of unity) C^α-smooth functions φ_i on V, such that

$$(+) \qquad g = \sum_{i=1}^{l} \varphi_i^2\, dx_i^2.$$

(A) **Corollary: Nash's Approximation Theorem.** *An arbitrary Riemannian C^α-metric g on V has a fine C^α-approximation by some C^α-metrics g' on V which admit isometric C^α-immersions $f'\colon (V, g') \to \mathbb{R}^{2l}$ for some $l = l(n) < \infty$, $n = \dim V$.*

Proof. According to $(**)$ each form $\varphi_i^2(dx_i)^2$ admits an ε-approximation by the form $(dy_{i,\varepsilon})^2 + (dz_{i,\varepsilon})^2$. If V is compact, then the sum $g' = g'_\varepsilon = \sum_{i=1}^{l} [(dy_{i,\varepsilon})^2 + (dz_{i,\varepsilon})^2]$ is ε-close to g'. As this g' is induced by the map $(y_{i,\varepsilon}, z_{i,\varepsilon})\colon V \to \mathbb{R}^{2l}$ the proof is accomplished for compact manifolds V. If V is non compact, then in addition we require every function x_i, $i = 1, \ldots, l$, to have its support in a disjoint union of compact subsets, say in $\bigcup_j U_{i_j} \subset V$. With such a function x_i we use in $(**)$ a function $\varepsilon = \varepsilon(v)$, $v \in V$ (instead of a constant $\varepsilon > 0$), such that $\varepsilon(v)$ is constant on every subset U_{ij}, $j = 1, 2, \ldots$, and such that the function $\varepsilon(v)$ is small in the *fine* C^∞-topology. Then the error $g - g'_\varepsilon$ is small in the *fine* C^α-topology. Q.E.D.

(A') **Remark.** If α is an integer, then g admits a C^α-approximation by some C^∞-metrics, and so the above g' and f' can be taken C^∞-smooth. If α is not an integer, then we have a C^β-approximation of g by C^∞-smooth metrics g', induced by imbeddings into $\mathbb{R}^{2l}$ for all $\beta < \alpha$, such that (V, g') admits a C^∞-immersion into $\mathbb{R}^l$.

Exercise. Let (V, g) be a complete simply connected surface of non-positive curvature $K(g) \leq 0$. Construct C^α-maps $f_\varepsilon: V \to \mathbb{R}^3$, $\varepsilon \to 0$, such that the induced metrics $f_\varepsilon^*(h)$, for $h = \sum_{i=1}^3 dx_i^2$, C^α-converge to g on compact subsets in V.

Hint. Use coordinates (u_1, u_2) on V for which $g = du_1^2 + \varphi(u_1, u_2)\, du_2^2$.

Remark. If $K < 0$, then every bounded domain $U \subset V$ admits an actual (not only approximate) isometric immersion $(U, g) \to \mathbb{R}^3$. But if $K(g) \leq -\chi^2 < 0$, then there is no global isometric C^2-immersion $(V, g) \to \mathbb{R}^3$ [see the survey by Poznyak (1973)]. This is the famous theorem of Hilbert-Efimov.

Isometric Imbeddings $V \to \mathbb{R}^q$. Start with a free C^∞-imbedding $f_0: V \to \mathbb{R}^{q_0}$ which is strictly short relative to a given C^α-metric g on V. Then the form $g - f_0^*(h_0)$, $h_0 = \sum_{i=1}^{q_0} dx_i^2$, is positive and so there exists a C^α-immersion $f': V \to \mathbb{R}^{2l}$ which is isometric relative to some C^α-metric g', such that the "error" $\delta = g - f_0^*(h_0) - g'$ is arbitrarily small in the fine C^α-topology. If $\alpha > 2$, then there exists a small perturbation of f_0, say $f_1: V \to \mathbb{R}^{q_0}$ of class C^α, such that $f_1^*(h_0) = f_0^*(h_0) + \delta$ (see 2.3.2). Thus, we obtain an *isometric* C^α-imbedding $(f_1, f'): (V, g) \to \mathbb{R}^{q_0 + 2l}$ as $g = f_0^*(h_0) + \delta + g'$. [Compare Moore-Schlafly (1980).]

We shall reduce in 3.1.7 the number $2l$ to $q_1 = (n + 2)(n + 3)/2$ and then we shall obtain the following

Imbedding Theorem. *Every Riemannian C^α-manifold, $2 < \alpha \leq \infty$, admits a free isometric C^α-imbedding $f: V \to \mathbb{R}^q$ for $q = n^2 + 10n + 3$.*

Remarks. If $\alpha > 4$, then the dimension of the ambient space can be reduced to $q = (n + 2)(n + 3)/2$ (see 3.1.7), but no such improvement is known for smaller α.

Isometric C^α-imbeddings $V \to \mathbb{R}^q$ for $\alpha = 3, 4, \ldots, \infty$ and $q = (n + 1)[\frac{3}{2}n(n + 1) + 4n]$ were obtained by Nash (1956). He starts with imbeddings of *compact* manifolds into $\mathbb{R}^q$ for $q = \frac{3}{2}n(n + 1) + 4n$, and then he reduces the non-compact case to the compact one with the following

(B) *Compact Decomposition.* Let V be a manifold without boundary, let $\bar{V}_i$, $i = 1, 2, \ldots$, be closed manifolds, $\dim \bar{V}_i = n = \dim V$, and let $\bar{f}_i: V \to \bar{V}_i$ be C^∞-maps with the following two properties:

(i) The supports of the differentials $D\bar{f}_i: T(V) \to T(\bar{V}_i)$ form a *locally finite* family: for every compact subset $U \subset V$ all maps $\bar{f}_i$, $i \geq i_0 = i_0(U)$ have $D\bar{f}_i | U = 0$.
(ii) There exists an open cover $\bigcup_{i=1}^\infty U_i = V$, such that the subset $U_i \subset V$ is mapped *diffeomorphically* by $\bar{f}_i$ onto an open subset $\bar{U} \subset \bar{V}_i$ for all $i = 1, \ldots$.

Then, as the reader will agree, every C^α-metric g on V admits a decomposition

$$g = \sum_i \bar{f}_i^*(\bar{g}_i),$$

for some Riemannian C^α-metrics $\bar{g}_i$ on $\bar{V}_i$.

(B′) **Corollary.** *Let $\bigcup_i U_i = V$ be an arbitrary locally finite cover of V by open subsets $U_i \subset V$. Then every C^α-metric g on V decomposes as*

$$g = \sum_j (dx_j)^2,$$

where x_j: $V \to \mathbb{R}$ are C^α-functions on V such that the support of x_j is contained in some subset U_i for $i = i(j)$ and $i(j) \to \infty$ for $j \to \infty$ (i.e. almost all functions x_j vanish on U_i for every fixed i)

Proof. We may assume (refine the cover if necessary), that the subsets U_i satisfy (ii) for some maps $\bar{f}_i$: $V \to \bar{V}_i \approx S^n$. Then $g = \sum_i \bar{f}_i^*(\bar{g}_i)$ and we obtain the required decomposition with isometric C^α-immersions $(S^n, \bar{g}_i) \to \mathbb{R}^q$ for all $i = 1, \ldots$.

Exercise. Assume, given n, q and α, there exists an isometric C^α-immersion $(S^n, g) \to \mathbb{R}^q$ for all C^α-metrics g on S^n. Show that an arbitrary n-dimensional Riemannian C^α-manifold V admits an isometric C^α-immersion into $\mathbb{R}^{(n+1)q}$. Prove the existence of such an immersion $V \to \mathbb{R}^{2q}$, provided V is diffeomorphic to S^n minus finitely many points, for example $V \approx \mathbb{R}^n$.

3.1.2 Isometric Immersions $V^n \to W^q$ for $q \geq (n+2)(n+5)/2$

The following proposition is proven on the basis of Lemmas (D) up to (F) at the end of this Sect. 3.1.2.

(A) **Extension Lemma.** *Let (V, g) and (W, h) be Riemannian C^∞-manifolds and let $f: (V, g) \to (W, h)$ be a free isometric C^∞-immersion. Consider the cylinder $(V \times \mathbb{R}^1, g \oplus dt^2)$ and let $U \subset V = V \times 0 \subset V \times \mathbb{R}^1$ be a compact subset in V. If the subset U is contractible in V and if $q \geq (n+1)(n+4)/2$, then the map $f|U: U \to W$ extends to a free isometric C^∞-map $\tilde{U} \to W$ for some neighborhood $\tilde{U} \subset V \times \mathbb{R}^1$ of U.*

(A′) **Corollary.** *Consider the product $(V \times \mathbb{R}^2, g \oplus dt_1^2 \oplus dt_2^2)$ and let $U \subset V = V \times 0 \subset V \times \mathbb{R}^2$ be a subset as in* (A). *If $q \geq (n+2)(n+5)/2$ then the map $f|U$ extends to a free isometric C^∞-map $\tilde{\tilde{U}} \to W$ for some neighborhood $\tilde{\tilde{U}} \subset V \times \mathbb{R}^2$ of U.*

Proof. First extend $f|U$ to $\tilde{U}$ and then apply (A) to $\tilde{U}$ in place of V.

One may equivalently express (A′) by saying that $f|U$ extends to a free isometric C^∞-map $U' \times D_\varepsilon^2 \to W$ for some neighborhood $U' \subset V$ of U and for some ε-ball $D_\varepsilon^2 \subset \mathbb{R}^2$, $\varepsilon > 0$. This leads to the following

(B) **Cylinder Lemma.** *If $q \geq (n+2)(n+5)/2$, then the map $f|U$ extends to a free isometric C^∞-maps $U' \times \mathbb{R}^1 \to W$.*

Proof. Use the above with any isometric C^∞-map ψ: $\mathbb{R}^1 \to D_\varepsilon^2$, $\psi(0) = 0$.

(B′) **Adding dx^2.** *Let $x = V \to \mathbb{R}^1$ be a C^∞-function with support in U. If $q > (n+2)(n+5)/2$, then there exists a free isometric C^∞-immersion $f_1: (V, g + (dx)^2) \to (W, h)$ which equals the immersion $f: V \to W$ outside the subset U.*

Proof. The graph of the function x is an *isometric* immersion

$$\Gamma_x: (V, g + (dx)^2) \to (V \times \mathbb{R}^1, g \oplus dt^2).$$

Hence, the composition of the map $\Gamma_x | U'$ with the above map $U' \times \mathbb{R}^1 \to W$ is a free isometric map $(U', g + (dx)^2) \to W$ which agrees with f outside U. Q.E.D.

(B″) **Remark.** One can make the map f_1 arbitrarily C^0-close to f by taking a sufficiently small ball $D_\varepsilon^2 \subset \mathbb{R}^2$.

(C) **Theorem.** *Let (V, g) and (W, h) be Riemannian C^∞-manifolds and let $f_0: V \to W$ be a strictly short map. If $q \geq (n+2)(n+5)/2$, then f_0 admits a fine C^0-approximation by free isometric C^∞-maps $f: (V, g) \to (W, h)$.*

Proof. Assume the map f_0 to be C^∞-smooth and free (see 1.1.4) and decompose the metric $g - f_0^*(h)$ [see (B′) in 3.1.1]

$$g - f_0^*(h) = \sum_j dx_j^2, \qquad j = 1, 2, \ldots,$$

for some locally finite sequence of C^∞-functions x_j on V. Then the above lemma (B′) provides free isometric C^∞-maps

$$f_k: (V, g_k) \to (W, h) \qquad \text{for } g_k = f_0^*(h) + \sum_{j=1}^{k} dx_j^2, \; k = 1, 2, \ldots,$$

which converge for $k \to \infty$ (in fact, the maps f_k stabilize on compact subsets in V) to the required isometric map $f: (V, g) \to (W, h)$.

(D) *The Cauchy Problem for Cylinders.* We study C^∞-maps $f: V \times \mathbb{R} \to (W, h)$ and use the local coordinates $u_1, \ldots, u_n, t$ in $V \times \mathbb{R}$. We denote by ∇_i, $i = 1, \ldots, n$, ∇_t, $\nabla_{ij} = \nabla_i \nabla_j$ etc., the covariant derivatives in (W, h) relative to the vector fields $\partial_i = \partial/\partial u_i$ and $\partial_t = \partial/\partial t$. Thus, the derivatives $\nabla_i f = D_f(\partial_i)$, $\nabla_{ij} f$ etc., are vectors fields in W along the mapped manifold $f: V \times \mathbb{R} \to W$, which are called, for brevity, *fields along* $V \times \mathbb{R}$. Such a field X is called *binormal* to $V \times t_0 \subset V \times \mathbb{R}$ for a given $t_0 \in \mathbb{R}$ if it is normal to the osculating spaces of the map $f | V \times t_0$ at all points $(v, t_0) \in V \times t_0$. Recall that the osculating space, called $T_{v,t_0}^2(f | V \times t_0) \subset T_w(W)$, $w = f(v, t_0)$, is spanned by the covariant derivatives $\nabla_i f$ and $\nabla_{ij} f$, $i, j = 1, \ldots, n$, at w, thus, the binormal fields X satisfy

$$\langle X, \nabla_i f \rangle = 0, \qquad \langle X, \nabla_{ij} f \rangle = 0,$$

for $\langle X, Y \rangle \overset{\text{def}}{=} h(X, Y)$.

(D′) **Lemma.** *A C^∞-map $f: V \times \mathbb{R} \to (W, h)$ induces the cylinder metric $g \oplus dt^2$ on $V \times \mathbb{R}$ for the metric $g = f^*(h) | V \times 0$, if and only if the field $\nabla_t f$ is binormal to $V \times t$ for all $t \in \mathbb{R}$ and $\langle \nabla_t f, \nabla_t f \rangle \equiv 1$.*

Proof. The isometric immersion equations for the metric $g \oplus dt^2$ are

$$\langle \nabla_i f, \nabla_j f \rangle = g_{ij} \overset{\text{def}}{=} g(\partial_i, \partial_j), \qquad i, j = 1, \ldots, n, \tag{1}$$

$$\langle \nabla_i f, \nabla_t f \rangle = 0, \tag{2}$$

$$\langle \nabla_t f, \nabla_t f \rangle = 1. \tag{3}$$

Since g_{ij} does not depend on t for the cylinder metric,

$$0 = \partial_t g_{ij} = \langle \nabla_{it} f, \nabla_j f \rangle + \langle \nabla_i f, \nabla_{jt} f \rangle. \tag{1'}$$

Then we differentiate (2) and alternate $i \leftrightarrow j$,

$$\begin{aligned} \langle \nabla_{ij} f, \nabla_t f \rangle + \langle \nabla_i f, \nabla_{ti} f \rangle &= 0 \\ \langle \nabla_{ji} f, \nabla_t f \rangle + \langle \nabla_j f, \nabla_{tj} f \rangle &= 0. \end{aligned} \tag{2'}$$

As $\nabla_{ij} = \nabla_{ji}$ and $\nabla_{it} = \nabla_{ti}$, we conclude with (1′) and (2′) to

$$\langle \nabla_t f, \nabla_{ij} f \rangle = 0. \tag{4}$$

The Eqs. (2), (3) and (4) show $\nabla_t f$ to be a binormal field of norm 1. This is the "only if" claim of the lemma and we obtain "if" by reversing the above calculation.

Janet's Equations. Differentiate (4) in t and write

$$\langle \nabla_{tt} f, \nabla_{ij} f \rangle + \langle X, \nabla_t \nabla_i X_j \rangle = 0 \tag{4'}$$

for $X = \nabla_t f$, $X_i = \nabla_i f$. Differentiate (3) in u_j and then in u_i,

$$\langle \nabla_t f, \nabla_{tj} f \rangle = 0, \tag{3'}$$

$$\langle \nabla_{ti} f, \nabla_{tj} f \rangle + \langle X, \nabla_i \nabla_t X_j \rangle = 0. \tag{3''}$$

The Eqs. (4′), (3″) and the derivatives of (2) and (3) in t result in the following system of $(n + 1)(n + 2)/2$ P.D.E. of second order

$$\begin{aligned} &\langle \nabla_{tt} f, \nabla_t f \rangle = 0, \\ &\langle \nabla_{tt} f, \nabla_i f \rangle = -\langle \nabla_t f, \nabla_{it} f \rangle, \\ &\langle \nabla_{tt} f, \nabla_{ij} f \rangle = \langle \nabla_{it} f, \nabla_{jt} f \rangle + \langle \nabla_t f, R(\nabla_t f, \nabla_i f, \nabla_j f) \rangle, \end{aligned} \tag{5}$$

where R denotes the curvature tensor of (W, h).

(D″) Lemma. *Every solution $f: V \times \mathbb{R} \to W$ of (5) which satisfies on $V \times 0 \subset V \times \mathbb{R}$ the initial conditions*

$$\begin{aligned} \langle \nabla_i f, \nabla_j f \rangle &= g_{ij} \\ \langle \nabla_t f, \nabla_t f \rangle &= 1 \\ \langle \nabla_t f, \nabla_i f \rangle &= 0 \\ \langle \nabla_t f, \nabla_{ij} f \rangle &= 0 \end{aligned} \tag{6}$$

is an isometric map $f: (V \times \mathbb{R}, g \oplus dt^2) \to (W, h)$.

Proof. By reversing the above calculation we obtain with (5)

$$\nabla_t \langle \nabla_t f, \nabla_t f \rangle = 0,$$
$$\nabla_t \langle \nabla_t f, \nabla_i f \rangle = 0,$$
$$\nabla_t \langle \nabla_t f, \nabla_{ij} f \rangle = 0,$$

and so the field $\nabla_t f$ satisfies the assumptions of (D′) on $V \times \mathbb{R}$. Q.E.D.

Thus, the extension of an isometric map $f|V \times 0\colon (V \times 0, g) \to (W, h)$ is reduced to the solution of the Cauchy problem for the system (5) with the initial data (6), where the equations (6) say that the map f is isometric on $V \times 0$, and that the field $X = \nabla_t f$ along $V \times 0$ is binormal to $V \times 0$ and has norm one.

(E) *Solution of* (6). Fix an isometric C^∞-map $f|V \times 0\colon (V \times 0, g) \to (W, h)$ and decide whether there exists a unit vector field X along $V \times 0$ which is binormal to $V \times 0$. If the map $f\colon V \times 0 \to W$ is free, then we can form the *binormal bundle* $BN \to V \times 0$ whose fiber $BN_v = T_w(W) \ominus T_v^2$, $v = (v, 0) \in V \times 0$, $w = f(v, 0)$, is the orthogonal complement to the osculating space $T_v^2 = T_{v,0}^2(f|V \times 0)$. This is possible because the dimension of T_v^2 is independent of $v \in V \times 0$ for free maps $f|V \times 0$, namely $\dim T_v^2 = [n(n+1)/2] + n$, $n = \dim V$. The fields X in question are unit sections $V \times 0 \to BN$, and so such an X exists in the following two cases,

(i) $\dim BN > n$, that is $\dim W = q > [n(n+1)/2] + 2n$

(ii) BN is a trivial bundle of dimension > 0. For example, if the manifold V is contractible and $q > [n(n+1)/2] + n$.

Exercise. Find a (non-free!) C^∞-immersion $\mathbb{R} \to \mathbb{R}^3$ which admits no continuous non-vanishing binormal field.

(E′) *Solution of* (5). Let the metric h on W be real analytic and let $f_0\colon V \to W$ be a free C^{an}-map which admits a unit binormal field X. *Then* f_0 *extends to an isometric* C^{an}*-map* $f\colon (\mathcal{O}\!p\, V, q \oplus dt^2) \to (W, h)$ *for a small neighborhood* $\mathcal{O}\!p\, V \subset V \times \mathbb{R}$, $V = V \times 0$, and for $g = f_0^*(h)$.

Proof. The system (5) is linear in $\nabla_{tt} f$ and so it can be C^{an}-resolved in $\nabla_{tt} f$ for linearly independent fields $\nabla_i f$, $\nabla_{ij} f$ and $\nabla_t f$ along $V = V \times 0$. This means the existence of a C^{an}-field Y along $V \times 0$ which satisfies (5) when substituted for $\nabla_{tt} f$. Then the system (5) can be C^{an}-resolved in $\nabla_{tt} f$ on a small neighborhood $\mathcal{O}\!p\, V \subset V \times \mathbb{R}$. This means the existence of a real analytic vector function Φ on $\mathcal{O}\!p\, V$ with the entries v, t, w, ∇_i, ∇_t, ∇_{ij}, ∇_{it}, such that $\Phi(v, t, f(v, t), \nabla_i f(v, t), \ldots)$ identically satisfies (5) (for all f) when substituted for $\nabla_{tt} f$. To obtain such a Φ on $\mathcal{O}\!p\, V$ (not only within a given coordinate neighborhood) we observe that the system (5) is invariant under coordinate changes in V and so (5) defines a differential relation $\mathcal{R}$ in the pertinent jet space $X^{(2)}$ over $V \times \mathbb{R}$ which splits as a Whitney sum, $X^{(2)} = X^\partial \oplus X^\perp$ (where the summand X^∂ corresponds to the derivative ∇_{tt} compare 2.4.1, 2.4.7). The initial data (f_0, X) on $V = V \times 0 \subset V \times \mathbb{R}$ define a section into $X^\perp$, say $\bar{X}\colon V \to X^\perp$, while the field Y, which resolves (5) over V, lifts this section to $\mathscr{R} \subset X^{(2)} \to X^\perp$. Call this lift $\bar{Y}: \bar{V} \to \mathscr{R}$ for $\bar{V} = \bar{X}(V) \subset X^\perp$ and extend $\bar{Y}$ to a section $\bar{\Phi}: \mathcal{O}p(\bar{V}) \to \mathscr{R}$ for some neighborhood $\mathcal{O}p(\bar{V}) \subset X^\perp$. Such a section $\bar{\Phi}$ gives us the required Φ as well as the correct global definition of Φ. Notice that the sections $\bar{Y}$ and $\bar{\Phi}$ (as well as Y and Φ) are not unique unless $q = [n(n+1)/2] + n + 1$.

Now, the Cauchy-Kovalevskaya theorem gives us a unique C^{an}-solution f of the system $\nabla_{tt} f = \Phi(v, t, f, \nabla_i(f), \ldots)$ on $\mathcal{O}p V$, which satisfies a given initial condition. Q.E.D.

Remark. It is unclear whether a similar extension theorem holds true for C^∞-immersions, except for $V = \mathbb{R}$ and $W = \mathbb{R}^3$, where such an extension clearly exists.

(F) *Free Cylinders* $V \times \mathbb{R} \to W$. Let $f_0: V \to W$ be a free map. A binormal field $X: V \to BN$ is called *regular* if the fields $\nabla_i f_0$, $\nabla_{ij} f_0$, X, $\nabla_i X$ are independent on V. This is equivalent to the regularity of the section $X: V \to BN$ for the connection in the bundle $BN \to V$ induced from the Riemannian connection of (W, h) (see 2.2.6).

Example. If $X = \nabla_t f$ for a *free* map $f: V \times \mathbb{R} \to W$, $f | V \times 0 = f_0$, then X is a regular field.

(F′) **Lemma.** *If the manifold V is parallelizable and if $BN \to V$ is a trivial bundle of dimension $\geq n + 2$, $n =$ dim V, then there exists a unitary regular field $V \to BN$. For example, a regular field exists if V is a contractible manifold and* dim $W = q \geq (n+1)(n+4)/2$.

This is an immediate consequence of the h-principle for regular fields (see 2.2.6).

Fix a free map $f_0: V \to W$ and a regular field X along V. Then a field Y along V is called *free* (relative to f_0 and X) if the fields $\nabla_i f_0$, $\nabla_{ij} f_0$, X, $\nabla_i X$, Y are independent on V.

Example. If $f: V \times \mathbb{R} \to W$ is a free map, then the field $\nabla_{tt} f | V \times 0$ is free relative to $f_0 = f | V \times 0$ and $X = \nabla_t f | V \times 0$. Conversely, if $f | V \times 0$ is a free map, the field $\nabla_t f | V \times 0$ is regular and $\nabla_{tt} f | V \times 0$ is free, then the map f is free on some neighborhood $\mathcal{O}p(V \times 0) \subset V \times \mathbb{R}$.

Let $f_0: V \to W$ be a free C^∞-map and X be a regular C^∞-field. Denote by $\tilde{T}^2 \to V$ the subbundle of the bundle $f_0^*(T(W)) \to V$, whose fiber $\tilde{T}_v^2$, $v \in V$, is spanned by the osculating space $T_v^2(V) \subset T_w(W)$, $w = f_0(v)$, and the vectors X and $\nabla_i X$ at w. Observe that dim $\tilde{T}_v^2 =$ dim $T_v^2 + n + 1 = [(n+1)(n+4)/2] - 1$, and that a field Y is free if and only if it is nowhere contained in the subbundle $\tilde{T}^2$. A free field exists, for example, if the orthogonal complement $f_0^*(T(W)) \ominus T^2$ is a trivial bundle of positive dimension. This is always the case for contractible manifolds V, provided $q \geq (n+1)(n+4)/2$.

(F″) **Proposition.** *Let $f_0: V \to W$ be a free C^∞-map, let X be a unit regular binormal C^∞-field along V and let Y_0 be a free field with respect to f_0 and X. Then the map f_0*

extends to a free C^∞-map $f\colon V \times \mathbb{R} \to W$ for $V = V \times 0$, for which the induced metric in a small neighborhood $\mathcal{O}p V \subset V \times \mathbb{R}$ is cylindrical,

$$f^*(h)|\mathcal{O}p V = f_0^*(h) \oplus dt^2,$$

and $\nabla_t f|V = X$.

Proof. First let f_0 and X_0 be real analytic. Then the existence of f without the freedom condition is established in (E′) with the aid of a field Y along V which serves for $\nabla_{tt} f|V$. To make f free we have to make Y free. To do this we first deform the given free field Y_0 to a free C^{an}-field Y_0' which is orthogonal to $\tilde{T}^2$, i.e. $Y_0'\colon V \to f_0^*(T(V)) \ominus \tilde{T}^2$. Then the field $Y + Y_0'$ satisfies (5) when substituted for $\nabla_{tt} f$, since Y_0' is normal to $X = \nabla_t f$, to $\nabla_i f$ and to $\nabla_{ij} f$. We multiply Y_0' by a large C^{an}-function $\lambda = \lambda(v)$, which makes the field $Y + \lambda Y_0'$ free, and then we solve (5) with this field $Y + \lambda Y_0'$ in place of Y. The new solution f has $\nabla_t f|V = X$ and $\nabla_{tt} f|V = Y + \lambda Y_0'$, and so this map $f\colon V \times \mathbb{R} \to W$ is free on $\mathcal{O}p V \subset V \times \mathbb{R}$.

Now, if f_0 and X are C^∞-smooth, then the formal part of the Cauchy-Kovalevskaya theorem provides a free C^∞-map $f\colon V \times \mathbb{R} \to W$, for which $f|V \times 0 = f_0$, $\nabla_t f|V \times 0 = X$ and $\nabla_{tt} f|V \times 0 = Y + \lambda Y_0'$, and which is *infinitesimally isometric* along $V = V \times 0$, i.e. $J_\delta^r|V \times 0 = 0$ for $\delta = f^*(h) - (f_0^*(h) \oplus dt^2)$ and for all $r = 0, 1, \dots$. Hence, a small C^∞-perturbation of f is isometric on $\mathcal{O}p V \subset V \times \mathbb{R}$ by the implicit function theorem (see 2.3.6). Q.E.D.

The Proof of Lemma (A). Lemma (F′) provides a regular binormal field along a small neighborhood $\mathcal{O}p U \subset V$. As $q \geq (n+1)(n+4)/2$ there is a free field along $\mathcal{O}p U$ as well. Hence (F″) applies to $f_0 = f|\mathcal{O}p U$. Q.E.D.

C^α-Immersions for $4 < \alpha < \infty$. Let (W, h) be a C^∞-manifold and let $f_0\colon V \to W$ be a free C^α-immersion. The map f_0 does not, in general, admit an extension to a C^α-map $f\colon V \times \mathbb{R} \to W$ which is isometric for the cylinder metric $f_0^*(h) \oplus dt^2$, because the last equation in (6) relates the derivative $\nabla_t f$ to $\nabla_{ij} f_0$ and so isometric cylinders $f\colon (V \times \mathbb{R}, f_0^*(h) \oplus dt^2) \to W$ which extend f_0 are at most $C^{\alpha-1}$-smooth for generic C^α-maps f_0.

Observe however that the metrics $g_0 = f_0^*(h)$ are somewhat better than just $C^{\alpha-1}$-smooth for C^α-maps f_0. Indeed, the curvature tensor R_{g_0} is $C^{\alpha-2}$-smooth by the Gauss theorema egregium, and so g_o is C^β-smooth for all $\beta < \alpha$ in appropriate (harmonic) local coordinates on V according to Jost-Karcher (1982).

Question. Does some neighborhood $\tilde{U} \subset V \times \mathbb{R}$ admit an isometric C^α-immersion $(\tilde{U}, f_0^*(h) \oplus dt^2) \to (W, h)$ for $q \geq (n+1)(n+4)/2$, for all free C^α-maps $f_0\colon V \to W$ and for a given $\alpha > 2$?

There is a similar regularity problem for *bendings* of a given free C^α-map $f_0\colon V \to \mathbb{R}^q$. Does there exists a C^α-continuous deformation $f_t\colon V \to \mathbb{R}^q$, $t \in [0,1]$, for which $f_t^*(h) = f_0^*(h)$ for all t (where $h = \sum_{i=1}^q dx_i^2$), such that the maps f_t for $t > 0$ are not congruent to f_0 by isometries of $\mathbb{R}^q$? (The implicit function theorem of 2.3.2 yields $C^{\alpha-1}$-bendings for $3 < \alpha < \infty$.)

Let us generalize Theorem (C) to C^α-manifolds (V, g) for $4 < \alpha < \infty$.

C^α-Extension Lemma [compare (A)]. *Let $f: (V, g) \to (W, h)$ be a free isometric C^α-map for which the metric $g = f^*(h)$ is C^α-smooth and let U be a (closed) n-dimensional ball in V, $n = \dim V$. If $\alpha > 4$ and $q \geq (n+1)(n+4)/2$, then there exists a C^α-immersion $\tilde{f}: V \times \mathbb{R} \to W$ with the following three properties:*

(i) *There is a neighborhood $\tilde{U} \subset V \times \mathbb{R}$ of $U \subset V = V \times 0 \subset V \times \mathbb{R}$ on which the map $\tilde{f}$ is free and the induced metric is cylindrical,*

$$\tilde{f}^*(h)|\tilde{U} = g \oplus dt^2,$$

(ii)
$$\tilde{f}^*(h)|V \times 0 = g,$$

(iii) *The map $\tilde{f}|V \times 0$ is as C^2-close to f as one wishes.*

Proof. Start with a fine C^β-approximation of f by a C^∞-map f_0 for some β in the interval $4 < \beta < \alpha$. Then we formally solve the systems (5) and (6) near U and thus we obtain a C^∞-map $\tilde{f}_0: V \times \mathbb{R} \to W$ whose $C^{\beta-1}$-norm is controlled (i.e. this norm is bounded on every compact subset $K \subset V \times \mathbb{R}$ by a constant $C = C(K, f)$) and such that the metric $\tilde{f}_0^*(h)|V \times 0$ is $C^{\beta-1}$-close to g and $\tilde{f}_0^*(h)|\mathcal{O}p\,U$ is $C^{\beta-2}$-close to $g \oplus dt^2$ for a small neighborhood $\mathcal{O}p\,U \subset V \times \mathbb{R}$. The implicit function theorem (see 2.3.2) now yields a $C^{\beta-2}$-small perturbation of $\tilde{f}_0$ to the required C^α-map $\tilde{f}$. Q.E.D.

The Proof of Theorem (C) *for C^α-Manifolds* (V, g). Let $f_0: V \to W$ be a strictly short map. Application of the implicit function theorem allows one to approximate f_0 by a free C^α-maps, say $f_0': V \to W$, such that the metric $g - f_0'(h)$ is C^∞-smooth and is therefore a sum

$$g - f_0'(h) = \sum_j dx_j^2$$

for some C^∞-*functions* x_j with small supports. Then the proof of (C) goes along with the above extension lemma.

C^1-Approximation. The proof of Theorem (C) shows that the isometric C^α-immersions $f: (V, g) \to (W, h)$, which C^0-approximate the strictly short map $f_0: V \to W$, are C^1-close to f_0, provided the map f_0 is C^1-smooth and the metric $f_0^*(h)$ is C^0-close to g. This equally applies to C^α-immersions, $4 < \alpha < \infty$, and so we have the following

(G) **Theorem.** *Let the metric h be C^∞-smooth and let g be C^α-smooth, $4 < \alpha < \infty$. Suppose $f_0: (V, g) \to (W, h)$ is an isometric C^1-map which admits a fine C^1-approximation by strictly short C^1-maps. If $q \geq (n+2)(n+5)/2$, then f_0 admits a fine C^1-approximation by isometric C^α-maps $(V, g) \to (W, h)$.*

Exercise. Let $f_0: V \to W$ be a C^2-immersion such that every point $w = f_0(v) \in W$, $v \in V$, admits a normal vector $\nu \in T_w(W) \ominus T_v(V)$ for which the second fundamental form $\Pi_\nu T_v(V)$ is positive definite. Obtain a fine C^2-approximation of f_0 by strictly short maps [for the metric $f_0^*(h)$]. Find a sharper infinitesimal criterion for such short approximations.

(H) *The Parametric h-Principle.* The proof of (C) applies to families of maps and then yields the parametric h-principle for free isometric maps relative to strictly short maps.

Exercises. Prove the parametric h-principle for free isometric C^α-map $f: V \to \mathbb{R}^q$, for $q \geq (n+2)(n+5)/2$, $n = \dim V$, for all C^α-smooth Riemannian manifolds V and for $\alpha > 4$.

Let $\partial_1, \ldots, \partial_k$ be C^∞-smooth linearly independent vector fields on V. Construct a C^∞-map $f: V \to \mathbb{R}^q$, $q = (k+2)(k+5)/2$, such that $\langle \partial_i f, \partial_j f \rangle = \delta_{ij}$ where $\delta_{ij} = 0$ for $i \neq j$ and $\delta_{ij} = 1$ for $i, j = 1, \ldots, k$.

Hint. Study the maps $V \to \mathbb{R}^q$ which are isometric and free (in the obvious sense) on the subbundle $\operatorname{Span}\{\partial_i\} \subset T(V)$.

3.1.3 Convex Cones in the Space of Metrics

Let (V, g) be a Riemannian C^∞-manifold of dimension n without boundary.

Proposition. *Let a Riemannian C^∞-metric δ on V satisfy $100^n\delta < g$ which means the positive definiteness of $g - 100^n\delta$. Then there exist C^∞-diffeomorphisms $p_i: V \to V$, $i = 1, \ldots, m = n + 8$, such that $\sum_{i=1}^m p_i^*(g) = mg + \delta$.*

Proof. Let (W, h) be the Cartesian product of m copies of (V, g) and let $f_0: V \to W$ be the *diagonal* imbedding for which $f_0^*(h) = mg$. Since $mn \geq (n+2)(n+5)/2$, there is a fine C^0-approximation of f_0 by a C^∞-map f, such that $f^*(h) = mg + \delta$. An inspection of the proof of Theorem (C) in 3.1.2. shows that for $\delta < 100^{-n}g$ the map f can be assumed so C^1-close to f_0, that the projections p_i of f to the factors V of W are immersions and, hence, C^∞-diffeomorphisms $V \to V$. Q.E.D.

(A)**Theorem.** *There is no (non-trivial) Diff-invariant convex cone in the space $\mathscr{G}_+^\infty$ of Riemannian C^∞-metrics on an arbitrary closed manifold V, where Diff stands for the group of C^∞-diffeomorphisms of V.*

Proof. Let $C \subset \mathscr{G}_+^\infty$ be a non-empty cone of metrics such that $g \in C$ implies $g + \delta \in C$ for any $\delta \in \mathscr{G}_+^\infty$ for which $\delta < \varepsilon_n g$ with a given fixed positive $\varepsilon_n > 0$. Then, obviously, $C = \mathscr{G}_+^\infty$ for compact manifolds V. The above proposition then applies to give the theorem.

Exercises. Extend the above theorem to C^0-metrics on V and also to C^{an}-metrics.

Let V be an *open* connected manifold without boundary of dimension $n \geq 2$. Show every Diff-invariant convex cone $C \subset \mathscr{G}_+^\infty(V)$ to be *stable*, i.e.

$$C + g \subset C \qquad \text{for all } g \in \mathscr{G}_+^\infty.$$

Complete metrics and also metrics of infinite volume provide interesting examples of such cones.

Consider a complete C^∞-metric $g = g_s$ on $\mathbb{R}^2$ which is S^1-symmetric around a fixed point $v_0 \in \mathbb{R}^2$ and such that the length of the circle of radius R about v_0 equals R^s for a given $s \in \mathbb{R}$ and for all $R \geq 1$. Let C_s denote the intersection of all Diff-invariant convex cones in $\mathscr{G}_+^\infty$ which contain g_s. Show that the cone C_s properly contains C_{s_1} for all $s_1 > s \leq 0$ and that $C_s = C_{s_1}$ for all $s_1 \geq s > 0$. Prove that C_s contains no (non-trivial) convex Diff-invariant subcones for any $s > 0$, and so C_s is a minimal cone in $\mathscr{G}_+^\infty(\mathbb{R}^2)$. Show that C_s, $s > 0$, is a *unique* minimal (convex, Diff-invariant) cone in $\mathscr{G}_+^\infty(\mathbb{R}^2)$. Find further examples of convex Diff-invariant cones in $\mathscr{G}_+^\infty(\mathbb{R}^2)$ and in $\mathscr{G}_+^\infty(V)$ for manifolds V of dimension >2.

3.1.4 Inducing Forms of Degree $d > 2$

Let Φ be a symmetric form of degree d on $\mathbb{R}^q$, for example, $\Phi = \sum_{i=1}^q x_i^d$. Consider a smooth map $f\colon V \to \mathbb{R}^q$ and let $\{\partial_1, \ldots, \partial_n\}$ be a frame of tangent vector fields on V. Then the induced differential form on V satisfies

$$f^*(\Phi)(\partial_{i_1}, \partial_{i_2}, \ldots, \partial_{i_d}) = \Phi(\partial_{i_1} f, \ldots, \partial_{i_d} f)$$

for all d-tuples of indices $i_1, \ldots, i_d = 1, \ldots, n$. For example, if $d = 3$, then

$$f^*(\Phi)(\partial_i, \partial_j, \partial_k) = \Phi(\partial_i f, \partial_j f, \partial_k f), \qquad i, j, k = 1, \ldots, n.$$

Let us write down the linearization $L = L_f(y)$ of the differential operator $f \mapsto f^*(\Phi)$ (compare 2.3.1). To simplify the notation, we assume $d = 3$ and we put $f_i = \partial_i f$ and $y_i = \partial_i y$. Then

$$L_f(y)(\partial_i, \partial_j, \partial_k) = \Phi(y_i, f_j, f_k) + \Phi(f_i, y_j, f_k) + \Phi(f_i, f_j, y_k).$$

In order to invert the operator $y \mapsto L_f(y)$ we consider to maps $y\colon V \to \mathbb{R}^q$ which are *Φ-normal* to $f_j \otimes f_k$ as follows

(1) $$\Phi(y, f_j, f_k) = 0, \qquad j, k = 1, \ldots, n.$$

Differentiate (1) and get with $f_{ij} = \partial_i \partial_j f$

(1′) $$\Phi(y_i, f_j, f_k) = -\Phi(y, f_{ij}, f_k) - \Phi(y, f_j, f_{ik}).$$

Hence, the solution of the P.D.E. system $L_f(y) = g$ is reduced to the solution of (1) and the following system

(2) $$\sum \Phi(y, f_{ij}, f_k) = -g(\partial_i, \partial_j, \partial_k), \qquad i, j, k = 1, \ldots, n,$$

where $\sum$ denotes the sum over the permutations of the indices i, j, and k. The systems (1) and (2) constitute $\binom{n+1}{2} + \binom{n+2}{3}$ linear algebraic equations in y, where $\binom{a}{b} = \dfrac{a!}{b!(a-b)!}$.

The map f is called *Φ-free* if it is C^2-smooth and if the total rank of the systems (1) and (2) equals $\binom{n+1}{2} + \binom{n+2}{3}$ at every point $v \in V$. More generally, if Φ is a form of any degree $d \geq 2$, then we associate to the map $f\colon V \to \mathbb{R}^q$ the system

$$\begin{aligned} &\Phi(y, f_{i_2}, \ldots, f_{i_d}) = 0 \\ &\sum \Phi(y, f_{i_1 i_2}, f_{i_3}, \ldots, f_{i_d}) = g_{i_1, \ldots, i_d}, \end{aligned} \tag{3}$$

where $\sum$ denotes the sum over all permutations of the given indices. Here we have $r = r(n, d) = \binom{n+d-2}{d-1} + \binom{n+d-1}{d}$ linear equations in y and we call the map *Φ-free* if the system (3) has rank $r = \binom{n+d-2}{d-1} + \binom{n+d-1}{d}$ at all points $v \in V$. This definition of freedom is independent of the choice of the frame $\partial_1, \ldots, \partial_n$ on V.

It is now clear that the differential operator $f \mapsto f^*(\Phi)$ is infinitesimally invertible at Φ-free maps $f\colon V \mapsto (\mathbb{R}^q, \Phi)$ for all forms Φ of any degree $d \geq 2$ (compare 2.3.1).

Again, let $\Phi = \sum_{i=1}^q x_i^d$ and let $\Sigma \subset X^{(2)}$, $X = V \times \mathbb{R}^q \to V$, denote the subset of the 2-jets of C^2-maps $f\colon V \to \mathbb{R}^q$ which are not Φ-free. A straightforward calculation gives $\operatorname{codim} \Sigma \geq q - r(n, d)$, and so *generic C^2-maps $f\colon V \to \mathbb{R}^q$ are Φ-free for $q \geq n + r(n, d)$* by Thom's transversality theorem.

Denote by F^∞ the space of Φ-free C^∞-maps $f\colon V \to \mathbb{R}^q$ with the fine C^∞-topology and let $\mathscr{G}^\infty$ denote the space of symmetric C^∞-forms on V of degree d. By the implicit function theorem (see 2.3.2) the differential operator $f \mapsto f^*(\Phi)$ sends F^∞ onto some *open* subset $\mathscr{G}^\infty(q) \subset \mathscr{G}^\infty$ which is non-empty for $q \geq n + r(n, d)$.

Positive Forms of Degree $d = 2k$. A symmetric form g on $\mathbb{R}^n$ of degree $2k$ defines a quadratic form on the symmetric power $(\mathbb{R}^n)^k$ by

$$g(x_1 \otimes x_2 \otimes \cdots \otimes x_k, y_1 \otimes y_2 \otimes \cdots \otimes y_k) = g(x_1, y_1, \ldots, x_k, y_k).$$

If this quadratic form is positive definite, then the form g on $\mathbb{R}^n$ is also called *positive*. Positive forms g on $\mathbb{R}^n$ constitute a unique (up to sign) minimal convex open cone (in the space of forms on $\mathbb{R}^n$) which is invariant under the linear group GL_n [see (B) in 2.4.9]. A symmetric differential form g of degree $2k$ on V is called positive if $g | T_v(V)$ is positive for all $v \in V$.

Theorem [compare (B′) in 3.1.1]. *An arbitrary positive C^∞-form g on V admits a decomposition*

$$g = \sum_j (dx_j)^{2k},$$

where each $x_j\colon V \to \mathbb{R}$ is a C^∞-function on V with support in some ball $U_j \subset V$, and such that every compact subset in V meets at most finitely many balls U_j.

Proof. Since the form g is positive, there exist C^∞-functions z_i on V, $i = 1, \ldots, l$ for some $l = l(n, \kappa)$ such that the form $g | T_v(V)$ is contained in the interior of the convex hull of the forms $(dz_i)^{2k} | T_v(V)$ for all tangent spaces $T_v(V)$, $v \in V$ [compare 3.1.1 and (B) in 2.4.9]. Thus we can obtain a decomposition

$$g = \sum_{i=1}^{l} \varphi_i^{2k} \, dz_i^{2k},$$

for some C^∞-functions φ_i, where the support of φ_i is the disjoint union of arbitrarily small subsets in V.

Let $x_\varepsilon = \varepsilon\varphi \sin \varepsilon^{-1} z$ and $y_\varepsilon = \varepsilon\varphi \cos \varepsilon^{-1} z$. Then (see 3.1.1)

$$[(dx_\varepsilon)^2 + (dy_\varepsilon)^2]^k \to \varphi^{2k}\, dz^{2k} \qquad \text{for } \varepsilon \to 0.$$

Next, we use the identity [compare (4) in 2.4.9] $(a^2 + b^2)^k = \sum_{j=1}^{2k} (\lambda_j a + \mu_j b)^{2k}$, where

$$\lambda_j = (\sin 2\pi/j)\left[\sum_{j=1}^{2k} (\sin 2\pi/j)^{2k}\right]^{-1/2k}$$

and

$$\mu_j = (\cos 2\pi/j)\left[\sum_{j=1}^{2k} (\cos 2\pi/j)^{2k}\right]^{-1/2k}.$$

For example, if $k = 2$, then

$$(a^2 + b^2)^2 = \tfrac{2}{3}(a^4 + b^4) + \tfrac{1}{6}[(a + b)^4 + (a - b)^4].$$

Thus we obtain

$$\sum_{j=1}^{2k} [d(\lambda_j x_\varepsilon + \mu_j y_\varepsilon)]^{2k} \underset{\varepsilon \to 0}{\longrightarrow} \varphi^{2k}\, dz^{2k}.$$

This approximation together with the openness of $\mathscr{G}^\alpha(q)$ makes the argument of 3.1.1 work for forms of arbitrary degree $2k \geq 2$. Q.E.D.

Exercises. Show that every positive C^α-form g of degree $2k$ on V, admits for $\alpha > 2$, a representation

$$g = \sum_{i=1}^{q} (df_i)^{2k}$$

for some C^α-functions $f_i\colon V \to \mathbb{R}$ and for

$$q = n + \binom{n + 2k - 2}{2k - 1} + (2k + 1)\binom{n + 2k - 1}{2k}.$$

Show that $g = \sum_{i=1}^{l} g_i^k$ for some positive quadratic forms g_i on V and for $l = \binom{n + 2k - 1}{2k}$, and thus reduce the above results on $2k$-forms to the case $k = 1$.

Let g be a C^α-form of *odd* degree $2k + 1$. Find an *algebraic* formula which makes

$$g = \sum_{i=1}^{q} (df_i)^{2k+1},$$

for some C^α-functions f_i on V, for

$$q = (2k + 1)\left[n + \binom{n + 2k - 1}{2k} + \binom{n + 2k}{2k + 1}\right]$$

and for all $\alpha \geq 1$ and $k \geq 1$. Then use the implicit function theorem to obtain

$g = \sum_{i=1}^{q} (df_i)^{2k+1}$ for

$$q = 2\left[n + \binom{n+2k-1}{2k} + \binom{n+2k}{2k+1}\right],$$

provided $\alpha > 2$.

Hint. See 2.3.7.

3.1.5 Immersions with a Prescribed Curvature

Consider a 4-form Φ on $\mathbb{R}^n$,

$$\Phi = \Phi(x_1, x_2, x_3, x_4), \qquad x_i \in \mathbb{R}^n,\ i = 1, \ldots, 4,$$

which is symmetric under the following permutations of the entries

$$x_1 \leftrightarrow x_2,\ x_3 \leftrightarrow x_4 \quad \text{and} \quad (x_1, x_2) \leftrightarrow (x_3, x_4)$$

This means Φ is a symmetric bilinear form on the symmetric square $(\mathbb{R}^n)^2$, and so the dimension of the space of all such forms Φ equals $\frac{n(n+1)}{4}\left(1 + \frac{n(n+1)}{2}\right)$. There is a canonical splitting of Φ into the sum $\Phi = \Phi^+ + \Phi^-$, where Φ^+ is the symmetric 4-form on $\mathbb{R}^n$ obtained by the complete symmetrization of Φ, and where $\Phi^- = \Phi - \Phi^+$ satisfies

$$\Phi^-(x_1, x_2, x_3, x_4) = \Phi(x_1, x_2, x_3, x_4) - \Phi(x_1, x_4, x_3, x_2)$$

The form Φ^- has the symmetry type of curvature tensors. These constitute a space of dimension

$$\frac{n(n+1)}{4}\left(1 + \frac{n(n+1)}{2}\right) - \frac{n(n+1)(n+2)(n+3)}{24} = \frac{n^2(n^2-1)}{12}.$$

Now, let $f\colon V \to W$ be an isometric C^2-immersion between Riemannian C^∞-manifolds $V = (V, g)$ and (W, h) of dimension n and q respectively. The map f then induces the form Φ of the above type on every tangent space $T_v(V)$, $v \in V$. Namely, take local coordinates u_i in V which are geodesic at $v \in V$ and define $\Phi = \Phi_f$ by

$$\Phi(\partial_i, \partial_j, \partial_k, \partial_l) = \langle \nabla_{ij} f, \nabla_{kl} f \rangle,$$

for $\partial_i = \partial f(v)/\partial u_i$ and for the covariant derivatives ∇_{ij} and their scalar products in W. The part Φ^- of $\Phi = \Phi_f$ depends only on the curvature tensor R of the induced metric $g = f^*(h)$ by the Gauss theorema egregium

$$R(g) = \Phi_f^-.$$

The remarkable feature of this formula is the absence of third derivatives of f in the expression for $R(f^*(h))$, where $h \mapsto f^*(h)$ is a first order differential operator and $g \mapsto R(g)$ is a second order operator. However, the composition of the two is a second (not third!) order operator.

The symmetric part Φ_f^+ also has a simple geometric interpretation. Let γ be a geodesic in V in the direction of some unit vector $\partial \in T_v(V)$. Then the value $\Phi_f^+(\partial, \partial, \partial, \partial)$ equals the (curvature)^2 of the curve $f(\gamma) \subset W$ at $w = f(v) \in W$.

Observe that the form Φ_f can be identified with the quadratic form on the symmetric square $(T(V))^2$ which is induced from h by the *second differential* D_f^2. This D_f^2 maps $(T(V))^2$ to the normal bundle $N_f \to V$ by sending $\partial_i \otimes \partial_j$ to $P_N(\nabla_{ij} f)$, for all bivectors $\partial_i \otimes \partial_j \in (T(V))^2$ and for the normal projection P_N: $T(W)|V \to N_f$.

Exercises. Show that the form Φ_f^+ is positive (in the sense of 3.1.6.) if and only if the map f is free.

Let V be a connected manifold and let f_i: $V \to W$, $i = 1, \ldots,$ be isometric C^2-immersions for which $\Phi_{f_i}^+ \leq \Phi_0^+$ [i.e. $\Phi_0 - \Phi_{f_i}^+ | T_v(V)$ lies in the closure of the cone of the positive forms for all $v \in V$] and for which the sequence $f_i(v_0) \in W$ has an accumulation point in W for a given point $v_0 \in V$. Show that some subsequence C^α-converges for all $\alpha < 2$ to an isometric map f: $V \to W$ which is twice differentiable almost everywhere and which a.e. satisfies $\Phi_f^+ \leq \Phi_0$.

(A) **Theorem.** *Let f_0: $V \to W$ be a free isometric C^∞-immersion and let Φ^+ be a continuous symmetric form on V such that $\Phi^+ - \Phi_{f_0}^+$ is a positive form. If $q \geq (n+2)(n+5)/2$, then the map f_0 admits a fine C^1-approximation by free isometric C^2-immersions f: $V \to W$ for which $\Phi_f^+ = \Phi^+$.*

Example. If $V = \mathbb{R}^n$ and $W = \mathbb{R}^q$ then the theorem implies the existence of C^2-solutions to the following system of $[n(n+1)/2] + [n(n+1)(n+2)(n+3)/24]$ P.D. equations in the unknown map f: $\mathbb{R}^n \to \mathbb{R}^q$.

$$\langle \partial_i f, \partial_j f \rangle = \delta_{ij}, \qquad 1 \leq i \leq j \leq n,$$

$$\langle \partial_{ij} f, \partial_{kl} f \rangle = \Phi_{ijkl}^+, \qquad 1 \leq i \leq j \leq k \leq l,$$

where $\delta_{ij} = 0$ for $i \neq j$ and $\delta_{ii} = 1$, and where Φ_{ijkl}^+ are arbitrary continuous functions on $\mathbb{R}^n$ for which the form $\Phi(\partial_i, \partial_j, \partial_k, \partial_l) = \Phi_{ijkl}^+$ is positive at all points $v \in \mathbb{R}^n$ [compare (B) in 2.4.9].

Proof [compare Nash (1954)]. Let $\Phi_0^+ \geq 0$ be a C^∞-form on V which equals $(dx)^4$ for some C^∞-function x: $V \to \mathbb{R}$ with the support in a ball $U \subset V$. Let us construct a family of free isometric C^∞-maps f_ε: $V \to W$ for small $\varepsilon > 0$, with the following four properties.

(i) f_ε equals f_0 outside U for all $\varepsilon > 0$;

(ii) the maps f_ε C^1-converge to f_0 for $\varepsilon \to 0$;

(iii) the C^2-distance between f_ε and f_0 is controlled by the C^0-norm of Φ_0^+, with some universal constant const $=$ const(q),

$$\|f_\varepsilon - f_0\|_2 \leq \text{const}\, \|\Phi_0^+\|_0,$$

for some fixed norms in the (linearized) space of maps close to f_0 and in the space of 4-forms on V;

(iv) the forms $\Phi_{f_\varepsilon}^+$ C^0-converge to $\Phi_{f_0}^+ + \Phi_0^+$ for $\varepsilon \to 0$.

First, we obtain by (A′) in 3.1.2 with the implicit function theorem (see 2.3.6) a family of C^∞-maps $\tilde{f}_\varepsilon$: $V \times \mathbb{R}^2 \to W$ with the following three properties

(a) $\tilde{f}_\varepsilon | V \backslash U = f_0 | V \backslash U$ for $V = V \times 0 \subset V \times \mathbb{R}^2$ and for all $\varepsilon > 0$;
(b) the family $\tilde{f}_\varepsilon$ is C^∞-continuous in $\varepsilon \geq 0$ and $\tilde{f}_0 | V = f_0$.
(c) the induced metric $\tilde{f}_\varepsilon^*(h)$ satisfies on some neighborhood $\tilde{U} \subset V \times \mathbb{R}^2$ of U,

$$\tilde{f}_\varepsilon^*(h) | \tilde{U} = g_\varepsilon \otimes dt_1^2 \otimes dt_2^2,$$

for $g_\varepsilon = g - \varepsilon^4\, dx^2$ and for all (small) $\varepsilon > 0$.
Then we put

$$f_\varepsilon(v) = \tilde{f}_\varepsilon(v, \varepsilon^2 \sin \varepsilon^{-1} x, \varepsilon^{-2}(1 - \cos \varepsilon^{-1} x)).$$

These maps f_ε are isometric and they obviously satisfy (i)–(iii). We check (iv) by evaluating $\Phi_{f_\varepsilon}^+$ in the geodesic coordinates $u_1, \ldots, u_n$ at some point $v \in V$. We have for $\nabla_i = \nabla_{u_i}$ and $\partial_i = \partial/\partial u_i$,

$$\nabla_i f_\varepsilon = \nabla_i \tilde{f}_\varepsilon + \varepsilon \partial_i \times [(\cos \varepsilon^{-1} x) \nabla_{t_1} \tilde{f}_\varepsilon - (\sin \varepsilon^{-1} x) \nabla_{t_2} \tilde{f}_\varepsilon],$$

and

$$\nabla_{ij} f_\varepsilon = \nabla_{ij} \tilde{f}_\varepsilon + (\partial_i x)(\partial_j x) z + 0(\varepsilon),$$

for $Z = -(\sin \varepsilon^{-1} x) \nabla_{t_1} \tilde{f}_\varepsilon - (\cos \varepsilon^{-1} x) \nabla_{t_2} \tilde{f}_\varepsilon$. The form Φ_{f_ε} at the point v is

$$\langle \nabla_{ij} f_\varepsilon, \nabla_{ij} f_\varepsilon \rangle = \langle \nabla_{ij} \tilde{f}_\varepsilon, \nabla_{ij} \tilde{f}_\varepsilon \rangle + (\partial_i x)(\partial_j x)(\partial_k x)(\partial_l x) + 0(\varepsilon),$$

as $\langle Z, Z \rangle = 1$ and $\langle \nabla_{ij} \tilde{f}_\varepsilon, Z \rangle = 0$ by (D″) in 3.2.1. Hence, $\Phi_{f_\varepsilon}^+ \to \Phi_{f_0}^+ + dx^4$.

Next, we use the decomposition $\Phi_0^+ = \sum_j dx_j^4$ for *positive* C^∞-forms Φ_0^+ on V (see 3.1.4) and thus we obtain maps f_ε with the properties (i)–(iv) for an *arbitrary positive* C^∞-form $\Phi_0^+ > 0$. Finally, we break the form $\Delta_0 = \Phi^+ - \Phi_{f_0}^+$ into the sum

$$\Delta_0 = \Phi_0^+ + \Delta_1,$$

where Φ_0^+ is a positive C^∞-form and Δ_1 is a small positive continuous form. We take the map f_{ε_1} with a small $\varepsilon_1 > 0$ for a new map f_1. For this f_1 the error $\Phi - \Phi_{f_1}^+$ is close to Δ_1. Then we pass to $f_2, f_3, \ldots$, for which the error $\Phi^+ - \Phi_{f_i}^+ \to 0$ for $i \to \infty$, and the property (iii) allows us to go to the limit $f = \lim_{i \to \infty} f_i$ which has $\Phi_f^+ = \Phi^+$. Q.E.D.

C^∞-Immersions with Given Curvature. The isometric immersion relation for maps $f: (V, g) \to \mathbb{R}^q$ prescribes the scalar products

$$\langle \partial_1 f, \partial_2 f \rangle = g(\partial_1, \partial_2) \tag{1}$$

for all vectorfields ∂_1 and ∂_2 in V. This implies

$$\langle \partial_1 \partial_2 f, \partial_2 f \rangle = \tfrac{1}{2} \partial_1 g(\partial_2, \partial_2)$$

and so

$$\langle \partial_2^2 f, \partial_1 f \rangle = \partial_2 g(\partial_1, \partial_2) - \tfrac{1}{2} \partial_1 g(\partial_2, \partial_2). \tag{1′}$$

Since commuting fields satisfy

(*) $$\partial_2\partial_3 = \tfrac{1}{4}[(\partial_2+\partial_3)^2 - (\partial_2-\partial_3)^2],$$

the Eqs. (1) express via (1′) the scalar products

(2) $$\langle \partial_2\partial_3 f, \partial_1 f\rangle$$

by combinations of derivatives of g. Next, for commuting fields ∂_i, $i = 1, \ldots, 4$,

(3) $$\partial_4\langle \partial_2\partial_3 f, \partial_1 f\rangle - \partial_2\langle \partial_3\partial_4 f, \partial_1 f\rangle = \langle \partial_2\partial_3 f, \partial_1\partial_4 f\rangle - \langle \partial_1\partial_2 f, \partial_3\partial_4 f\rangle,$$

and so we express (3) by combinations of second derivatives of g. This amounts to Gauss' formula $\Phi_f^- = R(g)$.

Now, let the curvature Φ_f^+ also be given. Then Φ^+ and Φ^- determine all of $\Phi = \Phi_f$ and so all scalar products

(4) $$\langle \partial_1\partial_2 f, \partial_3\partial_4 f\rangle$$

are expressed by Φ^+ and by derivatives of g. Abbreviate this expression to $\langle 12, 34\rangle$. Then

(5) $$\partial_1\langle 11, 22\rangle - \partial_2\langle 11, 12\rangle + \tfrac{1}{2}\partial_1\langle 12, 12\rangle = \langle \partial_1^3 f, \partial_2^2 f\rangle.$$

The scalar products (5) determine with (*) and with a similar expression for $\partial_1\partial_2\partial_3$ (by a combination of cubes of linear combinations of ∂_1, ∂_2 and ∂_3) all scalar products

(6) $$\langle \partial_1\partial_2\partial_3 f, \partial_4\partial_5 f\rangle$$

Furthermore, the derivatives of (2) give us the scalar products

(7) $$\langle \partial_1\partial_2\partial_3 f, \partial_4 f\rangle,$$

and so the forms $g = f^*(h)$ (for $h = \sum_{i=1}^q dx_i^2$ in $\mathbb{R}^q$) and Φ_f^+ uniquely determine the normal projections of the third derivatives of f to the (second) osculating space $T_v^2(f) \subset \mathbb{R}^q$ for all $v \in V$. In particular, if $\dim T_v^2(f) = q$ for all $v \in V$, then the third derivatives of f become functions of the first and second derivatives and so the system

(8) $$f^*(h) = g, \qquad \Phi_f^+ = \Phi^+$$

is completely integrable (see 2.3.8) in this case. In particular, (8) admits a C^∞-solution if and only if the forms g and Φ^+ satisfy the compatibility condition which expresses the symmetry

(9) $$\partial_1\partial_2\partial_3 f = \partial_2\partial_1\partial_3 f = \partial_2\partial_3\partial_1 f$$

in terms of g and Φ^+. This compatibility condition is called *the Codazzi equation.* Furthermore, Frobenius' theorem [see (F) in 2.3.8] implies the following *rigidity* of C^∞-maps $V \to \mathbb{R}^q$,

(B). *If $f_1^*(h) = f_2^*(h)$ and $\Phi_{f_1}^+ = \Phi_{f_2}^+$, and if the osculating space of the map f_1 has $\dim T_v^2(f_1) = q$ for all $v \in V$, then f_2 is congruent to f_1 by a rigid motion of $\mathbb{R}^q$, provided the manifold V is connected.*

The condition $q = \dim T_v^2(f_1)$ implies $q \le n(n+3)/2$. But the system (8) contains p equations for

$$p = \frac{n(n+1)}{2} + \frac{n(n+1)(n+2)(n+3)}{24},$$

and so one may expect a similar rigidity theorem for generic C^∞-maps $V \to \mathbb{R}^q$ for $q < p$. On the other hand, the Theorem (A) shows the complete breakdown of the rigidity for C^2-maps $f\colon V \to \mathbb{R}^q$, $q \ge (n+2)(n+5)/2$.

Now let $q \le q_0 = (3 - \sqrt{3})(n^2 - 4n)/6$. Then the dimension d of the space of quadratic form on $(\mathbb{R}^n)^2 = \mathbb{R}^{n(n+1)/2}$ of rank $r \le q - n$ satisfies

$$d = \frac{(q-n)[(n+1)^2 - q]}{2} \le \frac{n^2(n^2-1)}{12}.$$

This suggests that the tensor Φ_f of a generic map $f\colon V \to \mathbb{R}^q$ is uniquely determined by Φ_f^-, as Φ_f depends (at every point $v \in V$) on d parameters, while Φ_f^- may control $n^2(n^2-1)/12$ parameters corresponding to the curvature tensor $R(f^*(h))$. One knows (see Kobayashi-Nomizu) that the curvature tensor $R(f^*(h))$ uniquely determines Φ_f for generic maps $f\colon V \to \mathbb{R}^q$ for $q \le n + (n/3)$ and so these maps are uniquely determined (up to an isometry of $\mathbb{R}^q$) by the metric $f^*(h)$. On the other hand, if $d < n^2(n^2-1)/12$, then a generic curvature tensor on V does not come from any map $V \to \mathbb{R}^q$ and so generic C^∞-manifolds (V, g) admit no isometric C^2-immersions into $\mathbb{R}^q$ for $q < q_0$ [see E. Berger (1981) and Berger-Bryant-Griffiths (1983) for deeper relations between Φ_f and $R(f^*(h))$].

Exercises. Show that the above Proposition (B) fails in general to be true for C^2-maps.

Give an example of a C^∞-metric g on $V \approx S^3$, for which the space of isometric C^∞-immersions $f\colon (V, g) \to \mathbb{R}^4$, normalized by $f(v_0) = 0$ and by $D_f T_{v_0}(V) = D_0$ for a given isometry $D_0\colon T_{v_0}(V) \to \mathbb{R}^4$, is homeomorphic to the Cantor set. Show that no C^{an}-metric on V possesses a similar property.

Construct a C^∞-metric g on $V \approx S^3$ which admits no C^2-isometric immersion into $\mathbb{R}^4$, but such that every unit ball in (V, g) admits an isometric C^∞-imbedding into $\mathbb{R}^4$. Show that no C^{an}-metric on S^3 has this property.

Show that every C^∞-immersion $f_0\colon V \to \mathbb{R}^q$, $q \ge n+2$, admits a fine C^0-approximation by a C^∞-immersion $f\colon V \to \mathbb{R}^q$, which admits an infinite dimensional family of C^∞-deformations $f_t\colon V \to \mathbb{R}^q$, $t \in T$, such that $f_t^*(h) = f^*(h)$ and

$$\Phi_{f_t} = \Phi_f \qquad \text{for all } t \in T.$$

Isometric Immersions of Order $k \ge 1$ (Allendoerfer 1937; Spivak 1979). Let $f\colon V \to \mathbb{R}^q$ be a C^∞-map. Denote by $T_v^k(f) \subset T_w(\mathbb{R}^q)$, $w = f(v)$, the kth osculating space which by definition is the span of the derivatives $\partial_1 \partial_2 \dots \partial_l f(v)$ for all $l \le k$ and for all l-tuples of vector fields $\partial_1, \dots, \partial_l$ in V. The kth order differential D_f^k maps the symmetric power $(T_v(V))^k$ for all $v \in V$ to the orthogonal complement $T_v^k(f) \ominus T_v^{k-1}(f)$ by sending every k-vector $\partial_1 \otimes \cdots \otimes \partial_k \in (T_v(V))^k$ to the normal projection of the derivative $\partial_1 \partial_2 \dots \partial_k f(v)$ to $T_v^k(f) \ominus T_v^{k-1}(f)$. Thus, the Euclidean metric in $\mathbb{R}^q$ induces a certain quadratic form on $(T(V))^k$, that is a $2k$-linear form, called $G_k(f)$ on $T(V)$. For example, $G_2(f)$ is our old curvature form Φ_f. Denote by $g_k(f)$ the symmetrization of $G_k(f)$. Thus,

$$g_1(f) = G_1(f) = f^*(h) \qquad \text{and} \qquad g_2(f) = \Phi_f^+ .$$

Exercises. Express the form $G_k(f)$ by $g_k(f)$ and by derivatives (of order $<k$) of the forms $G_l(f)$ for $l < k$. Thus the forms $g_l(f)$, $l = 1, \dots, k$, uniquely determine the forms $G_l(f)$, $l = 1, \dots, k$.

Let $\dim T_v^k(f) = q$ for all $v \in V$. Show that f is uniquely determined (up to isometries of $\mathbb{R}^q$) by the forms $g_1(f), \dots, g_k(f)$ [compare (B)].

Now, let g_l, $l = 1, \dots, k$, be arbitrary symmetric differential $2l$-forms on V. *The isometric immersion problem* for $(V, g_1, \dots, g_k)$ is that of finding a C^α-maps $f\colon V \to \mathbb{R}^q$, for a given $\alpha \geq k$, such that $g_l(f) = g_l$ for $l = 1, \dots, k$. If such an f exists, then $g_l \geq 0$ for $l = 1, \dots, k$, and if $g_l > 0$, then the map f is necessarily k^{th}-order free.

Exercises. Show that the differential operator (of order k) $f \mapsto (g_1(f), \dots, g_k(f))$ is infinitesimally invertible (see 2.3.1) at all $(k+1)^{\text{th}}$ order free maps $f\colon V \to \mathbb{R}^q$ and apply the implicit function theorem to these maps f.

3.1.6 Extension of Isometric Immersions

We prove in this section the h-principle for extensions of free isometric immersions from a given submanifold $V_0 \subset V$ to a small neighborhood $\mathcal{O}\!\mathit{p}\, V_0 \subset V$. First, let V_0 be a single point $v_0 \in V$ and let $f_i, f_{ij}, \dots$ denote the covariant derivatives $\nabla_i f$, $\nabla_j \nabla_i f$, $\dots$ of a map $f\colon V \to (W, h)$ for fixed local coordinates u_i, $i = 1, \dots, n$, in V. We write the isometry condition for $f\colon (V, g) \to (W, h)$ as

$$\langle f_i, f_j \rangle = g_{ij}, \qquad 1 \leq i \leq j \leq n, \tag{1}$$

and then we differentiate,

$$\langle f_{ik}, f_j \rangle + \langle f_i, f_{jk} \rangle = \partial_k g_{ij}, \tag{2}$$

where $\partial_k = \partial/\partial u_k$. The Eqs. (2) are equivalent to

$$\langle f_i, f_{jk} \rangle = A^{ijk}, \tag{2'}$$

for

$$A^{ijk} = \tfrac{1}{2}(\partial_k g_{ij} + \partial_j g_{ik} - \partial_i g_{jk}).$$

We differentiate (2′),

$$\langle f_i, f_{jkl} \rangle + \langle f_{il}, f_{jk} \rangle = \partial_l A^{ijk}, \tag{3}$$

and then obtain

$$\langle f_{il}, f_{jk} \rangle - \langle f_{ik}, f_{jl} \rangle = B^{ijkl} + R^{ijkl}, \tag{4}$$

for

$$B^{ijkl} = \partial_l A^{ijk} - \partial_k A^{ijl}$$

and

$$R^{ijkl} = \langle f_i, (R(f_k, f_l), f_j) \rangle,$$

where R denotes the curvature tensor in (W, h).

The Eqs. (3) are linear in the third derivatives f_{jkl} and the Eqs. (4) give a complete consistency condition for (3) because the vectors f_i are linearly independent. Hence, the system (3) is solvable in $f_{jkl}(v_0) \in T_{w_0}(W)$, $w_0 = f(v_0)$, if the vectors $f_{ij}(v_0) \in T_{w_0}(W)$ satisfy (4). Moreover, the space of solutions of the system (1) + (2) + (3) at $v_0 \in V$ has the same homotopy type as the space of solutions of (1) + (2) + (4).

The Eqs. (4) are non-linear and their space of solutions (in a fixed tangent space $T_{w_0}(W)$) may be quite complicated. However, the space of *free* solutions, that are defined as p-tuples, $p = n + [n(n+1)/2]$, of linearly independent vectors f_i and f_{ij} in $T_{w_0}(W)$ which satisfy (1) + (2) + (4), has a fairly simple structure.

(A) **Lemma.** *The space of free solutions of* (1) + (2) + (4) *at every point* $w_0 = f(v_0)$ *is homotopy equivalent to the Stiefel manifold* $\mathrm{St}_p(T_{w_0}(W)) = \mathrm{St}_p \mathbb{R}^q$. *Hence, the space of free solutions of* (1) + (2) + (3) *is also homotopy equivalent to* $\mathrm{St}_p \mathbb{R}^q$.

Proof. The equations in question are invariant under coordinate changes in V and so one may use geodesic coordinates at v_0 for which $A^{ijk}(v_0) = 0$. Then the vectors f_i and f_{ij} become orthogonal in $T_{w_0}(W)$ and we must only determine $[n(n+1)/2]$-tuples of independent vectors $f_{ij} \in \mathbb{R}^{q-n}$ which satisfy (4). These tuples are given by the injective linear maps F of the symmetric square $(\mathbb{R}^n)^2 = \mathbb{R}^{[n(n+1)/2]}$ into $\mathbb{R}^{q-n}$ such that the Φ^--part of the induced form Φ_F on $\mathbb{R}^n$ is given by $B^{ijkl} + R^{ijkl}$ (see 3.1.5). If $\Phi = \Phi_F$ is positive definite (as a quadratic form) on $(\mathbb{R}^n)^2$, then the space of isometric maps $[(\mathbb{R}^n)^2, \Phi] \to \mathbb{R}^p$ equals $\mathrm{St}_{p-n} \mathbb{R}^{q-n}$. On the other hand, those forms Φ which are positive definite on $(\mathbb{R}^n)^2$ and which satisfy $\Phi^- = R$ for a given R, form a *nonempty* convex subset in the Euclidean space of all 4-forms Φ. This subset is nonempty, since $\Phi^- + \lambda\Phi^+$ is positive on $(\mathbb{R}^n)^2$ for every positive symmetric 4-form Φ^+ on $\mathbb{R}^n$ and for all sufficiently large $\lambda \geq \lambda_0 = \lambda_0(\Phi^-)$. Hence, the space of solutions of (4) is homotopy equivalent to $\mathrm{St}_{p-n} \mathbb{R}^{q-n}$ and the lemma follows.

(A′) **Corollary.** *The space of free isometric C^∞-immersion $\mathcal{O}p(v_0) \to W$* (compare 1.1.5) *is weakly homotopy equivalent to the Stiefel bundle* $\mathrm{St}_p(W)$ *of p-frames* $\mathrm{St}_p T_v(W)$, $v \in V$, *for* $p = n + [n(n+1)/2]$.

Proof. If a smooth map $f\colon \mathcal{O}p(v_0) \to W$ satisfies (1) + (2) + (3) at $v_0 \in V$, then f is infinitesimally isometric of second order at v_0, that is $J^2_{f^*(h)}(v_0) = J^2_g(v_0)$. By the implicit function theorem (see 2.3.2) a small perturbation of f is isometric on $\mathcal{O}p(v_0)$. Q.E.D.

Now, let $e\colon V_0 \subsetneq V$ be an arbitrary submanifold of codimension k_0 in V and let $f\colon V_0 \to W$ be a free isometric immersion. Let $N_e \to V_0$ denote the normal bundle

$T(V)\ominus T(V_0)$ and let $T(W)|V_0$ stand for the induced bundle $f^*(T(W))\to V_0$. We study homomorphisms $X\colon N_e\to T(W)|V_0$ over small (coordinate) neighborhoods $U_0\subset V_0$ by taking frames of independent fields $\partial_i\colon U_0\to N_e$, $i=1,\dots,k_0$, and then by expressing $X|U_0=\{X_i\}$ for the fields $X_i=X(\partial_i)$ along U_0 (compare 3.1.2). Consider a smooth map $\tilde f\colon \mathcal{O}p\,V_0\to W$ which extends $f_0=\tilde f|V_0$ and let $X=D_{\tilde f}|V_0$. If the differential $D_{\tilde f}\colon T(V)\to T(W)$ is isometric on $T(V)|V_0$ then X is an isometric homomorphism of N_e to the normal bundle $N_f=T(W)\ominus T(V_0)$. This is expressed by the equations

$$\langle X_i, X_j\rangle = g_{ij}, \tag{5}$$

$$\langle X_i, f_\mu\rangle = 0, \tag{6}$$

where f_μ denote the first covariant derivatives $\nabla_\mu f = D_f(\partial/\partial u_\mu)$ in (W,h) for some local coordinates u_μ in V_0, $\mu=1,\dots,n-k_0$. If the map $\tilde f$ is infinitesimally isometric of the first order along V_0 [i.e. $J^1_{\tilde g}|V_0 = J^1_g|V_0$ for $\tilde g=\tilde f^*(h)$], then the Eqs. (2′) above imply

$$\langle f_{\mu\nu}, X_i\rangle = A^{i\mu\nu}, \tag{7}$$

for $1\le\mu,\ \nu\le n-k_0$, $i=1,\ \dots,\ k_0$. The Eqs. (7) admit the following invariant description. The bundle $T(W)|V_0$ splits into the orthogonal sum,

$$T(W)|V_0 = T(V_0)\oplus(T(V_0))^2\oplus BN_f,$$

where the tangent bundle $T(V_0)$ is isometrically imbedded into $T(W)|V_0$ by the differential D_f and the symmetric square $(T(V_0))^2$ is imbedded into $T(W)|V$ by the second differential D_f^2, while BN_f denotes the binormal bundle of $f\colon V\to W$ (compare 3.1.2). Denote by P^1 and P^2 the normal projections of $T(W)|V_0$ on the subbundles $T(V_0)$ and $(T(V_0))^2$ respectively. Then the Eqs. (6) say

$$P^1\circ X = 0 \tag{6′}$$

and the Eqs. (7) become

$$P^2\circ X = A, \tag{7′}$$

for the homomorphism $A\colon N_e\to(T(V_0))^2$ which is defined by $A^{i\mu\nu}$, such that $\langle A\partial_i, f_{\mu\nu}\rangle = A^{i\mu\nu}$. Consider the adjoint homomorphism $A'\colon (T(V_0))^2\to N_e$, for which

$$\langle A\partial,\delta\rangle = \langle\partial, A'\delta\rangle$$

for the scalar products induced by the imbeddings $D_f^2\colon (T(V_0))^2\to T(W)|V_0$ and $X=D_{\tilde f}|N_e\to T(W)|V_0$ from the scalar product in (W,h). Since the covariant derivatives in V equal normal projections to V of covariant derivatives in W, the homomorphism A' equals the second differential $D_e^2\colon (T(V_0))^2\to N_e$ of the imbedding $e\colon V_0\hookrightarrow V$. The homomorphism A' also equals the normal projection of $(T(V_0))^2\subset T(W)|V_0$ to $D_fT(V_0)\subset T(W)|V_0$, and so A' is a *short homomorphism*. This means the positive semi definiteness of the form $\Phi_f-\Phi_e$ which is viewed here as a quadratic form on $(T(V))^2$ (compare 3.1.5). *Hence, the inequality $\Phi_f\ge\Phi_e$ between the curvatures of the maps f and e gives us a necessary condition for the existence of an isometric extension $\tilde f\colon V\to W$.* Furthermore, if the map $\tilde f$ is free, then the subbundles $T(V)$

and $T^2(f) = T(V_0) \oplus (T(V_0))^2$ are linearly independent in $T(W)|V_0$, and so the homomorphism A' is *strictly short*, which amounts to the strict inequality $\Phi_f > \Phi_e$.

The fields $X_i = \nabla_i \tilde{f}$, for infinitesimally isometric maps $\tilde{f}\colon \mathcal{O}p\, V_0 \to W$, also satisfy the following *differential* equations on V_0

$$\langle X_i, \nabla_\mu X_j \rangle = A^{i\mu j}, \qquad 1 \le i,j \le k_0, 1 \le \mu \le n - k_0, \tag{8}$$

as Eqs. (2′) show. These Eqs. (8) say that the connection induced by the homomorphism $X\colon N_e \to T(W)|V_0$ equals the normal connection of the subbundle $N_e \subset T(V)|V_0$. Observe that for a *free* map $\tilde{f}$ the homomorphism X is *regular* (compare 2.2.6 and 3.1.2). This amounts to linear independence of the vectors $P^3(\nabla_\mu X_j)$, $1 \le \mu \le n - k_0$, $1 \le i \le k_0$ where P^3 denotes the normal projection of $T(W)|V_0$ onto the binormal bundle $BN_f \subset T(W)|V_0$. The regularity of X is clearly sufficient for the strict inequality $\Phi_f > \Phi_e$.

(B) **Lemma.** *If* $q \ge m_0 + [m_0(m_0 + 1)/2] + k_0(m_0 + 2)$ *for* $m_0 = n - k_0 = \dim V_0$, *then regular homomorphisms* $X\colon N_e \to T(W)|V_0$, *which satisfy* (5) + (6) + (7) + (8), *abide by the h-principle.*

Proof. If the submanifold V is totally geodesic then the Eqs. (7) become

$$\langle f_{\mu\nu}, X_i \rangle = A^{i\mu\nu} = 0,$$

and so the homomorphisms X in question are regular isometric homomorphisms $N_e \to BN_f$ which induce a given connection in N_e. The h-principle for these is proven in 2.2.6. In order to apply the argument in 2.2.6 to $A^{i\mu\nu} \ne 0$, we must check the microflexibility of the sheaf of regular solutions of (5) + (6) + (7) + (8). We assume as in 2.2.6 the fields $\partial_i\colon V_0 \to N_e$ to be orthonormal and then we bring together the equations which relate the field X_l to X_i for $i < l$ and for a given $l \le k_0$.

$$\begin{aligned} \langle X_l, X_i \rangle &= 0 \\ \langle X_e, X_l \rangle &= 1 \\ \langle X_l, f_\mu \rangle &= 0 \\ \langle X_l, f_{\mu\nu} \rangle &= A^{i\mu\nu} \\ \langle X_l, \nabla_\mu X_i \rangle &= A^{l\mu i} \\ \langle \nabla_\mu X_l, X_i \rangle &= A^{i\mu l} \end{aligned} \tag{9}$$

All these equations but the last one are algebraic in X_l. Furthermore,

$$\langle \nabla_\mu X_l, X_i \rangle = \nabla_\mu \langle X_l, X_i \rangle - \langle X_l, \nabla_\mu X_i \rangle = \langle X_l \nabla_\mu X_i \rangle,$$

and so the last equation is redundant. If by induction we assume the homomorphism X to be regular on $\mathrm{Span}\{\partial_1, \ldots, \partial_l\} \subset N_e$, then the vector X_i, f_μ, $f_{\mu\nu}$, $\nabla_\mu X_i$, for $1 \le i \le l - 1$, $\mu = 1, \ldots, n - k_0$, are linearly independent and so the solutions X_l of (9) for which the homomorphism X is regular on $\mathrm{Span}\{X_1, \ldots, X_l\}$ form a d-dimensional sphere over every point $v \in V_0$ for $d = q - l - m_0 - [m_0(m_0 + 1)/2] - (l - 1)m_0$. The solutions X_l now become regular sections of the resulting sphere bundle and the argument of 2.2.6 applies.

(B′) Corollary. *If $\Phi_f > \Phi_e$ and $q \geq m_0 + [m_0(m_0+1)/2] + k_0(m_0+2)$, then the space of regular soutions $X = \{X_1, \ldots, X_k\}$ of (5) + (6) + (7) + (8) is weakly homotopy equivalent to the space of all injective homomorphisms of the bundle $N_e \oplus N_e \otimes T(V_0)$ to the binormal bundle BN_f.*

Proof. As $\Phi_f > \Phi_e$, the homomorphism $A'\colon (T(V_0))^2 \to N_e$ is strictly short and so the adjoint homomorphism $A\colon N_e \to (T(V_0))^2 \subset T(W)|V_0$ is also strictly short, which means the positive definiteness of the form $g' = g - A^*(h)$ on N_e, where $A^*(h)$ is the form induced by A.

Since $P^1 \circ X = 0$ and $P^2 \circ X = A$ by $(6') + (7')$, the homomorphism X is uniquely determined by $Y = P^3 \circ X\colon N_e \to BN_f$. The Eqs. (5) are equivalent to

$$\langle Y_i, Y_j \rangle = g'_{ij}, \tag{10}$$

and the Eqs. (8) become

$$\langle Y_i, \nabla'_\mu Y_j \rangle = C^{i\mu j}, \tag{11}$$

where the following notations are adopted. The derivative ∇' is the covariant derivative in the bundle BN_f, which means $\nabla'_\mu Y_j = P^3 \circ \nabla_\mu Y_j$, and so

$$\langle Y_i, \nabla'_\mu Y_j \rangle = \langle Y_i, \nabla_\mu Y_j \rangle,$$

where Y_i stands for $Y(\partial_i)$ for a given frame of sections $\partial_i\colon V_0 \to N_e$, $i = 1, \ldots, k_0$. Then we have with $X_i^0 = A(\partial_i) = X_i - Y_i$,

$$C^{i\mu j} = A^{i\mu j} - \langle X_i^0, \nabla_\mu X_j^0 \rangle + \langle \nabla_\mu X_i^0, Y_j \rangle - \langle \nabla_\mu X_j^0, Y_i \rangle.$$

It is now clear (compare 2.2.6) that the space of the 1-jets of regular homomorphisms $N_e \to BN_f$ which satisfy (10) + (11) is fiberwise homotopy equivalent to the Stiefel bundle $\mathrm{St}_r(BN_f)$ for $r = k_0(m_0+1)$. Q.E.D.

Second Derivatives $X_{ij} = \nabla_{ij}\tilde{f}$. Let $\{u_\mu, u_i\}$, $1 \leq \mu \leq m_0 = n - k_0$, $1 \leq i \leq k_0$, be local coordinates in V, such that u_μ extend local coordinates in V_0, u_i are constant on V_0 and the fields $\partial_i = (\partial/\partial u_i)|V_0$ are normal to V_0. If the extension $\tilde{f}$ of f is infinitesimally isometric of second order along V_0, then the fields $X_{ij} = \nabla_{ij}\tilde{f}|V_0$ satisfy the following algebraic equations according to formulae (2′) and (4).

$$\langle X_{ij}, f_\mu \rangle = A^{\mu ij} \tag{12}$$

$$\langle X_{ij}, X_k \rangle = A^{kij} \tag{13}$$

for $1 \leq i, j, k \leq k_0$, and $1 \leq \mu \leq m_0 = n - k_0$ and

$$\langle X_{ij}, f_{\mu\nu} \rangle = D^{\nu\mu ij} + \langle \nabla_\nu X_i, \nabla_\mu X_j \rangle \tag{14}$$

$$\langle X_{ij}, \nabla_\mu X_k \rangle - \langle X_{jk}, \nabla_\mu X_i \rangle = D^{ij\mu k} \tag{15}$$

$$\langle X_{ij}, X_{kl} \rangle - \langle X_{il}, X_{kj} \rangle = D^{iklj}, \tag{16}$$

where $1 \leq i, j, k, l \leq k_0$, $1 \leq \mu, \nu \leq m_0$, and D^{abcd} stands for $B^{abcd} + R^{abcd}$.

Let the map f be free and the homomorphism $X = \{X_i\}$ be regular. Then the fields f_μ, $f_{\mu\nu}$, X_i and $\nabla_\mu X_i$ a span certain subbundle $T^2(f, X) \subset T(W)|V_0$ of dimension $m_0 + [m_0(m_0+1)/2] + k_0 + m_0 k_0$. Denote by $BN(f, X)$ the orthogonal complement $T(W) \ominus T^2(f, X)$ and orthogonally split $X_{ij} = X'_{ij} + X''_{ij}$ for $X'_{ij}: V_0 \to T^2(f, X)$ and $X''_{ij}: V_0 \to BN(f, X)$. If we substitute $X'_{ij} + X''_{ij}$ for X_{ij} in the Eqs. (12)–(15), then we obtain similar equations with X'_{ij} in place of X_{ij},

$$(12') \qquad \langle X'_{ij}, f_\mu \rangle = A^{\mu ij}$$

$$(15') \qquad \langle X'_{ij}, \nabla_\mu X_k \rangle - \langle X'_{ij}, \nabla_\mu X_i \rangle = D^{ij\mu k},$$

while the Eqs. (16) become

$$(16') \qquad \langle X''_{ij}, X''_{kl} \rangle - \langle X''_{il}, X''_{kj} \rangle = E^{iklj},$$

where

$$E^{iklj} = D^{iklj} + \langle X'_{il}, X'_{kj} \rangle - \langle X'_{ij}, X'_{kl} \rangle.$$

The Eqs. (12')–(15') form a non-singular system of linear equations in X'_{ij} and so the solutions X'_{ij} of these equations are just sections of an affine subbundle in $T^2(f, X)$.

The Eqs. (16') are quadratic in X''_{ij}. However, *free* solutions X''_{ij} of (16') have very simple structure [compare Lemma (A)], where the freedom means linear independence of X''_{ij} which is equivalent to the freedom of the map $\tilde{f}$. Namely, for every point $v \in V_0$, and for every $E(v) = E^{iklj}(v)$, where $E^{iklj}(v)$ is a function if $f_\mu(v)$, $f_{\mu\nu}(v)$, $X_i(v)$ and $\nabla_\mu X_i(v)$, the space of free solutions of (16') in the fiber $BN(f, X)|V$ is a smooth manifold, say $G = G(E(v))$, which is homotopy equivalent to the Stiefel manifold of $[k_0(k_0 + 1)/2]$-tuples of independent vectors in $BN(f, X)|v$. Furthermore, this manifold continuously depends upon v and $E(v)$ such that the union $\bigcup G(E(v))$ over all $v \in V_0$ and all possible values of $E(v)$ form a fibration over the space of pairs $(v, E(v))$. This analysis and Lemma (B) give us the following description of systems of vector fields X_i and X_{ij} along V, where X_i satisfy (5) + (6) + (7) + (8) and the regularity condition, while the fields X_{ij} satisfy (12) + (13) + (14) + (15) + (16) and the freedom condition (i.e. the vectors X''_{ij} are independent).

(C) Lemma. *The above systems of fields satisfy the h-principle. This h-principle shows that the space of these systems (X_i, X_{ij}) is weakly homotopy equivalent to the space of injective homomorphisms*

$$\mathscr{I}: N_e \oplus (N_e \otimes T(V_0)) \oplus (N_e)^2 \to BN_f.$$

Third derivatives $X_{ijk} = \nabla_k \tilde{f}_{ij}$. An extension $\tilde{f}$ of f is infinitesimally isometric of second order along V_0 if and only if the first derivatives $\nabla_i \tilde{f} = X_i$ satisfy the Eqs. (5)–(8), the second derivatives $\nabla_{ij} \tilde{f} = X_{ij}$ satisfy (12)–(16) and the third derivatives $X_{ijk} = \nabla_k \nabla_{ij} \tilde{f}$ satisfy [see (3) and the following discussion]

$$(17) \qquad \begin{aligned} \langle X_{ijk}, f_\mu \rangle &= \partial_k A^{\mu ij} - \langle X_{ij}, \nabla_\mu X_k \rangle \\ \langle X_{ijk}, X_l \rangle &= \partial_k A^{lij} - \langle X_{ij}, X_{kl} \rangle, \end{aligned}$$

for $1 \le i, j, k, l \le k_0$, $1 \le \mu \le m_0 = n - k_0$. Since these equations are linear in X_{ijk} and since the Eqs. (14)–(16) give us the complete consistency condition for (17), these equations (17) are solvable in some vector fields X_{ijk} along V_0 and the space of these solutions is contractible. Hence, Lemma (C) yields the following

(C′) **Corollary.** *Free extensions $\tilde{f}\colon \mathcal{O}p V_0 \to W$, for $\tilde{f}|V_0 = f$, which are infinitesimally isometric of second order along V_0 satisfy the h-principle. The space of these extensions has the weak homotopy type of the space of the above injective homomorphisms $\mathcal{I}$.*

Now, we apply the implicit function theorem [compare (A′)] and obtain the following solution to the free isometric extension problem.

(D) **Theorem.** *Let V and W be Riemannian C^∞-manifolds, let $e\colon V_0 \subsetneq V$ be a C^∞-submanifold and let $f\colon V_0 \to W$ be a free isometric C^∞-immersion. Then free isometric C^∞-immersions $\mathcal{O}p V_0 \to W$ (for an "infinitely small" neighborhood $\mathcal{O}p V_0 \subset V$, see 1.4.1) which extend $f = \tilde{f}|V_0$ satisfy the h-principle. Such an extension $\tilde{f}$ exists if and only if the following two conditions are satisfied.*

(i) *the (relative) curvature of the immersion f is strictly greater than the curvature of $e\colon V_0 \subsetneq V$, that is the form $\Phi_f - \Phi_e$ is positive definite on the symmetric square $(T(V_0))^2$;*

(ii) *there exists an injective homomorphism*

$$\mathcal{I}\colon N_e \oplus (N_e \otimes T(V_0)) \oplus N_e^2 \to BN_f,$$

where N_e is the normal bundle of e and BN_f is the binormal bundle of f.

(D′) **Remark.** The above considerations show that the restriction map $\tilde{f} \mapsto f = \tilde{f}|V_0$ is a Serre fibration of the space of free isometric C^∞-immersion over the space of those free C^∞-immersions $f\colon V_0 \to W$ which satisfy the condition (i).

Exercises. Extend the Theorem (D) to C^{an}-immersions.

Let the manifolds V and W be parallelizable and let $V_0 \subset V$ be a totally geodesic submanifold. Show that every free isometric C^∞-immersion $f\colon V_0 \to V$ extends to a free isometric C^∞-immersion $\tilde{f}\colon \mathcal{O}p V_0 \to W$, for $q = \dim W \geq n(n+3)/2$, $n = \dim V$.

Let V and W be Riemannian C^{an}-manifolds, let $e\colon V_0 \subsetneq V$ be a 1-codimensional C^{an}-submanifold and let $f\colon V_0 \to W$ be a free isometric C^{an}-immersion for which $\Phi_f > \Phi_e$. Show that f extends to a (not necessarily free) isometric C^{an}-immersion $\tilde{f}\colon \mathcal{O}p V_0 \to W$ if and only if there exists an injective homomorphism $N_e \to BN_f$.

Hint. Use the Cauchy-Kovalevskaya theorem in place of the implicit function theorem.

The h-Principle for Isometric Immersions $\mathcal{O}p V_0 \to W$. The results in 2.1.2 and the proof of the Theorem (A) of 3.1.5 show that an arbitrary strictly short map f_0: $V_0 \to W$ admits a C^0-approximation by isometric C^∞-immersions $f\colon V_0 \to W$ with an arbitrarily large curvature form Φ_f^+, provided $\dim W = q \geq (m_0 + 2)(m_0 + 5)/2$ for $m_0 = n - k_0 = \dim V_0$. If the form Φ_f^+ is large, then the quadratic form Φ_f on $(T(V))^2$ also is large and for $\Phi_f > \Phi_e$ the Theorem (D) allows one to extend the immersion f to $\mathcal{O}p V_0 \subset V$. Moreover, we obtain with the Remark (D′) the h-principle for free isometric C^∞-immersions $\mathcal{O}p V_0 \to W$ relative to strictly short maps for $q \geq (m_0 + 2)(m_0 + 5)/2$. This inequality is automatically satisfied for $k_0 = \operatorname{codim} V_0 \geq 2$ and so we have the following

(E) **Theorem.** *Let $V_0 \subset V$ be a C^∞-submanifold of codimension $k_0 \geq 2$ in an n-dimensional Riemannian C^∞-manifold V and let $f_0\colon V_0 \to W$ be a strictly short map, where W is a Riemannian C^∞-manifold of an arbitrary dimension q. Then the following condition (i) is necessary and sufficient for the existence of a free isometric C^∞-map $f\colon \mathcal{O}p\, V_0 \to W$ whose restriction $f|V_0$ is arbitrarily C^0-close to f_0.*

(i) *There exists an injective homomorphism of the bundle $T(V) \oplus (T(V))^2 | V_0$ to the induced bundle $f_0^*(T(W))$.*

(E′) **Corollary.** *If the manifold V is parallelizable, then some neighborhood $U \subset V$ of an arbitrary C^∞-*submanifold *$V_0 \subset V$ of* codim *$V_0 = k_0 \geq 2$ admits a free isometric C^∞-immersion $f\colon U \to \mathbb{R}^q$ for $q = n(n+3)/2$ where $n =$* dim *V.*

Observe that no neighborhood $U \subset V$ admits a free map $U \to \mathbb{R}^q$ for $q < n(n+3)/2$.

Exercises. Extend (E) and (E′) to the case $k_0 = 1$ and $q \geq (n+1)(n+4)/2$.

Let the manifolds V, W and $e\colon V_0 \subsetneq V$ be real analytic and parallelizable, let $k_0 = \operatorname{codim} V_0 \geq 3$ and let the normal bundle $N_e \to V_0$ admit a 1-dimensional subbundle (e.g. $k_0 > m_0 = \dim V_0$). Show for $q \geq n(n+1)/2$ the existence of an isometric (possibly non free) C^{an}-immersion of some neighborhood $U \subset V$ of V_0 into W. Then assume $k_0 = 2$ and prove the existence of such an immersion $U \to W$ for $q \geq n(n+3)/2$.

3.1.7 Isometric Immersions $V^n \to W^q$ for $q \geq (n+2)(n+3)/2$

Let $V = (V, g)$ and $W = (W, h)$ be Riemannian C^∞-manifolds. We use *special* coordinates $u_1, u_2, \ldots, u_n$ in V, where $u_2, u_3, \ldots, u_n$ are global cyclic coordinates of some C^∞-imbedded normally oriented torus $T^{n-1} \subsetneq V$ and where u_1 is the normal coordinate in some tubular neighborhood $U \subset V$ of T^{n-1}, such that $u_1 = 0$ is the equation for $T^{n-1} \subset U \subset V$. Then we use coordinates $\{u_1, \ldots, u_n, t, \theta\}$ in $V \times \mathbb{R}^2$ for the Euclidean coordinates $\{t, \theta\}$ in $\mathbb{R}^2$.

Let $F\colon V \times \mathbb{R}^2 \to W$ be a C^∞-immersion and let $x = x(u_1)$ be a C^∞-function with compact support in the real variable u_1. Then we consider the functions $t(u_1) = \varepsilon \sin \varepsilon^{-1} x$ and $\theta(u_1) = \varepsilon(1 - \cos \varepsilon^{-1} x)$ for $\varepsilon > 0$ and we define the following C^∞-map $f = f_\varepsilon\colon V \to W$ for all $\varepsilon > 0$,

$$f(v) = f(u_1, \ldots, u_n) = F(u_1, \ldots, u_n, t(u_1), \theta(u_1)), \qquad \text{for } v \in U,$$

and

$$f(v) = F(v, 0) \qquad \text{for } v \in V \backslash U.$$

A straightforward computation gives the following formulae for the covariant derivative of f in the Riemannian manifold $W = (W, h)$.

$$\nabla_i f = \nabla_i F \quad \text{and} \quad \nabla_{ij} f = \nabla_{ij} F \tag{1}$$

for $2 \leq i, j \leq n$;

(2) $$\nabla_1 f = \nabla_1 F + x' Y^1,$$

where $x' = dx(u_1)/du_1$ and $Y^1 = c\nabla_t F + s\nabla_\theta F$, for $c = \cos \varepsilon^{-1} x$ and $s = \sin \varepsilon^{-1} x$;

(3) $$\nabla_{1i} f = x_{1i} F + x' \nabla_i Y^1$$

for $2 \le i \le n$;

(4) $$\nabla_{11} f = \nabla_{11} F + x'' Y^1 + \varepsilon^{-1}(x')^2 Y^2 + x' Y^3,$$

where

$$Y^2 = -s\nabla_t F + c\nabla_\theta F$$

and

$$Y^3 = 2(c\nabla_{1t} F + s\nabla_{1\theta} F) + x'(c^2 \nabla_{tt} F + 2cs\nabla_{t\theta} F + s^2 \nabla_{\theta\theta} F).$$

Here $\nabla_{1t} = \nabla_{u_1} \nabla_t$ and so on.

If the map

$$F: (V \times \mathbb{R}^2, g \oplus dt^2 \oplus t\theta^2) \to (W, h)$$

is infinitesimally isometric of infinite order along $V = V \times 0 \subset V \times \mathbb{R}^2$, then the induced metric $f_\varepsilon^*(h)$ C^∞-converges to $g + (dx)^2$ for $\varepsilon \to 0$. Moreover the jet of the difference satisfies for $\varepsilon \to 0$

$$\varepsilon^{-k} J^r(f_\varepsilon^*(h) - g - dx^2) \to 0,$$

for all $k, r = 1, 2, \ldots$. Indeed, the map $x \mapsto (\varepsilon \sin \varepsilon^{-1} x, \varepsilon(1 - \cos \varepsilon^{-1} x)$ *isometrically* sends $(\mathbb{R}, dx^2)$ into $(\mathbb{R}^2, dt^2 + d\theta^2)$, and so the argument of 3.1.2 applies.

Now, we test the freedom of the maps $f = f_\varepsilon$, $\varepsilon \to 0$, for possibly non free maps F. We assume the restriction $F|V$ to be a free map $V \to W$ and we denote by $BN \subset T(W)|V$ the binormal bundle of the map $F|V$. Put

$$Y(\alpha) = (\cos \alpha)\nabla_t F + (\sin \alpha)\nabla_\theta F,$$

for all $\alpha \in [0, 2\pi)]$, and let $Z_i(\alpha)$, $i = 2, \ldots, n$, denote the normal projections of the covariant derivatives $\nabla_i Y(\alpha)|U$ to BN for the coordinates domain $U \subset V$.

(A) **Lemma.** *If the map F is infinitesimally isometric of order $r \ge 1$ along $U = U \times 0 \subset V \times \mathbb{R}^2$, and if the fields $Z_i = Z_i(\alpha)$, $i = 2, \ldots, n$, are linearly independent for every $\alpha \in [0, 2\pi]$, then the map $f = f_\varepsilon: V \to W$ is free for all sufficiently small $\varepsilon > 0$.*

Proof. The Eqs. (1)–(4) in 3.1.2 show that the fields $Y_1|U$ and $Y_2|U$ are mutually orthogonal, have unit norm and that they are orthogonal to the fields $\nabla_i F|U$, $\nabla_{ij} F|U$ and $\nabla_k Y^1$ for $1 \le i, j \le n$ and $2 \le k \le n$. Since the fields $\nabla_i F|V$ and $\nabla_{ij} F|V$ are linearly independent, since the binormal projections Z_k of $\nabla_k Y^1$ also are independent and since the field Y^3 remains bounded for $\varepsilon \to 0$, the following $n + [n(n+1)/2]$ fields also are independent along U for all small $\varepsilon > 0$,

$$\nabla_i F, \quad \nabla_{ij} F, \nabla_{ij} + x' \nabla_i Y_1,$$

$$\nabla_1 F + x' Y^1, \quad \nabla_{11} F + x'' Y^1 + \varepsilon^{-1}(x')^2 Y^2 + x' Y^3 \quad \text{for } 2 \le i, j \le n,$$

and the lemma follows.

Remark. The above argument shows the maps f_ε to be uniformly free for $\varepsilon \to 0$, which means the uniform bound from below for the norm of the Hessian of the derivatives $\nabla_i f_\varepsilon$ and $\nabla_{ij} f_\varepsilon$, $1 \leq i, j \leq n$.

This remark and the relation $(*)$ allows one to apply the maps f_ε the implicite function theorem for operators of polynomial growth (see 2.3.6) and thus to obtain the following generalization of the adding dx^2 lemma [see (B′) in 3.1.2].

(B) **Lemma.** *Let the map F be infinitesimally isometric of infinite order along V and let the fields Z_i, $i = 2, \ldots, n$, be independent. Then there exists a small C^∞-perturbation f^1 of the map f_ε, for any sufficiently small $\varepsilon > 0$, such that $f^1(h) = g + (dx)^2$.*

Let us show that the assumptions of this lemma can be met for $q \geq (n+2)(n+3)/2$. Consider a free isometric C^∞-immersion $f_0 \colon (V, g) \to (W, h)$ and let the special coordinate domain $T^{n-1} \times \mathbb{R} \approx U \subset V$ lie in some topological ball in V.

(C) **Lemma.** *If $q \geq (n+2)(n+3)/2$, then there exists a C^∞-maps $F \colon V \times \mathbb{R}^2 \to W$ which satisfies the assumptions of* (B) *and for which $F|V = f_0$.*

Proof. First we construct an orthonormal 2-frame of binormal fields X_1 and X_2 along $U \subset V \to W$, such that

(i) $\langle \nabla_i X_k, X_l \rangle = 0$, $1 \leq i \leq n$, $1 \leq k, l \leq 2$,
(ii) the field X_1 is regular,
(iii) the frame (X_1, X_2) is semi regular (see 2.2.6) along the hypersurface $u_1 =$ const in U for all const $\in \mathbb{R}$. This means the linear independence of the orthogonal projections of the derivatives $\nabla_i(cX_1 + sX_2)$, $i = 2, \ldots, n$, to the binormal bundle $BN = BN_{f_0} \subset T(W)|V$ for every pair of constants (c, s), $c^2 + s^2 = 1$. The existence of these fields X_1 and X_2, which are sections $U \to BN$, follows for $\dim BN = q - [n(n+3)/2] \geq n + 3$, from the h-principle for semi regular bundle homomorphisms (see 2.2.6).

Next we construct (compare 3.1.2) a free infinitesimally isometric (along U) map $F^1 \colon (U \times \mathbb{R}, g \oplus dt^2) \to W$, for which $F^1|U = f_0|U$ and $\nabla_t F^1|U = X_1$, such that the field $\nabla_{tt} F^1|U$ is normal to X_2.

Finally, we extend F^1 to the desired infinitesimally isometric map F: $(U \times \mathbb{R}^2, g \oplus dt^2 \oplus dt^2 \oplus d\theta^2) \to W$ with $\nabla_\theta F|U = X_2$. Q.E.D.

Now, we are able to produce isometric immersions $V \to W$ with the best known lower bound on $q = \dim W$.

(D) **Theorem.** *Let $f_0 \colon V \to W$ be a strictly short map between Riemannian C^∞-manifolds of dimensions n and q such that $q \geq (n+2)(n+3)/2$. Then f_0 admits a fine C^0-approximation by free isometric C^∞-maps $f \colon V \to W$.*

Proof. We only need to show (compare 3.1.2) that every Riemannian C^∞-metric on V decomposes into a locally finite sum $\sum_j dx_j^2$, where every function x_j is of the form

$x_j = x_j(u_1)$ for some special coordinate system in a small domain $U = U_j \approx T^{n-1} \times \mathbb{R}$ in V. If V is compact, then the existence of such a (finite) decomposition directly follows from Theorem (A) in 3.1.3. The compact decomposition [see (B) in 3.1.1] reduces the non-compact case to the compact one. Q.E.D.

Exercises. Extend the C^1-approximation theorem (G) in 3.1.2, the parametric h-principle (H) in 3.1.2 and the curvature theorem (A) in 3.1.5 to the dimensions q in the interval $(n+2)(n+3)/2 \leq q < (n+2)(n+5)/2$.

Generalize the above to C^α-immersions of C^α-manifolds for $\alpha \geq 100$.

3.1.8 Isometric Cylinders $V^n \times \mathbb{R} \to W^q$ for $q \geq (n+2)(n+3)/2$

If $q \geq (n+3)(n+4)/2$, then the above Theorem (D) gives us a complete hold of free isometric immersions $(V \times \mathbb{R}, g \oplus dt^2) \to (W, h)$. We shall see presently how *the proof* of that theorem yields such an immersion for $q \geq (n+2)(n+3)/2$ for *compact* (possibly with boundary) manifolds V.

Consider a C^∞-immersion $F: (V \times \mathbb{R}^2, g + dt^2 + d\theta^2) \to (W, h)$ which is infinitesimally isometric of first order along V and whose restriction $F|V: V \to W$ is free. Then the differential D_F maps the normal bundle $N = V \times \mathbb{R}^2 \to V$ of $V = V \times 0 \subset V \times \mathbb{R}^2$ into the binormal bundle $BN \subset T(W)|V$ of the map $F|V$ by sending the fields $\partial/\partial t$ and $\partial/\partial\theta$ in $V \times \mathbb{R}^2$ to the orthonormal fields $\nabla_t F$ and $\nabla_\theta F$ along V.

Consider the circle $S_\varepsilon \subset \mathbb{R}^2$ given by $t^2 + \theta^2 = \varepsilon^2$ and concentrate on the map $F|V \times S_\varepsilon: V \times S_\varepsilon \to W$.

(A) Lemma. *If the manifold V is compact and if the homomorphism $D_F: N \to BN$ is semi regular* (see 2.2.6), *then the map $F|V \times S_\varepsilon$ is free for all sufficiently small $\varepsilon > 0$. Moreover the maps $F|V \times S_\varepsilon$ are uniformly free for $\varepsilon \to 0$.*

Proof. The semi regularity assumption amounts to linear independence of the binormal components of the vectors $\nabla_{u_i}(c\nabla_t F + s\nabla_\theta F)$, $i = 1, \ldots, n$, along V for every pair $(c, s) \in \mathbb{R}^2$, $c^2 + s^2 \neq 0$ and for every system of local coordinates $u_1, \ldots, u_n$ in V. We parametrize $S_\varepsilon \subset \mathbb{R}^2$ by $t = \varepsilon \sin \varepsilon^{-1} x$ and $\theta = \varepsilon \cos \varepsilon^{-1} x$ for $x \in \mathbb{R}$ and we have the following formulae for the covariant derivatives of the map $\tilde{f}_\varepsilon = F|V \times S_\varepsilon$

$$\nabla_i \tilde{f}_\varepsilon = \nabla_i F$$
$$\nabla_{ij} \tilde{f}_\varepsilon = \nabla_{ij} F,$$

for $1 \leq i, j \leq n$, and

$$\nabla_x \tilde{f}_\varepsilon = c\nabla_t F - s\nabla_\theta F$$
$$\nabla_{ix} \tilde{f}_\varepsilon = \nabla_i(c\nabla_t F - s\nabla_\theta F)$$
$$\nabla_{xx} \tilde{f}_\varepsilon = -\varepsilon^{-1}(s\nabla_t F + c\nabla_\theta F) + c^2\nabla_{tt} F - 2cs\nabla_{t\theta} F + s^2\nabla_{\theta\theta} F,$$

where $s = \sin \varepsilon^{-1} x$ and $c = \cos \varepsilon^{-1} x$. These derivatives are independent for $\varepsilon \to 0$ [compare (A) in 3.1.7] and the lemma is proven.

If the map F is infinitesimally isometric of infinite order along V, then there is a small perturbation of the map $\tilde{f}_\varepsilon$ which is a free *isometric* map

$$\tilde{f}'_\varepsilon\colon (V \times S^1_\varepsilon, g \oplus dx^2) \to (W, h)$$

by the implicit function theorem (compare 3.1.7). We can even find a perturbation, such that $\tilde{f}'_\varepsilon | V \times s_0 = F | V \times 0$, for a given point $s_0 \in S^1$. This is done by translating $S^1_\varepsilon \subset \mathbb{R}^2$ to the circle $S^1_\varepsilon(s_0)$ around s_0 and then by perturbing the map $F|V \times S^1_\varepsilon(s_0)$ without changing it on $V \times 0 \subset V \times S^1_\varepsilon(s_0)$.

(A′) **Corollary.** *Let V be a compact parallelizable manifold and let $f\colon V \to W$ be a free isometric C^∞-immersion for which the induced bundle $f^*(T(W))$ is trivial. If $q \geq [(n+2)(n+3)/2] + 1$, then there exists a free isometric C^∞-immersion $\tilde{f}_\varepsilon\colon V \times S^1_\varepsilon \to W$ for all $\varepsilon > 0$, such that $\tilde{f}_\varepsilon | V \times s_0 = f$ for a given point $s_0 \in S^1_\varepsilon$.*

Proof. The binormal bundle of f satisfies

$$BN_f = f^*(T(W)) \ominus [T(V) \oplus (T(V))^2],$$

and so BN_f is trivial in our case. Furthermore

$$\dim BN_f = q - \frac{n(n+3)}{2} \geq n + 4,$$

and so there is a semi regular homomorphism $N = V \times \mathbb{R}^2 \to BN_f$ as in (B′) of 2.2.6. This homomorphism extends to an infinitesimally isometric map F as in 3.1.7. Thus, we obtain the required immersion $V \times S^1_\varepsilon \to W$ for short circles S^1_ε and by taking cyclic coverings of $V \times S^1_\varepsilon$ we make the length of the circle any number we want. Q.E.D.

(A″) **Example.** If (V, g) is isometric to a flat torus, then $V \times S^1$ is also a flat torus and so we obtain by applying (A′) and the Theorem (D) of 3.1.7 a free isometric C^∞-immersion of the standard flat torus T^{n+1} into any given C^∞-manifold W, provided $\dim W \geq [(n+2)(n+3)/2] + 1$. The universal covering of T^{n+1} gives us a free isometric immersion $\mathbb{R}^{n+1} \to W$.

(A‴) **Remark.** If the manifolds V and W and the map f are real analytic, then by the Cauchy-Kovalevskaya theorem there exists an isometric (non-free) map of a small neighborhood $\mathcal{O}p\,V \subset V \times \mathbb{R}^2$ into W for all $q \geq (n+2)(n+3)/2$ [see (E′) in 3.1.2] and thus we obtain (non-free) isometric C^{an}-maps $\mathbb{R}^{n+1} \to W$ for $\dim W = (n+2)(n+3)/2$.

Exercise. Prove the above result for C^∞-manifolds and maps.

Hint. See 2.3.6 and 3.1.9.

Cylinders Between Isometric Immersions. We consider two homotopic isometric C^∞-immersions f_0 and $f_1\colon V \to W$ and ask ourselves if there exists *an isometric C^∞-cylinder between f_0 and f_1*, which by definition is an isometric C^∞-map $\bar{f}\colon (V \times [0, l], g \oplus dt^2) \to W$, for some $l \in (0, \infty)$, such that $\bar{f} | V \times 0 = f_0$ and $\bar{f} | V \times l =$

f_1. An obvious obstruction for transforming a given homotopy between f_0 and f_1 to a cylinder may come from the "longness" of the homotopy and so we assume the existence of a *strictly short* homotopy f_t, $t \in [0, 1]$, between f_0 and f_1 for which the map f_t is strictly short for $0 < t < 1$. Next we assume the maps f_0 and f_1 to be free. If $q \geq (n + 2)(n + 3)/2$, then the techniques of the previous section allow us to stretch a strictly short homotopy to a free isometric C^∞-homotopy. This is also denoted by f_t but now the map $f: V \times [0, 1] \to W$ is C^∞-smooth and $f_t = f|V \times t: V \to W$ is a free isometric immersion for all $t \in [0, 1]$. Finally, let the manifold V be compact and parallelizable and assume the induced bundle $f_0^*(T(W))$ to be trivial.

(B) Theorem. *If $q \geq [(n + 2)(n + 3)/2] + 1$, then the maps f_0 and f_1 can be joined by a free isometric C^∞-cylinder $V \times [0, l] \to W$. If $q = (n + 2)(n + 3)/2$, then a (possibly non-free) isometric C^∞-cylinder exists, provided the Riemannian manifolds are real analytic (with no analyticity assumption on the maps f_0 and f_1).*

Proof. Take the product $V \times S^1$ for $S^1 = S_1^1 \subset \mathbb{R}^2$ given by $t^2 + \theta^2 = 1$ and apply (an obvious generalization of) the proof of (A′) to the free isometric maps f_t, $t \in [0, 1]$. Thus, we obtain a C^∞-family of free isometric maps $\tilde{f}_t: V \times S^1 \to W$, such that $\tilde{f}_0|V \times 0 = f_0$ and $\tilde{f}_1|V \times 0 = f_1$, where the circle S^1 is parametrized by $[0, 2\pi]$. Since free isometric maps $V \times S^1 \to W$ are microflexible (see 1.4.2 and 2.3.2), there is a small perturbation $\tilde{f}_t'$ of $\tilde{f}_t$ such that

(i) the maps $\tilde{f}_t': V \times S^1 \to W$ are free isometric C^∞-immersions for all $t \in [0, 1]$ and $\tilde{f}_t' = \tilde{f}_t$ for $t = 0$ and $t = 1$;

(ii) there exist some points $0 = t_0 < \cdots < t_i < \cdots < t_k = 1$ and some points s_i and s_i' in S^1, such that

$$\tilde{f}_{t_{i-1}}'|\mathcal{O}p(V \times s_i) = \tilde{f}_{t_i}'|\mathcal{O}p(V \times s_i'), \qquad i = 1, \ldots, k,$$

for small neighborhoods $\mathcal{O}p(V \times s_i) \subset V \times S^1$. Thus we obtain an isometric map $\tilde{\tilde{f}}: V \times \tilde{S} \to W$ for the non-Hausdorff manifold $\tilde{S}$ which is obtained from $k + 1$ copies of S^1 by gluing together the small intervals $\mathcal{O}p(s_{i-1}) \subset S^1 = S_{i-1}^1$ and $\mathcal{O}p(s_i') \subset S^1 = S_i^1$, $i = 0, \ldots, k$. There is an obvious isometric C^∞-map of some interval $[0, l]$ into $\tilde{S}$ which joins the points $0 \in S^1 = S_0^1 \subset \tilde{S}$ and $0 \in S^1 = S_k^1 \subset \tilde{S}$, say $\varphi: [0, l] \to \tilde{S}$. Then the composed map $\tilde{\tilde{f}} \circ (\mathrm{Id} \times \varphi): V \times [0, l] \to W$ is the required cylinder.

The case $q = (n + 2)(n + 3)/2$. Suppose for a while that the maps f_0 and f_1 are real analytic. Then the maps f_t can be made real analytic for all $t \in [0, 1]$ and next they extend to a family of isometric C^{an}-maps $F_t: V \times D^2 \to W$, for a small disk $D^2 = D_\varepsilon^2 \subset \mathbb{R}^2 = \{t, \theta\}$ around the origin, such that the map F_t is free on $V \times D \cap (\mathbb{R}^1 \times 0)$ for all $t \in [0, 1]$. Thus, one obtains a family of isometric maps $\tilde{f}_t: V \times S^1 \to W$, such that the map $\tilde{f}_t|\mathcal{O}p(V \times 0)$ is free for all $t \in [0, 1]$. This partial freedom is sufficient for the above microflexibility argument to apply and the case of analytic maps f_0 and f_1 is concluded. In fact, this argument provides a C^∞-cylinder $\bar{f}: V \times [0, l] \to W$, such that the restrictions $\bar{f}|V \times [0, \varepsilon]$ and $\bar{f}|V \times [l - \varepsilon, l]$ for small $\varepsilon > 0$ are *given* free isometric C^{an}-maps, say $\bar{f}_0$ and $\bar{f}_1: V \times [0, \varepsilon] \to W$, where the interval $[l - \varepsilon, l]$ is identified with $[0, \varepsilon]$. (The maps $\bar{f}_0$ and $\bar{f}_1$ are, of course, not quite arbitrary. They are assumed to extend f_0 and f_1 respectively and they must be homotopic, which means the existence of a family of free isometric maps $\bar{f}_t$: $V \times [0, \varepsilon] \to W$, $t \in [0, 1]$, between $\bar{f}_0$ and $\bar{f}_1$.)

Now, let the maps f_0 and f_1 be C^∞-smooth. Then they can be extended to free isometric C^∞-maps $\bar{f}_0$ and $\bar{f}_1$ of $V \times [0, \varepsilon] \to W$. Since the Riemannian manifolds V and W are real analytic, one can make these extensions real analytic outside $V = V \times 0 \subset V \times [0, \varepsilon]$ by the implicit function theorem (see 2.3.2 and 2.3.6). Then the C^{an}-maps $\bar{f}_0 | V \times [\varepsilon/2, \varepsilon]$ and $\bar{f}_1 | V \times [\varepsilon/2, \varepsilon]$ are joined by an isometric cylinder and thus the maps f_0 and f_1 are also joined by a cylinder. Q.E.D.

Example. Let f_1 and f_2 be free isometric immersions $S^1 \to \mathbb{R}^q$. If $q \geq 6$, then there is an isometric C^∞-cylinder between f_1 and f_2 which can be assumed free for $q \geq 7$.

Question. Does an isometric cylinder between f_0 and f_1 exist for $q = 4$ and $q = 5$?

Exercises. Prove for C^{an}-manifolds V and W that every free isometric C^∞-cylinder between C^{an}-maps f_0 and f_1: $V \to W$ admits a C^∞-approximation by an isometric C^{an}-cylinder between f_0 and f_1.

Hint. See 2.3.6.

Let $V = (V, g)$ be a compact parallelizable Riemannian C^∞-manifold and let f_0: $V \times S^1 \to (W, h)$ be a strictly short map, such that the induced bundle $f_0^*(T(W))$ is trivial. Prove for $\dim W \geq [(n + 2)(n + 3)/2] + 1$ the existence of an approximation of f_0 by a free C^∞-map f: $V \times S^1 \to (W, h)$, for which the induced metric $f^*(h)$ equals $g \oplus \lambda\, dx^2$, where the (large) positive constant λ may depend on the required precision of the approximation.

Stretching a Cylinder. Let f: $V \times [0, l] \to W$ be an isometric cylinder. Suppose that the map $f | V \times [a, b]$, for some subinterval $[a, b] \subset [0, l]$, extends to an isometric map $\tilde{f}$: $V \times [a, b] \times [0, \varepsilon] \to W$ for some $\varepsilon > 0$. Show the existence of an arbitrary long isometric cylinder $V \times [0, l_1] \to W$ for all $l_1 \geq l$ which equals f near the ends of the cylinder. Study the freedom of this long cylinder. (An interesting case is when f is free but $\tilde{f}$ may be non-free). Study extensions $\tilde{f}$ which are only infinitesimally isometric (along $V \times [a, b]$) and give a sufficient condition for the existence of a stretching for $q \geq (n + 2)(n + 3)/2$, $n = \dim V$.

Folded Cylinders. A *folded C^∞-cylinder* is a continuous map f: $V \times [0, l] \to W$ with the following property. There is a subdivision of $[0, l]$ into closed subintervals

$$[l_{i-1}, l_i] \qquad \text{for } i = 0, \ldots, k \quad \text{and} \quad l_0 = 0,\, l_k = l,$$

such that the map f is C^∞-smooth on each segment $[l_{i-1}, l_i]$ and

$$f(v, l_i - \delta) = f(v, l_i + \delta),$$

for all $v \in V$, $i = 1, \ldots, k - 1$, and for all sufficiently small $\delta > 0$.

Let f_t: $V \times [0, \varepsilon] \to W$, $t \in [0, 1]$, be a C^∞-family of free isometric C^∞-maps. Prove, in case V is compact, the existence of a folded C^∞-cylinder $\bar{f}$ between $f_0 | V \times 0$ and f_1: $| V \times 0$, which is free at every segment $[l_{-1}, l_i]$ where $\bar{f}$ is C^∞-smooth.

Unfolding the Folds. Let $f\colon V \times [0, l] \to W$ be a folded C^∞-cylinder and suppose $f | V \times [l_i, l_i + \delta]$ admits an extension to an isometric C^∞-map

$$\tilde{f}\colon V \times [l_i, l_i + \delta] \times [0, \varepsilon] \to W,$$

for all folding points $l_i \subset [0, l]$, and for some positive δ and ε. Prove the existence of an isometric C^∞-cylinder (without any folds) between $f | V \times 0$ and $f | V \times l$. Study the freedom of this smooth cylinder and analyse the possibilities of infinitesimally isometric extensions $\tilde{f}$. Give an alternative proof of the Theorem (B) by using folded (and then unfolded) cylinders.

Let (T^2, g) be a torus with an arbitrary flat metric and let $f_0\colon T^n \to (W, h)$ be an arbitrary continuous map such that the induced bundle $f_0^*(T(W))$ is trivial. Prove for a C^{an}-manifold W of dimension $q \geq (n+1)(n+2)/2$ the existence of a C^0-approximation of f_0 by a C^∞-map $f\colon T^n \to W$ for which the induced metric $f^*(h)$ equals λg for some constant $\lambda = \lambda(f) > 0$. Show for $q \geq [(n+1)(n+2)/2] + 1$ the existence of a similar *free* approximation f.

Consider a manifold $V = V_0 \times [0, 1]$ with an arbitrary (noncylindrical) Riemannian metric g on V. Observe that the metric $g + \lambda\, dt^2$ is "nearly cylindrical" for the projection $t\colon V_0 \times [0, 1] \to [0, 1]$ and for large constants $\lambda > 0$. Combine this observation with the implicit function theorem and with the above discussion on cylindrical immersions and prove the following

Proposition. *If a manifold $V \approx V_0 \times [0, 1]$ is compact and parallelizable and if a strictly short map $f_0\colon (V, g) \to (W, h)$ induces the trivial bundle $f_0^*(T(W))$ on V, then for* $\dim W \geq [(n+1)(n+2)/2] + 1$, $n = \dim V$, *there is a free C^∞-approximation $f\colon V \to W$ of f_0 for which the induced metric $f^*(h)$ equals $g + \lambda\, dt^2$ for some constant $\lambda = \lambda(f) > 0$.*

Generalize this proposition to nonsplit (compact or not) manifolds V by taking an arbitrary Morse function $t\colon V \to \mathbb{R}$ and thus show the existence of a free isometric C^∞-immersion of (V, g), (for all Riemannian metrics g on V) into the pseudo-Riemannian manifold $(W \times \mathbb{R}, h \oplus (-dt^2))$, where W is an arbitrary Riemannian manifold of dimension $q \geq [(n+1)(n+2)/2] + 1$. For example, every parallelizable Riemannian C^∞-manifold V of dimension n admits a free isometric immersion into the pseudo-Euclidean space $\mathbb{R}^{q+1} = \mathbb{R}^{q,1}$ with the metric

$$\sum_{i=1}^{q} dx_i^2 - dx_{q+1}^2, \qquad \text{for } q = \frac{(n+1)(n+2)}{2} + 1.$$

Give examples of n-dimensional (non-parallelizable!) manifolds V which admit no free maps into $\mathbb{R}^q$ for $q < n(n+4)/2$.

3.1.9 Non-free Isometric Maps

The study of non-free maps faces two major problems. First, the algebra-geometric structure of jet spaces of isometric maps becomes quite complicated unless some regularity (freedom like) condition is imposed on these maps. The second difficulty arises from the failure of the implicit function theorem for non-free C^∞-maps. This problem is less severe for C^{an}-maps due to Cauchy-Kovalevskaya theorem. (Jacobowitz 1982; Bryant-Griffiths-Yang 1983).

A Weakened Freedom Condition. Take a hyperplane $H \subset T_v(V)$, $v \in V$, and let $V_0 \subset V$ be a hypersurface through the point v which has $T_v(V_0) = H$ and which is geodesic at v. We call an isometric C^2-map $f: V \to W$ *free along* H, if the map $f | V_0$ is free in some neighborhood $U \subset V_0$ of $v \in V_0$. The map f is called *weakly free* if it is free along some hyperplane $H \subset T_v(V)$ for all points $v \in V$.

The structure of local isometric weakly free maps $\mathcal{O}p(v_0) \to W$, $v_0 \in V$, is very simple in the C^{an}-category. Indeed, every free isometric C^{an}-map $f_0: V_0 \to W$ extends near $v_0 \in V_0 \subset V$ to an isometric C^{an}-map of a small neighborhood $\mathcal{O}p(v_0) \subset V$ into W, provided the binormal space $BN_{f_0}(V_0, v_0) \subset T_{w_0}(W)$, $w_0 = f_0(v_0)$ has dimension $b \geq 1$. If $b = 1$, then the extension is unique with a choice of an orientation in BN_{f_0}. If $b \geq 2$ (which amounts to $\dim W \geq [n(n+1)/2] + 1$), then the space of these extensions is homotopy equivalent to the sphere S^{b-1}. This follows from our analysis in 3.1.6.

Example. Consider a surface $V = V^2$ with a Riemannian C^{an}-metric g on V. One obtains all weakly free isometric C^{an}-immersions $\mathcal{O}p(v_0) \to \mathbb{R}^3$ by taking any geodesic $V_0 \subset V$ through $v_0 \in V$ and then by construction a free isometric C^{an}-immersion of a germ of V_0 at v_0 into $\mathbb{R}^3$. The extension to V is then unique up to a choice of a binormal to V_0 in $\mathbb{R}^3$. If the Gauss curvature K_g does not vanish at $v \in V$, then every isometric immersion $\mathcal{O}p(v_0) \to \mathbb{R}^3$ is free on some geodesic $V_0 \subset V$, and so the above description gives us *all* isometric C^{an}-immersions $\mathcal{O}p(v_0) \to \mathbb{R}^3$. This is not true anymore if $K_g(v_0) = 0$. Moreover, Hopf and Schilt (1937) found the following example of an isometric C^{an}-immersion $f: \mathcal{O}p(v_0) \to \mathbb{R}^3$ which cannot be isometrically C^2-deformed to any weakly free isometric immersion. To see an obstruction to such a deformation, we look at the tangential (Gauss) map G_f: $\mathcal{O}p(v_0) \to S^2$, whose Jacobian equals the Gauss curvature K of the surface. If $v_0 \in V$ is an isolated zero of $K = K(v)$, then the map G_f is a covering of the punctured neighborhood $\mathcal{O}p(v_0) \backslash v_0$ onto $\mathcal{O}p(s_0) \backslash s_0$ for $s_0 = G_f(v_0) \subset S^2$, $\mathcal{O}p(s_0) \subset S^2$. The local degree of this map $\deg(G_f, v_0)$ clearly is invariant under isometric deformations of the map f. If f is a weakly free map, then obviously, $|\deg(G_f, v_0)| = 1$, and so no analytic surface in $\mathbb{R}^3$ with $|\deg| > 1$ can be isometrically deformed to a weakly free surface.

Exercises. (a) Show for $K \geq 0$, that $\deg(G_f, v_0) = 1$ for the above surfaces V. Then consider a smooth function $\varphi: \mathbb{R}^2 \to \mathbb{R}$ with an isolated critical point $v_0 \in \mathbb{R}^2$ and prove Kronecker's inequality

$$\operatorname{index}(\operatorname{grad} \varphi, v_0) \leq 1.$$

(b) Construct for a given integer $-d \leq 1$ a C^{an}-disk $V \subset \mathbb{R}^3$ with an isolated zero $v_0 \subset V$ of the Gauss curvature and with $\deg(G, v_0) = -d$ for the tangential map $G: V \to S^2$.

Let $f\colon V \to \mathbb{R}^3$ be a C^{an}-immersion for which the point $v_0 \in V$ is an isolated zero of the Gauss curvature $K_g(v)$ of the induced metric g and let $|\deg(G_f, v_0)| = d \geq 2$. Then the map f is *d-flat at* v_0. Namely, the dth-osculating space $T^d_{v_0}(f) \subset T_{w_0}(\mathbb{R}^3)$, $w_0 = f(v_0)$, coincides with the tangent space $T^1_{v_0}(f) \doteq D_f(T(V)) \subset T_{w_0}(\mathbb{R}^3)$. Suppose that the map f is *generic* in the space of d-flat maps. Then $T^{d+1}_{v_0}(f) \neq T^d_{v_0}(f)$ and small perturbations of f, preserving the d-flatness at v_0 also have at v_0 an isolated zero of the induced curvature, while the degree $\deg(G_f, v_0)$ is constant under such perturbations.

Theorem (Efimov 1949). *The above generic d-flat C^{an}-immersion $f\colon V \to \mathbb{R}^3$ is rigid for $d = B$. Namely, f admits no C^{an}-family of isometric maps $f_t\colon (V, g) \to \mathbb{R}^3$, $f_0 = f$, for which $D_{f_t} | T_{v_0}(V) = D_f | T_{v_0}(V)$.*

Proof. The basic algebraic ingredient of the theorem is

Efimov's Lemma. *Let $F = F(x, y)$ be a homogeneous polynomial (form) of degree $d + 1$ in two variables and let F_{11}, F_{12} and F_{22} denote the second derivatives of F. If $d = 8$ and the form F is generic, then the linear differential operator*

$$\Phi \mapsto F_{11}\Phi_{22} - 2F_{12}\Phi_{12} + F_{22}\Phi_{11}$$

is injective on the space of formal power series $\Phi = \Phi(x, y)$ which start with terms of degree $d + 1$,

$$\Phi(x, y) = a_0 x^9 + a_1 x^8 y + \cdots + a_9 y^9 + b_0 x^{10} + \cdots.$$

To prove the lemma, one should find one single form F for which the above operator is injective. Let us show this to be the case for the form $F(x, y) = \frac{1}{6}x^3 y^6$. The corresponding operator

$$\Phi \mapsto x y^6 \Phi_{22} - 6x^2 y^5 \Phi_{12} + 5x^2 y^4 \Phi_{11}$$

maps the space $\{\Phi^m\}$ of forms of degree m into the space $\{\Phi^{m+4}\}$ and the matrix of the operator $\{\Phi^m\} \to \{\Phi^{m+4}\}$ is diagonal in the monomial basis. The entries on the diagonal are

$$P(k, l) = l(l - 1) - 6kl + 5k(k - 1),$$

for $k = 0, \dots, m$ and $l = m - k$. We must show that the equation $P(k, l) = 0$ admits no integral solution $k, l \geq 0$ for $k + l \geq 9$. Observe that

$$P(k, l) = (l - k)(l - 5k) - 5k - l.$$

Put $p = l - k$ and $q = l - 5k$. Then the equation $P(k, l) = 0$ becomes

$$(*) \qquad 2pq = 5p - 3q.$$

The integral solutions of $(*)$ are

$$p = \frac{5p' - 3q'}{2q'} \quad \text{and} \quad q = \frac{5p' - 3q'}{2p'},$$

where

$$p' = \pm 1, \pm 3 \quad \text{and} \quad q' = \pm 1, \pm 5.$$

Since $l + k = \frac{1}{2}(3p - q) \geq 9$, we must have

$$\frac{(5p' - 3q')(3p' - q')}{p'q'} \geq 36,$$

which is satisfied by no choice of the above p' and q'. Q.E.D.

We prove the theorem by studying the derivatives of the isometric immersion equations (2′) and (4) of 3.1.6, which now become

(1) $$\langle f_i, f_{jk} \rangle = A^{ijk}, \qquad 1 \leq i, j, k \leq 2,$$

and

(2) $$f_{11}f_{22} - f_{12}^2 = B.$$

If we differentiate (1) $m - 2$ times, then we express the scalar products $\langle f_i, \partial^m f \rangle$ by functions in (derivatives of) the metric g and in the derivatives $\partial^k f$ of order $k < m$. Hence, the orthogonal projection of every derivative $\partial^m f(v_0) \in T_{w_0}(\mathbb{R}^3)$ onto the tangent space $D_f(T_{v_0}(V)) \subset T_{w_0}(\mathbb{R}^3)$ is uniquely determined by the induced metric g on V and by $\partial^k f(v_0)$ for $k < m$. Since the map f is d-flat, the *normal projection* of $\partial^m f(v_0)$ to $N_{v_0}(f) = T_{w_0}(\mathbb{R}^3) \ominus D_f T_{v_0}(V)$, called $\overline{\partial^m f(v_0)}$, vanishes for $m = 2, 3, \ldots, d$, while the normal component $\overline{\partial^{d+1} f(v_0)}$ of $\partial^{d+1} f(v_0)$ defines a form $F = F(x, y) = \sum_{k+l=d+1} a_{kl} x^k y^l$, where x and y correspond to the local coordinates u_1 and u_2 on V and where $a_{kl} = \binom{d+1}{k} \dfrac{\overline{\partial^{k+} f(v_0)}}{du_1^k \, du_1^l}$. If we differentiate the Eqs. (2) $2(d - 1)$ times and then take the normal projection of the derivatives of f, then we see that all terms with $\overline{\partial^m f(v_0)}$ for $m > d + 1$ appear in products with $\overline{\partial^m f(v_0)}$ for $m < d + 1$. Therefore, all these terms vanish and so we left with some quadratic equations in a_{kl}. These equations can be compactly expressed by

(2′) $$F_{11}F_{22} - F_{12}^2 = B_d(v_0),$$

where B_d is a function in (derivatives of) the metric g and in the derivatives $\partial^m f(v_0)$ for $m \leq d$. Since f is a generic d-flat map, the form F is also generic and so for $d = 8$ there is no deformation of F which keeps the Hessian $F_{11}F_{22} - F_{12}^2$ of F unchanged. Indeed, the linearized deformation equation is

$$F_{11}\Phi_{22} - 2F_{12}\Phi_{12} + F_{22}\Phi_{11},$$

and so Efimov's Lemma applies.

Now, for all $m > d + 1$ we consider the form $\Phi^m(x, y) = \sum_{k+l=m} a_{kl} x^k y^l$, where

$$a_{kl} = \binom{m}{k} \frac{\overline{\partial^{k+l} f(v_0)}}{du_1^k \, du_1^l}.$$

This form Φ^m for $m > d + 1$, unlike $\Phi^{d+1} = F$, is not invariant under coordinate changes in V but it does satisfy the following equation in our fixed coordinate system.

$$F_{11}\Phi^m_{22} - 2F_{12}\Phi^m_{12} + F_{22}\Phi^m_{11} = B_m,$$

where B_m is a function in g and in $\partial^n f$ for $n < m$. This is seen by differentiating (2) $(m + d - 3)$ times and then by looking at the normal components of the derivatives. The key observation here (like for $m = d + 1$) is the vanishing of the terms which contain derivatives $\partial^n f$ for $n > m$, as they appear in products with derivatives $\partial^l f$ for $l \leq d$. Efimov's lemma implies the rigidity of the forms Φ^m for all $m = d + 2$, $d + 3$, and thus the rigidity of f [compare (G) in 2.3.8]. Q.E.D.

Remarks and Exercises. Efimov's lemma (and hence, the theorem) is likely to be true for all $d \geq 3$. But the lemma obviously is false for $d = 2$, and generic 2-flat immersions may be non-rigid.

Questions. Let V be a surface with a C^{an}-metric g on V. What is the number $N = N(g)$ of connected components in the space of isometric C^{an}-maps $(\mathcal{O}p(v_0), g) \to \mathbb{R}^3$, $v_0 \in V$? Efimov's theorem gives examples of $N(g) \geq 2$ (we identify immersions which are congruent by reflections of $\mathbb{R}^3$). Do any metrics have $N(g) = 3$? Is there a universal upper bound $N(g) \leq N_0$ independent of g?

Exercises. Take a generic form $F = F(x, y)$ of degree $d \geq 4$ and show that the differential operator

$$\Phi \mapsto F_{11}\Phi_{11} + F_{12}\Phi_{12} + F_{22}\Phi_{11}$$

is injective on the space of forms Φ of degree m for all $m \geq 4$.

Consider a quadratic differential operator D on functions $f: \mathbb{R}^n \to \mathbb{R}$,

$$D: f \mapsto \sum_{i,j=1}^{n} a_{ij}(\partial_i f)(\partial_j f),$$

where a_{ij} are generic C^{an}-functions on $\mathbb{R}^n$. Take a generic function f in the space of the C^{an}-functions f on $\mathbb{R}^n$ which are d-flat at the origin, that is $J^d_f(0) = 0$, and show for $d \geq (100)^n$ this f admits no C^{an}-deformation f_t for which $D(f_t) = D(f)$.

Generalize this rigidity to systems of homogeneous polynomial differential operators of arbitrary order and degree ≥ 2.

Questions. Take a generic C^{an}-immersion f in the space of immersions $V^n \to \mathbb{R}^q$ which are d-flat at a given point $v_0 \in V^n$. Is f C^{an}-rigid for $q \leq n(n + 1)/2$ and for d large, say for $d \geq (100)^q$? This rigidigy means non-existence of non-trivial isometric C^{an}-deformations in the space of d-flat immersions $V \to \mathbb{R}^q$. Unfortunately, there is no generalization (?) of the theorem of Hopf-Schilt to codimension ≥ 2 which would guarantee d-flatness of all isometric deformations.

If $q < n(n + 1)/2$, then the rigidity is expected with no flatness assumption (see 2.3.8), but for $q = n(n + 1)/2$ the flatness plays a crucial role in making the isometric immersion equation overdetermined. This is seen by differentiating m times the Eqs. (2′) of 3.1.6 and by differentiating $m + d - 1$ times the Eq. (4) of 3.1.6. The resulting algebraic equations for $\partial^n f(v_0)$ contain no terms with $n > m + 2$, as those derivatives are coupled with the vanishing derivatives $\partial^k f(v_0)$, $k = 2, \ldots, d$, and so the equations in $\partial^n f(v_0)$, $n = 2, \ldots, m + 2$, are formally overdetermined for large d.

The local rigidity of a C^{an}-immersion $V \to \mathbb{R}^q$ at some point $v_0 \in V$ obviously implies the global C^{an}-rigidity. This idea is used by Efimov (1949) to prove the C^{an}-rigidigy of the standard round torus $T^2 \subset \mathbb{R}^3$.

Exercise. Let the first Stiefel-Whitney class of a connected manifold $V = V^n$ satisfy $(w_1)^3 \neq 0$. Show that every C^{an}-immersion $f\colon V \to \mathbb{R}^{n+1}$ is rigid. (This is only interesting for *open* manifolds V as all closed hypersurfaces $V \subset \mathbb{R}^{n+1}$ are rigid for $n \geq 3$.)

Hint. If the Gauss map $G_f\colon V \to P^n\mathbb{R}$ has rank $G_f \geq 3$ at some point $v_0 \in V$ then f is rigid near v_0.

Question. Let the first rational Pontryagin class of V satisfy $(p_1)^3 \neq 0$. Then is every C^{an}-immersion $f\colon V \to \mathbb{R}^{n+2}$ rigid?

Making Non-free Maps Free. The Euclidean space $\mathbb{R}^q$ obviously admits large families of isometric C^{an}-deformations in the ambient spaces $\mathbb{R}^N \supset \mathbb{R}^q$. This allows one to make some maps $f\colon V^n \to \mathbb{R}^q$ free by deformating in a space $\mathbb{R}^N$ of sufficiently high dimension N.

Exercises. Show that every isometric C^∞-immersion $f_0\colon (V,g) \to \mathbb{R}^q \subset \mathbb{R}^N$ admits an isometric deformation to a *free* isometric immersion $f\colon V \to \mathbb{R}^N$ for $N \geq q + [n(n+3)/2]$.

Hint. Apply Thom's transversality theorem to deformations of V which come from isometric deformations of $\mathbb{R}^q$ in $\mathbb{R}^N$.

Let V be a compact Riemannian manifold and let $f_0\colon V \to \mathbb{R}^q \subset \mathbb{R}^N$ be an isometric immersion. Construct for $N \geq q + [(n+1)(n+4)/2]$ an isometric C^∞-cylinder $f\colon V \times \mathbb{R} \to \mathbb{R}^N$, such that $f|V \times 0 = f_0$ and such that the map f is free near some submanifold $V \times t_0 \subset V \times \mathbb{R}$, $t_0 \in \mathbb{R}$. Then prove that any two isometric C^∞-immersions f_0 and $f_1\colon V \to \mathbb{R}^q$ can be joined by an isometric cylinder $V \times [0, l] \to \mathbb{R}^N \supset \mathbb{R}^q$ for $N \geq q + [(n+1)(n+4)/2]$ (compare 3.1.8).

Show that every isometric C^∞-immersion $f_0\colon V \to \mathbb{R}^q \subset \mathbb{R}^N$ admits an isometric deformation to an isometric *C^∞-imbedding* $f\colon V \to \mathbb{R}^N$, provided

$$N \leq \min(2q + n + 1, q + 3n + 1).$$

(This estimate for N, probably, can be improved.)

3.2 Isometric Immersions in Low Codimension

There is no general theory of isometric immersions $V^n \to W^q$ for $q < (n+2)(n+3)/2$ but various facts are known for special manifolds. What follows is a brief (and not at all complete) discussion on some results. For further information we refer to the surveys by Gromov-Rochlin (1970), Posnjak (1973), Posnjak-Sokolov (1977), Aminov (1982) and to the book by Burago-Zalgaller (1980).

3.2.1 Parabolic Immersions

Consider a smooth immersion f of V into a Riemannian manifold (W, h). A point $v \in V$ is called *parabolic* for f if the induced metric $g = f^*(h)$ has the same sectional curvatures at v as the metric h at $w = f(v) \in W$:

$$R_g(X, Y; X, Y) = R_h(X, Y; X, Y),$$

for all vectors X and Y in $T_v(V) \subset T_w(W)$. This is equivalent by the Gauss theorema egregium to the symmetry of the 4-form $\Phi_f | T_v(V)$ (see 3.1.5), which means

$$\Phi_f^- = \Phi_f - \Phi_f^+ = 0.$$

We call an immersion f *parabolic* if all points $v \in V$ are parabolic. For example, *isometric* immersions between manifolds of *constant* sectional curvature $\kappa = K(V, g) = K(W, h)$ are parabolic

Flat Directions. A tangent vector (direction) $X \in T_v(V) \subset T(V) \subset T(W)|V$ is called *flat* for f if $\nabla_X \tau \in T_v(V)$ for the covariant derivative ∇_X in W of an arbitrary vectorfield τ tangent to V. This is equivalent to the inclusion $X \in \operatorname{Ker} \Phi_f | T_v(V)$, where $\operatorname{Ker} \Phi$ consists, by definition, of the vectors X, such that $\Phi(X, X_1, X_2, X_3) = 0$ for all triples of vectors (X_1, X_2, X_3) in $T_v(V)$.

If W is a complete simply connected manifold of *constant* curvature then the space of flat directions, denoted by $Fl = \operatorname{Ker} \Phi_f \subset T(V)$, clearly equals the kernel of the differential of the obvious tangential map $\tilde{G}_f\colon V \to \tilde{G}r_n W$, where $\tilde{G}r_n W$ stands for the space of the (complete connected) totally geodesic submanifolds X of dimension n in W. Furthermore, if $W = \mathbb{R}^q$, then the space $\tilde{G}r_n \mathbb{R}^q$ canonically projects onto the Grassmann manifold, $\pi\colon \tilde{G}r_n \mathbb{R}^q \to Gr_n \mathbb{R}^q$ with the fibers $\approx \mathbb{R}^{q-n}$, and the composition $\pi \circ \tilde{G}_f$ equals the ordinary tangential map $G_f\colon V \to Gr_n \mathbb{R}^q$. It is clear that the differential of π is injective on the image $D\tilde{G}_f(T(V)) \subset T(\tilde{G}r_n \mathbb{R}^q)$. Therefore,

$$Fl = \operatorname{Ker} D\tilde{G}_f = \operatorname{Ker} DG_f.$$

We need a formula for the curvature of the natural Euclidean connection ∇ (see 2.2.6) in the canonical n-dimensional bundle $H \to Gr_n \mathbb{R}^q$. We regard the points $P \in Gr_n \mathbb{R}^q$ as orthogonal projectors $P\colon \mathbb{R}^q \to \mathbb{R}^q$ of rank n and we view the vectors $X \in H_P \subset H$ as vectors $X \in \mathbb{R}^q$ for which $P(X) = X$. If $X = X(u_1, \ldots, u_{n(q-n)})$ is a section of H expressed in some local coordinates u_i, $i = 1, \ldots, n(q-n) = \dim Gr_n \mathbb{R}^q$, then

$$(\nabla_i X)(P) = P(\partial_i X) \qquad \text{for } \partial_i = \partial / \partial u_i,$$

by the definition of ∇. The curvature operator Ω of ∇ is defined in the coordinates u_i by the operators $\Omega_{ij} = -\Omega_{ji} = (\nabla_i \nabla_j - \nabla_j \nabla_i)\colon H \to H$.

(A) **Lemma.** *The operators* Ω_{ij} *satisfy*

$$\Omega_{ij}(X) = (P_i P_j - P_j P_i) X,$$

where $P_i\colon \mathbb{R}^q \to \mathbb{R}^q$ *denotes the partial derivative* $\partial_i P$ *of the projector* $P = P(u_1, \ldots, u_{n(q-n)})\colon \mathbb{R}^q \to \mathbb{R}^q$.

Proof. Since $P^2 = P$ and $PX = X$, we have

$$\begin{aligned}\nabla_i\nabla_j X &= P\partial_i P\partial_j X = \partial_i(P\partial_j X) - P_i P\partial_j X = P_i(\partial_j X - P\partial_j X) + P\partial_{ij}X\\ &= P_i(\partial_j X - \partial_j(PX) + P_j X) + P\partial_{ij}X = (P_i P_j + P\partial_{ij})X,\end{aligned}$$

and the lemma follows.

Take an r-dimensional linear subspace $I \subset T_P(Gr_n\mathbb{R}^q)$ and choose local coordinates u_i such that the vectors $\partial_i | T_P(Gr_n\mathbb{R}^q)$ lie in I for $i = 1, \ldots, r$. We call I an *Ω-isotropic* subspace if $\Omega_{ij} = 0$ for $i, j = 1, \ldots, r$. This can be expressed in the invariant language by saying that the (operator valued) curvature form vanishes on I, that is $\Omega | I = 0$.

(A′) **Lemma.** *The dimension of every Ω-isotropic subspace I abides*

$$r = \dim I \leq q - n.$$

Proof. Let $S \subset \mathrm{Hom}(\mathbb{R}^q \to \mathbb{R}^q)$ be the span of the operators $P_i = \partial_i P$ and let $A = P(\mathbb{R}^q) \subset \mathbb{R}^q$. Then the following three conditions are satisfied.

(i) The operators $s \in S$ are symmetric.
(ii) The vector $s(a)$ is normal to A for all $s \in S$ and $a \in A$.
(iii) The operators $s \in S$ commute on A,

$$s_1 s_2(a) = s_2 s_1(a) \quad \text{for all } s_1, s_2 \in S \quad \text{and} \quad a \in A.$$

Indeed, the operators s are symmetric as linear combinations of derivatives of a symmetric operator. Furthermore, the identity $P^2 = P$ differentiates to $PP_i + P_i P = P_i$ which yields (ii). Finally, the condition $\Omega | I = 0$ is equivalent to (iii).

Now, the proof is immediate with the following

Sublemma. *Let linear subspaces $A \subset \mathbb{R}^q$ and $S \subset \mathrm{Hom}(\mathbb{R}^q \to \mathbb{R}^q)$ satisfy* (i)–(iii). *Then* $\dim A + \dim S \leq q$.

Proof. Take a non-zero vector $a_0 \in A$ and let $S_0 \subset S$ be the subspace of those operators s for which $s(a_0) = 0$. Then the linear subspace $S(a_0) = \{s(a_0) | s \in S\} \subset \mathbb{R}^q$ has $\dim S(a_0) = \dim S - \dim S_0$. Furthermore, all $s_0 \in S_0$ satisfy

$$\langle s_0(x), s(a_0)\rangle = \langle x, s_0 s(a_0)\rangle = \langle x, s\, s_0(a_0)\rangle = 0,$$

for all $x \in \mathbb{R}^q$ and so the images of the operators $s_0 \in S_0$ are normal to $S(a_0)$. Take the orthogonal complement $R_0 \subset \mathbb{R}^q$ to the span of $S(a_0)$ and a_0 and let $A_0 = A \cap R_0 \subset R_0$. Since $q_0 = \dim R_0 < q$, we may assume by induction the inequality $\dim A_0 + \dim S_0 \leq q_0$ which implies what we need,

$$(\dim A - 1) + (\dim S - \dim S(a_0)) \leq q - \dim S(a_0) - 1.$$

(A″) **Example.** The curvature Ω on the oriented Grassmannian $Gr_2\mathbb{R}^q$ is an ordinary (real valued) 2-form which is non-singular (symplectic) by (A′). Hence the Euler class (of the canonical 2-bundle) $[\Omega] \in H^2(Gr_2\mathbb{R}^q; \mathbb{R})$ satisfies $[\Omega]^{q-2} \neq 0$.

Exercise. Show that the orthogonal projection $\mathbb{C}^q \to \mathbb{R}^q \subset \mathbb{C}^q$ induces a diffeomorphism of the quadric $\sum_{i=1}^q z_i^2 = 0$ in $\mathbb{C}P^{q-1}$ onto $Gr_2(\mathbb{R}^q)$, such that the simplectic Kähler form on the quadric goes to the form Ω.

(A‴) **Corollary** (Tompkins 1939). *If $v \in V$ is a parabolic point of an immersion f: $V \to W$, then the subspace $Fl_v = Fl_v(f) \subset T_v(V)$ of flat vectors has* $\dim Fl_v \geq 2n - q$.

Proof. The form $\Phi_f | T_v(V)$ for $T_v(V) \subset T_w(W)$, $w = f(v)$, is invariant under the map $\exp^{-1}$: $U \to T_w(W)$, where exp: $T_w(W) \to W$ is the (geodesic) exponential map and $U \subset W$ is a small neighborhood of $w \in W$. Therefore, one may assume $W = \mathbb{R}^q = T_w(W)$ and use the map G_f: $V \to Gr_n \mathbb{R}^q$. The parabolicity of the point $v \in V$ now is equivalent to the vanishing of the induced curvature,

$$G_f^*(\Omega) | T_v(V) = 0,$$

and so the image $DG_f(T_v(V)) \subset T_x(Gr_n \mathbb{R}^q)$, $x = G_f(v)$, is Ω-isotropic. Hence,

$$\dim Fl_v = n - \operatorname{rank} DG_f | T_v(V) \geq 2n - q.$$

Remark. Let Φ be an arbitrary *symmetric* 4-form on $\mathbb{R}^n$, such that the associated quadratic form $\bar{\Phi}$ on the symmetric square $(\mathbb{R}^n)^2$ is positive semidefinite. The following inequality is an algebraic equivalence of (A‴),

$$\operatorname{rank} \Phi \overset{\text{def}}{=} n - \dim \operatorname{Ker} \Phi \leq \operatorname{rank} \bar{\Phi}.$$

Examples. If $\dim W = \dim V + 1$, then the inequality $\dim Fl_v \geq n - 1$, obviously, is equivalent to the parabolicity of the point v. In particular, parabolic immersions f: $V \to \mathbb{R}^{n+1}$ are characterized by the inequality $\operatorname{rank}_v G_f \leq 1$ for all $v \in V$, which (obviously) is invariant under *projective* transformations of $\mathbb{R}^{n+1}$. If we take, for instance, an *isometric* immersion f of the disk ($B^2 = \{x^2 + y^2 \leq 1\}$, $g_0 = dx^2 + dy^2$) into $\mathbb{R}^3$ and then compose this f with some projective map p: $\mathbb{R}^3 \to \mathbb{R}^3$ [which must be regular on $f(B) \subset \mathbb{R}^3$], we obtain a *new flat metric* $g = (p \circ f)^*(h)$ on B^2 (where $h = \sum_{i=1}^3 dx_i^2$). Thus, the metric g_0 generates an interesting (?) class of flat metrics g on B^2 as f and p vary.

Fix a projective map p of the open unit ball $B^{n+1} \subset \mathbb{R}^{n+1}$ onto the hyperbolic space H^{n+1} (of constant curvature $\kappa = -1$). Then parabolic immersions f: $V \to B^{n+1}$ go to parabolic immersions $p \circ f$: $V \to H^{n+1}$ which induce in $V = V^n$ metrics of constant curvature $\kappa = -1$. This gives many examples of *complete* parabolic C^∞-hypersurfaces in H^{n+1} which correspond to *proper* parabolic imbeddings f: $V \to B^{n+1}$.

Exercise. Classify for all $n \geq 2$, all complete parabolic hypersurfaces in H^{n+1} with free non-Abelian fundamental groups.

(B) *Integrability of* $Fl \subset T(V)$. Denote by $i_0 = i_0(f)$, for a given immersion f: $V \to W$, the minimum of $\dim Fl_v$ over all $v \in V$ and let $\tilde{\Sigma}^{i_0} \subset V$ be the subset where this minimum is assumed. This $\tilde{\Sigma}^{i_0}$ is an open subset in V and the spaces Fl_v, $v \in \tilde{\Sigma}^{i_0}$, form an i_0-dimensional vector bundle $Fl^{i_0} \to \tilde{\Sigma}^{i_0}$. If the manifold W has *constant curvature*, then the bundle Fl^{i_0} is integrable because $Fl = \operatorname{Ker} D\tilde{G}_f$. Furthermore, if

W is complete and simply connected, then the integral leaf $\mathscr{L}_v \subset \tilde{\Sigma}^{i_0}$ through $v \in \tilde{\Sigma}^{i_0}$ is the path connected component of the point v in $\tilde{\Sigma}^{i_0} \cap \tilde{G}_f^{-1}(x)$ for $x = \tilde{G}_f(v) \in \tilde{G}r_n W$. The points $v \in \tilde{\Sigma}^{i_0}$, for which $c = \tilde{G}_f(v)$ lie in the image $\tilde{G}_f(V \backslash \tilde{\Sigma}^{i_0}) \subset \tilde{G}r_n W$, have zero measure in $\tilde{\Sigma}^{i_0}$, because rank $\tilde{G}_f | V \backslash \tilde{\Sigma}^{i_0} <$ rank $\tilde{G}_f | \tilde{\Sigma}^{i_0} = n - i_0$ and Sard's theorem applies. Hence, *generic leaves* $\mathscr{L}_v$ *are smooth* i_i*-dimensional properly imbedded submanifolds in* V.

Lemma. *Let* X *and* Y *be smooth vector fields in* $Fl \subset T(V) \subset T(W)|V$. *If the manifold* W *has constant curvature, then the covariant derivative* $Z = \nabla_X Y$ *(in* W*) also lies in* Fl.

Proof. To show $\nabla_Z \tau \in T(V)$ for all fields $\tau \in T(V)$ we observe that the Lie bracket $[Z, \tau] = \nabla_Z \tau - \nabla_\tau Z$ always lies in $T(V)$ (for any two fields tangent to V) and so we must prove that $\nabla_\tau Z = \nabla_\tau \nabla_Z Y$ lies in $T(V)$. Since W has constant curvature, the operator $\nabla_\tau \nabla_X - \nabla_X \nabla_\tau = R(X, \tau)$ preserves the tangent bundle $T(V) \subset T(W)|V$. Hence the field

$$\nabla_X \nabla_\tau Y = \nabla_X \tau' + \nabla_{\tau''} Y,$$

where $\tau' = \nabla_\tau Y \in T(V)$ and $\tau'' = [\tau, X] \in T(V)$, lies in $T(V)$. Q.E.D.

Corollary (O'Neil 1962). *Let* W *be a complete simply connected manifold of constant curvature and let* $\dim Fl_v(f) > 0$ *for all* $v \in V$. *Then there exists a complete totally geodesic submanifold* $\mathscr{L} \subset V$ *of dimension* $i_0 > 0$, *such that the map* f *is flat on* $\mathscr{L}$, [*i.e.* $f(\mathscr{L}) \subset W$ *is contained in some* i_0*-dimensional totally geodesic submanifold* $X^{i_0} \subset W$] *and the map* $\tilde{G}_f$ *has rank* i_0 *at all points* $v \in \mathscr{L}$.

Proof. The generic leaves $\mathscr{L} = \mathscr{L}_v$ discussed above are totally geodesic and $f|\mathscr{L}$ is flat by the lemma.

Theorem (O'Neil 1962). *Every isometric map* $f\colon \mathbb{R}^n \to \mathbb{R}^{2n-1}$ *is flat on a straight line and so the image* $f(\mathbb{R}^n) \subset \mathbb{R}^{2n-1}$ *is unbounded.*

Proof. We have, by (A‴), $\dim Fl_v > 0$, and the map f is isometrically flat on some affine subspace $\mathscr{L} \subset \mathbb{R}^n$. Q.E.D.

Theorem (Ferus 1975). *Every isometric* C^∞*-immersion* $f\colon S^n \to S^{2n-1}$ *is flat, if both spheres are assumed to have constant curvature* $\kappa = 1$.

Proof. By our earlier considerations there is an open subset $U \subset \tilde{\Sigma}^{i_0}$ which is foliated by great spheres $\mathscr{L}_v \subset U \subset S^n$ of dimension i_0 such that the map f is flat on these spheres. Since great spheres are invariant under the central involution $s \mapsto -s$ of S^n, the subset $U \subset S^n$ is also invariant and the map $f|U\colon U \to S^{2n-1}$ is symmetric:

$$f(-u) = -f(u), \qquad \text{for all } u \in U.$$

Since the map $f|U$ is isometric as well as symmetric, every great circle $S^1 \subset U$ goes by f onto a great circle in S^{2n-1}. Since U is open, the map f is flat on U and so $i_0 = n$. Q.E.D.

Exercises. (a) Let $f: S^n \to S^q$ be a short (for example, isometric) map. Show that f is flat isometric under either of the following two conditions,

(i) There are points $s_i \in S^n$, $i = 1, \ldots, n$, in general position (i.e. contained in no equator $S^{n-1} \subset S^n$), such that $f(-s_i) = -f(s_i)$, for $i = 1, \ldots, n$.
(ii) There is a single point $s \in S^n$ for which $f(-s) = -f(s)$ and such that the differential $D_f: T_s(S^n) \to T_{f(s)}(S^q)$ exists and it is isometric (for example, f is C^1-isometric near $s \in S^n$).

(a′) Let $U \subset S^n$ be an open subset whose complement $C = S^n \backslash U$ is a smooth k-dimensional submanifold which contains no pairs of opposite points. Show that every isometric C^∞-immersion $f: U \to S^q$ is flat in the following two cases,

(i) $n \geq k + 2$ and $q \leq 2n - 2$;
(ii) $n \geq 2k + 2$ and $q \leq 2n - 1$.

(b) Let $f: \mathbb{R}^n \to \mathbb{R}^q$ be a short map which isometrically and bijectively sends the ray $\{x_1 \geq 0, x_i = 0, \text{for } i = 2, \ldots, n\}$ in $\mathbb{R}^n$ onto the ray $\{x_1 \geq 0, x_i = 0 \text{ for } i = 2, \ldots, q\}$ in $\mathbb{R}^q$. Show that the half-space $\{x_i \geq 0\} \subset \mathbb{R}^n$ goes into $\{x_1 \geq 0\} \subset \mathbb{R}^q$.

(b′) Assume that the above short map f is flat isometric on the line $\{x_i = 0$ for $i = 2, \ldots, n\} \subset \mathbb{R}^n$ and show that the map f splits in the following sense. Every hyperplane $\{x_1 = \text{const}\} \subset \mathbb{R}^n$ goes into the hyperplane $\{x_1 = \text{const}\} \subset \mathbb{R}^q$ and every normal line to $\{x_1 = \text{const}\} \subset \mathbb{R}^n$ isometrically and bijectively goes onto a normal line to $\{x_1 = \text{const}\} \subset \mathbb{R}^q$. Use this and obtain an orthogonal splitting of an arbitrary isometric C^∞-immersion $f: \mathbb{R}^n \to \mathbb{R}^{2n-1}$ [compare Stiel (1965)].

(c) Let V be a complete manifold of non-negative sectional curvature and let $f: V \to \mathbb{R}^q$ be a short map which isometrically and bijectively sends some geodesic $\gamma \in V$ onto the line $\{x_i = 0 \text{ for } i = 2, \ldots, q\} \subset \mathbb{R}^q$. Show that the manifold V splits isometrically, $V = V' \times \mathbb{R}$, where $\gamma = v_0' \times \mathbb{R}$ for some $v_0' \in V'$, and the map f also splits: The hypersurfaces $V' \times r \subset V$ go into hyperplanes $\{x_1 = \text{const}\} \subset \mathbb{R}^q$, and the geodesics $v' \times \mathbb{R}$ go onto normal lines to these hyperplanes [compare Hartman (1970)].

(c′) Generalize (c) to manifolds V of non-negative *Ricci curvature* by applying the splitting theorem of Cheeger-Gromoll (1971).

(d) Study short maps of an open hemisphere $S^n_+ \subset S^n$ into S^q which are flat isometric on some geodesic arc of length π in S^n_+.

(C) **Remark.** In fact, all leaves $\mathscr{L}_v$, $v \in \tilde{\Sigma}^{i_0}$, (not only generic ones) are *properly* imbedded into V. Indeed, consider the normal bundle $L^\perp \to \mathscr{L}$ of a given leaf $\mathscr{L} = \mathscr{L}_{v_0} \subset V$ and let $h_v: L^\perp_{v_0} \to L^\perp_v$, $v \in \mathscr{L}$, denote the holonomy map of the foliation, which is

$$h_v = D_v^{-1} D_{v_0} \qquad \text{for } D_v = D\tilde{G}_f | L_v^\perp: L_v^\perp \to T_x(\tilde{G}r_n W),$$

where $x = \tilde{G}_f(v) = \tilde{G}_f(v_0)$. Since the foliation is totally geodesic in V, the bundle $L^\perp$ continuously extends to the closure $\bar{\mathscr{L}} \subset V$ of $\mathscr{L}$ in V and the holonomy maps h_v continuously extend to all $v \in \bar{\mathscr{L}}$. Then, by continuity, $D_{v_0} = D_v h_v$ for all $v \in \bar{\mathscr{L}}$, which implies

$$\operatorname{rank} D_v \geq \operatorname{rank} D_{v_0} = n - i_0,$$

and so $\bar{\mathscr{L}} = \mathscr{L}$. Q.E.D.

(D) *Geodesic Laminations.* Consider an isometric C^1-immersion $f: V \to W$ between manifolds of constant sectional curvature $\kappa = K(V) = K(W)$ and let us define a partition of V into totally geodesic leaves $\mathscr{L}_v \subset V$ for *all* $v \in V$. We first assume that W is the *standard* (i.e. complete simply connected) manifold, $W = M^q[\kappa]$ of curvature κ, and that V is an ε-ball in such a manifold, $V = B^n(\varepsilon) \subset M^n[\kappa]$, where we assume $4\kappa\varepsilon^2 \leq \pi^2$. Then we define $\mathscr{L}_v \subset V$ to be the subset of those points $v' \in V$ for which $\operatorname{dist}(f(v'), f(v)) = \operatorname{dist}(v', v)$. Since $\kappa(V) = \kappa(W)$ and since the map f is smooth, the subset $\mathscr{L}_v \subset V$ is geodesically convex for all $v \in V$ and so $\mathscr{L}_{v'} = \mathscr{L}_v$ for all $v' \in \mathscr{L}_v$. Hence, V is *partitioned* into closed convex subsets called *leaves*, $\mathscr{L}_v \subset V$. If the map f is C^∞-smooth, then this partition agrees on the subset $\Sigma^{i_0} \subset V$ with the above foliation $\Sigma^{i_0} = \{\mathscr{L}_v\}$. Furthermore, let a point $v \in V$ lie in the *interior* of the subset $\Sigma^i \subset V$ where the rank of the map $\tilde{G}_f$ equals $i \geq i_0$. The interior of Σ^i is foliated, like Σ^{i_0}, into i-dimensional leaves which are subsets of the convex leaves $\mathscr{L}_v$. Since the union $\bigcup_{i=i_0}^n \operatorname{Int} \Sigma^i$ is dense in V, every convex leaf $\mathscr{L}_v$ is locally a limit of i-dimensional leaves of (local) geodesic foliations and so $\mathscr{L}_v$ satisfies the following three properties

(i) $\dim \mathscr{L}_v \geq i_0 \geq 2n - q$ for all $v \in V$.

(ii) If $i_0 \geq 1$, then the (convex!) leaf $\mathscr{L}_v \subset V$ has no extreme points *inside* $V = B^n(\varepsilon)$, and so $\mathscr{L}_v$ equals the convex hull $\operatorname{Conv}(\mathscr{L}_v \cap \partial V)$.

(iii) If $i_0 \geq r + 1$ for $r \geq 0$, then (ii) holds true for the intersection of $\mathscr{L}_v$ with every complete totally geodesic subspace $X^{n-r} \subset M^n[\kappa] \supset V$ of codimension r,

$$\mathscr{L}_v \cap X^{n-r} = \operatorname{Conv}(\mathscr{L}_v \cap X^{n-r} \cap \partial V).$$

(D′) **Corollary.** *The intersection $\mathscr{L}_v \cap X^{n-r}$ have*

$$\operatorname{Diam}(\mathscr{L}_v \cap X^{n-r}) \geq \operatorname{dist}(v, \partial V),$$

for all totally geodesic subspaces $X^{n-r} \subset M^n[\kappa]$ through v and for all $v \in V$.

Exercises. (a) Let v_0 be the center of the ball $V = B^n(\varepsilon) \subset M^n[\kappa]$. Prove that

$$\operatorname{Diam} \mathscr{L}_{v_0} = \operatorname{Diam} V = 2\varepsilon, \qquad \text{for } 2i_0 > n,$$

and that

$$\operatorname{Diam} \mathscr{L}_{v_0} \geq \operatorname{Diam} \Delta^n, \qquad \text{for } i_0 \geq 1,$$

where Δ^n is a regular geodesic n-simplex in V with the vertices on the boundary ∂V.

(b) Let V be the unit ball $B^n(1) \subset \mathbb{R}^n$. Prove that every isometric C^{an}-immersion $f: V \to \mathbb{R}^{2n-1}$ has $\operatorname{Diam} f(V) = 2$.

Construct for given numbers $n = 2, 3, \ldots$, and $\delta > 0$ an isometric C^∞-immersion $f: V \to \mathbb{R}^{2n-1}$ for which $\operatorname{Diam} f(V) \leq \sqrt{3} + \delta$.

(c) Let V be a *closed* hemisphere in $S^n = M^n[1]$ and let $v_0 \in V$ be the center of $V = B^n(\frac{\pi}{2}) \subset S^n$. Show that the leaf $\mathscr{L}_{v_0} \subset V$ equals V, provided $i_0 \geq 1$, and hence every isometric C^∞-immersion $V \to S^{2n-1}$ is flat. Find non-flat isometric C^∞-immersions of the *interior* $\operatorname{Int} V$ into S^{n+1} for all $n = 2, 3, \ldots$.

Denote by Q the quotient space of the partition of the ball $V = B^n(\varepsilon) \subset M^n[\kappa]$ into leaves $\mathscr{L}_v$. If V is a *closed* ball, then Q is a compact space.

(D″) **Lemma.** *The topological dimension of Q satisfies*

$$\dim Q \le n - i_0.$$

Proof. There exists, by (D′), a countable collection of geodesic subspaces $X_j^{n-i_0} \subset M^n[\kappa]$, $j = 1, 2, \ldots$, such that the images of the balls $B_j^{n-i_0} = V \cap X_j^{n-i_0}$ under the quotient map $V \to Q$ cover the space Q and so the problem is reduced to the following

Sublemma. *Let $\alpha: B^{n-i_0} \to Q$ be a continuous map for which the pullback $\alpha^{-1}(x) \in B^{n-i_0}$ is a convex subset in B^{n-i_0} for all $x \in Q$. Then*

$$\dim \alpha(B^{n-i_0}) \le n - i_0.$$

Proof. We assume by induction

$$\dim \alpha(B') \le n - i_0 - 1,$$

for all convex subsets $B' \subset B^{n-i_0}$ of dimensional $n - i_0 - 1$. Then for an arbitrary neighborhood $U \subset Q$ of a given point $x \in Q$ we take a neighborhood $U' \subset \alpha^{-1}(U) \subset B^{n-i_0}$ of the subset $\alpha^{-1}(x) \subset B^{n-i_0}$, such that the boundary $\partial U' \subset B^{n-i_0}$ is a finite union of convex subsets B' of dimension $n - i_0 - 1$. Since the pullbacks $f^{-1}(x')$ are convex and hence, connected for all $x' \in Q$, there exists a neighborhood $U'' \subset U$ of x, such that the boundary $\partial U'$ is contained in the image $\alpha(\partial U')$. As $\dim \alpha(\partial U') \le n - i_0 - 1$, the sublemma follows from the very definition of $\dim Q$.

Now let V and W be arbitrary manifolds of constant curvature κ and let $f: V \to W$ be an isometric C^∞-immersion. Then every small ball in V is partitioned into convex leaves $\mathscr{L}_v$ and these local partitions define a global partition, called a *lamination* [compare Thurston (1978)], of V into locally (possibly noncomplete) geodesic submanifolds $\mathscr{L}_v \subset V$. This lamination has codimension $n - i_0$ by (D″) and the map f is isometrically flat on every leaf $\mathscr{L}_v \subset V$. If the manifold V is compact (possibly with boundary) and if the receiving space W is simply connected, then all leaves are closed subsets in V and the quotient space Q is a compact Hausdorff space of dimension $n - i_0$.

(E) *The Holonomy Dimension $Hd(V)$.* Let V be an arbitrary n-dimensional manifold of constant curvature $\kappa = K(V)$. The *holonomy covering* of V is a unique minimal Galois' covering $\tilde{h}: \tilde{V} \to V$, such that $\tilde{V}$ admits an isometric immersion $\tilde{d}: \tilde{V} \to M^n[\kappa]$, called *the developing* map. Denote by Γ the Galois group of the covering $\tilde{h}$, let $K(\Gamma, 1)$ be the corresponding Eilenberg-Mac Lane space and let $H: V \to K(\Gamma, 1)$ be the *classifying* map which isomorphically maps the Galois group $\Gamma = \pi_1(V)/\pi_1(\tilde{V})$ onto $\pi_1(K(\Gamma, 1)) \approx \Gamma$. Denote by $Hd(V)$ the minimal integer k for which there exists a k-dimensional polyhedron K and continuous maps $F: V \to K$ and $G: K \to K(\Gamma, 1)$, such that the composed map $G \circ F: V \to K(\Gamma, 1)$ is homotopic to H.

Examples. If V is a *closed* manifold, then $Hd(V) = \dim V$ unless V is the sphere S^n of curvature $\kappa > 0$.

Let U be an open subset in a complete manifold V of dimension n. If the inclusion homomorphism on the k-dimensional homology, $H_k(U) \to H_k(V)$ does not vanish, then $Hd(U) \ge k$, unless $U = V = S^n$, $n = k$.

Remark. The developing map $\tilde{d}: \tilde{V} \to M^n[\kappa]$ isomorphically maps Γ onto a subgroup $\tilde{d}_*(\Gamma)$ in the isometry group $\mathrm{Is}(M^n[\kappa])$ such that the map $\tilde{d}$ is Γ-equivariant.

Exercise. Show for $n \geq 3$, that every finitely presented subgroup in $\mathrm{Is}(M^n[\kappa])$ is the image $\tilde{d}_*(\Gamma)$ for some compact manifold (with a boundary!) of constant curvature κ.

(E′) **Theorem.** *Let V be an n-dimensional manifold of constant curvature κ, such that the group $\tilde{d}_*(\Gamma)$ acts freely on the image $\tilde{d}(\tilde{V}) \subset M^n[\kappa]$. If V admits an isometric C^∞-immersion $f: V \to M^q[\kappa]$, then*

$$q \geq \dim V + Hd(V).$$

Proof. Since the map f is flat isometric on every leaf $\mathscr{L}_v \subset V$ and since the group $\tilde{d}_*(\Gamma)$ is free on the image $\tilde{d}(\mathscr{L}_v)$, the holonomy covering $\tilde{V} \to V$ is trivial over every leaf $\mathscr{L}_v \subset V$. Therefore, the classifying map $H: V \to K(\Gamma, 1)$ extends to the mapping cylinder C of the quotient map $\alpha: V \to Q$ (which collapses each leaf to a point in Q) and so

$$Hd(V) \leq \dim Q \leq n - i_0 \leq q - n.$$

(F) **Exercises and Open Questions.** (a) Find, for given n and q satisfying $n \leq q < 2n$, a compact manifold $V = V^n$ of constant curvature $\kappa = -1$, which admits an isometric C^∞-immersion $V \to M^q[\kappa]$ (here $M^q[\kappa]$ is the hyperbolic space) but has no such immersion into $M^{q-1}[\kappa]$.

(a′) Do similar manifolds V exist for $\kappa = 1$? What happens to $q = 2n$ *for* $\kappa = \pm 1$?

(b) Let K be a k-dimensional piecewise smooth subpolyhedron in an n-dimensional manifold V of constant curvature κ. What is the minimal $q = q(n, k, \kappa)$, such that some small neighborhood $U \subset V$ of K admits an isometric C^∞-immersion $U \to M^q[\kappa]$ for all V and K? No estimate better than $n + k \leq q \leq (n+2)(n+3)/2$ is known for $k \geq 1$.

(b′) Show that $q(n, 1, \kappa) = n + 1$.

(b″) Let U_ε be the ε-neighborhood of a closed geodesic in an n-dimensional manifold of constant curvature κ. Prove for small $\varepsilon > 0$ the existence of an isometric C^{an}-immersion $U_\varepsilon \to M^{n+1}[\kappa]$. Study the boundary ∂U_ε for $n = 3$ and show that every complete flat 2-dimensional manifold admits an isometric immersion into $M^4(\kappa)$ for all κ. For example the 2-torus with an arbitrary flat metric (as well as every flat Klein bottle) admits an isometric C^{an}-immersion into $\mathbb{R}^4$.

(b‴) Observe the following Jackobowitz [1976] that every flat 2-torus admits a *flat* isometric immersion into a flat *split* 3-torus which is an isometric product of three circles. [In general, every flat n-torus goes to a split $\frac{n(n+1)}{2}$-torus, see Jackobowitz (1976).] Use this to show that every flat 3-torus admits an isometric C^∞-immersion into $\mathbb{R}^8$.

(c) Let T^n be the flat torus which is obtained by identification or opposite faces of the cube $\{|x_i| \leq 1, i = 1, \ldots, n\} \subset \mathbb{R}^n$, and let $B(r) \subset T^n$ be the ball $\{\sum_{i=1}^n x_i^2 \leq r^2\}$ for some $r < 1$. Show that the complement $U_r^n = T^n \backslash B(r)$ admits no isometric C^∞-immersion into $\mathbb{R}^{2n-1}$ for $2(\sqrt{n} - r) \geq \pi r$. Construct such an immersion $U_r^2 \to \mathbb{R}^3$ for $r = 0.9$.

(c′) Let $V(\varepsilon)$ denote the complement to the square $\{|x_i| \leq \varepsilon, i = 1, 2\}$ in T^2. Show that $V(\varepsilon)$ admits an isometric C^∞-immersion into $\mathbb{R}^3$ if and only if $\varepsilon > \frac{1}{2}$. Show that $V(0.9)$ admits an isometric C^{an}-immersion into $\mathbb{R}^3$ but $V(0.51)$ has no such immersion.

(c″) Find a flat metric on S^2 minus three open disks which admits no isometric C^∞-immersion into $\mathbb{R}^3$. {No such metric is known on the cylinder $S^1 \times [0, 1]$, but many are known on the Möbius band, see Halpern-Weaver (1977).}

(d) Show that the compact disk B^2 with an *arbitrary* flat metric admits an isometric C^∞*-imbedding* into $\mathbb{R}^3$. [No nontrivial imbedding result is known for flat metrics on B^n for $n \geq 3$.]

(d′) (Michel Katz, unpublished). Construct infinitely many pairwise topologically non-isotopic isometric C^∞-imbeddings of the flat cylinder $S^1 \times [0, \varepsilon]$ into $\mathbb{R}^3$ for length $(S^1) = 1$ and for $\varepsilon = 10^{-6}$.

(d″) Find a complete (nonflat!) C^∞-surface V which admits uncountably many pairwise non-isotopic proper isometric C^∞-imbedding $V \to \mathbb{R}^3$.

(e) Let $U \subset \mathbb{R}^n$ be an arbitrary open subset. Define $\mathrm{Rad}_k(U \subset \mathbb{R}^n)$ to be the lower bound of those numbers $R \geq 0$ for which there exists a continuous map $\alpha\colon U \to \mathbb{R}^n$ with the following two properties

(i) the topological dimension of the image satisfies

$$\dim f(U) \leq k,$$

(ii) $\mathrm{dist}(u, f(u)) \leq R$ for all $u \in U$.

See Appendix 1 in Gromov (1983) for basic properties of Rad_k.

Examples. If U contains a $(k + 1)$-dimensional cycle $C \subset U \subset \mathbb{R}^n$ which is not homologous to zero in the ε-neighborhood $U_\varepsilon(C) \subset \mathbb{R}^n$, then $\mathrm{Rad}_k(U) \geq \varepsilon$.

If the complement $\mathbb{R}^n \backslash U$ is a union of disjoint compact subsets in $\mathbb{R}^n$, then $\mathrm{Rad}_{n-2}(U) = \infty$.

If the complement $\mathbb{R}^n \backslash U$ can be covered by compact subsets $K_j \subset \mathbb{R}^n$, $j = 1, 2, \ldots$, such that $\mathrm{Diam}\, K_j \leq \mathrm{const} < \infty$ and such that no l subsets among K_j intersect, then $\mathrm{Rad}_{n-l}(U) = \infty$.

(e′) Let U admit an isometric C^∞-immersion into a ball $B^{n+k}(R) \subset \mathbb{R}^{n+k}$ of radius R. Prove that

$$\mathrm{Rad}_k(U) \leq \sqrt{n}R.$$

(e″) Let $K \subset V$ be a properly imbedded smooth submanifold of dimension k. What is the minimal q such that some small neighborhood $U \subset \mathbb{R}^n$ of K admits a *bounded* isometric C^∞-immersion $\mathrm{f}\colon U \to \mathbb{R}^q$? The above (e′) implies the estimate $q \leq n + k$ for "sufficiently spread" submanifolds K and the upper bound $q \leq n + 1$ is obvious for $k = 1$. Nothing beyond this is known.

(f) Let V be a complete n-dimensional manifold of constant curvature $\kappa \leq 0$ which admits an isometric C^∞-immersion $f\colon V \to M^q[\kappa]$. Show that V is homotopy equivalent to a $(q - n)$-dimensional polyhedron.

Additional References. Dajczer-Gromoll (1984), Laroubi (1984), Zeghib (1984).

3.2.2 Hyperbolic Immersion

Consider a smooth immersion $f\colon V \to (W,h)$ and compare sectional curvatures of (W,h) and (V,g) for the induced metric $g = f^*(h)$. Namely, set

$$\Delta = \Delta(\partial_1,\partial_2) = R_g(\partial_1,\partial_2;\partial_1,\partial_2) - R_h(\partial_1,\partial_2;\partial_1,\partial_2),$$

for all pairs of orthonormal tangent fields ∂_1 and ∂_2 in $T(V) \subset T(W)|V$. The immersion f is called *hyperbolic* if $\Delta \le 0$ and it is called *strictly hyperbolic* if $\Delta < 0$ for all orthonormal fields ∂_1 and ∂_2 on V. (The equality $\Delta \equiv 0$ amounts to the parabolicity of f.)

Denote by Π_ν the *second quadratic form* (on V) for a normal vector $\nu \in N_v = N \subset T(W) \ominus T(V)|V$. Recall that

$$\Pi_\nu(\partial_1(v),\partial_2(v)) = \langle \nabla_{\partial_1}\nabla_{\partial_2} f(v), \nu\rangle,$$

where the covariant derivatives are understood in (W,h). Let

$$Di_\nu(\partial_1,\partial_2) = \Pi_\nu(\partial_1,\partial_1)\Pi_\nu(\partial_2,\partial_2) - (\Pi_\nu(\partial_1,\partial_2))^2$$

be the discriminant of Π_ν. Then Gauss theorema egregium claims

$$\Delta(\partial_1(v),\partial_2(v)) = \sum_{j=1}^{q-n} Di_{\nu_j}(\partial_1(v),\partial_2(v)),$$

for an arbitrary orthonormal basis $(\nu_1,\ldots,\nu_{q-n})$ in N_v.

Denote by $\alpha\colon Gr_2 V \to V$ the Grassman bundle of tangent 2-planes in V and let $L \to Gr_2 V$ be the canonical oriented 2-bundle. Take a plane $L_x \subset L$ over some point $x \in Gr_2 T_v(V) \subset Gr_2 V$ which is spanned by some orthonormal vectors $\partial_1(v)$ and $\partial_2(v)$ in $T_v(V)$. Then the discriminant of the form $\Pi_\nu L_x$ equals the product of the eigenvalues α_1 and α_2 of the form $\Pi_\nu|L_x$. Let us define another form $\bar{\Pi}_\nu = \bar{\Pi}_\nu^x$ on L_x as follows

$$\bar{\Pi}_\nu = (\Pi_\nu - \tfrac{1}{2}(\alpha_1 + \alpha_2)g)|L_x.$$

The form $\bar{\Pi}_\nu$ clearly has zero trace and $\bar{\Pi}_\nu = 0$ if and only if $\alpha_1 = \alpha_2$.

We call the immersion $f\colon V \to W$ *umbilic* on L_x if $\bar{\Pi}_\nu^x = 0$ for all normal vectors $\nu \in N_v$. The umbilicity of f on L_x obviously implies $\Delta|L_x \ge 0$ and so strictly hyperbolic immersions are nowhere umbilic.

Denote by $\tilde{N} \to Gr_2 V$ the lift of the normal bundle, $\tilde{N} = \alpha^*(N)$, and let $\bar{L}^2 \to Gr_2 V$ be the 2-bundle whose fiber $\bar{L}_2^x$ consists of quadratic forms on L_x with zero trace. The correspondance $\nu \mapsto \bar{\Pi}_\nu$ is a homomorphism of bundles,

$$\bar{\Pi}\colon \tilde{N} \to \bar{L}^2,$$

whose zeros $x \in Gr_2 V$ correspond to umbilic planes $L_x \subset T(V)$. Thus we conclude to the following

(A) Lemma. *If $f\colon V \to W$ is a strictly hyperbolic immersion, then the homomorphism $\bar{\Pi}\colon \tilde{N} \to \bar{L}^2$ does not vanish and so the Euler class $\bar{\chi}$ of the bundle* $\mathrm{Hom}(\tilde{N} \to \bar{L}^2)$ *is zero.*

(A′) **Corollary** (Liber-Chern-Kuiper-Otsuki-Springer). *There is no strictly hyperbolic immersions $f: V \to W$ for*

$$\dim W = q \leq 2n - 2, \qquad n = \dim V.$$

Proof. The oriented 2-bundles L and $\bar{L}^2$ can be viewed as complex line bundles and they obviously satisfy the relation $\bar{L}^2 = L^{-2}$, which implies

$$\chi(\bar{L}^2) = -2\chi(L)$$

Then the restriction of the class $\bar{\chi}$ to the Grassman manifold

$$Gr_2 \mathbb{R}^n = Gr_2 T_v(V) \subset Gr_2 V,$$

where the bundle $\tilde{N}$ is trivial, satisfies for all $v \in V$

$$\bar{\chi} | Gr_2 \mathbb{R}^n = (-2\chi(L))^k | Gr_2 \mathbb{R}^n,$$

for $k = \dim \tilde{N} = q - n$. The class $(\chi(L))^k \in H^{2k}(Gr_2 \mathbb{R}^n; \mathbb{R})$ does not vanish for $2k \leq \dim Gr_2 \mathbb{R}^n = 2(n-2)$ by (A″) of 3.2.1 and the Lemma is proven.

(A″) **Corollary.** *Let some normal Pontryagin class $p_l(N) \in H^{4l}(V, \mathbb{R})$ be non zero while $p_i(N) = 0$ for $i \neq l$. Then the hyperbolicity of the immersion $f: V \to W$ implies $q > 2n + 2l - 2$.*

Proof. Since $p_i = 0$ for $i \neq l$ the Euler class $\bar{\chi}$ of $\mathrm{Hom}(\tilde{N} \to \bar{L}^2)$ is expressed in terms of the induced class $p_l(\tilde{N}) = \alpha^*(p_l(\tilde{N}))$ by the well-known formula

$$\bar{\chi} = (-2\chi((L))^m \alpha^*(p_l(N)),$$

for $m + 2l = \dim N = q - n$. To conclude the proof we must show that

$$p \neq 0 \Rightarrow (\chi(L))^{n-2} \alpha^*(p) \neq 0 \qquad \text{for all } p \in H^*(V; \mathbb{R}).$$

The class $(\chi(L))^{n-2} \in H^{2n-2}(Gr_2 V)$ is given by a $(2n-2)$-form ω whose restriction to $GrT_v(V) = Gr_2 \mathbb{R}^n$ is a normalized volume form [see (A″) in 3.2.1]. Hence, the integrals of the forms π on V over cycles $C \subset V$ satisfy

$$\int_C \pi = \int_{\tilde{C}} \omega \wedge \alpha^*(\pi)$$

for the pull-back $\tilde{C} = \alpha^{-1}(C) \subset Gr_2 V$. If $p \neq 0$, then p is given by a form π such that $\int_C \pi \neq 0$ for some cycle C in V and the above implication follows. Q.E.D.

Example. Let us construct for given n and $l \leq n/4$ a manifold $V = V^n$, such that the normal bundle N of every immersion $f: V \to \mathbb{R}^q$ has $p_l(N) \neq 0$ and $p_i(N) = 0$ for $i \neq l$. It is enough to consider the case $n = 4l$ and $q > 2n$. *Take a* $(q-n)$-dimensional vector bundle $M \to S^{4l}$ for which $p_l(M) \neq 0$. As $q - n > n$, the total space of the bundle M has the same proper rational homotopy type as the trivial bundle $\mathbb{R}^{q-n} \times S^4$ (this is a standard corollary of Serre's theorem on finiteness of the stable homotopy groups of spheres) and so there is a proper map $\varphi: \mathbb{R}^{q-n} \times S^{4l} \to M$ of non-zero degree. Make this φ smooth and transversal to the zero section $S^4 \subset M$ and then take the pullback $V = \varphi^{-1}(S^{4l}) \subset \mathbb{R}^{q-n} \times S^{4l}$.

Corollary. *There is no strictly hyperbolic immersion* $V \times \mathbb{R}^k \to \mathbb{R}^{q+2k}$ *for all* $k = 0, 1, \ldots,$ *for* $q \le \frac{5}{2}n - 2$ *and for the above manifold* V *of dimension* $n = 4l$.

Exercises. (a) Fill in the details in the above argument.

(b) Express the condition $\bar{\chi} \neq 0$ *in terms of characteristic classes of the bundles* $T(V)$ and N over V (without any simplifying assumptions like $p_i = 0$ for $i \neq l$).

Asymptotic Directions. A non-zero tangent vector (direction) $X \in T_v(V)$ is called *asymptotic* for an immersion $f\colon V \to W$ if $\Pi_\nu(X, X) = 0$ for all normal vectors $\nu \in N_v$. This is equivalent to the system of the following $q - n$ homogeneous quadratic equations

$$(*) \qquad \Pi_{\nu_j}(X, X) = 0, \qquad j = 1, \ldots, q - n,$$

where $\{\nu_1, \ldots, \nu_{q-n}\}$ is some basis in N_v.

(B) **Lemma** (T. Springer and T. Otsuki 1953). *If* $q \le 2n - 1$ *and if* f *is hyperbolic, then there is an asymptotic direction* X *in every tangent space* $T_v(V)$. *Furthermore, if* f *is strictly hyperbolic, then* $q = 2n - 1$ [compare (A')] *and there are exactly* 2^n *distinct hyperbolic directions in* $T_v(V)$, *which, moreover, continuously depend on* $v \in V$.

Proof. If $n > q - n$, then the system $(*)$ has, according to Bezout's theorem, a non-zero complex solution $Z = X + Y\sqrt{-1}$, for X, $Y \in T_v(V)$. The equations $\Pi_{\nu_j}(Z, Z) = 0$ amounts to the system

$$(**) \qquad \begin{aligned} \Pi_{\nu_j}(X, X) &= \Pi_{\nu_j}(Y, Y) \\ \Pi_{\nu_j}(X, Y) &= 0. \end{aligned}$$

Since f is hyperbolic,

$$\Delta(X, Y) = \sum_j \Pi_{\nu_j}(X, X)\Pi_{\nu_j}(Y, Y) - (\Pi_{\nu_j}(X, Y))^2 \le 0$$

for all X and Y in $T_v(V)$. Therefore, the solutions X and Y of $(**)$ also satisfy $(*)$ and at least one of them is non-zero, which is the required asymptotic direction.

If f is strictly hyperbolic, then $\Delta(X, Y) < 0$ for all linearly independent vectors X and Y in $T_v(V)$. It follows that all non-zero solutions X of $(*)$ are simple and isolated (in the real projective space P^{n-1}). Indeed, if $X + \varepsilon X'$ is an infinitesimal deformation of a solution X of $(*)$, then $\Pi_{\nu_j}(X, X') = 0$, and the inequality $\Delta(X, X') \le 0$ implies $X' = \lambda X$, $\lambda \in \mathbb{R}$. For the same reason all solutions of $(**)$ are real, $Z = (a + b\sqrt{-1})X$, and so all complex solutions of $(*)$ are simple and isolated (in $\mathbb{C}P^{n-1}$) as well as real. It follows, with Besout's theorem, that $q - n \ge n - 1$ and the number of solutions (in P^{n-1}) equals 2^n. As these solutions are simple, they are continuous (in fact, smooth) in $v \in V$ by the implicit function theorem. Q.E.D.

Remark. If f is a *parabolic* immersion, then every asymptotic direction X is flat since $\Delta(X, Y) = 0$, $Y \in T_v(V)$. Thus we obtain another proof of (A''') in 3.2.1 for $q = 2n - 1$. In fact, the general case $q \le 2n - 1$ also follows from the above lemma.

(B′) **Corollary.** (Borisenko 1977). *Let $f\colon V^n \to W^q$ be a strictly hyperbolic immersion for $q = 2n - 1$. If V^n is a closed manifold, then the Euler characteristic $\chi(V^n) = 0$.*

Proof. Indeed every asymptotic direction $X \in T_v(V)$, $\|X\| = 1$, extends to a unique normal asymptotic vector field on some finite covering of V^n.

Remark. If V^n and W^q have constant sectional curvature, then a finite covering of V is parallelizable, since there are n out of 2^n asymptotic directions which are linearly independent on V (see Moor 1972; Borisenko 1977). However, one does not yet know whether there are (isometric) hyperbolic immersions between round spheres, $S^n \to S^{2n-1}$ for $n = 3, 7$.

Let us slightly generalize (A′). Take an arbitrary immersion $f\colon V \to W$ and consider a unit normal vector $\nu \in N_v$ at some point $v \in V$.

(B″) **Lemma.** *If the form Π_ν is positive definite on some k-dimensional subspace $L \subset T_v(V)$ and if*

$$q - n = \dim W - \dim V < k,$$

then the immersion f is not hyperbolic. Moreover, there are orthonormal vectors X and Y in L for which

$$\Delta(X, Y) \geq Di_{\nu_j}(X, Y) = \Pi_\nu(X, X)\Pi_\nu(Y, Y) - (\Pi_\nu(X, Y))^2 > 0.$$

Proof. Take an orthonormal basis $\{\nu_j\}$ in N_v, $j = 1, \ldots, q - n$ with $\nu_1 = \nu$. The proof of (A′) yields orthonormal vectors X and Y in L such that $\sum_{j=2}^{q-n} Di_{\nu_j}(X, Y) \geq 0$ and so

$$\Delta(X, Y) = \sum_{j=1}^{q-n} Di_{\nu_j}(X, Y) \geq Di_\nu(X, Y) > 0.$$

Exercises. (a) Prove (B″) with the argument of (B).

(b) (Chern-Kuiper 1952; Jacobowitz 1973). Let $f\colon V^n \to \mathbb{R}^{2n-1}$ be a smooth immersion. Show that f cannot be hyperbolic if the manifold is closed. Assume, moreover, that the image $f(V)$ lies in the ball of radius R, say $B_x(R)$, $x \in \mathbb{R}^q$, and prove the existence of a 2-plane τ in $T_{v_0}(V)$ for some $v_0 \in V$ at which the sectional curvature $K(\tau) \geq R^{-2}$.

Hint. Consider a maximum point $v_0 \in V$ of the function $v \mapsto \operatorname{dist}(v, x)$ and apply (B″) to $\nu = (f(v) - x)/\|f(v) - x\|$.

(b′) Prove the existence of τ with $K(\tau) \geq R^{-2}$ under the following weaker assumption: the orthogonal projection of $f(V) \subset \mathbb{R}^{2n-1}$ into some hyperplane $\mathbb{R}^{2n-2} \subset \mathbb{R}^{2n-1}$ is contained in a ball of radius R. Derive the following relation between $D = \operatorname{Diam} V$ and $K_+ = \sup_\tau K(\tau)$ for those Riemannian manifolds V which admit isometric C^2-immersions into $\mathbb{R}^{2n-1}$,

$$(*) \qquad K_+ \geq D^{-2}\frac{2n-1}{n-1}.$$

Verify (*) for the real projective space P^3 of constant curvature and prove that there is no isometric C^2-immersion $P^3 \to \mathbb{R}^5$. Generalize this by showing that no metric on P^n whose sectional curvature is pinched between 0.99 and 1 admits an isometric C^2-immersion into $\mathbb{R}^{n+2}$ for all $n = 3, 4, \ldots$.

(b″) (A. Gray 1969). Let $f\colon V^n \to W^q$ be a proper C^2-immersion where W^q is a complete simply connected manifold of non-positive sectional curvature. Show V to be homotopy equivalent to a k-dimensional polyhedron for some $k \le q - n$, provided one of the following two conditions is satisfied.

(i) f is hyperbolic,
(ii) the manifold W^q has *constant* negative curvature and the induced metric in V^n has non-positive curvature.

Hint. Study the critical points of the function $v \to \operatorname{dist}(w_0, v)$ for some $w_0 \in W^q$.

(C) *Flat Directions of Hyperbolic Immersions*. Let $f\colon V \to W$ be a hyperbolic immersion. Then the quadratic form Π_ν has rank $\Pi_\nu \le 2q - 2n$ for all normal vectors $\nu \in T_v(V)$, $v \in T$. This is an immediate corollary of (B″). In particular, the forms Π_ν are singular for $q < \frac{3}{2}n$. In order to exploit this for manifolds W of *constant curvature* we recall classical facts on the following

Geometric Legendre Transform. Let $P = P^q$ be the real projective space and let P^* be the dual space. Points $y \in P^*$ by definition are hyperplanes in P, called $y^* \subset P$. Observe that $P^{**} = P$ and so points $x \in P$ correspond to hyperplanes $x^* \subset P^*$. Observe that $x \in y^* \Leftrightarrow y \in x^*$ and denote by $Q \subset P \times P^*$ the set of those pairs (x, y) for which $x \in y^*$. Denote by π and π^* the projections of Q to P and to P^* respectively. Observe that the bundle $\pi\colon Q \to P$ is canonically isomorphic to the Grassmann bundle of hyperplanes in $T(P)$ called $Gr_{q-1} P \to P$, while the bundle $\pi^*\colon Q \to P^*$ is canonically isomorphic to $Gr_{q-1} P^* \to P^*$.

The manifold Q carries a natural hyperplane field (a contact structure) $\theta \subset T(Q)$, where the hyperplane $\theta_z \subset T_z(Q)$, $z = (x, y) \in Q$, is defined as the span of the tangent spaces to the fibers $\pi^{-1}(x) \subset Q$ and $(\pi^*)^{-1}(y) \subset Q$ at $z \in Q$. The tangent spaces to the fibers, say T_z^x and T_z^y of dimension $q - 1$ in $T_z(Q)$, have $T_z^x \cap T_z^y = 0$ and so they span a *hyperplane* in $T_z(Q)$ for all $z \in Q$.

Take an arbitrary smooth submanifold $V \subset P$ of positive codimension and denote by $\tilde{V} \subset Q$ the subset of those pairs $(x, y) \in Q \subset P \times P^*$ for which $x \in V$ and the hyperplane $y^* \subset P$ is tangent to V at the point x. For example, if V consists of a single point $v \in P$, then $\tilde{V} = \pi^{-1}(v)$; if V is a hyperplane, $V = y^*$ for some $y \in P^*$, then $\tilde{V} = (\pi^*)^{-1}(y)$.

A submanifold $Z \subset Q$ is called *Legendre* if $\dim Z = q - 1$ and if the tangent space $T_z(Z) \subset T_z(Q)$ lies in θ_z for all $z \in Z$.

Lemma (Legendre). *The above submanifold $\tilde{V} \subset Q$ is Legendre for all $V \subset P$. Furthermore, let $Z \subset Q$ be Legendre, let the projection $\pi|Z$ have constant rank $= n$ and let the image $V = \pi(Z) \subset P$ be a smooth submanifold in P. Then $\tilde{V} \supset Z$. Moreover, if the intersection $Z \cap \pi^{-1}(x)$ is a closed manifold for all $x \in P$, then $\tilde{V} = Z$.*

The (well known and easy) proof is left to the reader.

Exercise. Let P be an *arbitrary* smooth manifold of dimension q and let $Q = Gr_{q-1} P$. Define $\theta_z \subset T(Q)$, $z = (x, H)$, for $x \in P$ and $H \subset T_x(P)$, to be the pull-back of the hyperplane H under the (differential of the) projection $Q \to P$. The resulting hyperplane field $z \mapsto \theta_z$ on Q is called *the canonical contact structure* on Q and the submanifolds $Z = Z^{q-1} \subset Q$ tangent to θ are called *Legendre*. Show this new field θ to be equal to the previous one for the *projective* spaces and extend Legendre's lemma to all smooth manifolds P.

Definition. Legendre's transform of a submanifold $V \subset P$ is the image $V^* = \pi^*(\tilde{V}) \subset P^*$.

Warning. This V^*, in general, is not a submanifold in P^* since the map $\pi^* | \tilde{V}$ may have non-constant rank.

Exercise. Let codim $V \geq 2$ and let the map $\pi^* | \tilde{V}$ have constant rank. Show that V is *flat*, i.e. V lies in a *projective subspace* $P^n \subset P^q$ for $n = \dim V$.

The above Legendre's lemma immediately implies the famous *Legendre duality theorem* which claims the identity $(V^*)^* = V$ for the submanifolds V of positive codimension in P. Since V^* may be singular, the identity $(V^*)^* = V$ applies only to those points $z \in \tilde{V}$ where the map $\pi^* | \tilde{V}$ has locally constant rank. Namely, if rank $\pi^* | \tilde{U}' \equiv k$ for some small neighborhood $\tilde{U}' \subset \tilde{V}$ of $z \in \tilde{V}$, then the image $\pi^*(\tilde{U}')$ is a k-dimensional submanifold, say $U^* \subset P$, such that $\tilde{U}^* \supset \tilde{U}'$. Recall that $\dim \tilde{U}^* = \dim \tilde{U}' = q - 1$.

Denote by $\tilde{\Sigma}_j^* \subset \tilde{V}, j = 0, \ldots, q-1$, the subset of those $z \in \tilde{V}$ where rank $\pi^* | \tilde{V} = j$. The (connected components of the) pull-backs of the map $\pi^* | \tilde{V}$ foliate the interior of the subset $\tilde{\Sigma}_j^*, j = 0, \ldots, q-1$, into $(q-1-j)$-dimensional leaves $\tilde{\mathscr{L}}_z \subset \operatorname{Int} \tilde{\Sigma}_j^*$, through the points $z \in \operatorname{Int} \tilde{\Sigma}_j^*$. Every such leaf projects under the map $\pi | \tilde{V} : \tilde{V} \to V \subset P$ onto a $(q-1-j)$-dimensional submanifold, called $\hat{\mathscr{L}}_z \subset V$, which is uniquely determined by the point

$$z = (x, y) \in \operatorname{Int} \tilde{\Sigma}_j^* \subset \tilde{V} \subset Q \subset P \times P^*,$$

where $x \in \hat{\mathscr{L}}_z \subset V \subset P$ and where the hyperplane $y^* \subset P^*$ is tangent to V at x. Since the hyperplanes $l^* \subset P^*$, $l \in \hat{\mathscr{L}}_z$, are tangent to $V^* \subset P^*$ at $y \in V^*$ by Legendre duality, the submanifold $\hat{\mathscr{L}}_z \subset P$ is flat for all $z \in \operatorname{Int} \tilde{\Sigma}_j^*, j = 0, \ldots, q-1$, and the hyperplane y^* is tangent to V at all points $l \subset \hat{\mathscr{L}}_z$. Furthermore, if j_0 is the greatest integer for which $\tilde{\Sigma}_{j_0}^* \neq \varnothing$, then the leaf $\hat{\mathscr{L}}_z$ is a *closed* subset in V for a generic point $z \in \pi(\tilde{\Sigma}_{j_0}^*) \subset V$ according to Sard's theorem [compare (B) in 3.2.1].

Finally, we observe that the kernel of the differential $D_z \pi^* | \tilde{V}$ *isomorphically* projects onto Ker Π_v under (the differential of) the map $\pi | V$, where $v \in T_x(P)$ is the unit normal to the hyperplane $y^* \subset P$ for all $z = (x, y) \in \tilde{V}$. (Here we use the standard metric of constant curvature $+1$ in the projective space P and we may take an arbitrary point $z \in \tilde{V}$.)

Exercise. Study the relations between the manifolds $\hat{\mathscr{L}}_z$ and the flat leaves $\mathscr{L}_v$ [see (B) in 3.2.1].

Since the above considerations are projectively invariant they apply to all manifolds $W = W^q$ of *constant* sectional curvature and so we have the following

Proposition. *If an isometric immersion $f\colon V \to W$ has rank $\Pi_v \leq k < n = \dim V$ for all normal vectors v, then there exists a totally geodesic submanifold $\mathscr{L} \subset V$ of dimension k which is (relatively) complete in V and such that the map $f|\mathscr{L}\colon \mathscr{L} \to V$ is isometric flat.*

Corollary. *If the manifolds V and W are complete and if the map f is hyperbolic, then the image $f(V) \subset W$ contains a complete totally geodesic submanifold of dimension $\geq 3n - 2q$.*

This result is due to Borisenko (1977) where the reader is referred to for a further study of hyperbolic immersions.

(D) Examples of Hyperbolic Immersions. There are various constructions of complete hyperbolic surfaces in $\mathbb{R}^3$ (see Rosendorn 1966). For example, let $K \subset \mathbb{R}^3$ be a piecewise linear one-dimensional polyhedron (graph), such that every vertex $k_0 \subset K$ lies in the *interior* of the convex hull of the neighbor vertices. Then the boundary of an appropriate small neighborhood of K is a *hyperbolic* surface which is complete, provided K contains no connected infinite chain of edges whose total length is finite.

Exercise. Make the above precise. Then construct a complete bounded hyperbolic immersion $\mathbb{R}^2 \to \mathbb{R}^3$. Observe that the direct product of hyperbolic immersions is hyperbolic and construct a complete bounded hyperbolic immersion $\mathbb{R}^{2n} \to \mathbb{R}^{3n}$, $n = 2, 3, \ldots$.

If W^q is a manifold of constant sectional curvature, then the hyperbolicity is quite a restrictive condition for complete immersions $V^n \to W^q$ for $q \leq 2n - 2$, and especially for $q < \frac{3}{2}n$. Yet, there is no geometric classification of these immersions. In fact, one does not even know which closed manifolds V^n admit non-flat hyperbolic immersions in a given manifold W^q for $n + 2 \leq q < \frac{3}{2}n$.

Clifford Torus. The product of n unit circles $S^1 \subset \mathbb{R}^2$ is called the *Clifford torus* $T^n \subset S^{2n-1} \subset \mathbb{R}^{2n}$, where

$$S^{2n-1} = \{x \in \mathbb{R}^{2n} \mid \|x\| = \sqrt{n}\}.$$

This torus is strictly hyperbolic in the sphere S^{2n-1}.

The cone from the origin over the Clifford torus is a (non-strictly) hyperbolic immersion $C\colon T^n \times (0, \infty) \to \mathbb{R}^{2n}$, which is non-complete at the origin. Furthermore, the map $(t, x) \mapsto (C(t, x), f(x))$ for $f(x) = x^{-1}$ is a complete hyperbolic imbedding of

$T^n \times (0, \infty)$ into $\mathbb{R}^{2n+1}$. Observe that the sectional curvature of the induced metric converges to zero for $x \to \infty$.

Question. Does the Klein bottle admit a (strictly) hyperbolic immersion into S^3?

Isometric Immersions of Hyperbolic Spaces. The hyperbolic plane H^2 (with the complete metric of constant curvature -1) admits no isometric C^2-immersion $f: H^2 \to \mathbb{R}^3$ by a famous theorem of Hilbert. In fact, no complete immersion $V^2 \to \mathbb{R}^3$ is *uniformly* strictly hyperbolic which means $K \leq -\kappa < 0$ for the Gauss curvature K of the induced metric (see Efimov 1964; T. Milnor 1972).

Question. How does the Hilbert-Efimov theorem generalize to manifolds of dimension ≥ 2?

It is unknown whether the hyperbolic plane admits a C^k-immersion into $\mathbb{R}^4$ for $k \geq 2$. However, there exist closed strictly (and hence, uniformly) hyperbolic surfaces in $\mathbb{R}^4$ (see Rosendorn 1961). The universal covering of such a surface is a complete bounded immersion $\mathbb{R}^2 \to \mathbb{R}^4$ which is strictly uniformly hyperbolic [see the survey by Poznjak (1973), for additional information].

Theorem (Rosendorn 1960). *There exists an isometric C^∞-immersion $H^2 \to \mathbb{R}^5$.*

Proof. Take (*horospherical*) coordinates t and u in H^2, $-\infty < t, u < \infty$, such that the hyperbolic metric equals $(dt)^2 + e^{2t}(du)^2$. Decompose, $e^{2t} = \varphi_1^2(t) + \varphi_2^2(t)$, for some C^∞-function $\varphi_i(t)$, $i = 1, 2$, such that the support of φ_i, $i = 1, 2$, is a disjoint union of closed subintervals in $\mathbb{R}$. Let $\varepsilon_i(t)$ be a positive function which is locally constant on the support of φ_i for $i = 1, 2$. Set

$$f_i = f_i(t, u) = \varepsilon_i(t)\varphi_i(t)\sin(u\varepsilon_i^{-1}(t))$$

and

$$f_i' = \varepsilon_i(t)\varphi_i(t)\cos(u\varepsilon_i^{-1}(t)),$$

for $i = 1, 2$. By Nash's formula (see 3.1.1)

$$\sum_{i=1,2} (df_i)^2 + (df_i')^2 = e^{2t}(du)^2 - \delta(t)(dt)^2,$$

for $\delta(t) = \sum_{i=1,2} (\varepsilon_i(t))^2 \left(\frac{d\varphi_i(t)}{dt}\right)^2$.

If the functions ε_i are chosen sufficiently small, then $1 - \delta(t) > 0$. Therefore, there is a C^∞-function $\alpha(t)$ such that $(d\alpha(t)/dt)^2 = 1 - \delta(t)$ and the five functions f_i, f_i' and α, for $i = 1, 2$, define the required isometric immersion $H^2 \to \mathbb{R}^5$.

An Isometric C^∞-Immersion $H^n \to \mathbb{R}^{4n-3}$. The metric (of constant curvature -1) on the hyperbolic space is $(dt)^2 + e^{2t}\sum_{j=1}^{n-1}(du_j)^2$, where $t, u_1, \ldots, u_{n-1}$ are the *horospherical coordinates* in H^n. Define with the above φ_i and ε_i the functions $f_{ij} = \varepsilon_i\varphi_i \sin u_j\varepsilon_i^{-1}$ and $f_{ij}' = \varepsilon_i\varphi_i \cos u_j\varepsilon_i^{-1}$ and observe that the map $(f_{ij}, f_{ij}', \alpha\sqrt{n-1})$: $H^n \to \mathbb{R}^{4n-3}$ for $i = 1, 2$, and $j = 1, \ldots, n-1$ is isometric for the above $\alpha = \alpha(t)$.

An Isometric C^{an}-Immersion $H^n_- \to \mathbb{R}^{2n-1}$ (F. Schur 1886). Denote by $H^n_- \subset H^n$ the *horoball*, $H^n_- = \{t, u_1, \ldots, u_n | t \leq 0\}$. Take $f_j = \varepsilon e^t \sin \varepsilon^{-1} u_j$ and $f_j' = \varepsilon e^t \cos \varepsilon^{-1} u_j$ for $j = 1, \ldots, n-1$ and for a small *constant* $\varepsilon > 0$, and let $\beta = \beta(t)$ satisfy for $t \leq 0$,

$$\frac{d\beta(t)}{dt} = 1 - (n-1)\varepsilon^2 e^{2t}.$$

Then the map $(f_j, f_j', \beta)\colon H^n_- \to \mathbb{R}^{2n-1}$ is C^{an}-isometric.

Corollary. *Every relatively compact open subset $U \subset H^n$ admits an isometric C^{an}-immersion into $\mathbb{R}^{2n-1}$.*

Immersion $H^n_- \to M^{2n-1}[\kappa]$. Consider a Riemannian C^{an}-manifold U and let $\gamma \subset U$ be an infinite geodesic. Let the isometry group $Is(U)$ be transitive on γ and let the isotropy subgroup I_w, $w \in \gamma$, contain the torus T^{n-1} as a subgroup. Then an obvious generalization of the above formulae provides an isometric C^{an}-immersion $H^n_- \to U$. In particular, there is an isometric C^{an}-immersion of H^n_- into the ε-neighborhood of an arbitrary (infinite or periodic) geodesic γ in any given space of dimension $\geq 2n-1$ of constant curvature κ for all $\kappa \in \mathbb{R}$ and all $\varepsilon > 0$.

Exercises. Consider the (Riemannian) product of $(n-1)$-copies of H^2 with the metric $\sum_{j=1}^{n-1} (dt_j)^2 + e^{2t}(du_j)^2$. Show that the submanifold $\{t_j, u_j | t_1 = t_2 = \cdots = t_{n-1}\}$ is isometric to H^n with constant curvature $-(n-1)$.

Consider coordinates x and y in H^2 in which the hyperbolic metric is $dx^2 + [\mathrm{ch}(x)]^2 dy^2$ for $\mathrm{ch}(x) = (e^x + e^{-x})/2$. Prove with an appropriate splitting $[\mathrm{ch}(x)]^2 - \frac{1}{2} = \varphi_1^2(x) + \varphi_2^2(x)$ the following theorem of Blanusa (1955).

There exists a proper isometric C^∞-imbedding $H^2 \to \mathbb{R}^6$ whose image is the graph of a C^∞-map $\mathbb{R}^2 \to \mathbb{R}^4$. Furthermore, there exists a proper isometric C^∞-imbedding $H^n \to \mathbb{R}^{6n-6}$ for all $n \geq 2$.

Find a small isometric C^∞-perturbation of the linear imbedding $\mathbb{R}^6 \to \mathbb{R}^{10}$ which makes the composed isometric map $H^2 \to \mathbb{R}^{10}$ free and then approximate this map by an isometric C^{an}-imbedding $H^2 \to \mathbb{R}^{10}$. Prove a similar result for $\mathbb{R}^8$ instead of $\mathbb{R}^{10}$ and thus show that *existence* of a *proper isometric C^{an}-imbedding $H^n \to \mathbb{R}^{8n-8}$*.

(D′) *An Application of the Theory of Sheaves.* The strict hyperbolicity condition for immersions $V \to W$ is clearly open and (Diff V)-invariant. Hence, the h-principle applies to *open* manifolds V (see 2.2.2), and so the construction of hyperbolic immersions reduces to the study of the corresponding space of jets.

Lemma. *Choose some vectors a_i in $\mathbb{R}^n$, $i = 1, \ldots, k$, put*

$$f_i(x) = f_{a_{i,\lambda}}(x) = \langle a_i, a_i \rangle \langle x, x \rangle - \lambda \langle x, a_i \rangle^2,$$

and let $F\colon \mathrm{Id} \oplus f_1 \oplus \cdots \oplus f_k\colon \mathbb{R}^n \to \mathbb{R}^{n+k}$. If the vectors a_i span a subspace of dimension $\geq n-1$ in $\mathbb{R}^n$ and if the number λ is sufficiently large, then the map F is strictly hyperbolic at the origin $0 \in \mathbb{R}^n$.

Proof. Let $f_i(x, y)$ be the symmetric forms corresponding to the (quadratic) function f_i. Then, obviously,

$$\Delta(x, y) = \sum_{i=1}^{k} f_i(x, x) f_i(y, y) - (f_i(x, y))^2 > 0$$

for all large λ and for all orthonormal vectors x and y in $\mathbb{R}^n$. Q.E.D.

Corollary. *Consider an immersion φ of an open manifold V into a Riemannian manifold W. Let $v_i \in N_\varphi \subset T(W)|V$, $i = 1, \ldots, k$, be mutually orthogonal normal (to V) field and let α: $N_\varphi \to T(V)$ be a homomorphism. If the tangent fields $a_i = \alpha(v_i)$ span a subspace of dimension $\geq n - 1$, at every point $v \in V$, then the map f is homotopic to a strictly hyperbolic immersion.*

Proof. Consider the above functions

$$f_i \quad \text{on} \quad \mathbb{R}^n = T_v(V), \qquad v \in V, \qquad \text{for } f_i = f_{a_i(v), \lambda(v)},$$

where $\lambda(v)$ is a sufficiently large continuous function in $v \in V$ and define

$$F_v = D_v \varphi + \sum_{i=1}^{k} f_i v_i \colon T_v(V) \to T_w(W), \qquad w = \varphi(v).$$

The maps F_v form a continuous in $v \in V$ family of maps $T_v(V) \to T_w(V)$ which are strictly hyperbolic near the zero section $V \to T(V)$. Hence, by the h-principle, the map φ can be deformed to a hyperbolic immersion.

Theorem. *If $q \geq 3n - 2$ then an arbitrary continuous map φ: $V^n \to W^q$ is homotopic to a hyperbolic immersion, provided V^n is open.*

Proof. We may assume φ to be an immersion. Construct normal fields v_i, $i = 1, \ldots, q - n$ by induction in i as follows. Let $\Sigma(i) \subset V$ be the subset where the fields $v_1, \ldots, v_i$ are independent. We construct v_{i+1} by first taking a *generic* normal field v'_{i+1} on the complement $V \setminus \Sigma(i)$ such that $\langle v'_{i+1}, v_j \rangle = 0$, $j = 1, \ldots, i$, and then by taking $v_{i+1} = \psi v'_{i+1}$ where ψ is a smooth function on V whose zero set equals $\Sigma(i)$. The subset $\Sigma^k \subset V$, where codim Span $\{v_i\}_{i=1,\ldots,q-n} \geq k$ clearly has codim $\Sigma^k \geq k$ for all $k = 0, 1, \ldots, q - n$. Therefore, the fields $a_i = \alpha(v_i)$ for a *generic* homomorphism α: $N_\varphi \to T(V)$ satisfy the following inequality over a small neighborhood of the $(n-1)$-skeleton of some triangulation of V.

$$\dim \operatorname{Span} \{a_i\} \geq q - 2n + 1.$$

Hence, the above corollary applies for $q \geq 3n - 2$.

(E) **Exercises and Open Questions.** (a) Replace the above genericity argument by a homotopy theoretic consideration.

(b) Prove that an arbitrary immersion of an open manifold V into a Riemannian manifold W can be regularly homotoped to an immersion f: $V \to W$ which is *ε-parabolic*, that is $|\Pi_v(X, Y)| \leq \varepsilon$ for all unit normal vectors v, and for all unit tangent vectors X and Y, where $\varepsilon > 0$ is any given number.

(b′) Show that the existence of a smooth immersion $V \to \mathbb{R}^m$ implies the existence of a hyperbolic immersion $V \to \mathbb{R}^{2m-1}$ for all open manifolds V.

(c) What is the minimal $q = q(n)$ for which the above existence theorem holds true?

(d) Let V be an arbitrary (possibly closed) n-dimensional manifold and let W have positive sectional curvature. Prove the h-principle for strictly hyperbolic immersions $V \to W$ for $\dim W \geq (n+2)(n+3)/2$ (compare 3.1.7).

(d′) Find counterexamples to the above h-principle for $\dim W = 2n - 1$. (No such example is known for $\dim W \geq 2n$).

(e) An immersion $V \to W$ is called strictly *elliptic* if the discriminant Δ (defined at the beginning of this section) satisfies $\Delta(X, Y) > 0$ for all orthonormal tangent vectors X and Y in $T(V)$. (If $W = \mathbb{R}^q$, this amounts to the positivity of the sectional curvature of the induced metric in V.) Prove that every strictly elliptic immersion admits a non-vanishing normal vector field. Prove the converse for *open* manifolds V: if an immersion $f\colon V \to W$ admits a non-vanishing normal field, then f can be regularly homotoped to a strictly elliptic immersion $V \to W$.

(e′) Recall Whitney's theorem: *no imbedding of the projective plane P^2 into $\mathbb{R}^4$* admits a normal field. Hence, no *imbedding $P^2 \to \mathbb{R}^4$* is strictly elliptic.

(e″) Does there exist an elliptic *immersion $P^2 \to \mathbb{R}^4$*?

(e‴) Construct an elliptic imbedding of $P^3 \approx SO(3)$ into $\mathbb{R}^6$.

3.2.3 Geometric Obstructions to Isometric C^2-Immersions $V^2 \to \mathbb{R}^3$

In this section we discuss several inequalities between geometric invariants of surfaces V in $\mathbb{R}^3$ which indicate a complete break-down of the h-principle for isometric C^2-immersions $V \to \mathbb{R}^3$.

(A) *A Lower Bound on the Curvature $K(v)$ of a Compact Surface V.* Let $V(a) = \{v \in V \mid K(v) \geq a^2\}$ for $a \geq 0$ and assume the existence of an isometric C^2-immersion $f\colon V \to \mathbb{R}^3$ such that $\|f(v)\| \leq R_0$ for all $v \in V$.

(i) If V has no boundary and if $0 \leq a \leq R_0^{-1}$, *then*

$$3a^2 \operatorname{Area} V(a) + \int_{V(a)} K(v)\,dv \geq 4\pi(1 - aR_0)^2.$$

In particular, $\int_{V(0)} K(v)\,dv = \int_V \max(K(v), 0)\,dv \geq 4\pi$.

(ii) *If for some $\alpha < 1$ the boundary points of V satisfy $\|f(v)\| \leq \alpha R_0$ for all $v \in \partial V$, then*

$$3a^2 \operatorname{Area} V(a) + \int_{V(a)} K(v)\,dv \geq (2\pi - 4 \arcsin \alpha)(1 - aR_0)^2.$$

In particular,

$$\int_V \max(K(v), 0)\,dv > 2\pi - 8\alpha.$$

Proof. Take the (interior) normal field ν of the immersion $f \mid V(a)$ for which the second fundamental form Π_ν has non-negative eigenvalues $0 \leq \lambda_1(v) \leq \lambda_2(v)$,

$v \in V(a)$, called the *principal curvatures* of f. Consider the (normal exponential) map e: $V(a) \times \mathbb{R} \to \mathbb{R}^3$, such that $e(v,0) = f(v)$, and which isometrically sends the line $v \times \mathbb{R}$ onto the straight line in $\mathbb{R}^3$ oriented by the normal vector $\nu(v)$ for all $v \in V$. The Jacobian of the map e is clearly $J(v,t) = (1 - \lambda_1(v)t)(1 - \lambda_2(v)t)$.

Let $W = \{(v,t) \in V(a) \times \mathbb{R} | \lambda_2(v) \leq t \leq a^{-1}\} \subset V(a) \times \mathbb{R}$. Then the image $e(W) \subset \mathbb{R}^3$ contains the ball $\{\|x\| \leq R_0 - a^{-1}\} \subset \mathbb{R}^3$. Indeed, each maximum point $v = v(x) \in V$ of the distance function $v \to \operatorname{dist}(f(v), x)$ satisfies $e(v(x), \operatorname{dist}(f(v), x)) = x$. Therefore,

$$\tfrac{4}{3}\pi(R_0 - a^{-1})^3 \leq \int_W J(v,t)\,dv\,dt \leq \int_W (1 + \lambda_1(v)\lambda_2(v)t^2)\,dv\,dt$$

$$= \int_W (1 + K(v)t^2)\,dv\,dt \leq a^{-1} \operatorname{Area} V(a) + \tfrac{1}{3}a^3 \int_{V(a)} K(v)\,dv,$$

which implies (i).

If V has a boundary, then $e(W)$ contains the subset $X_r - \{x \in \mathbb{R}^3 | \|x\| \leq \min(R_0 - a^{-1}, \|x - f(v)\| - \alpha R_0\}$ for all $v \in V$. If $\|f(v)\| = R_0$, then $\operatorname{Vol} X_r \geq \frac{1}{3}(2\pi - 4 \arcsin \alpha)(1 - aR_0)^2$ which implies (ii). Q.E.D.

(A′) **Exercises and Generalizations.** (a) Prove (A) for $a = 0$ by applying the Gauss-Bonnet formula to the boundary of the convex hull $\operatorname{Conv} f(V) \subset \mathbb{R}^3$.

(b) Consider a complete 3-dimensional Riemannian manifold W^3 whose sectional curvatures are pinched between $+1$ and -1 and whose injectivity radius is bounded from below by a constant $\rho_0 > 0$. Assume the existence of an isometric C^2-immersion f: $V \to W^3$ and find a lower bound on $\int_{V(a)} K(v)\,dv$ in terms of ρ_0, a and $d = \operatorname{Diam} f(V)$. Show in particular that

$$\int_V \max(K(v), 0)\,dv \geq 4\pi - \delta$$

for closed surfaces V, where $\delta = \delta(\rho_0, d) \to 0$ for $d \to 0$.

(b′) Generalize (b) by allowing the manifold W^3 to have a locally convex boundary.

(c) Consider an n-dimensional manifold V for $n \geq 3$ and denote by $K^+(v)$ the upper bound of sectional curvatures on the 2-planes in the tangent space $T_v(V)$. Assume the existence of an isometric C^2-immersion f: $V \to \mathbb{R}^{n+1}$ and estimate from below the integral $\int_V (\max(K^+(v), 0))^{n/2}\,dv$. Find a similar estimate for immersions f: $V \to W^{n+1}$.

Question. Does the existence of an isometric C^2-immersion $V^n \to \mathbb{R}^q$, $n + 2 \leq q \leq 2n - 1$, imply some integral inequality for the curvature of the (closed) manifold V^n?

(B) *An Upper Bound on* Rad V. Assume the boundary ∂V non-empty and set $\operatorname{Rad} V = \sup_{v \in V} \operatorname{dist}(v, \partial V)$. Observe that manifolds of positive sectional curvature $K(V) \geq \kappa^2 > 0$ have $\operatorname{Rad} V \leq \pi\kappa^{-1}$.

If $K(V) > 0$ and if V admits an isometric C^2-immersion into the ball $B(R_0) = \{\|x\| \leq R_0\} \subset \mathbb{R}^{n+1}$, $n = \dim V$, then $\operatorname{Rad} V \leq \pi R_0$.

Proof. The image $f(V) \subset \mathbb{R}^{n+1}$ is a locally convex hypersurface. Take the (exterior) normal ray r_v to $f(V)$ at $f(v)$ with respect to which the second fundamental form at v is negative definite and let $s = s(v)$ be the intersection point of r_v with the sphere $\partial B(R_0)$. The resulting map $V \to \partial B(R_0)$ for $v \mapsto s(v)$ is clearly infinitesimally enlarging (see 1.2.4), that is the induced metric of constant curvature R_0^{-2} in V is greater than the original metric in V. Hence, Rad $V \leq \pi R_0$ by the above remark.

Exercises. (a) Let V_0 be a closed Riemannian manifold whose fundamental group is finite. Prove the existence of a constant $R_0 = R_0(V_0) < \infty$, such that no manifold V with Rad $V \geq R_0$ and with dim V = dim V_0 admits an infinitesimally enlarging (for example, isometric) map into V_0.

(b) Give an upper bound on Rad V for a locally convex hypersurface V in a Riemannian manifold W with a locally convex boundary ∂W.

(c) Construct, for a given $\varepsilon > 0$, a C^∞-metric $g = g_\varepsilon$ on the 2-dist D^2 with the following three properties

(1) The curvature of g everywhere is positive ≥ 1.
(2) The manifold (D^2, g) admits an isometric S^1-action with a fixed point $v_0 \in D^2$, such that $\mathrm{dist}_g(v_0, \partial D^2) > 1$.
(3) The unit ball B in (D^2, g) around v_0 has $\int_B K(v; g)\, dv \leq \frac{3}{2}\pi$ and $\mathrm{Diam}_g \partial B <$ $\mathrm{dist}(\partial B, \partial D^2)$ + length $\partial D^2 \leq \varepsilon$.

Prove for $\varepsilon < \frac{1}{8}$ that no isometric C^2-immersion $(D^2, g) \to \mathbb{R}^3$ exists. Then consider an arbitrary closed Riemannian manifold W^3 and show that no isometric C^2-immersion $(D^2, \partial g) \to W^3$ exists for all small positive $\delta = \delta(W^3)$. Estimate δ in terms of W.

(C) *An Upper Bound on* $R(V)$ *for Non-Elliptic Surfaces* V. Let us construct a *projective* map P_+ of the unit ball $B^3 \subset \mathbb{R}^3$ into the unit sphere $S^3 \subset \mathbb{R}^4$ as follows. First move S^3 to a new position in $\mathbb{R}^4$ where it is tangent to B^3 at the center $0 \in B^3$ and then radially project B^3 to the moved sphere from the center of this sphere. In a similar way one obtains a projective map $P_-: B^3 \to H^3$ for the hyperbolic space H^3 of curvature -1.

Consider an immersion $f: V \to B^3$ which induces a metric g on V with the sectional curvature $K = K(v)$. Denote by $g_\pm$ the metric induced by the map $P_\pm \circ f$ and let $K_\pm$ denote the curvature of $g_\pm$.

(C′) **Lemma.** *There exists a universal constant* $c_0 > 0$ (*which is, in fact,* ≤ 100), *such that*

(∗)
$$C_0^{-1} g \leq g_\pm \leq C_0 g$$

and

(∗∗)
$$C_0^{-1} K \leq K_\pm \mp 1 \leq C_0 K.$$

Proof. The inequality (∗) is obvious for a constant C_0 which majorizes the norms of the differentials $DP_\pm$ and $(DP_\pm)^{-1}$. To prove (∗∗) we observe that projective maps send planes in $\mathbb{R}^3$ to (totally geodesic) planes in S^3 (and in H^3). Hence, the discriminants of the second fundamental forms of the maps f and $P_\pm \circ f$ satisfy with some constant $C_0 > 0$,

$$(***) \qquad C_0^{-1} \operatorname{Discr} \Pi \le \operatorname{Discr} \Pi_\pm \le C_0 \operatorname{Discr} \Pi,$$

and (**) follows [compare (E′) in 2.4.4].

(C″) **Corollary.** *If a surface $V = (V, g)$ has curvature $K \ge R_0^{-2}$ for some $R_0 \ge 0$ and if V admits an isometric C^2-immersion into the ball $B(R) \subset \mathbb{R}^3$ for $R \le R_0/\sqrt{2C_0}$, then* $\operatorname{Rad} V \le \pi R \sqrt{2C_0}$.

Proof. We may assume $R = 1$. Then $K_+ \ge \frac{1}{2}$ by (**). Hence, $\operatorname{Rad}(V, g_+) \le \pi\sqrt{2}$, and $\operatorname{Rad}(V, g) \le \pi/\sqrt{2C_0}$ by (*).

(C‴) **Exercises and Generalizations.** (a) Let V be a complete non-compact surface, such that $\liminf K(v)[\operatorname{dist}(v, v_0)]^2 = 0$ for some $v_0 \in V$. Prove every isometric C^2-immersion $V \to \mathbb{R}^3$ to be unbounded.

(a′) Let a compact surface V with a *connected* boundary satisfy

$$K(V) \ge -R_0^{-2}, \qquad \text{for some } R_0 \ge 0,$$

$$2\pi - \int_V \max(K(v), 0)\, dv \ge \beta > 0,$$

and

$$\text{length } \partial V \le R_0 \beta / 30\sqrt{2C_0}.$$

(For example, V is flat.) Assume $\operatorname{Rad} V > \pi r \sqrt{2C_0}$ for $r = 30\beta^{-1}$ length ∂V and show that no isometric C^2-immersion $V \to \mathbb{R}^3$ exists. Apply this to a closed surface with $K \le 0$ minus a small ball.

(b) Geralize (C′) to C^2-*hypersurfaces* V in the unit ball $B^{n+1} \subset \mathbb{R}^{n+1}$, $n \ge 3$. Let the sectional curvature of V satisfy $K(V) \ge -(2C_0)^{-1}$ for the above C_0 and establish the following properties of V.

(b_1) $\operatorname{Rad} V \le \pi\sqrt{2C_0}$.

(b_2) If V is closed, then the homology group $H_i(V; \mathbb{R}) = 0$ for $1 \le i \le n-1$.

Hint. Use an estimate by A. Weinstein (1970) on the curvature operator of manifolds $V^n \subset \mathbb{R}^{n+2}$ [see Aminov (1975), Baldin-Mercuri (1980) and Moore (1978) for further results].

(c) Let V be a complete non-compact surface with a finitely generated fundamental group. Let $\limsup_{v \to \infty} K(v) \le 0$ and let V admit a bounded isometric C^2-immersion $f\colon V \to \mathbb{R}^3$. Show the balls $B(R) \subset V$ around a fixed point $v_0 \in V$ to satisfy

$$(*) \qquad \liminf_{R \to \infty} R^{-1} \log \operatorname{Area} B(R) \ge \varepsilon > 0$$

under one of the following three conditions.

(i) Area $V = \infty$,
(ii) $\liminf_{v \to \infty} K(v) \geq -\text{const} > -\infty$,
(iii) $K(v) < 0$ outside a compact subset in V.

Hint. Use the map P_- to prove (i) and (ii). Consult Verner (1970) for (iii).

Questions. (a) Are the conditions (i)–(iii) essential? How does one generalize (*) to hypersurfaces $V^n \subset \mathbb{R}^{n+1}$ for $n \geq 3$?

(b) Does (C″) generalize to immersions $V \to W^3$, where the Riemannian manifold W^3 is not projectively flat? Is there a generalization of (C″) to immersions $V^n \to \mathbb{R}^q$ for $n + 2 \leq q \leq 2n - 1$? Under what conditions on (the dimensions of) V^n and W^q all isometric (or C^2-nearly isometric) C^2-immersions $f: V^n \to W^q$ satisfy $\operatorname{Diam} f(V^n) \geq \text{const} = \text{const}(V^n, W^q)$.

(D) *Isoperimetric Inequalities for Surfaces.* Let V be a compact connected surface and let $\omega_+ = \int_V \max(K(v) + 1, 0)\, dv$. Then the area $A = A(V)$ admits the following bound in terms of the length $L = L(\partial V)$, of the Euler characteristic $\chi = \chi(V)$ and of the integral ω_+.

Basic Inequality (Fiala 1941; Ionin 1969).

$$L^2/2 \geq A(A/2 - \omega_+ + 2\pi\chi). \tag{*}$$

Proof. Let $V(t) = \{v \in V | \operatorname{dist}(v, \partial V) \geq t\} \subset V$. Then

$$L(t) \overset{\text{def}}{=} L(\partial V(t)) = -\frac{dA(V(t))}{dt}. \tag{1}$$

If the boundary curve $\partial V(t)$ is smooth, then by Gauss-Bonnet,

$$\frac{dL(t)}{dt} = \int_{V(t)} K(v)\, dv - 2\pi\chi(V(t)).$$

For non-smooth curves one still has the inequality

$$\frac{dL(t)}{dt} \leq \int_{V(t)} K(v)\, dv - 2\pi\chi(V(t)).$$

Since $\chi(V(t)) \geq \chi(V)$ for $t \geq 0$, and since $\int_{V(t)} K(v)\, dv \leq \omega_+ - A(V(t))$, we get

$$\frac{dL(t)}{dt} \leq \omega_+ - A(V(t)) - 2\pi\chi. \tag{2}$$

Now, we obtain (*) by multiplying (1) with (2) and then by integrating in t over $[0, \infty]$.

Exercise. Fill in the detail in this proof.

A Modification of (*). *Let* $\omega_- = \int_V \max(K(v), -1)\, dv$ *and let* $\varepsilon = 2\pi\chi - \omega_-/A$.
(a) *If* $\varepsilon \geq 1$, *then* $L \geq A$.
(b) *If* $0 \leq \varepsilon \leq 1$, *then* $L \geq 2\pi\chi - \omega_-$.

Proof. Since $\int_{V(t)} K(v)\,dv \leq \omega_- + A - A(V(t))$, we obtain with the above

$$-\int_0^T L(t)\frac{dL(t)}{dt} \geq \int_0^T (\omega_- - 2\pi\chi + A - A(t))\frac{dA(V(t)}{dt}.$$

We get (a) with $T = \infty$ and we obtain (b) with $T = T_0$, such that $A(T_0) = (1-\varepsilon)A$.

Now, let V admit an isometric C^2-immersion f into the unit ball $B^3 \subset \mathbb{R}^3$. Then V satisfies the following (refinement of a special case of)

Inequality of Burago (1968). *Let* $\omega_0 = \int_V \max(K(v), 0)\,dv$. *Then*

$$(**) \qquad L^2 \geq A(C_1 A - C_2\omega_0 + C_3\chi),$$

where $C_1 = C_0^{-2}$, $C_2 = 2C_0^3$ *and where* $C_3 = C_3(V,f)$ *satisfies* $4\pi C_0^{-2} \leq C_3 \leq 4\pi C_0^2$ *for the constant* C_0 of Lemma (C′).

Proof. We get with (C′) the inequality $\omega_+(g_-) \leq C_0^2\omega_0$ and we apply $(*)$ to the length and the area of (V, g_-),

$$L_-^2/2 \geq A_-(A_-/2 - C_0^2\omega_0 + 2\pi\chi).$$

Then $(**)$ reduces to the following inequalities which are immediate with (C′).

$$C_0^2 L^2 \geq L_-^2 \quad \text{and} \quad C_0^{-1}A \leq A_- \leq C_0 A.$$

Corollary. *If* $\chi(V) \geq 0$ *and* $K(V) \leq 0$. *Then* Area $V \leq C_1^{-1/2}$ length ∂V.

Exercises. (a) Let a surface V in the unit ball $B^3 \subset \mathbb{R}^3$ have $\chi(V) \geq 0$ and $K(V) \leq (2C_0)^{-1}$. Show that Area $V \leq C$ length ∂V for some universal constant $C > 0$. [Compare Jorge-Xavier (1981).]

(b) Find a sequence of smooth imbedded surfaces $V_i \subset B^3$, $i = 1, \ldots$, of *non-positive* curvature with the following two properties,

(1) The boundaries ∂V_i, $i = 1, \ldots$, are equal to a fixed connected smooth curve $\partial = \partial V_1 = \partial V_2 = \cdots$,
(2) $\chi(V_i) \to -\infty$ for $i \to \infty$ and Area $V_i \geq \sqrt{-\chi(V_i)}$ for $i = 1, 2, \ldots$.

(c) Let $V' = \{v \in V \mid K(v) \geq (2C_0)^{-1}\}$ and $\omega' = \int_{V'} K(v)\,dv$. Prove $(**)$ with ω' in place of ω_0 (and with some constants C_i' instead of C_i).

(d) Let the boundary ∂V be connected, let $\omega_0 \leq \pi$ and let V admit an isometric C^2-immersion $V \to \mathbb{R}^3$. Show that $10L^4 \geq A(C_1 A - C_2\pi L^2 + C_3\chi L^2)$ and find for all $\chi < 0$ examples of (non-immersible into $\mathbb{R}^3$!) surfaces V which violate this inequality.

(E) *Conformal Maps and Isoperimetric Inequalities*. A C^1-map between two oriented Riemannian manifolds, $f\colon V^n \to W^n$ is called (orientation preserving) *conformal* if the Jacobian of f is related to the differential of f by the equality $J(v) = \|D_v\|^n$ for all $v \in V$. A C^0-map f is called *quasiconformal* if the (distribution) differential locally has $\int \|Df\|^n\,dv < \infty$ and if $J(v) \geq C\|D_v\|^n$ for some constant $C > 0$ and for almost all $v \in V$.

Consider a proper function $x\colon V \to \mathbb{R}_+$ and let $V(t) = x^{-1}[0,t] \subset V$. Such an increasing family of compact subsets $V(t) \subset V$ is called an *exhaustion* of V. Any monotone change of the parameter $t \to \Theta \in \mathbb{R}_+$ gives another labelling of the exhaustion $V(t)$. Let us assume V to carry a Riemannian metric and let us parametrize (label) a given exhaustion $V(t)$ with the following

Conformal and Subconformal Measures in $\mathbb{R}_+$. A measure $d\Theta$ for a monotone increasing function $\Theta = \Theta(t)$, $t \in \mathbb{R}_+$, is called *subconformal* [relative to $V(t)$] if for every two points t_1 and $t_2 > t_1$ in $\mathbb{R}_+$, and for an arbitrary continuous function $h\colon x^{-1}[t_1, t_2] \to \mathbb{R}_+$ there exists a point $t_3 \in [t_1, t_2]$ such that

$$\int_{x^{-1}[t_1,t_2]} h^n \geq (\Theta(t_2) - \Theta(t_1)) \left(\int_{x^{-1}(t_3)} h^{n-1} \right)^{n/(n-1)},$$

where the function h^n is integrated with respect to the (n-dimensional) Riemannian measure on V and h^{n-1} is integrated over the $(n-1)$-dimensional measure on the fiber $x^{-1}(t_3)$. Here and below we assume the fibers $x^{-1}(t)$ to be rectifiable $(n-1)$-dimensional subsets in V. Moreover, to avoid a trivial mess we assume the integral $\int_{x^{-1}(t)} F$ to be continuous in t for all continuous functions F on V.

Measures majorized by subconformal measures are subconformal. Therefore, there exists a unique *maximal* subconformal measure $d\Theta$ which is called the *conformal measure*.

Example. Let V be complete and let $V(t)$ be the ball of radius t around a fixed point in V. Then the measure $d\Theta = (\operatorname{Vol} \partial V(t))^{1/(1-n)}\, dt$ obviously is conformal. (The boundary $\partial V(t)$ may be *properly* contained in the sphere $\{v \in V | \operatorname{dist}(v_0, v) = t\}$, but we assume this does not happen for our manifold V).

Exercises. (a) Assume the function $x\colon V \to \mathbb{R}_+$ to be smooth, put $\varphi(t) = (\int_{x^{-1}(t)} \|dx\|^{n-1})^{1/(1-n)}$ and show that $\varphi(t)\, dt$ is the conformal measure.

(b) Prove the conformal measure to be invariant under conformal changes of the Riemannian metric in V.

(c) Let $x\colon V \to W$ be a smooth proper map, such that $\dim x^{-1}(w) = m$, $w \in W$. Take a covector $l \in T_w^*(w)$, pull back by $D_f^*\colon T_w^*(W) \to T_v^*(V)$, $v \in x^{-1}(w)$, and set $\|l\| = (\int_{x^{-1}(w)} \|D_f^* l\|^m)^{1/m}$. Prove that the (Finsler) metric in W defined by the dual norm on $T(W)$ is a conformal invariant of V. Study similar metrics on submanifolds $Z \subset W$ for $x | x^{-1}(Z)\colon x^{-1}(Z) \to Z$ and use these to generalize the definition to non-smooth maps x.

Subconformal Exhaustions. An arbitrary exhaustion $V(t)$ of V can be reparametrized by $\Theta = \int d\Theta$ for some subconformal (e.g. the conformal) measure $d\Theta$. Then $V(\Theta) = V(t(\Theta))$ is called a *subconformal exhaustion* and the range of Θ is denoted by $(\Theta_-, \Theta_+) \subset \mathbb{R}$.

Let W, $\dim W = \dim V$, be a manifold with a Riemannian metric g, whose (oriented) volume form is denoted by ω, and let $f\colon V \to W$ be a conformal map. Set $A(\Theta) = \operatorname{Vol}_g V(\Theta) \overset{\text{def}}{=} \int_{V(\Theta)} f^*(\omega)$ and let $L(\Theta) = \operatorname{Vol}_g f | \partial V(\Theta)$ which by definition is the $(n-1)$-dimensional volume of $f(\partial V(\Theta)) \subset W$ counted with the geometric multiplicity.

Lemma. *If the exhaustion $V(\Theta)$ is subconformal, then*

$$(+)\qquad \int_{\Theta_0}^{\Theta_+} (L(\Theta)/A(\Theta))^{n/(n-1)}\, d\Theta \leq \text{const} < \infty,$$

for all $\Theta_0 \in (\Theta_-, \Theta_+)$ and for const $=$ const(Θ_0).

Proof. Let $h(v) = \|D_v f\|$. Since the map f is conformal,

$$A(\Theta) = \int_{V(\Theta)} h^n \quad \text{and} \quad L(\Theta) = \int_{\partial V(\Theta)} h^{n-1}.$$

Since the exhaustion is subconformal,

$$\int_{V(\Theta)} h^n \geq \int_{\Theta_-}^{\Theta} d\Theta \left(\int_{\partial V(\Theta)} h^{n-1}\right)^{n/(n-1)},$$

and so

$$y(T) = \int_{\Theta_-}^{T} (L(\Theta))^{n/(n-1)} d\Theta \leq A(T)$$

for all $T \in (\Theta_-, \Theta_+)$. Therefore,

$$\int_{\Theta_0}^{\Theta_+} (L(\Theta)/A(\Theta))^{n/(n-1)}\, d\Theta \leq \int_{\Theta_0}^{\Theta_+} y^{-n/(n-1)} \frac{dy}{d\Theta} \leq (n-1)(y(\Theta_0))^{1/(1-n)} < \infty.$$

Parabolic Exhaustions. An exhaustion is called *parabolic* if $\Theta_+ = \infty$ for the conformal parameter of this exhaustion.

Example. If the concentric spheres $S^{n-1}(R)$ around some point in a complete manifold V have

$$\int_0^{\infty} [\operatorname{Vol} S^{n-1}(R)]^{1/(1-n)}\, dR = \infty,$$

then the exhaustion by the balls $B^n(R)$ is parabolic.

Now the Lemma yields the following

Corollary. *For parabolic exhaustions $A(t)$,*

$$(++)\qquad \liminf_{t\to\infty} L(t)/A(t) = 0.$$

The inequalities $(+)$ and $(++)$ show that the integral $A(t) = \int_{V(t)} f^*(\omega)$ is rather stable under reasonable changes of ω. Namely, let ω' be a form *equivalent* to ω in the sense that the difference $\omega - \omega'$ is a differential of a *bounded* measurable form on V, say $\omega - \omega' = d\sigma$. Then $A'(t) = \int_{V(t)} f^*(\omega')$ satisfies by Stokes' theorem

$$|(A(t) - A'(t))/A(t)| \leq \|\sigma\| L(t)/A(t),$$

and so

$$\liminf_{t\to\infty} |(A(t) - A'(t))/A(t)| = 0$$

in the parabolic case. In particular, if ω is equivalent to zero (i.e. $\omega = d\sigma$), then every conformal map $f\colon V \to W$ is constant, provided (some exhaustion of) V is parabolic.

Exercises. (a) Show, by making $\omega = d\sigma$, that the following manifolds W receive no non-constant conformal maps from parabolic manifolds without boundary (e.g. from $\mathbb{R}^n$),

(i) W is compact connected and has a non-*empty* boundary.
(ii) W is complete, the sectional curvature of W is everywhere negative ≤ -1 and the fundamental group $\pi_1(W)$ is solvable.
(iii) W is a (Riemannian) normal covering with a free non-Abelian Galois group of an open subset $U \subset W_0$, where W_0 is a closed Riemannian manifold and the complement $\Sigma = W_0 \backslash U$ is a smooth simplicial subcomplex in W_0. For example, U equals S^2 minus three points.

(b) Generalize $(+)$, $(++)$ and the above (a) to quasi-conformal maps.

Distribution of Values $w = f(v) \in W$. The above bound on $|A(t) - A'(t)|$ may fail for forms $\omega' = \omega + d$ if the (distribution) form σ is not bounded. Consider, for example, the singular form (current) δ_w on W for some $w \in W$ which (viewed as a measure) consists of a single atom at w of mass one, $\int_W \delta_w = 1$. Then the integral $N_w(t) = \int_{V(t)} f^*(\delta_w)\,dv$ equals the number of solutions $v \in V(t)$ to the equation $f(v) = w$. Observe that $\int_W N_w(t)\,dw = A(t)$ for all t. Assume $\operatorname{Vol} W < \infty$ and introduce the error (often called the *defect*) $d_w(t) = 1 - (\operatorname{Vol} W) N_w(t)/A(t)$. If the map f misses w, then $d_w(t) = 1$ and so $|A - A'| = |A - N_w| = A$.

Now, let $\dim W = \dim V = 2$. Put $K(t) = \int_{V(t)} K(f(v)) f^*(\omega)$ for the curvature $K(w)$ of W. In other words $K(t)$ is the integral curvature of $V(t)$ for the metric induced from W outside the singularities of the map f. If $K(w) \geq -1$, $w \in W$, then the modified inequality $(*)$ implies $L(t) \geq \min(A(t), 2\pi\chi(t) - K(t))$ for $\chi(t) = \chi(V(t))$ for all t. In fact, since the map is conformal, the singularities of the induced metric g in V correspond to the branching points of f (the curvature of the induced singular metric has the atomic mass $-2\pi m$ at every branch point where the differential D_f vanishes with order m) and these singularities may be removed by a smooth approximation g' of g. Hence, the isoperimetric inequalities of (D) apply to metrics with isolated singularities (with an obvious definition of the curvature at the singular points). Now, we have with $M^2(t) = [\max(0, 2\pi\chi(t) - K(t)]^2$ the following relation for the conformal parameter Θ,

$$(+++) \qquad \int_{\Theta_0}^{\Theta_+} \min(1, M^2(\Theta)/A^2(\Theta))\,d\Theta < \infty.$$

We apply this inequality to the following very special class of metrics g in W: The metric g has constant negative curvature -1 outside a discrete subset $\{w_i\}$, $i = 1, 2, \ldots$, in W and the singularity at $w_i \in W$ carries an atom of positive curvature, say $p_i \delta_{w_i}$ for $0 < p_i < 2\pi$. Every such metric is locally isometric to the sector of the angle $2\pi - p$ in the hyperbolic plane with the two sides identified. For example, the boundary W of every convex polyhedron in the hyperbolic space H^3 carries such a metric whose (intrinsic!) singularities are the vertices of the polyhedron.

If W is a closed surface with our special metric, then

$$\operatorname{Area} W = \sum_i p_i - 2\pi\chi(W)$$

and

$$K(t) = -A(t) + \sum_i p_i N_{w_i}(t).$$

Therefore, *the inequality* $(+++)$ *provides a non-trivial relation between the errors (defects)* $d_{w_i}(t)$ *for* $t \to \infty$, *in case the exhaustion* $V(t)$ *is parabolic.*

Finally, observe that the defects $d_{w_i}(t)$ are rather unsensible [like $A(t)$ see above] to a choice of a (singular or non-singular) metric in W and so the inequality $(+++)$ for singular metrics may yield defect relations for smooth ones. More precise statements and further results are in the following references and exercises.

References. Our brief exposition of conformal maps followed Ahlfors' geometric approach to Nevanlinna's value distribution theory. This theory and various generalizations can be found in Hayman (1964), Griffiths (1974), Cowen-Griffiths (1976), Rickman (1983), Mattila-Rickman (1979).

Exercises. (a) Let W be the torus T^2 with a smooth metric and let $V = \mathbb{R}^2$, exhausted by concentric balls $V(t) = \{\|v\| \le t\}$. Let $d_w^\varepsilon(t) = \max(0, d_w(t) - \varepsilon)$ and show that

$$\int_1^\infty (d_w^\varepsilon(t))^2\, d\log t < \infty,$$

for all $w \in W$, for all $\varepsilon > 0$ and for all non-constant quasi-conformal maps $f\colon V \to W$.

Hint. Use a metric on T^2 whose curvature mass at the only singular point $w \in T^2$ equals $2\pi - \varepsilon$.

(a′) Prove a similar defect relation for quasi-conformal maps $\mathbb{R}^2 \to S^2$ by using metrices on S^2 with $k > 2 = \chi(S^2)$ singular points.

(b) Let V be an infinite cyclic covering of a closed surface V_0. Show that every non-constant quasi-conformal map $f\colon V \to S^2$ misses at most two points.

(b′) Exhaust V by concentric balls $V(t)$ (for the Riemannian metric induced from V_0), consider a quasi-conformal map $f\colon V \to T^2$, for which $A(t)/t \to \infty$, and show that $\int_1^\infty (d_w^\varepsilon(t))^2\, d\log t < \infty$ for the above d_w^ε. Modify this for maps $V \to S^2$.

(c) Let V and W be complete Riemannian manifolds with the following properties.

(i) W is a regular (Riemannian) covering of a compact manifold W_0, such that the Galois group contains a free non-Abelian subgroup.
(ii) The unit ball $B_v(1) \subset V$ satisfies $\operatorname{Vol} B_v(1) \le C < 1$ for all $v \in V$.

Prove each quasiconformal map $f: V \to W$ to be *uniformly* continuous. Moreover, $\mathrm{dist}(f(v_1), f(v_2)) \le \mathrm{const}(W_0, C, d)$ for $d = \mathrm{dist}(v_1, v_2)$.

(d) Let V be a surface with a parabolic exhaustion $V(t)$ and let $f: V \to \mathbb{R}^3$ be a conformal immersion, that is the induced metric g on V equals $h^2 g_0$ for the original metric g_0 on V and for some positive function h on V. Let $A(t) = \mathrm{Area}_g V(t)$, $D(t) = \mathrm{Diam}\, f(V(t))$, $\chi(t) = \chi(V(t))$ and $K_+(t) = \int_{V(t)} \max(0, K_g(v))(dv)_g$. Show that

$$\limsup_{t\to\infty} D^2(t)(K_+(t) + |\chi(t)|)/A(t) \ge \varepsilon > 0$$

for some universal $\varepsilon > 10^{-8}$.

3.2.4 Isometric C^∞-Immersions $V^2 \to \mathbb{R}^q$ for $3 \le q \le 6$

Let us recall the classical result of Alexandrov (1944) and Pogorelov (1969) on

Elliptic Isometric Imbeddings $S^2 \to M^3[\kappa]$. *If a $C^\infty(C^{\mathrm{an}})$-metric on the sphere S^2 has curvature $K(s) > \kappa$ then there exists an isometric $C^\infty(C^{\mathrm{an}})$-imbedding $f: (S^2, g) \to M^3[\kappa]$, where $M^3[\kappa]$ is the standard (spherical for $\kappa > 0$, Euclidean for $\kappa = 0$ and hyperbolic $\kappa < 0$) 3-space of constant curvature κ. (This imbedding is unique up to isometries of $M^3[\kappa]$.) In particular, every metric g admits an isometric imbedding into some hyperbolic 3-space.*

The proof (see Pogorelov 1969) is based on the *continuity method*, where the starting point is the following fact: *the space of metrices g on S^2 for which $K_g > \kappa$ is connected for all $\kappa \in \mathbb{R}$.*

Proof. The claim is obvious for $\kappa < 0$. To handle $\kappa \ge 0$ we use the conformal representation $g = e^{2\varphi} g_0$ for the standard metric g_0 on S^2. A straight-forward computation gives $K_g = e^{-2\varphi}(1 - \Delta\varphi)$ for the Laplace operator $\Delta = \Delta_g$. Hence, the relation $K_g > \kappa$ reduces to $\kappa e^{2\varphi} < 1 - \Delta\varphi$. Since the function $e^{2\varphi}$ is convex, the space of solution φ of this inequality is a convex subset in the space of functions. Q.E.D.

Exercises. (a) Let V be an arbitrary (possibly non-compact) manifold of dimension $n \ge 2$ and let $G(a,b)$, $-\infty \le a < b \le \infty$, denote the space of complete metrics g on V whose sectional curvatures satisfy $a < K_g < b$. Show the space $G(a,b)$ to be weakly contractible (i.e. non-empty, path connected, with zero homotopy groups π_i for $i \ge 1$) in case $a < 0 < b$.

(b) Show the group $\mathrm{Diff}\, S^2$ to be weakly homotopy equivalent to the orthogonal group $O(3)$.

Hint. Compare the spaces $G'/\mathrm{Diff}\, S^2$, where G' is the (contractible!) space of metrics g on S^2 with no symmetries, and $\Phi'/\mathrm{Conf}\, S^2$, where Φ' is the space of functions φ on S^2 for which $e^{2\varphi} g_0 \in G$ and Conf denotes the group of conformal automorphisms of S^2.

(b′) Determine the weak homotopy type of Diff V for an arbitrary surface V. *Hint*. Use the conformal representation $g = e^{2\varphi}g_0$ for all metrics g on V, where g_0 is a metric of constant curvature.

Hyperbolic Immersions $D^2 \to \mathbb{R}^3$. Take an isometrically imbedded interval $Y = [y_1, y_2]$ in a surface V and assume the normal exponential map exp: $Y \times \mathbb{R} \to V$ to be a smooth imbedding on $Y \times [x_1, x_2]$. Then the metric g of V is expressed in the coordinates x and y by $g = dx^2 + \varphi^2\,dy^2$ for some function $\varphi = \varphi(x, y)$ (which is determined by (V, g) and $Y \subset V$). Let f_ε: $D^2 \to \mathbb{R}^3$, for $D^2 = \exp(Y \times [x_1, x_2])$, be defined by the functions x, $\varepsilon\varphi \sin \varepsilon^{-1}y$ and $\varepsilon\varphi \cos \varepsilon^{-1}y$. Then the induced metric g_ε on D^2 is $g + \varepsilon^2(d\varphi)^2$ (see 3.1.1).

Example. Let V be a complete simply connected surface of non-positive curvature and let Y be a double infinite geodesic in V. Then the map exp: $Y \times \mathbb{R} \to V$ is a diffeomorphism and so the metric g is *globally* $dx^2 + \varphi^2\,dy^2$. Therefore, every compact disk $D^2 \subset V$ admits an ε-approximate isometric immersion $D^2 \to \mathbb{R}^3$ for all $\varepsilon > 0$.

It is unknown for general C^∞-manifolds V whether the immersion f_ε for small $\varepsilon > 0$ admits an approximation by *isometric* C^∞-immersions. However, such an approximation was obtained by Posnjak (1973), in the following two cases,

(a) the curvature $K(V)$ is negative,
(b) the surface (V, g) is real analytic.

Hence, in these two cases the disk $D^2 \subset V$ admits an isometric $C^\infty(C^{\mathrm{an}})$-immersion into $\mathbb{R}^3$.

Counterexample. The above approximation result may fail for C^2-metrics. Moreover, Pogorelov (1971) found a C^2-metric g on V whose curvature $K(v)$ is Lipschitz, but no neighborhood of some fixed point $v_0 \in V$ admits an isometric C^2-immersion into $\mathbb{R}^3$.

Isometric Immersions into Warped Products. Consider a manifold V wirth a metric g_0. Let $\delta(t)$ and $\psi(t)$ for $t_1 < t < t_2$ be C^{an}-functions, where $\psi(t) > 0$ for all $t \in (t_1, t_2)$. Then we define the *warped* metric $\psi(t)g_0 + \delta(t)\,dt^2$ on $W = V \times (t_1, t_2)$, which is positive for $\delta > 0$.

(A) **Examples.** (a) Let g_0 be the metric of constant curvature $+1$ on S^n. Then

(i) the metric $t^2 g_0 + \delta\,dt^2$, $0 < t < \infty$, equals the standard flat metric on $\mathbb{R}^{n+1}$ minus the origin for all *constant* $\delta > 0$;
(ii) the metric $(\sin t)^2 g_0 + dt^2$, $0 < t < \frac{\pi}{2}$, is the standard metric of curvature $+1$ on the hemisphere S_+^{n+1} minus the center;
(iii) the metric $(\sinh t)^2 g_0 + dt^2$, $0 < t < \infty$, is the standard metric of curvature -1 on the hyperbolic space H^{n+1} minus a point, where $\sinh t = \frac{1}{2}(e^t - e^{-t})$.

(b) Let g_0 be the standard flat metric on $\mathbb{R}^n$. Then $e^{2t}g_0 + \delta\,dt^2$ is the metric of constant curvature $-\delta$ for all constants $\delta > 0$.

(c) Let g_0 be the metric of constant curvature -1 on $H^n = M^n[-1]$. Then the metric $(\cosh t)^2 g_0 + dt^2$, $-\infty < t < -\infty$ is isometric to that on H^{n+1}, where $\cosh t = \frac{1}{2}(e^t + e^{-t})$.

Metrics on Graphs. Consider a C^1-function $h\colon V \to (t_1, t_2)$ and observe that the graph of h, called $\Gamma_h\colon V \to (W, \psi(t) g_0 + \delta(t)\, dt^2)$, induces on V the metric $g_0^h = \psi(h(v)) g_0 + \delta(h(v))(dh)^2$. This generalizes the basic (twisting) formula in 3.1.1 for the flat metric $t^2 g_0 + \varepsilon^2\, dt^2$ on $S^n \times (0, \infty) = \mathbb{R}^{n+1} \backslash 0$.

Let the derivative $\psi'(t)$ be nonvanishing on some subinterval $[\bar{t}_1, \bar{t}_2] \subset (t_1, t_2)$ and let $\psi(t_0) = 1$ for some $t_0 \subset [t_1', t_2']$. Let us parametrize $[\bar{t}_1, \bar{t}_2]$ by $\Theta = \frac{1}{2}\log\psi(t)$, set $\Theta_i = \Theta(\bar{t}_i)$, $i = 1, 2$, and let the following two conditions be satisfied for some $A > 1$.

(1) $|\Theta_i| > A$, $i = 1, 2$;
(2) the C^5-norm of the function $\bar{\delta}(\Theta) = \delta(t(\Theta))(dt/d\Theta)^2 \Theta \in [\Theta_1, \Theta_2]$, satisfies $\|\bar{\delta}(\Theta)\|_{C^5} \leq A^{-1}$.

Now, let V be a connected compact surface (with or without boundary) and let $\{g_x\}$ be a smooth family of C^∞-metrics on V where the parameter x runs over some Euclidean space $\mathbb{R}^p$. Recall that the conformal equivalence classes of the metrics on V close to g_0 are parametrized by a finite dimensional ball B^r [in the Teichmüller space for some choice of the basis in $\pi_1(V)$] according to Poincarés uniformization theorem. Let us relate to each metric g_x its conformal class $[g_x] \in B^r$ for x close to zero and assume that the map $x \mapsto [g_x]$ is a diffeomorphism on a small neighborhood $U \subset \mathbb{R}^p$ around the origin. (This implies, of course, $p = r$.) Finally, take a C^∞-metric g in the conformal class $[g_0]$, which means the existence of a function h_0 on V for which $(V, e^{2h_0} g_0)$ is isometric to (V, g).

(B) Proposition. *If the above constant $A = A(\psi, \delta)$ is sufficiently large, $A \geq \varepsilon^{-1}$ for a small number $\varepsilon = \varepsilon(V, g, \{g_x\}) > 0$, then there exists a C^∞-function h on V such that the metric $g_x^h = \psi(h(v)) g_x + \delta(h(v))(dh)^2$ is isometric to g for some $x \in \mathbb{R}^p$ close to zero. Furthermore, if g and g_x are real analytic, then the function h and the implied isometry $(V, g) \to (V, g_x^h)$ are also real analytic.*

Proof. Switch to the parameter $\Theta = \frac{1}{2}\log t$. Then $g_x^h = e^{2h} g_0 + \bar{\delta}(h)(dh)^2$, where $\|\bar{\delta}(\Theta)\|_{C^5} \leq \varepsilon$ for $\Theta_1 \leq \Theta \leq \Theta_2$. Let $g(\varphi) = e^{2\varphi} g - e^{2\varphi}\bar{\delta}(-\varphi)(d\varphi)^2$ and assume the existence of an isometry $f\colon (V, g(\varphi)) \to (V, g_x)$. Then the metric g_x^h for $h(v) = -\varphi(f^{-1}(v))$ is isometric to g. Thus our problem is reduced to finding an isometry f for some $x \in \mathbb{R}^p$ and for some function φ on V.

Let us write $\hat{g}$ for the volume form of g in case the surface V is oriented and let $\omega(g)$ denote the curvature *form* of g that is $K_g \hat{g}$ of g for the curvature (function) K_g of g. By a classical formula,

$$(*) \qquad \omega(e^{2\varphi} g) = \omega(g) + \Delta(\varphi \hat{g}),$$

for the Laplace operator $\Delta = \Delta_g$ on 2-forms. Since the metric $g(\varphi)$ is induced from a warped product, the curvature of $g(\varphi)$ is a differential expression of second (not third!) order in φ by Gauss theorema egregium. Hence,

$$(**) \qquad \omega(g(\varphi)) = \omega(g) + \Delta(\varphi\hat{g}) + D_\varepsilon(\varphi)$$

where D_ε is a non-linear differential operator of second order whose coefficients depend on g and $\bar{\delta}$, such that $\|D_\varepsilon\|_{C^3} \to 0$ for $\varepsilon \to 0$. (This is the ε which bounds $\|\bar{\delta}\|_{C^5}$.)

Fix a function φ_0 for which $e^{2\varphi_0}g$ is isometric to g_0 and let φ be C^2-close to φ_0. Then there exists a unique $x = x(\varphi)$ such that $g(\varphi)$ is conformally equivalent to g_x. The conformal equivalence $f\colon (V, g(\varphi)) \to (V, g_x)$ may be not unique [for $\chi(V) \geq 0$] but we agree on some canonical choice of $f = f_\varphi$. Denote by $I(\varphi)$ the induced 2-form $f_\varphi^*(\omega(g_x))$ for $x = x(\varphi)$.

Lemma. *If $\varepsilon = \varepsilon(V, g, \{g_x\}) > 0$ is sufficiently small, then the functional equation*

$$\omega(g) + \Delta(\varphi\hat{g}) + D_\varepsilon(\varphi) = I(\varphi)$$

is $C^\infty(C^{an})$-solvable in φ. Moreover, this solution exists and is unique with the following normalization,

(a) $\int_V e^{2\varphi}\hat{g} = \int_V \hat{g}_x$, *in case V has no boundary;*
(b) *The 2-forms $f_\varphi^*(\omega(g_x))$ and $e^{2\varphi}\hat{g}$ are equal at the points $v \in \partial V$, in case $\partial V \neq 0$.*

Proof. We treat only (a) and leave (b) to the reader. $B_\varepsilon\colon \varphi \to \omega(g) + \Delta\varphi\hat{g} + D_\varepsilon(\varphi) - I(\varphi)$ sends functions φ normalized by (a) to 2-forms ω for which $\int \omega = 0$. If a function φ is of Hölder class $C^{i,\alpha}$, $i = 2, 3, \ldots, 0 < \alpha < 1$, then $B_\varepsilon(\varphi)$ is $C^{i-2,\alpha}$-smooth. Moreover, the operator $B_\varepsilon C^{i,\alpha} \to C^{i-2,\alpha}$ is differentiable. Its linearization (differential) at $\varphi_0 \in C^{i,\alpha}$ is a small perturbation of the Laplace operator Δ which sends $C^{i,\alpha}$-functions φ with $\int_V \varphi\hat{g} = 0$ to $C^{i-2,\alpha}$-forms $\omega = \Delta\varphi\hat{g}$ with $\int_V \omega = 0$. Since Δ is an invertible operator, the operator B_ε is invertible near φ_0 by the implicit function theorem. Thus the lemma is proven.

Exercise. Fill in the details in the above argument.

If φ satisfies $B_\varepsilon(\varphi) = 0$ then the conformal map $f_\varphi\colon (V, g(\varphi)) \to (V, g_x)$ is curvature preserving. Therefore the Jacobian ψ of this map satisfies $\Delta \log \psi = 0$ for $\Delta = \Delta_{g(\varphi)}$ and so $\psi \equiv 1$ under conditions (a) and (b). This completes the proof for orientable surfaces V. The non-orientable case is treated by passing to the oriented double covering of V.

(B′) **Isometric Immersions $V \to W^3$ for $K(W^3) = -\varepsilon^{-2}$.** *Let V be a compact connected surface with a $C^\infty(C^{an})$-metric. Then for every positive $\varepsilon < \varepsilon_0 = \varepsilon_0(V) > 0$ there exists a 3-dimensional manifold W of constant curvature $-\varepsilon^{-2}$ which receives an isometric $C^\infty(C^{an})$-immersion $f\colon V \to W$. Moreover, one can take W and f, such that*

(i) *If V is (homeomorphic to) S^2, then W^3 is the hyperbolic space H^3 (of curvature $-\varepsilon^{-2}$) and f is an imbedding.*
(ii) *If V is P^2 then W is obtained from H^3 by dividing H^3 minus a point $w_0 \in H^3$ by the geodesic symmetry in w_0. This W is homeomorphic to $P^2 \times \mathbb{R}$ and it is not complete (at w_0). The map f is an imbedding.*

(iii) *If V is a closed surface with $\chi(V) \leq 0$, then W is homeomorphic to $V \times \mathbb{R}$ and complete. The map f is an imbedding.*
(iv) *If V is orientable with a non-empty boundary, then $W = H^3$, but one cannot (?) guarantee an imbedding.*
(v) *If V is non-orientable and ∂V is non-empty, then one may take the above $(H^3 \backslash w_0)/\mathbb{Z}_2$ for W. One may also take a complete W which is homeomorphic to (Klein bottle) $\times \mathbb{R}$. In neither case one guaranties an imbedding.*

Proof. Let V be closed and let g_x be the metrics of constant curvature $+1, 0$ or -1 on V parametrized by the Teichmüller space X. Thus every metric g is conformal to some g_x and this g_x is unique if we specify the homotopy class of the conformal equivalence. Now, we observe [compare (A)] that the warped product $W = (v \times \mathbb{R}, \varepsilon^2\psi^2 g_x + \varepsilon^2\, dt^2)$, where $\psi(t) = \sinh t$ for $K(g_x) = +1$, $\psi(t) = e^t$ for $K(g_x) = 0$ and $\psi(t) = \cosh t$ for $K(g_x) = -1$, has constant negative curvature $-\varepsilon^{-2}$. One can easily verify that $\bar{\delta}(\Theta)$ is small for ε small, and (B) applies.

Now let $V = (V, g)$ have a boundary and let $V_0 = (V_0, g_0)$ be a complete flat manifold which is non-orientable in case V is non-orientable. Then by the uniformization theory for surfaces with boundaries there exists a family of C^{an}-immersions $f_x\colon V \to V_0$, such that the induced flat metrics g_x on V satisfy the assumptions of (B). [If $V = D^2$, for example, we need a single conformal immersion $(V, g) \to V_0$ which exists by the Riemann mapping theorem.] Thus we obtain with (B) an isometric immersion of V into $W = (V_0 \times \mathbb{R}, \varepsilon^2 e^{2t} g_0 + \varepsilon^2\, dt^2)$. In a similar way one may use the projective plane P^2 with $K(P^2) = 1$ for V_0.

(B″) **Exercises.** (a) Let $V = (V, g)$ be a non-compact surface and let $V_0 = (V_0, g_0)$ be a closed non-orientable surface for which $\chi(V_0) \geq 0$ (i.e. V_0 is either P^2 or the Klein bottle). Prove the existence of a conformal immersion $V \to V_0$. [Compare Gunning-Narasimhan (1967).]

(b) Let $\tilde{V} \subset H^3$ be an immersed elliptic (i.e. locally convex) non-compact connected surface for which the orthogonal projection to a hyperbolic plane, say $P\colon \tilde{V} \to H_0 \subset H^3$, is a *proper* map such that $\mathrm{dist}(v, P(v)) \leq \mathrm{const} < \infty$ for all $v \in \tilde{V}$. Show that, in fact, $\tilde{V}$ is embedded, the projection P is a homeomorphism and V lies in a halfspace, say in $H^3_+ \subset H^3$, bounded by H_0.

(b′) Let the curvature of the induced metric in $\tilde{V}$ satisfy $-1 \leq a \leq \tilde{K} \leq b < 0$ and show that $0 \leq C_a \leq \mathrm{dist}(v, P(v)) \leq C_b$ for all $V \in \tilde{V}$, where C_κ denotes the constant for which the hypersurface $\{w \in H^3 | \mathrm{dist}(w, H_0) = C_\kappa\}$ has curvature κ. In particular, if $\tilde{K} \equiv \mathrm{const} = \kappa$, then $\mathrm{dist}(v, P(v)) \equiv \mathrm{const}' = C_\kappa$.

(b″) Let $V = (V, g)$ be a closed surface of genus ≥ 2. Show the warped product manifold W of curvature $-\varepsilon^{-2}$ which isometrically contains V [see (iii)] to be unique up to isometry for given V and ε. Then show the isometric immersion $V \to W$ to be also unique up to an isometry of W. Prove a similar uniqueness of isometric immersions of flat tori into hyperbolic manifolds.

(c) Let $W_+(x) = (V \times \mathbb{R}_+, (\cosh t)^2 g_x + dt^2)$ be the warped product of curvature -1, where the metrics g_x on V of curvature -1 are parametrized by the Teichmüller space $X \ni x$ [which is homeomorphic to the Euclidean space of dimension

$6(\text{genus } V) - 6]$. Let $\mathscr{X}$ be the space of pairs $(x \in X, f: V \to W_+(x))$ where f is a strictly elliptic immersion whose projection to $V \times 0 \subset W_+(x)$ is homotopic to the identity. Show the differential operator $\mathscr{D}: \mathscr{X} \to G$ assigning to each (x, f) the metric $g \in G$ on V induced by f to be a *Fredholm* map for appropriate Banach manifold structures in $\mathscr{X}$ and in G.

(c′) Show the Fredholm index of $\mathscr{D}$ (that is the index of the linearization of $\mathscr{D}$ at any point in $\mathscr{X}$) to be zero.

(c″) Show with (b″) every metric $g_0 \in G$ on V of *constant* (negative) curvature to be a *regular* value of $\mathscr{D}$ and observe that $\mathscr{D}^{-1}(g_0)$ consists of a single point in $\mathscr{X}$.

(d) Prove [with Pogorelov's (1969) a priori estimates of the extrinsic curvatures and their derivatives of convex surfaces in H^3] the operator $\mathscr{D}$ to be a *proper* map (in suitable topologies) of $\mathscr{X}$ into the subspace $G_+ \subset G$ of metrices g with $K(g) > -1$.

(d′) Prove with (c″) and (d) the map $\mathscr{D}: \mathscr{X}^\infty \to G_+^\infty$ to be of Fredholm degree $= 1$ and, hence, onto where ∞ refers to the subspaces of C^∞-maps and C^∞-metrics in $\mathscr{X}$ and in G_+.

(d″) Sharpen (B′) by showing $\varepsilon_0(V) \geq \inf_{v \in V}(-K_v(V, g))^{-2}$ for *closed* surfaces V of genus ≥ 1 (compare the existence claim in the Alexandrov-Pogorelov theorem of genus $= 0$)

(d‴) Show with (B) in 3.2.3, that no such bound on ε_0 is possible for metrics g on the disk D^2.

Question. Let V be a closed surface with finitely many punctures (or a more general open surface) and let $W(x)$ be some family of hyperbolic 3-manifolds. What are the properties of the pertinent operator $\mathscr{D}$?

Isometric Immersions $V \to M^4[\kappa]$. There is no single known obstruction for isometric C^∞-immersions of surfaces into Riemannian 4-manifolds. On the positive side one has the following

(C) **Theorem.** *Let $V = (V, g)$ be a C^∞ (C^{an})-surface which is either homeomorphic to the torus T^2 or where V is compact orientable with a non-empty boundary. Then V admits an isometric C^∞ (C^{an}) immersion into an arbitrary small 4-ball with constant curvature κ for all $\kappa \in \mathbb{R}$.*

Proof. First let $\partial V \neq \varnothing$. Then there is a family of C^{an}-immersion $f_x: V \to \mathbb{R}^2$ such that the induced metrics g_x satisfy the assumption of (B) (compare the previous proof). Therefore, (V, g) isometrically immerses into $W_\varepsilon = ((\mathbb{R}^2 \times \mathbb{R}, (\varepsilon^{-1}t)^2 g_0 + dt^2)$ for the flat metric g_0 on $\mathbb{R}^2$ and for all small $\varepsilon > 0$. Then we take the isometric C^{an}-immersion of W_ε into $\mathbb{R}^4 = M^4[0]$ which is the cone over the obvious isometric immersion of $\mathbb{R}^2$ into the ε-sphere $S_\varepsilon^3 \subset \mathbb{R}^4$. The case $\kappa \neq 0$ is treated in a similar way. Furthermore, the case $V = T^2$ reduces via (B) to the following

Lemma (B. Lawson). *Let V be a flat torus. Then there exist isometric C^{an}-immersions $V \to S_{\varepsilon_i}^3$ for some sequence $\varepsilon_i \to 0$.*

Proof. Consider the Hopf fibration $S^3 \to S^2$ and observe that the pull-back of an arbitrary closed curve in S^2 is an (intrinsically) flat torus in $S^3 = S_1^3$. One finds among these tori and their finite covers the homothety class of each flat metric on T^2 (an easy exercise) and the lemma follows.

Exercise. Use an approximation of isometric immersions $(V, g) \to S_\varepsilon^3$ by nearly isometric *imbeddings* $(V, g) \to S_\varepsilon^3$, and show that *every compact connected oriented surface V with boundary admits an isometric $C^\infty(C^{an})$-imbedding into $\mathbb{R}^4$.*

Question. Let a compact surface $V = (V, g)$ (with or without boundary) admit a non-zero vector field. Does there exist a (nearly) isometric C^2-immersion $V \to S_\varepsilon^3$ for small $\varepsilon \le \varepsilon_0(V, g)$?

(C′) **Isometric Immersions $V \to M^4[-\varepsilon^{-2}]$.** *An arbitrary compact $C^\infty(C^{an})$-surface $V = (V, g)$ admits an isometric $C^\infty(C^{an})$-immersion into the hyperbolic space $M^4[-\varepsilon^{-2}]$ of curvature $-\varepsilon^{-2}$ for all positive $\varepsilon \le \varepsilon_0 = \varepsilon_0(V) > 0$.*

Proof. In order to apply (B) to the warped product $M^4[-\varepsilon^{-2}] = (\mathbb{R}^3 \times \mathbb{R}, \varepsilon^2 e^{2t} g_0 + \varepsilon^2 dt^2)$ we need a family of C^{an}-immersions $f_x \colon V \to \mathbb{R}^3$ whose induced metrics g_x uniquely represents all conformal classes of metrics on V near $[g]$. To construct f_x we start with an arbitrary family g_x' whose conformal classes run over the Teichmüller space X. Then we have by the Nash-Kuiper immersion theorem a C^1-map $F \colon V \times X \to \mathbb{R}^3$ which is isometric on each fiber $(V = V \times x, g_x')$. Finally we C^1-approximate F by a C^{an}-map $f \colon V \times X \to \mathbb{R}^3$. The conformal classes of the induced metrics g on $V = V \times x$ define a C^{an}-map of X into itself, $x \mapsto [g_x] \in X$, which is C^0-close to the identity. Since X is a manifold, every small (in the fine C^0-topology) perturbation of the identity is a surjective map. We cannot always guarantee this map to be a global diffeomorphism, but it is a local diffeomorphism over a fixed class $[g] \in X$ provided f is generic. Q.E.D.

Exercises. Construct isometric $C^\infty(C^{an})$-*imbeddings* $V \to M^4[-\varepsilon^{-2}]$ for surfaces V with a non-empty boundary and for closed ones with *even* Euler characteristics [if $V = P^2$ and $K(V) > -\varepsilon^{-2}$, then no isometric C^2-imbedding $V \to M^4[-\varepsilon^{-2}]$ is possible, see (e′) in (E) of 3.2.2].

Prove (all forms of) the h-principle for conformal $C^\infty(C^{an})$-immersions $V \to W$, $\dim W > \dim V = 2$ for arbitrary $C^\infty(C^{an})$-metrics on V and W. Study conformal *imbeddings* $V \to W$.

Let X be a (possibly singular) real analytic space (e.g. an affine real algebraic variety). Show that every C^0-small continuous perturbation of the identity map $X \to X$ sends X *onto* X.

Question. Let τ be an (integrable or not) 2-dimensional subbundle of $T(V)$, $\dim V = n > 2$. Does the h-principle hold true for those C^∞-maps $V \to W$, $\dim W > 2$, which are conformal on τ for given Riemannian metrics on V and on W?

Isometric Immersions $V \to M^5[\kappa]$. *In the following three cases a compact surface V admits an isometric $C^\infty(C^{an})$-imbedding into the space $M^5[\kappa]$ of constant curvature for all $\kappa \in \mathbb{R}$.* [See (D) below for a more general result.]

(i) *V is homeomorphic to S^2, T^2 or to the Klein bottle.*
(ii) *V is connected with a non-empty boundary.*
(iii) *V is orientable and there exists an isometric orientation reversing involution $I: V \to V$ whose fixed point set is non-empty. In this case there is an equivariant isometric imbedding $V \to M^5[\kappa]$ which is symmetric in a hyperplane in $M^5[\kappa]$.*

Proof. The sphere S^2 with any metric imbeds into $M^3[-\varepsilon^{-2}]$ and each ball in $M^3[-\varepsilon^{-2}]$ isometrically immerses into $M^5[\kappa]$ [see (D) in 3.2.2]. The resulting immersion can be $C^\infty(C^{an})$-approximated by an isometric *imbedding*. (This approximation is left to the reader.)

The case of $V = T^2$ was studied in (C). The proof of (C) reduces the case of the Klein bottle and (ii) to isometric C^{an}-immersions of the *flat* Klein bottles K^2 into $S^4 = M^4[\varepsilon^{-2}]$. To obtain these we take the δ-neighborhood U_δ of the projective line in $P_\varepsilon^4 = S_\varepsilon^4/\mathbb{Z}_2$ whose boundary is a flat K^2. If δ is small compared with ε, then U_δ admits an isometric C^{an}-immersion into S_ε^4 and we find each flat K^2 among finite covers of ∂U_δ [compare the Exercise (b″) in (F) of 3.2.1]. Thus we get isometric immersions into $M^5[\kappa]$. Passing to isometric *imbeddings* is an exercise for the reader.

Now, in case (iii), we consider the quotient $V_0 = V/I$ which is a smooth surface with a boundary. Then for an arbitrary symmetric metric g on V one can easily find a C^{an}-function ψ on V with the following three properties,

(a) $\psi(I(v)) = -\psi(v)$, $v \in V$.
(b) The form $g - (d\psi)^2$ is positive definite outside the curves fixed by I.
(c) There exist a $C^\infty(C^{an})$-metric g_0 on V_0 and a $C^\infty(C^{an})$-map $\psi_0: V \to V_0$, such that $\psi_0^*(g_0) = g - (d\psi)^2$ and $\psi_0(Iv) = \psi_0(v)$.

Finally, we take an isometric imbedding $f_0: (V_0, g_0) \to \mathbb{R}^4$ and take $(\psi, \psi_0 \circ f_0)$: $V \to \mathbb{R}^5$ for the required imbedding for $\kappa = 0$. The case $\kappa \neq 0$ is left to the reader.

Corollary. *There exists a closed C^{an}-surface $V \subset \mathbb{R}^5$ with a metric of constant curvature -1. Therefore, there exists an isometric C^{an}-immersion $H^n \to \mathbb{R}^{5n-5}$ for $H^n = M^n[-1]$ and for all $n = 2, 3, \ldots$.* [Compare (D) in 3.2.2.]

Immersions $V \to M^6[\kappa]$. *An arbitrary (possibly non-compact) $C^\infty(C^{an})$-surface (V, g) admits an isometric $C^\infty(C^{an})$-imbedding into $M^6[\kappa]$ for all $\kappa \in \mathbb{R}$.*

Proof. If V is compact the proof is immediate with (B). Indeed, we have a family of C^{an}-immersions $f_x: V \to \mathbb{R}^3$, $x \in X$, for which the conformal classes of the induced metrics g_x on V cover a neighborhood of the class $[g]$ in the Teichmüller space X. Then we use the isometric immersion $I_\varepsilon: \mathbb{R}^3 \to S^5 \subset \mathbb{R}^6$, $\varepsilon > 0$, given by $\varepsilon \sin \varepsilon^{-1} y_i$ and $\varepsilon \cos \varepsilon^{-1} y_i$ for the coordinates y_i, $i = 1, 2, 3$ in $\mathbb{R}^3$. Since g is conformal to g_{x_0} for some $x_0 \in X$, we may write $g = \psi^2 g_{x_0}$. We observe that the immersion $F_\varepsilon = \psi(I_\varepsilon \circ f_{x_0})$: $V \to \mathbb{R}^6$ is ε-isometric for $\psi^2 g_{x_0}$, as the induced metric is $\psi^2 g_{x_0} + 3\varepsilon^2 (d\psi)^2$. Finally, we perturb F_ε to an isometric immersion $F: (V, \psi^2 g_{x_0}) \to \mathbb{R}^6 = M^6[0]$ by applying (B). We leave to the reader the case $\kappa \neq 0$ as well as the elimination of double points of F to obtain imbeddings.

Now let $V = (V, g)$ be non-compact and let $\Sigma_1 \subset U$ be a disjoint union of simple closed double-sided curves in V, such that every component of the complement $V \backslash \Sigma_1$ is relatively compact in V. Let Σ_2 and Σ_3 in V be small deformations of Σ_1 such that some tubular neighborhoods $U_i \subset V$ of Σ_i, $i = 1, 2, 3$ do not intersect.

Sublemma. *There exist C^∞-functions y_i and ψ_i on V, $i = 1, 2, 3$, such that $g = \sum_{i=1}^3 \psi_i^2 (dy_i)^2$ and such that the support of ψ_i equals $V \backslash U_i$ for $i = 1, 2, 3$.*

Proof. The equation $\sum_{i=1}^3 \psi_i^2\, dy_i^2 = g$ is easily solvable on the union $U = U_1 \cup U_2 \cup U_3 \subset V$. Indeed, U is a didjoint union of cylinders $S^1 \times [0, 1]$ and so there is a conformal C^∞-immersion $U \to \mathbb{R}^2$ which amounts to a C^∞-decomposition $g|U = e^\rho(dz_1^2 + dz_2^2)$. On the other hand, the convex integration applies to y_i outside U for $\psi_i > 0$ being fixed. Thus one obtains a global decomposition $g = \sum_{i=1}^3 \varphi_i^2\, dz_i^2$ where the functions φ_i are C^∞ but z_i, $i = 1, 2, 3$ are only C^1-smooth. Passing to a C^∞-decomposition is immediate in the case where V is simply connected since for any C^∞-triple (y_1, y_2, y_3) which is C^1-close [in the fine C^1-topology to (z_1, z_2, z_3)] the metric $\Sigma \varphi^2\, dy^2$ is conformal to g. Hence, (V, g) is isometric to $(V, e^\alpha \Sigma \varphi_i^2\, dy_i^2)$ for some C^∞-function α on V. Furthermore, if $\pi_1(V)$ is finitely generated, then the Teichmüller space X of the conformal classes of metrics on V is finite dimensional and the approximation argument applies to families $g_x = \Sigma \varphi_i^2\, dz_{i,x}^2$ for which $[g_x] = x \in X$. Finally, the desired approximation is obtained on an arbitrary surface V by applying the above to an increasing family of compact smooth domains $V_j \subset V$, $j = 1, 2, \ldots$, which exhaust V.

Exercise. Fill in the details in the argument. Then prove (all forms of) the h-principle for C^∞-solutions ψ_i and y_i to the equation $\sum_{i=1}^3 \psi_i^2\, dy_i^2 = g$.

Now, let $g = \sum_{i=1}^3 \psi_i^2\, dy_i^2$ and let $\varepsilon_i = \varepsilon_i(v)$, $v \in V$, $i = 1, 2, 3$, be small positive functions on V which rapidly decay for $v \to \infty$ and such that each ε_i is locally constant outside U_i. Then the map $F_\varepsilon = (s_i, c_i)\colon V \to \mathbb{R}^6$ for $s_i = \varepsilon_i \psi_i \sin \varepsilon_i^{-1} y_i$ and $c_i = \varepsilon_i \psi_i \cos \varepsilon_i^{-1} y_i$, $i = 1, 2, 3$, induces the metric $g + \sum_{i=1}^3 \varepsilon_i^2\, d\psi_i^2$ on V which can be made arbitrarily C^∞-close (in the *fine* C^∞-topology) to g for small ε_i. We claim the existence of all C^∞-perturbations $\tilde{\psi}_i$ and $\tilde{y}_i$ of ψ_i and y_i respectively for which $(V, \sum_{i=1}^3 \tilde{\psi}_i^2\, d\tilde{y}_i^2 - \varepsilon_i^2\, d\tilde{\psi}^2)$ is isometric to $(V, g = \sum_{i=1}^3 \psi_i^2\, dy_i^2)$ where each function ε_i is assumed C^∞-small and ψ_i is required to vanish on the subset where ε_i is not locally constant. This amounts [compare the proof of (B)] to the existence of a small perturbation $(y_{1,x}, y_{2,x}, y_{3,x})$ of (y_1, y_2, y_3) such that there exists a conformal map $f\colon (V, g_x - \sum_{i=1}^3 \varepsilon_i^2\, d\psi_i^2) \to (V, g)$, where $g_x = \sum_{i=1}^3 \psi_i^2\, dy_{i,x}^2$, which preserves the curvature forms. This can be expressed by the following equation for the conformal factor e^{2h},

$$\Delta h \hat{g}_x + \omega(g_x) + D_\varepsilon(h) = f_x^*(\omega(g)).$$

This equation is similar to the one used in the proof of (B) and the solution is obtained on compact subsets in V by the argument in (B). The solution on all of V comes with the standard compact exhaustion trick. Thus a C^∞-immersion $(V, g) \to \mathbb{R}^6$ is constructed.

Exercises. Fill in the details in this proof. Extend the argument to immersions $V \to M^6[\kappa]$ for $\kappa \neq 0$. Make the proof work for C^{an}-immersions of C^{an}-surfaces. Eliminate possible double points and thus obtain an isometric $C^\infty(C^{an})$-imbedding $(V, g) \to M^6[\kappa]$ for all $\kappa \in \mathbb{R}$.

References. Our basic analytic tool, Proposition (B), is a modification of H. Weil's implicit function theorem which claims the existence of isometric immersions $(S^2, g) \to \mathbb{R}^3$ for the metrics g which are C^∞-close to the standard one (see Pogorelov's 1969-book for an account of the theory of isometric immersions $(S^2, g) \to M^3[\kappa]$). Isometric C^∞-immersions $(D^2, g) \to \mathbb{R}^4$ due to Posnyak (1973). Conformal immersions $V^2 \to \mathbb{R}^3$ were obtained by A. Garsia (1961).

(D) *Immersions* $V^2 \to \mathbb{R}^5$. We construct here an isometric $C^\infty(C^{an})$-imbedding $(V, g) \to \mathbb{R}^5$ for all *compact* surfaces V. We start with the study of the following *$(k + l)$-decompositions*,

$$g = \sum_{i=1}^{k} dx_i^2 + \sum_{i=k+1}^{k+l} \varphi_i \, dx_i^2, \tag{$*$}$$

where the map $X = (x_1, \ldots, x_{k+l})\colon V \to \mathbb{R}^{k+l}$ is a C^∞-immersion and where φ_i, $i = k+1, \ldots, k+l$, are *positive* C^∞-functions on V. (The case $l = 0$ amounts to isometric C^∞-immersions $X\colon V \to \mathbb{R}^k$.) Let Φ stand for the vector functions $\Phi = (\varphi_i)$, $i = 1, \ldots, k + l$, with positive components φ_i where $\varphi_i \equiv 1$ for $i = 1, \ldots, k$. The expression $(*)$ defines the differential operator

$$D\colon (X, \Phi) \mapsto g = \sum_{i=1}^{k+l} \varphi_i \, dx_i^2.$$

Definitions. A $(k + l)$-decomposition defined by a pair (X, Φ) is called *stable* if the operator D is infinitesimally invertible at (X, Φ) in the variables x_i and φ_i. That is the linearized equation $L(\tilde{X}, \tilde{\Phi}) = \tilde{g}$, for the linearization $L = L_{X,\Phi}$ of D admits a solution $(\tilde{X}, \tilde{\Phi}) = (\tilde{x}_i, \tilde{\varphi}_i)$, $i = 1, \ldots, k + l$ where $\tilde{\varphi}_i \equiv 0$ for $i = 1, \ldots, k$. Moreover, there is a differential operator $M = M_{X,\Phi}$ which sends metrics $\tilde{g}$ to the above pairs $(\tilde{X}, \tilde{\Phi})$, such that $L \circ M = \mathrm{Id}$.

A $(k + l)$-decomposition is called *strongly stable* if the operator D is infinitesimally invertible in x_i, $i = 1, \ldots, k + l$ and in φ_i, now for $i = k + 2, \ldots, k + l$. This means the existence of a linearized solution $(\tilde{X}, \tilde{\Phi})$ with $\tilde{\varphi}_i \equiv 0$ for $i = 1, \ldots, k + 1$.

(D′) **Lemma.** *If a metric g admits a strongly stable $(k + l)$-composition, then g also admits a stable $(k' + l')$-decomposition for $k' = k + 2$ and $l' = l - 1$.*

Proof. If a decomposition $g = \sum_{i=1}^{k+l} \varphi_i \, dx_i^2$ is strongly stable, then for every small $\varepsilon > 0$ there exist C^∞-small perturbations x_i' of x_i and φ_i' of φ_i, such that $\varphi_i' \equiv \varphi_i$ for $i = 1, \ldots, k + 1$, and such that

$$\sum_{i=1}^{k+l} \varphi_i'(dx_i')^2 = g - \varepsilon^2 (d\sqrt{\varphi_{k+1}})^2.$$

This follows from Nash's implicit function theorem (see 2.3.2). Furthermore, the decomposition $\sum_{i=1}^{k+l} \varphi_i'(dx_i')^2$ is also strongly stable since the infinitesimal invertibility is *defined* (see 2.3.1) in some neighborhood of the pair (X, Φ). Now let $s = \varepsilon\sqrt{\varphi_{k+1}} \sin \varepsilon^{-1} x_{k+1}'$ and $c = \varepsilon\sqrt{\varphi_{k+1}} \cos \varepsilon^{-1} x_{k+1}'$. Then

$$\sum_{i=1}^{k} (dx_i')^2 + ds^2 + dc^2 + \sum_{i=k+2}^{k+l} \varphi_i'(dx_i')^2 = g.$$

Finally, an easy argument (which is left to the reader) reduces the stability of this decomposition to the strong stability of the above $\sum_{i=1}^{k+l} \varphi_i'(dx_i')^2$.

Exercise. Generalize (D′) to manifolds V^n for *all* $n = 1, 2, 3, \ldots$.

(D″) **Lemma.** *Let g admit a stable decomposition, $g = \sum_{i=1}^{k+l} \varphi_i \, dx_i^2$ for $\varphi_i \equiv 1$, $i = 1, \ldots, k$. If $k + 2l \geq 5$, then there exists a strongly stable decomposition $g = \sum_{i=1}^{k+l} \varphi_i'(dx_i')^2$ for some C^∞-small perturbation (φ_i', x_i') of (φ_i, x_i), where $\varphi_i' \equiv 1$ for $i = 1, \ldots, k$.*

Remark. The isometric C^∞-immersion problem $(V, g) \to \mathbb{R}^5$ is reduced with (D′) and (D″) to the construction of a stable decomposition $g = dx_1^2 + \varphi_1 \, dx_2^2 + \varphi_3 \, dx_3^2$.

Proof. The linearized equation $L(\tilde{X}, \tilde{\Phi}) = \tilde{g}$, where $\tilde{X} = (\tilde{x}_i)$, $i = 1, \ldots, k + l$ and $\tilde{\Phi} = (\tilde{\varphi}_i)$ $i = k + 2, \ldots, k + l$ (where $\tilde{\varphi}_i \equiv 0$ for $i = 1, \ldots, k + 1$) contains $k + 2l - 1$ unknown functions and so it is underdetermined for $k + 2l \geq 5$, as the metric $\tilde{g}$ has three components, say $g_{\mu\nu}$, $1 \leq \mu \leq \nu \leq 2$, in some local coordinates (u_1, u_2) on V. Hence, the results in 2.3.8 suggest the strong stability of $\sum_{i=1}^{k+l} \varphi_i \, dx_i^2$ for *generic* functions x_i, $i = 1, \ldots, k + l$ and φ_i, $i = k + 1, \ldots, k + l$. The study of the equation $L(\tilde{X}, \tilde{\Phi}) = \tilde{g}$ is greatly simplified with the following auxiliary (unknown) functions

$$(+) \qquad \tilde{y}_\mu = \langle \tilde{X}, Y_\mu \rangle, \qquad \mu = 1, 2,$$

where Y_μ denotes the vector-functions, $Y_\mu = (\varphi_i \partial_\mu x_i)$, $i = 1, \ldots, k + l$, for $\partial_\mu = \partial/\partial u_\mu$, and where $\langle \ , \ \rangle$ stands for the (Euclidean) scalar product. Next, we denote by $Y_{\mu\nu}$, $1 \leq \mu$, $\nu \leq 2$, the vector function with the components $-2(\varphi_i \partial_{\mu\nu} x_i + (\partial_\nu \varphi_i)(\partial_\mu x_i))$, $i = 1, \ldots, k + l$, and we denote by $Z_{\mu\nu}$ the vector function whose first $k + 1$ components are zero and the following $l - 1$ are $(\partial_\mu x_i)(\partial_\nu x_i)$, $i = k + 2, \ldots, k + l$. Then we introduce the following three equations in the unknowns $\tilde{X}$, $\tilde{\Phi}$ and $\tilde{y}_\mu$,

$$(++) \qquad \langle X, Y_{\mu\nu} \rangle + \langle \tilde{\Phi}, Z_{\mu\nu} \rangle = \tilde{g}_{\mu\nu} - \partial_\mu \tilde{y}_\nu - \partial_\nu \tilde{y}_\mu.$$

A straightforward computation (compare 2.3.1 and 2.3.8) leads to the following conclusion: if for some functions $\tilde{y}_1$ and $\tilde{y}_2$ the linear algebraic equations (+) and (++) are consistant and, hence, satisfied by some $\tilde{X}$ and $\tilde{\Phi}$, then $L(\tilde{X}, \tilde{\Phi}) = g$. The five linear equations (+) and (++) contain $k + 2l - 1$ unknowns (these are $\tilde{x}_i$, $i = 1, \ldots, k + l$, and $\tilde{\varphi}_i$, $i = k + 2, \ldots, k + l$) and in the *generic case* the consistency condition is expressed by m equations in $\tilde{y}_\mu$ and $\partial_\nu \tilde{y}_\mu$ for $m \leq 6 - k - 2l$. Since $k + 2l \leq 5$, we are left with at most one differential equation of first order for which the solution is especially easy (see 2.3.8). Recall that the operator $L_{X,\Phi}$ on $\tilde{x}_i$, $i = 1, \ldots, k + l$ and $\tilde{\varphi}_i$, $i = k + 2, \ldots, k + l$ is invertible (by some $M = M_{X,\Phi}$), provided the infinite order jet of the given pair (X, Φ) misses a certain subset Σ in the space

J^∞ of jets of all such pairs. (We use J^∞ instead of J^r for $r < \infty$ since the order of the operator M plays no role in our discussion. In fact, we always deal with operators of a fixed order and so the space J^∞ stands for J^r for some fixed large integer $r < \infty$ which is not specified here.) If we apply the considerations in 2.3.8 to the Eqs. $(+)$ and $(++)$, we immediately conclude to the inequality $\operatorname{codim} \Sigma \geq 1$. But we need the following sharper inequality.

Let $\tilde{J}_g^\infty \subset J^\infty$ consist of the jets of stable infinitesimal solutions (X, Φ) to the equation $\sum_{i=1}^{k+l} \varphi_i \, dx_i^2 = g$. In other words, $\tilde{J}_g^\infty$ is (the lift to J^∞ of) the differential relation which governs stable solutions of this equation. We denote by $\tilde{\Sigma} \cap \tilde{J}_g^\infty$ the intersection $\Sigma \cap \tilde{J}_g^\infty$ and we claim the subset $\tilde{\Sigma} \subset \tilde{J}_g^\infty$ to have $\operatorname{codim} \tilde{\Sigma} \geq 3$ for all C^∞-metrics g on V. (Here again ∞ means some integer $r < \infty$.) The proof is straightforward with the Eqs. $(+)$ and $(++)$ (compare 2.3.8) and we leave this to the reader.

Now the transversality theorem for infinitesimally invertible operators (see 2.3.2) implies (D″).

Exercise. Generalize (D″) to n-dimensional manifolds, $n \geq 2$, for $k + 2l \geq n(n+3)/2$.

Our construction of isometric C^∞-immersions $(V, g) \to \mathbb{R}^5$ is concluded with the following

(D‴) **Proposition.** *An arbitrary C^∞-metric g on V admits a stable decomposition,*

$$g = dx_1^2 + \varphi_2 \, dx_2^2 + \varphi_3 \, dx_3^2. \tag{**}$$

Proof. Let $\mathbb{R}^3$ be the (x_1, x_2, x_3)-space and let $H \subset \mathbb{R}^3$ be the (x_2, x_3)-plane $\{x_1 = 0\}$. For an immersion $X = (x_1, x_2, x_3)\colon V \to \mathbb{R}^3$ we denote by $\Sigma^1 \subset V \subsetneq \mathbb{R}^3$ the singularity of the normal projection $V \to H$. Throughout the following discussion we require the following two properties of X.

(i) The singularity Σ^1 is generic. Namely, Σ^1 consists of finitely many smooth closed curves and the projection $V \to H$ folds along $\Sigma^1 \backslash \Sigma^{11}$, where Σ^{11} is the finite set of non-degenerate cuspidal points. Furthermore, the critical points of the functions $x_2 | \Sigma^1$ and $x_3 | \Sigma^1$ are non-degenerate. We denote by $\Sigma_1^1 \subset \Sigma^1$ the union of these points. Observe that the set Σ_1^1 is finite and $\Sigma^{11} \subset \Sigma_1^1$.

(ii) The function x_i has $\|\mathrm{grad}_g x_1\| < 1$ outside Σ^1. [If (**) is satisfied with some φ_1 and φ_2 at a point $v \in \Sigma^1$, then $\|\mathrm{grad}_g x_1(v)\| = 1$].

If the Eq. (**) is satisfied with some continuous functions φ_1 and φ_2, on an open subset in V, then [due to (i)] the functions φ_1 and φ_2 are uniquely determined by x_i, $i = 1, 2, 3$, and by g. Hence, there exists a unique maximal open subset, say $U = U(X) \subset V$, on which g admits a stable decomposition (**) for a given X and some (C^∞-smooth and positive) functions φ_2 and φ_3. We want $U = V$ and we construct an appropriate X in six steps.

Step 0. We start with an arbitrary map X which satisfies (i) and (ii). If the surface V is orientable, we may assume $\Sigma^{11} = 0$ and then the following step may be omitted.

Step 1: Making $U \supset \Sigma^{11}$. Consider smooth positive functions $\overline{\varphi_2}$ and $\overline{\varphi_3}$ on $\mathbb{R}^3$. and introduce the metric $\bar{g} = dx_1^2 + \overline{\varphi_2} \, dx_2^2 + \bar{\varphi}_3 \, dx_3^2$ on $\mathbb{R}^3$. This generalizes the warped products in (A). In particular, such a metric may have any given constant curvature

κ at a given point in $\mathbb{R}^3$. If the map $X: (V, g) \to (\mathbb{R}^3, \bar{g})$ is isometric, then (**) is satisfied with $\varphi_2(v) = \overline{\varphi_2}(X(v))$ and $\varphi_3(v) = \overline{\varphi_3}(X(v))$. Furthermore, if $v \in \Sigma^{11}$, then the point v is hyperbolic and the vertical (i.e. normal to H) line through v is tangent to V along an asymptotic direction in $T_v(V)$. To arrive at this situation, we let, with some choice of $\overline{\varphi_2}$ and $\overline{\varphi_3}$, the metric $\bar{g}$ have large positive curvature κ near $X(v) \in \mathbb{R}^3$. Then we construct an infinitesimally isometric immersion X': $(\mathcal{O}p(v), g) \to (\mathbb{R}^3, \bar{g})$, such that $X'(v) = X(v) \in \mathbb{R}^3$. If $\kappa > K_g(v)$, then this X' is hyperbolic and we can make one of the two asymptotic directions vertical at v. Furthermore, with appropriate small perturbations of $\overline{\varphi_2}$, $\overline{\varphi_3}$ and X' we arrive at a *generic* situation for which the point v is a non-degenerate cusp and for which the corresponding decomposition (**) is stable near v. The stability allows us to apply Nash's implicit function theorem and thus to arrive at an immersion, which is still called $X: V \to \mathbb{R}^3$, for which the projection $V \to H$ has non-singular cusps at the points of a given finite subset $\Sigma^{11} \subset V$ and such that $U \supset \Sigma^{11}$. Of course, the projection $V \to H$ may now have cusps besides Σ^{11}. However, maps $V \to H$ without cusps satisfy (all forms of) the h-principle (see 1.3.1) and so it is not hard (and left to the reader) to arrange cusps at Σ^{11}, such that the required extension outside $\mathcal{O}p\Sigma^{11}$ exists.

Step 2: Making $U \supset \Sigma_1^1$. Let $v \in \Sigma_1^1$ be a non-cuspidal point. First we deform X outside $\mathcal{O}p\Sigma^{11}$ in order to make $x_1 = 0$ on a small segment $S \subset \Sigma^1$ around $v \in \Sigma^1$. We can also assume the segment $S \subset \Sigma^1 \subset V$ to be geodesic in (V, g) and then choose a metric $\bar{g} = dx_1^2 + \overline{\varphi_2} + \overline{\varphi_3}\, dx_3^2$ in $\mathbb{R}^3$ near $X(v) \in \mathbb{R}^3$ for which isometric immersions

$$X': (V, g) \to (\mathbb{R}^3, \bar{g}) \quad \text{with } X'|S = X|S$$

would fold along S after projecting to H like X. Such an immersion X' is actually obtained near S by first achieving the stability (for perturbed X and $\bar{g}$) and then by applying Nash's implicit function theorem. The extension outside S is obvious and thus we let $U \ni v$ for all $v \in \Sigma_1^1$.

Step 3: Making $U \supset \Sigma^1$. One might do this by generalizing the above argument, but we take a different route to show additional features of the picture. First we regularly homotope the immersion $X = (x_1, x_2, x_3)$: $V \to \mathbb{R}^3$ without changing x_2 and x_3 at all, also keeping fixed $x_1|\mathcal{O}p\Sigma_1^1$, and such that the homotoped function x_1, called x_1', satisfies

(i) $\|\mathrm{grad}_g x_1'\| = 1$ on Σ^1, and Σ^1 is a non-degenerate maximum set for $\|\mathrm{grad}_g x_1'\|^2$. That is $1 - \|\mathrm{grad}_g x_1'\|^2 \geq \varepsilon(\mathrm{dist}(v, \Sigma^1))^2$ for some $\varepsilon > 0$;
(ii) The field $\mathrm{grad}_g x_1'|\Sigma^1$ is transversal to Σ^1 outside Σ^{11}.

Next we consider a metric $g' = (dx_1')^2 + \varphi_2\, dx_2^2 + \varphi_3\, dx_3^2$ on V for some positive C^∞-function φ_1 and φ_2, such that $g'|\mathcal{O}p\Sigma_1^1 = g|\mathcal{O}p\Sigma_1^1$, and observe that $\mathrm{grad}_{g'} x_1'$ also satisfies (i) and (ii). It follows that the quadratic forms $g - (dx_1')^2$ and $g' - (dx_1')^2$ on $T(V)$ are equivalent under some fiber-wise linear diffeomorphism α: $T(V) \to T(V)$. Moreover, there is an α which keeps Σ^1 (or rather the zero section $\Sigma^1 \to T(V)$) fixed, such that $\alpha^*(g') = g$, and such that $\alpha^*(\mathrm{grad}_g x_1') = \mathrm{grad}_{g'} x_1'$. Furthermore, we may require this α to be fixed over $\mathcal{O}p\Sigma_1^1$ and to be isotopic (in the category of fiber-wise linear diffeomorphisms which are fixed over $\mathcal{O}p\Sigma_1^1$) to the identity. Consider the linear forms $l_i' = \sqrt{\varphi_i}\, dx_i$, $i = 2, 3$, let $l_i = \alpha^*(l_i')$ and observe that $(dx_1')^2 + l_2^2 + l_3^2 = g$. Since the forms l_i do not vanish on $T(\Sigma^1) \subset T(V)$ outside Σ_1^1

there exist functions x_i' and $\varphi_i' > 0$, $i = 2, 3$, on a small neighborhood $\mathcal{O}p\Sigma^1 \subset V$, such that $l_i = \sqrt{\varphi_i'}\, dx_i'$ and $x_i'|\Sigma^1 = x_i$, $i = 1, 2$. Moreover, these x_i' and φ_i' are unique on $\mathcal{O}p\Sigma^1$ and they are equal to x_i and φ_i respectively on $\mathcal{O}p\Sigma_1^1 \subset V$. The functions x_i' extend to all of V, such that $\Sigma^1(x_2', x_3') = \Sigma^1 = \Sigma^1(x_2, x_3)$. Furthermore we achieve the stability of the decomposition $[(dx_1')^2 + \varphi_2'(dx_2')^2 + \varphi_3'(dx_3')^2]|\mathcal{O}p\Sigma^1 = g|\mathcal{O}p\Sigma^1$ by choosing *generic* x_1' and α. (Checking this is left to the reader.) Then we return to the notation x_1, φ_i.

Step 4: Satisfying (**) *outside* Σ^1 *with* C^1*-functions* x_1', x_2', x_3'. We seek a C^1-diffeomorphism $\beta'\colon V \to V$, which is the identity near Σ^1, and a C^1-function x_1' on V, for which $x_1'|\mathcal{O}p\Sigma^1 = x_1|\mathcal{O}p\Sigma^1$, such that the functions x_i' where $x_i' = x_i \circ \beta'$ for $i = 2$, 3, satisfy (**) with some *continuous* functions $\varphi_i' > 0$ on V. This can be expressed outside Σ^1 by the following differential relation (of the first order) between x_i'.

(***) $$\langle \operatorname{grad}_{\bar g} x_2', \operatorname{grad}_{\bar g} x_3' \rangle_{\bar g} = 0,$$

for $\bar g = g - (dx_1')^2$. A straightforward check up (left to the reader) shows the convex integration (see 2.4.7) to apply to (***) and then the existence of β' and x_1' is obvious.

Step 5: Making x_i' C^∞*-smooth outside* Σ^1. Take a compact region with a smooth boundary in V, say $V_0 \subset V$, which does not meet Σ^1 but comes close to it. Namely, $V_0 \cup \mathcal{O}p\Sigma^1 = V$. Take C^∞-smooth functions x_i, $i = 1, 2, 3$ and φ_i, $i = 2, 3$ on V_0 such that $x_i|\mathcal{O}p\partial V_0 = x_i'$ $\mathcal{O}p\partial V_0$ and $\varphi_i|\mathcal{O}p\partial V_0 = \varphi_i'|\mathcal{O}p\partial V_0$. Then the C^∞-metric $g_1 = g - dx_1^2$ on V_0 is C^0-close to $g_2 = \varphi_2\, dx_2^2 + \varphi_3\, dx_3^2$ [since $g = (dx_1')^2 + \varphi_2'(dx_2')^2 + \varphi_3'(dx_3')^2$ on V] and $g_1|\mathcal{O}p\partial V_0 = g_2|\mathcal{O}p\partial V_2$. If V_2 is simply connected, then there exists a conformal mapping $A\colon (V_0, g_1) \to (V_0, g_2)$ which is C^0-close to the identity. In fact, such an A exists for all V_0 provided the approximating functions x_i and φ_i are chosen such that the conformal classes $[g_1]$ and $[g_2]$ are equal. Recall that the conformal classes of metrics on V_0 form a finite dimensional (Teichmüller) space and so the required adjustment of x_i and φ_i is easy with the convex integration applied to *families* at the previous step [compare the proof in (C′)]. The induced metric $A^*(g_2)$ on V_0 equals $\psi^{-1}g_1$ for some C^∞-function $\psi > 0$ and so the functions $x_i^* = x_i \circ A$ and $\varphi_i^* = \varphi_i \circ A$, $i = 2, 3$, satisfy on V_0,

(*○) $$dx_1^2 + \psi\varphi_2(dx_2)^2 + \psi\varphi_3(dx_3)^2 = g.$$

This solution of (**) on V_0 may not agree with what we had before on $\mathcal{O}p\partial V_0 \subset \mathcal{O}p\Sigma^1$. However, the metrics g_1 and g_2 are equal on $\mathcal{O}p\partial V_0$ and they are C^∞-smooth. Hence, the C^0-closeness of A to the identity yields the C^∞-closeness between A and Id on $\mathcal{O}p\partial V_0$. This implies the C^∞-closeness of x_i^* to x_i' and of $\psi\varphi_i^*$ to φ_i' on V_0 for $i = 2$, 3. Finally, the *stable* solutions of (**) on $\mathcal{O}p\partial V_0$ are microflexible (see 2.3.2) and so a small C^∞-deformation of our solutions on V_0 and on $\mathcal{O}p\Sigma^1$ makes them agree on $\mathcal{O}p\partial V_0$. To complete the proof we need the decomposition (*○) to be stable on V_0. This is achieved by introducing a small generic perturbation (which keeps $[g_1] = [g_2]$) into our approximating functions x_i and φ_i. The check up of this genericity is left to the reader.

Thus the proof of (D‴) is concluded and we have a C^∞-immersion $(V, g) \to \mathbb{R}^5$. Furthermore, the stability of our solution to the (isometric immersion) equation $g = \sum_{i=1}^5 dx_i^2$ makes possible (see 2.3.2) the elimination of possible double points and the C^{an}-approximation. Hence, we obtain the desired $C^\infty(C^{an})$-imbedding of each compact $C^\infty(C^{an})$-surface into $\mathbb{R}^5$.

Exercise. Construct an isometric $C^\infty(C^{an})$-imbedding $(V, g) \to M^5[\kappa]$ for all $\kappa \in \mathbb{R}$.

Remarks on Immersions $(V, g) \to \mathbb{R}^5$ *for Non-Compact Surfaces* V. It is not hard to obtain a stable decomposition $g = dx_1^2 + \varphi_2\,dx_2^2 + \varphi_3\,dx_3^2$ for non-compact surfaces V. Unfortunately, this does not lead to an isometric immersion since the form $\varepsilon(d\sqrt{\varphi})^2$ is never small in the *fine* C^∞-topology (here $\varepsilon > 0$ and the support of φ is non-compact) to which Nash's implicit function theorem applies. However, one may expect the above proof to generalize to complete non-compact surfaces with a *bounded geometry*. This means a uniform lower bound on the injectivity radius, $\operatorname{Inj\,Rad}_v(V, g) \geq \delta > 0$ for all $v \in V$, and an upper bound on the norms of some covariant derivatives of the curvature. For example, $\|\nabla^r K_g(v)\| \leq \text{const} < \infty$ for the derivatives ∇^r of the order $r = 0, 1, \ldots, 10$.

Another class of metrics g may be handled with a C^∞-decomposition

$$g = dx_1^2 + \varphi_2^2\,dx_2^2 + \varphi_3^2\,dx_3^2,$$

where the functions φ_i, $i = 2, 3$, vanish on a small neighborhood $U_i \subset V$ of a disjoint union of simple closed curves in V, such that the connected components of the support $\operatorname{Supp}\varphi_i$ are relatively compact in V. Such decompositions exist (an exercise for the reader) for all (V, g) and they allow (non-constant!) functions $\varepsilon(v)$ for Nash's twisting. This implies the density of the immersible into $\mathbb{R}^5$ metrics in the space of metrics on V with the *fine* C^∞-topology. The presence of zeros of φ_i obstructs the stability, but one may try something else instead of Nash's implicit function theorem. The problem boils down to perturbing the functions x_1, x_2 and $\varphi_2 > 0$ in a given decomposition $g = dx_1^2 + \varphi_2^2\,dx_2^2$ (near a closed curve in V) to x_1', x_2' and φ_2', such that $(dx_1')^2 + (\varphi_2')^2(dx_2')^2 = g - \varepsilon(d\varphi_2')^2$ for a given arbitrary small $\varepsilon > 0$. This is likely to work if the curvature $K(v)$ does not vanish (near the curve in question) or if the metric g is C^{an}. Thus one expects an isometric $C^\infty(C^{an})$-imbedding $(V, g) \to \mathbb{R}^5$ to exist if $K(v)$ does not vanish outside a compact subset in V or if g is real analytic outside a compact subset.

Further Questions. Does every C^∞-metric g on a (compact) surface V admit the following C^∞-decomposition?

(a) $$g = dx_1^2 + \varphi^2(dx_2^2 + dx_3^2),$$

(b) $$g = dx_1^2 + dx_2^2 + \varphi_3^2\,dx_3^2.$$

The decomposition (b) amounts (at least locally) to an isometric immersion $(V, g) \to (\mathbb{R}^3, \bar{g})$ for $\bar{g} = dx_1^2 + dx_2^2 + \bar{\varphi}^2\,dx_3^2$. Here x_i are the coordinates in $\mathbb{R}^3$ and $\bar{\varphi} = \bar{\varphi}(x_1, x_2, x_3)$. If $\bar{\varphi} > 0$ (and thus $\bar{g}$ is positive definite), one defines *the ellipticity* of the decomposition (b) as the ellipticity (i.e. the local convexity) of the immersion $(V, g) \to (\mathbb{R}^3, \bar{g})$. If a metric g on a *compact* surface V admits an *elliptic* decomposition (b) then there is an isometric C^∞-immersion $(V, g) \to \mathbb{R}^4$. (This is an exercise for the reader). For example, let (V_0, g_0) be a complete simply connected surface without conjugate points (for instance, $K_{g_0} \leq 0$). Then every closed convex C^∞-hypersurface $V \subset (V_0 \times \mathbb{R}, \bar{g} = g_0 + dx^2)$ admits an isometric C^∞-immersion into $\mathbb{R}^4$. [If $K_{g_0} \leq 0$, then metric spheres V in $(V_0 \times \mathbb{R}, \bar{g})$ are convex.]

Exercise. Let some metric g on a *closed* surface V admit an elliptic decomposition (b). Show V to be homeomorphic to S^2.

Let g be a positive *semi*definite quadratic C^∞-form on a closed connected surface V, which degenerate (i.e. has rank <2) on a non-empty closed (connected or not) curve $\Sigma^1 \subset V$. Under what condition can one C^∞-decompose g by

(c) $$g = \varphi_1\, dx_1^2 + \varphi_2\, dx_2^2?$$

or even better, by

(d) $$g = \varphi(dx_1^2 + dx_2^2)?$$

Do the C^∞-decomposition (c), (d) abide by the h-principle?

The existence of (c) is related to the following question. Consider a Riemannian C^∞-manifold W which is diffeomorphic to the product $W_0 \times [0, 1]$ for a closed manifold W_0. The gradient field of each C^∞-function f on W, which has no critical points and is constant on the boundary of W, defines a diffeomorphism, say D_f: $(W_0 \times 0) \to (W_0 \times 1)$. The question is whether *every* diffeomorphism D: $W_0 \times 0 \to W_0 \times 1$, which is isotopic to some D_f, is, in fact, *equal* to $D_{f'}$ for another function f' on W.

Exercise. Two positive semi-definite C^∞-forms on $\mathbb{R}^n$, say g_1 and g_2, are called equivalent (near the origin) if there is a fiberwise linear diffeomorphism α: $T(\mathbb{R}^n) \to T(\mathbb{R}^n)$ which keeps the origin $0 \in \mathbb{R}^n$ fixed and for which $\alpha^*(g_1) = g_2$ near $0 \in \mathbb{R}^n$. Study this equivalence relation between the forms on $\mathbb{R}^n$ which are induced by (generic) C^∞-maps $\mathbb{R}^n \to \mathbb{R}^q$ from positive definite forms g on $\mathbb{R}^q$.

(E) *$(k + l)$-Decompositions and Isometric C^∞-Immersion $(V, g) \to \mathbb{R}^{k+2l}$ for* $\dim V = n \geq 3$. The decomposition $(*)$ of (D),

$(*)$ $$g = \sum_{i=1}^{k} dx_i^2 + \sum_{i=k+1}^{k+l} \varphi_i\, dx_i^2$$

makes sense for manifolds V of any dimension n. We insist, as earlier, on C^∞-smoothness of the functions x_i and φ_i, and on the positivity of φ_i. The decomposition $(*)$ is called non-singular at $v \in V$ if the quadratic forms dx_i^2, $i = k + 1, \ldots, k + l$ span the space of the quadratic forms on the tangent space $T_v(V)$. If $l < n(n + 1)/2$, then $(*)$ is singular at all points $v \in V$, but for $l \geq n(n + 1)/2$ the pertinent singularity $\Sigma = \Sigma(x_i) \subset V$, $i = k + 1, \ldots, k + l$, has $\operatorname{codim} \Sigma \geq l - [n(n + 1)/2] + 1$ for *generic* functions x_i.

If a form g_0 admits a non-singular (i.e. nowhere singular) $(k + l)$-decomposition, then, obviously, all C^∞-forms g which are C^0-close to g also admit such a decomposition. Furthermore, each non-singular decomposition can be perturbed (for $n \geq 2$) into a stable one by applying the "generic" argumentation of (D). This gives [compare (D)] an isometric C^∞-immersion $(V, g) \to \mathbb{R}^{k+2l}$ for the above metrics g which are C^0-close to g_0 and for all compact manifolds V. This fact has certain merit only for $k + 2l < (n + 2)(n + 3)/2$, where the results of 3.1 do not apply, and so the interesting dimensions here are $n = 3$ and 4, for which $2n(n + 1) < (n + 2)(n + 3)$.

If V is a closed manifold, then non-singular decompositions do not exist at all for $l \leq n(n + 1)/2$, because of the critical points of the functions x_i, $i = k + 1, \ldots, k + l$. However, for $l > n(n + 1)/2$ these decompositions enjoy the h-principle (this is an exercise on the convexe integration) and so they are not hard to come by. Moreover, one has the following

Proposition. *An arbitrary C^∞-metric g on V decomposes into $g = dx_1^2 + \sum_{i=2}^{l} \varphi_i\, dx_i^2$, for $l = n(n + 1)/2$, such that this decomposition is stable near the singularity $\Sigma = \Sigma(x_i) \subset V$, $i = 2, \ldots, l$.*

The proof is a generalization of the steps (0)–(4) in (D‴). Namely, we start with generic x_i, $i = 2, \ldots, 4$. Then, by a local argument near $\Sigma = \Sigma(x_i) \subset V$ we achieve a stable decomposition near Σ. Finally, we extend this with a convex integration outside Σ. Working everything out is left to the reader.

Corollary. *An arbitrary compact Riemannian 3-dimensional $C^\infty(C^{\mathrm{an}})$-manifold (V, g) admits an isometric $C^\infty(C^{\mathrm{an}})$-imbedding $(V, g) \to \mathbb{R}^{13}$.* (The general theorem in 3.1.7 gives an immersion into $\mathbb{R}^{15}$.)

Exercise. Show that the C^∞-metrics g on the ball B^4 which are C^0-close to a flat metric g_0 admit isometric C^∞-imbeddings $(B^4, g) \to \mathbb{R}^{20}$. Generalize this to all compact parallelizable manifolds V^n with a non-empty boundary.

Question. Does every metric g admit a decomposition $g = \sum_{i=1}^{l} \varphi_i\, dx_i^2$ for $l = n(n + 1)/2$ which is stable near the singular set Σ? If this is so, then there are isometric C^∞-immersions $(V^3, g) \to \mathbb{R}^{12}$ and $(V^4, g) \to \mathbb{R}^{20}$.

Singular Decompositions. Take a compact normally oriented hypersurface H in a Riemannian manifold (V, g), $\dim V = n$. If the induced metric in H, say g', admits a *nonsingular* decomposition, $g' = \sum_{i=1}^{l} \varphi_i'(dx_i')^2$, then the metric g on a small tubular neighborhood $U \subset V$ of H also decomposes,

$$(**) \qquad g|U = dx_0^2 + \sum_{i=1}^{l} \varphi_i\, dx_i^2.$$

Indeed, let $P\colon U \to H$ be the normal projection and let $x_0(u) = \pm\mathrm{dist}(u, H)$, where the $\pm$ sign distinguishes the two components of $U \backslash H$. Take $x_i = x_i' \circ P$ and observe that the functions φ_i' on H extend to some φ_i on U such that $\sum_{i=1}^{l} \varphi_i\, dx_i^2 = g - dx_0^2$. Since every metric g' locally decomposes with $l = n(n - 1)/2$, $n - 1 = \dim H$, we get $(**)$ with this l near each point $v \in V$. In case the decomposition $(**)$ is stable, the genericity consideration suggests for $4l > n(n + 1)$ the existence of an isometric C^∞-immersion $V \to \mathbb{R}^{2l+1}$. This may be interesting for $n \leq 7$ where $n(n - 1) + 1 < (n + 2)(n + 3)/2$. Observe that the stability (and the due genericity) can sometimes be achieved by adding extra terms dx_i^2. Furthermore, the actual size of the neighborhood where $(**)$ exists can be estimated from below in terms of the geometry of V. Specific statements are contained in the following

Exercise. Let (V, g) be a complete Riemannian manifold with an absolute bound on the sectional curvature, $|K(V,g)| \leq C^2$ for some $C > 0$. Let all geodesic loops at a given point $v_0 \in V$ have length $\geq \delta$ for some $\delta > 0$ and let $B = B_{v_0}(R)$ be the ball in V of radius R around v_0. Prove for $R \leq \min(\delta/4, (10C)^{-1})$ the existence of a decomposition

$$g|B = dx_0^2 + \sum_{i=1}^{l} \varphi_i \, dx_i^2, \qquad l = n(n-1)/2.$$

Prove the existence of a *stable* decomposition

$$g|B = \sum_{i=1}^{k} dx_i^2 + \sum_{i=k+1}^{k+l} \varphi_i \, dx_i^2, \qquad k = 2n-2, \, l = n(n-1)/2.$$

Prove the existence of an isometric $C^\infty(C^{\mathrm{an}})$-imbedding $(B, g) \to \mathbb{R}^{k+2l+1}$ for the above k and l. (It is likely that a smaller k will do.)

Final Remarks and Questions. The $(k+l)$-decompositions seem interesting enough in their own right. One may study them by the techniques of 3.1. This will probably give the global existence theorem for all k and l, such that $k + 2l \geq (n+2)(n+3)/2$. One also expects a $(k+l)$-decomposition to exist for $k + 2l \geq n(n+1)/2$ on a small neighborhood $U \subset V$ of a given point in case the metric is C^{an}. Then (for $k = 0$) one would try to construct an isometric C^{an}-imbedding of (U, g) into the ε-ball $B_\varepsilon \subset \mathbb{R}^{2l}$ for all $l \geq n(n+1)/4$ and for *all* $\varepsilon > 0$. An especially interesting decomposition is $g = \sum_{i=1}^{3} \varphi_i \, dx_i^3$ on non-closed 3-dimensional manifolds (V, g), but little is known about its existence.

3.3 Isometric C^∞-Immersions of Pseudo-Riemannian Manifolds

Let V be an n-dimensional manifold with a quadratic C^∞-form g. The dimension of the zero part of g, called $[g]_0$, may be a non-constant function on V, denoted $n_0 = n_0(v) = [g|T_v(V)]_0 = n - \mathrm{rank}_v g$, and so the positive and negative parts of g may also have variable dimensions, denoted by $n_+(n) = [g_v]_+$ and $n_-(v) = [g_v]_- = n - n_0(v) - n_+(v)$.

We denote by $W = (W, h)$ a C^∞-smooth *pseudo-Riemannian* manifold which means $[h_0] \equiv 0$. Hence the dimensions $q_\pm = [h]_\pm$ are constant and $q_+ + q_- = q$.

Our analysis of the isometric immersion equation $f^*(h) = g$ for C^∞-maps f: $V \to W$ combines the Riemannian immersions techniques of 3.1 with the theory of sheaves (compare 2.2.2). Some results are even stronger for pseudo-Riemannian manifolds as we are aided in many constructions by the presence of *isotropic directions* in W on which the metric h is zero. For example, we construct global immersions $V \to W$ for $q \geq [n(n+3)/2] + 2$ for some V and W, (see 3.3.5), where similar Riemannian immersions require (at the present state of knowledge) $q \geq (n+2)(n+3)/2$. However, the h-principle is still unknown for free isometric immersions $V \to W$ for $\min(q_+, q_-) \leq n(n+1)/2$.

3.3.1 Local Pseudo-Riemannian Immersions

We start with two simple properties of submersions (which are not necessarily fibrations).

(A) *Let B be a finite dimensional polyhedron and let α: $A \to B$ be a microfibration* (see 1.4.2). *If the fiber $\alpha^{-1}(b)$ is k-connected for all $b \in B$ then the space of sections $B \to A$ is l-connected for $l \geq k - \dim B$ [(-1)-connected means non-empty]. Furthermore, if $\dim B \leq k + 1$, then each partial section $B_0 \to A$ extends to B for all sub-polyhedra $B_0 \subset B$.*

Proof. The sheaf Φ of sections $A \to B$ is flexible (see 1.4.2, 2.2.1) and the germs $\Phi(b)$, $b \in B$, are k-connected since α is a microfibration. The flexibility allows the induction by skeletons which reduces the problem to microfibrations over the m-cube $[0, 1]^m$, $m \leq \dim B$. Then the induction in m (compare 2.2.1) reduces the problem further to $m = 1$ where the proof is straightforward and left to the reader.

(A′) *Let B be a smooth manifold stratified by submanifolds $\Sigma_i \subset B$, $i = 0, 1, \ldots,$* (compare 1.3.2) *such that* $\operatorname{codim} \Sigma_i \geq i$. *Let α: $A \to B$ be a smooth submersion such that the fiber $\alpha^{-1}(b)$ is $(k - i)$-connected for all $b \in \Sigma_i$ and for all $i = 0, 1, \ldots$. Then the induced homomorphism on the homotopy groups α_*: $\pi_j(A) \to \pi_j(B)$ is an isomorphism for $j \leq k$ and an epimorphism for $j = k + 1$.*

Proof. To lift an element $\sigma \in \pi_j(B)$ to A we realize it by a generic smooth map x: $S^j \to B$, such that $\dim x^{-1}(\Sigma_i) \leq j - i$. Then we lift x to A by applying (A) over the strata $x^{-1}(\Sigma_i) \subset S^j$, $i = 0, 1, \ldots$. A similar lift applies to the balls B^j, $\partial B^j = S^{j-1}$ which implies the injectivity of α_* as well as the surjectivity.

Exercise. Show every submersion with (non-empty!) contractible fibers to be a Serre fibration.

(B) The study of the second jets of free isometric immersion reduces [see (C) below] to the following quasi-linear algebraic equations in $(m - n)$-tuples of vectors $X_\mu \in \mathbb{R}^{q_+ + q_-}$, $\mu = n + 1, \ldots, m$. Fix linearly independent vectors $X_1, \ldots, X_n$ in $\mathbb{R}^{q_+ + q_-}$ and take a subset, say M, of pairs (μ, ν) of integers $\mu = n + 1, n + 2, \ldots$, and $\nu = 1, 2, \ldots$, such that $\mu > \nu$.

Consider the equations

$$(*) \qquad \langle X_\mu, X_\nu \rangle = A_{\mu\nu}, \qquad (\mu, \nu) \in M,$$

where each $A_{\mu\nu}$ is a smooth function in $X_1, X_2, \ldots, X_\nu$. Denote by $R(m) \subset \mathbb{R}^{q(m-n)}$ for $m = n + 1, n + 2, \ldots$, the space of those *linearly independent* m-tuples of vectors $X_1, \ldots, X_n, X_{n+1}, \ldots, X_m$ in $\mathbb{R}^q$, where $X_1, \ldots, X_n$ are the fixed vectors and where the vectors $X_{n+1}, \ldots, X_m$ satisfy the above equations for $\mu \leq m$. This $R(m)$ is a smooth submanifold in $\mathbb{R}^{q(m-n)}$ and the natural projection Π_m: $R(m) \to R(m - 1)$ is a submersion whose fibers have dimension $s(m)$ which is the number of the pairs (m, ν) in M.

The space $R(m)$ is stratified by the rank of the pseudo-Euclidean form $h(X, X) = \langle X,X\rangle = \sum_{j=1}^{q_+} x_j^2 - \sum_{j=q_++1}^{q} x_j^2$ on the spans of m-tuples in $R(m)$ as follows. The stratum $R_r(m) \subset R(m)$ consists of those m-tuples $(X_1,\ldots,X_m)$ where m-rank $= r$ which is equavalent to $[h|\mathrm{Span}\{X_1,\ldots,X_m\}]_0 = r$. The projection Π_m sends $R_r(m)$ to the union of $R_{r-1}(m-1)$, $R_r(m-1)$ and $R_{r+1}(m-1)$ in $R(m-1)$. Then one can easily see the following inequality for the codimension $\mathrm{cd}_r(m) = \mathrm{codim}\, R_r(m)$.

$$(+) \qquad cd_r(m) \geq \min(cd_{r-1}(m-1) + m - s(m), cd_r(m-1), cd_{r+1}(m-1)),$$

which by induction implies

$$(++) \qquad \mathrm{codim}\, R_r(m) \geq r - n_0$$

for $n_0 = [h|\mathrm{Span}\{X_1,\ldots,X_n\}]_0$.

Next, the pullback $\Pi_m^{-1}(y)$ for $y \in R_r(m)$ is the difference $L\backslash L'$ of affine subspace L and L' in $\mathbb{R}^{q_+,q_-}$ such that $\dim L = q - s(m)$ and $\dim L' = r$. Hence, this pullback is $(q-m-r-1)$-connected. Thus (A′) applies to Π_m and an obvious induction on m shows the space $R(m)$ to be at least $(q - m - n_0 - 1)$-connected.

Denote by $R(m, A_+, A_-) \subset R(m)$, for some non-negative integers A_+ and A_-, the subspace of those m-tuples of vectors for which the numbers $N_0 = [h|\mathrm{Span}\{X_1,\ldots,X_m\}]_0$ and $N_\pm = [h|\mathrm{Span}\{X_1,\ldots,X_m\}]_\pm$ abide the inequalities $N_0 + N_+ \leq A_+$ and $N_+ + N_- \leq A_-$. Assume [in order to have $R(m, A_+, A_-)$ non-empty] $A_\pm \geq n_\pm + n_0$ for $n_\pm = [h|\mathrm{Span}\{X_1,\ldots,X_n\}]_\pm$ and $A_0 = A_+ + A_- - m \geq 0$. Take a point $\sigma = \{X_1,\ldots,X_{m-1}\} \in R_r(m-1)$ and let $N'_\pm = N'_\pm(\sigma) = [h|\mathrm{Span}\{X_1,\ldots,X_{m-1}\}]_\pm$. Since the map Π_m sends $R(m, A_+, A_-)$ into $R(m-1, A_+, A_-)$, the intersection $R^\sigma = \Pi_m^{-1}(\sigma) \cap R(m, A_+, A_-)$ is empty unless $r \leq A_0 + 1$ and $N'_\pm + r \leq A_\pm$. If $N'_\pm + r \leq A_\pm - 1$, then $R^\sigma = \Pi_m^{-1}(\sigma)$ and so R^σ is $(q - m - r - 1)$-connected. If $r \leq A_0$, then either $N'_+ + r \leq A_+ - 1$ or $N'_- + r \leq A_- - 1$. In the case where $N'_+ \leq A_+ - 1$ and $N'_- + r = A_-$, the space R^σ is homotopy equivalent to the sphere S^k for

$$k = q_+ - N'_+ - r - 1 = q_+ - m - N'_- = q_+ + A_- - m - r.$$

Similarly, if $N'_+ + r = A_+$, then R^σ is homotopy equivalent to S^k for $k = q_- + A_+ - m - r$. Hence the space R^σ is $(p' - r - m - 1)$-connected for

$$p' = \min(q, q_+ + A_-, q_- + A_+),$$

and for all $\sigma \in R_r(m-1) \cap R(m-1, A_+, A_-)$, where $r \leq A_0$.

Since $\mathrm{codim}\, R_r(m-1) \geq r - n_0$ we conclude (by induction in m) with (A′) to the p-connectivity of the space $R(m, A_+, A_-)$ for

$$p = \min(A_0 - n_0 - 1, p' - n_0 - m - 1).$$

(B′) **Exercise.** Let S be the space of linear embeddings $s\colon \mathbb{R}^m \to \mathbb{R}^{q_+,q_-}$ and set $S_i = \{s \in S | [s^*(h)]_0 = i\}$. Show that $\mathrm{codim}\, S_i = i(i+1)/2$. Then let

$$S[A_+, A_-] = \{s \in S | [s^*(h)]_0 + [s^*(h)]_\pm \leq A_\pm\},$$

and show this space $S[A_+, A_-]$ to be p-connected for

$$p = \min\left(\frac{A_0(A_0+3)}{2}, q - m - 1, q_+ + A_- - m - 1, q_- + A_+ - m - 1\right).$$

where $A_0 = A_+ + A_- - m$. Finally, prove the $[A_0(A_0+3)/2]$-connectedness of the space of those quadratic forms g on $\mathbb{R}^m$ for which $[g]_0 + [g]_\pm \le A_\pm$.

(C) Fix local coordinates $u_1, \ldots, u_n$ in a neighborhood of a point $v \in V$ and consider (germs of) isometric C^∞-immersions $f\colon (V,g) \to (W,h)$ with a fixed value $w = f(v)$. Then the first covariant derivatives $X_i = \nabla_i f(v) \in T_w(W) = \mathbb{R}^{q_+, q_-}$ satisfy for $\langle\ ,\ \rangle = \langle\ ,\ \rangle_h$

(1) $$\langle X_i, X_j \rangle = g_{ij}(v), \qquad i,j = 1, \ldots, n$$

and the second derivatives

$$X_{kl} = \nabla_{kl} f(v) \in T_w(W) = \mathbb{R}^{q_+, q_-}, \qquad 1 \le k \le l \le n,$$

satisfy the equations

(2) $$\begin{aligned} \langle X_{kl}, X_i \rangle &= A^{kli}(v) \\ \langle X_{kl}, X_{ij} \rangle &= D^{klij}(v) + \langle X_{kj}, X_{il} \rangle, \end{aligned}$$

for the pairs (k,l) and (i,j) for which $i \le j < k \le l$, where $A^{kli} = \frac{1}{2}(\partial_i g_{kl} + \partial g_{ki} - \partial_k g_{li})$ and where $D^{klij} = \partial_l A^{kij} - \partial_j A^{kil} + \langle X_k \tilde{R}((X_j, X_l), X_i) \rangle$ for the curvature tensor $\tilde{R}((\ldots))$ in (W,h) (compare 3.1.6). Denote by R the space of those solutions $\sigma = \{X_i, X_{kl}\}$ to (1) + (2) for which the vectors X_i and X_{kl} are linearly independent. Let $R_r \subset R$ consist of those σ for which $N_0(\sigma) = [h|\operatorname{Span}\sigma]_0 = r$ and let $R[A_+, A_-] \subset R$ be the subset of those σ for which $N_0(\sigma) + N_\pm(\sigma) \le A_\pm$. Observe that $\operatorname{Span}\sigma \subset \mathbb{R}^{q_+,q_-} = T_w(W)$ equals the osculating space $T_v^2(v) \subset T_w(W)$ of the map f and so the independence of X_i, X_{ij} reflects the freedom of f at the point v. (Compare 3.1.6.) Since the space of independent solutions $\{X_1, \ldots, X_n\}$ to the system (1) (obviously) is $(p-1)$-connected for $p \le q_\pm - n_\pm(v) - n_0(v)$, the discussion in (B) implies with (A) the following two properties of the system (1) + (2).

(C′) *If $p \le q_\pm - n_\pm(v)$ as well as $p \le q - [n(n+3)/2] - n_0(v)$, then the space R is $(p-1)$-connected. Furthermore, if $A_\pm \ge n_\pm + n_0$ and*

$$p \le \min(A_0 - n_0, q_+ + A_- - [n(n+3)/2] - n_0, q_- + A_+ - [n(n+3)/2] - n_0),$$

for $A_0 = A_+ + A_- - [n(n+3)/2]$, then the space $R(A_+, A_-)$ is $(p-1)$-connected.

(C″) *The subspace $R_r \subset R$ has*

$$\operatorname{codim} R_r \ge r - n_0.$$

(D) Let I be the space of (germs at v of) free isometric C^∞-immersions $f\colon \mathcal{O}p(v) \to W$, such that $f(v) = w$. The second jet defines a map $J\colon I \to R$ by $f \mapsto J_f^2(v)$. This J is a weak homotopy equivalence by the argument in 3.1.6. Moreover, the same argument shows the map $J\colon I(A_+, A_-) \to R(A_+, A_-)$ for $I(A_+, A_-) = J^{-1}(R(A_+, A_-))$ to be a *w.h.* equivalence as well. Hence, *the spaces I and $I(A_+, A_-)$ have the connectivity indicated in* (C″). *In particular, the space I is non-empty, provided $q_\pm \ge n_\pm + n_0$ and $q \ge [n(n+3)/2] + n_0$.*

Next, we combine (C″) with the transversality theorem in 2.3.2 and conclude to the following propositions.

(D′) *Let $f_0\colon (V, g) \to (W, h)$ be a free isometric C^∞-immersion. Then there is an arbitrary C^∞-small isometric deformation of f_0 to a free isometric C^∞-immersion f for which the restriction of the form h on the osculating bundle of f satisfies $[h|T_f^2(v)]_0 \leq n_0 + n$ at all points $v \in V$.*

(D″) **Exercises.** (a) Prove the local isometric C^α-immersibility into W of C^α-manifolds (V, g) for the Hölder classes C^α, $\alpha > 2$.

(b) Show that codim $R_r \geq n_0 + n^{(1/2)+\varepsilon}$ for every fixed $\varepsilon > 0$ and for $n \geq n(\varepsilon)$. Improve with this the connectivity estimate for $R(A_+, A_-)$.

Question. What is the first non-trivial homotopy group of the space $R(A_+, A_-)$? (The answer depends on g_{ij}. The simplest case is $g_{ij} = 0$, $A^{kli} = 0$ and $D^{klij} = 0$).

(E) *Local δ-Extensions.* Let $f_0\colon (v, g) \to (W, h)$ be a free isometric C^α-immersion. A field X in W along V (compare 3.1.2) is called a *δ-field* for $\delta = \pm 1$ or 0 if it is nowhere contained in the osculating bundle $T^2(V) \subset T(W)|V$ of f_0 and if it satisfies the following equations

$$\langle X, X \rangle = \delta$$

$$\langle X, \nabla_i f_0 \rangle = 0 \qquad (3)$$

$$\langle X, \nabla_{ij} f_0 \rangle = 0.$$

The first equation prescribes the length δ to X and the remaining equations express the *binormality* of X that is the normality to the osculating bundle.

If the form h is non-singular on $T^2(V)$, that is $N_0 = [h|T^2(V)]_0 \equiv 0$, then every non-zero solution X to (3) is nowhere contained in $T^2(V)$, but for singular forms $h|T^2(V)$ the "nowhere contained" condition is non-vacuous. The inequality $N_0(v) + N_+(v) = [h|T_v^2(V)]_0 + [h|T_v^2]_+ < q_+$ clearly is necessary and sufficient for the existence of a 1-field on a small neighborhood $\mathcal{O}p(v) \subset V$, while the inequality $N_-(v) + N_0(v) < q_-$ provides (-1)-fields. If both inequalities are fulfilled, $N_0(v) + N_\pm(v) < q_\pm$, then there is a 0-field near v.

An extension of f_0 to an immersion $f\colon V \times \mathbb{R} \to W$, for $V = V \times 0 \subset V \times \mathbb{R}$, is called *δ-cylindrical* if this f is isometric for the form $g \oplus \delta\, dt^2$ on $V \times \mathbb{R}$. The "δ-cylindrical" condition implies the Eqs. (3) for $X = \nabla_t f$ (compare 3.1.2). Conversely,

If the manifold (W, h) and the map f_0 are real analytic and if there is a continuous δ-field X along V, then there exists a real analytic δ-cylindrical extension f of f_0 to a small neighborhood $\mathcal{O}p V \subset V \times \mathbb{R}$.

Proof. Continuous δ-fields can be approximated by real analytic ones and so X is assumed C^{an} to start with. Next, we recall Janet's equations (see 3.1.2) for the field $X' = \nabla_{tt} f$,

$$\langle X', \nabla_t f \rangle = 0$$

$$\langle X', \nabla_i f \rangle = P_i \qquad (4)$$

$$\langle X', \nabla_{ij} f \rangle = P_{ij},$$

where $P_i = -\langle \nabla_t f, \nabla_{it} f \rangle$ and

$$P_{ij} = \langle \nabla_{it} f, \nabla_{jt} f \rangle + \langle \nabla_t f, R(\nabla_t f, \nabla_i f, \nabla_j f)$$

for the curvature tensor R of (W, h). If $\nabla_t f | V = X$, then the Eqs. (4) are (algebraically) solvable in X' on V. Thus the system (4) reduces to the Cauchy-Kovalevskaya form and, hence, it is solvable (in f) near $V \subset V \times \mathbb{R}$ with the initial conditions $f | V = f_0$ and $\nabla_t f | V = X$. These solutions $f\colon \mathcal{O}\!p\, V \to W$ are exactly (and only) δ-cylindrical extensions by the argument in 3.1.2. Q.E.D.

Exercise. Generalize the above to isometric non-cylindrical extensions and thus prove the existence of a local isometric C^{an}-immersion of $\mathcal{O}\!p(v) \subset V$ into W, provided $q_\pm \geq n_\pm(v) + n_0(v)$ and $q \geq [n(n+1)/2] + n_0(v)$. [Compare Friedman (1961).]

Assume $(V, g) = (V_0 \times \mathbb{R}^m, g_0 \oplus 0)$ for $\dim V_0 = n - m$ and let $q < [n(n+1)/2] + m$. Show that no isometric C^∞-immersion $\mathcal{O}\!p(v) \to W$ exists for *generic* C^{an}-metrics g_0 on V_0 and h on W. [One does not know the conditions which would guarantee isometric C^{an}-immersions $\mathcal{O}\!p(v) \to W$ for $n(n+1)/2 \leq q \leq [n(n+1)/2] + n_0(v) - 1$.]

A δ-field X is called *regular* (compare 3.1.2) if the vectors X, $\nabla_i X$, $\nabla_i f_0$ and $\nabla_{ij} f_0$ are linearly independent along V, which means the independence of the fields X, $\nabla_i X$ from the osculating bundle $T^2(V) \subset T(W)|V$. A solution X' of (4) along V is called *free* (relative to a given regular δ-field X) if X' is nowhere contained in $\operatorname{Span}\{T^2(V), X, \nabla_1 X, \ldots, \nabla_n X\}$.

Lemma. *If the immersion f_0 admits a C^1-smooth regular δ-field X and a continuous free solution X' of* (4) *then there exists a free δ-cylindrical C^∞-extension of f_0 to $\mathcal{O}\!p(V) \subset V \times \mathbb{R}$.*

Proof. One can assume the fields X and X' to be C^∞-smooth and then one obtains an infinitesimally isometric extension f to $V \times \mathbb{R}$, such that $\nabla_t f | V = X$ and $\nabla_{tt} f | V = X'$ (compare 3.1.2). The regularity of X and the freedom of X' imply the freedom of f near $V \subset V \times \mathbb{R}$ and then f can be perturbed to an isometric extension near V (see 2.3.6).

Corollary. *Let f_0 admit at some point $v \in V$ binormal vectors $Y, Y_1, \ldots, Y_n, Y'$ in $T_w(W) \supset T_v^2(V)$, $w = f_0(v)$, which are linearly independent from T_v^2 and such that $\langle X, X \rangle = \delta$. Then there exists a free δ-cylindrical C^∞-extension of f_0 to $\mathcal{O}\!p(v) \subset V \times \mathbb{R}$.*

Proof. Let us differentiate (3) and obtain the following equations for the derivatives $\nabla_k X$,

$$\begin{aligned} \langle \nabla_k X, X \rangle &= 0 \\ \langle \nabla_k X, \nabla_i f_0 \rangle &= -\langle X, \nabla_{ik} f_0 \rangle \\ \langle \nabla_k X, \nabla_{ij} f_0 \rangle &= -\langle X, \nabla_{ijk} f_0 \rangle. \end{aligned} \tag{5}$$

Observe that adding a binormal field to $\nabla_k x$ does not affect these equations. Hence, a field X for which $X(v) = Y$ admits a perturbation to a δ-field Z, such that $\nabla_k Z(v) = \nabla_k X(v) + \alpha Y$ for $k = 1, \ldots, n$ and for any real number α. If this α is sufficiently large, then the vectors $\nabla_k Z(v)$ are independent from $Z(0)$ and T_v^2 which implies the regularity of Z on $\mathcal{O}p(v) \subset V$. Next, the solutions X' of (4) on V are also unsensitive to adding binormal vectors and so the lemma applies to Z and $Z' + \alpha Y'$.

(E′) **Lemma.** *If $N_0(v) + N_\pm(v) < q_\pm$ and $q > [(n+1)(n+4)/2] + N_0(v)$, then there is a free 0-cylindrical C^∞-extension of f_0 to $\mathcal{O}p(v) \subset V \times \mathbb{R}$.*

Indeed the above Y, Y_i and Y' obviously exist in this case.

(E″) **Corollary.** *Let $\tilde{g}$ be an arbitrary C^∞-form on $V \times \mathbb{R}$ which extends g on $V \subset V \times \mathbb{R}$. Then the above inequalities imply the existence of a free isometric C^∞-immersion of $\mathcal{O}p(v) \subset V \times \mathbb{R}$ to W which extends f_0.*

Proof. Let $f \colon V \times \mathbb{R} \to W$ be a free 0-cylindrical immersion and consider the diffeomorphisms $P_\varepsilon \colon V \times \mathbb{R} \to V \times \mathbb{R}$ given by $P_\varepsilon(v, t) = (v, \varepsilon t)$ for $\varepsilon > 0$. If $\varepsilon \to 0$ then the induced metric $\tilde{g}_\varepsilon = P_\varepsilon^*(\tilde{g})$ on $V \times \mathbb{R}$ C^∞-converges to the 0-cylindrical metric $\tilde{g}_0 = g \oplus 0 = P_0^*(\tilde{g})$ as $\tilde{g} | V = g$. Hence, there is a small C^∞-perturbation f_ε of f which is an isometric immersion of $\mathcal{O}p\, v \subset (V \times \mathbb{R}, \tilde{g}_\varepsilon)$ to W (see 2.3.6). Of course, one uses for a non-compact manifold V a fast decaying function $\varepsilon = \varepsilon(v)$ rather than a constant). Then the composed map $f \circ P_\varepsilon^{-1}$ is isometric on $(\mathcal{O}p\, V, \tilde{g})$ and thus the isometric immersion problem for $\tilde{g}$ is reduced to that for the 0-cylindrical metric of the Lemma.

3.3.2 Global Immersions

We start with an open manifold V which admits a submersion $V \to V_0$ with open fibers of a positive dimension, such that the form g on V is induced from some C^∞-form g_0 on V_0. Then free isometric C^∞-immersions (V, g) W satisfy the h-principle (see 2.2.2), and we obtain with (A), (D) and (D′) of 3.3.1 the following

(A) **Theorem.** *If $q_\pm \geq n_\pm + n_0 + n + p$ and $q \geq [n(n+5)/2] + n_0 + p$ then the space of free isometric C^∞-immersions $V \to W$ is p-connected. In particular, these inequalities with $p = -1$ imply the existence of a free isometric C^∞-immersion $f \colon V \to W$ in every homotopy class of maps $V \to W$. Moreover, one can choose this f, such that $N_0 = [h | T_f^2(V)]_0 \leq n + n_0$ and $N_0 + N_\pm \leq A_\pm$, provided the numbers A_+ and A_- satisfy $A_\pm \geq n_\pm + n_0$, $A_+ + A_- \geq [n(n+5)/2] + n_0 - 1$, $q_+ + A_- \geq [n(n+5)/2] + n_0 - 1$ and $q_- + A_+ \geq [n(n+5)/2] + n_0 - 1$.*

Remark. We have the p-connectedness rather than the $(p-1)$-connectedness since the pertinent section of V into the space of jets is needed only over the $(n-1)$-skeleton of some triangulation of V.

(A′) **Exercises.** (a) Assume the submersion $V \to V_0$ has contractible fibers and prove the existence of a free isometric C^∞-immersion $f \colon V \to W$ for $q_\pm \geq n_\pm + n_0 + \dim V_0$ and $q \geq [n(n+3)/2] + n_0 + \dim V_0$.

(b) Assume $n_0(v)$ to be constant in $v \in V$. Then the tangent bundle $T(V)$ splits into the Whitney sum, $T(V) = T_0 \oplus T_+ \oplus T_-$, where $T_0 = \ker g$ and the bundles T_+ and T_- (non-uniquely) represent the positive and the negative parts of g. The bundle $T(V)$ is called *g-trivial* if the bundles T_0, T_+ and T_- are trivial, and this definition obviously extends to an arbitrary bundle with a quadratic form of constant rank. Now, assume $T(V)$ to be g-trivial and consider a continuous map $f_0: V \to W$, for which the induced bundle $f_0^*(T(W)) \to W$ is $f_0^*(h)$-trivial. (Both conditions are satisfied for *contractible* manifolds V.) Prove f_0 to be homotopic to a free isometric C^∞-immersion $f: V \to W$, provided $q_+ \geq n_+ + n_0$, $q_- \geq n_+ + n_0$ and $q \geq [n(n+3)/2] + n_0$.

(c) Let $\tilde{g}$ be an *arbitrary* quadratic C^∞-form on V, let $V_0 \subset V$ be a smooth simplicial subcomplex of positive codimension and let $U \subset V$ be a regular neighborhood of V_0. Construct a smooth map $P: U \to U$, such that $P(U) = V_0$, and for which the induced form $\tilde{g}_0 = P_0^*(g)$ on U has the following properties.

(i) Free isometric C^∞-immersions $(U, \tilde{g}_0) \to (W, h)$ satisfy the h-principle for all manifolds (W, h).

(ii) Every free isometric C^∞-immersion $f_0: (U, \tilde{g}_0) \to (W, h)$ admits a perturbation to a free isometric C^∞-immersion $f: (\mathcal{O}\!p\, V_0, \tilde{g}) \to (W, h)$ [compare (E″) in 3.3.1]. Prove with the above the existence of a small neighborhood $\mathcal{O}\!p\, V_0 \subset V$ which admits a free isometric C^∞-immersion $(\mathcal{O}\!p\, V_0, \tilde{g}) \to (W, h)$, provided $q_\pm \geq [\tilde{g}]_\pm + [\tilde{g}]_0 + n$ and $q \geq [n(n+5)/2] + [\tilde{g}]_0$.

(A″) **Remark.** The above theorem applies to the product $(V \times \mathbb{R}, g \oplus 0)$ for an *arbitrary* manifold (V, g). Thus one obtains a *free isometric C^∞-immersion $(V, g) \to W$ for all (V, g), provided $q_\pm \geq n_\pm + n + n_0 + 1$ and $q \geq [n(n+7)/2] + n_0 + 3$. Moreover, under these assumptions an arbitrary continuous map $f_0: V \to W$ admits a fine C^0-approximation by free isometric C^∞-immersions.* [This follows from the theory of sheaves in 2.2.2. See (B′) below and 3.3.5 for sharper results.]

Exercise. Let the bundles $T(V)$ and $f_0^*(T(V))$ be respectively g- and $f^*(h)$-trivial. Obtain the above approximation for $q_\pm \geq n_\pm + n_0 + 1$ and $q \geq [n(n+5)/2] + n_0 + 3$.

(B) Let us give a geometric interpretation of the second jets of immersions $V \to W$. To do this we introduce the (abstract) *osculcting bundle* $T^2(V) \to V$ whose sections $s: V \to T^2(V)$ are second order differential operators on functions ρ on V, say $\varphi \mapsto s(\varphi)$, such that $s(\text{const}) = 0$. If $u_1, \ldots, u_n$ are local coordinates in V, then every such operator s is a combination $s = \sum_{i=1}^n s_i \partial_i + \sum_{i,j=1}^n s_{ij} \partial_{ij}$, for some functions s_i and s_{ij}, where $\partial_i = \partial/\partial u_i$ and $\partial_{ij} = \partial^2/\partial u_i \partial u_j$. Hence, the operators ∂_i and ∂_{ij} at $v \in V$ form a basis in $T_v^2(V)$ which transforms according to the chain rule with a change of coordinates. The bundle $T^2(V)$ is a natural bundle (see 2.3.7) but it is not a tensor bundle. However, the tangent bundle (which is spanned by ∂_i) imbeds into $T^2(V)$ and the quotient $T^2(V) | T(V)$ is canonically isomorphic to the symmetric square $(T(V)^2$. Furthermore, every Riemannian (or pseudo-Riemannian) metric on V defines a splitting $T^2(V) = T(V) \oplus (T(V))^2$ by assigning the *covariant* derivatives ∇_{ij} for a basis in $(T(V))^2 \subset T^2(V)$.

Each smooth map $f: V \to (W, h)$ defines the (full) *second differential* D_f^2:

$T^2(V) \to T(W)$ which sends $T^2(V)$ onto the (geometric) osculating sub-bundle $T_f^2 \subset T(W)|V$. Namely, $\partial_i \mapsto \nabla_i f$ and $\partial_{ij} \mapsto \nabla_{ij} f$ for the covariant derivatives ∇_i and ∇_{ij} in (W, h).

If V is endowed with a quadratic differential form g, then a linear map L: $T_v^2(V) \to T_w(W)$ is called *skew-isometric* if the vectors $X_i = L(\partial_i)$ and $X_{ij} = L(\partial_{ij})$ satisfy the equations (1) + (2) of (C) in 3.3.1.

The parametric h-principle for free isometric immersions $f: (V, g) \to (W, h)$ claims, with this terminology, the map $f \mapsto D_f^2$ to be a *weak homotopy equivalence* of the space of free isometric immersions to the space of fiberwise injective skew-isometric homomorphisms $T^2(V) \to T(W)$. This h-principle does hold true for the forms g induced by submersions $V \to (V, g_0)$ [see in (A)]. Furthermore,

(B′) *If* $q_\pm \geq [n(n+1)/2] + n_\pm + n_0$ *and* $q \geq n(n+3) + n_0 + 3$, *then the above h-principle holds true for all manifolds* (V, g).

Indeed, under these conditions each free isometric C^∞-immersion $(\mathcal{O}p(v), g) \to W$ extends to a free 0-cylinder [see (E′) in 3.3.1] and the results in 2.2.4 apply.

It follows for the above q and $q_\pm$ that *the space of free isometric C^∞-immersions $V \to W$ is $(p-1)$-connected for all forms g provided*

$$p = \min\left(q_\pm - n_+ - n_0, q - \frac{n(n+5)}{2} - n_0\right).$$

[Compare (A).]

Unfortunately, this does not yield isometric immersions $V \to W$ for $q < n(n+3) + 3$ [compare (A″)]. These are obtained with the following analysis of skew-symmetric homomorphisms φ: $T^2(V) \to T(W)$. Let $\varphi^*(h)$ denote the induced form in $T^2(V)$ and let $N_0 = N_0(\varphi, v) = [\varphi^*(h)|T_v^2(v)]_0$ and $N_\pm = N_\pm(\varphi, v) = [\varphi^*(h)|T_v^2(v)]_\pm$. Then with the proof of (B′) we obtain the following

Lemma. *Let a continuous map f_0: $V \to W$ lift to a fiberwise injective skew-isometric homomorphism φ: $T^2(V) \to T(W)$, such that $N_0 + N_\pm < q_\pm$ and $N_0 < q - [(n+1)(n+4)/2]$. Then f_0 admits a C^0-approximation by free isometric C^∞-immersions f: $V \to W$.*

Warning. If a free isometric immersion f: $V \to W$ is extended to a 0-cylinder $\tilde{f}$: $V \times \mathbb{R} \to W$ to which a diffeomorphism d: $V \times \mathbb{R} \to V \times \mathbb{R}$ commuting with the projection $V \times \mathbb{R} \to V$ is applied, then the resulting (free isometric) immersion $f' = \tilde{f} \circ d | V$ may have the dimensions $N_\pm$ and N_0 different from those of f. Thus, there is no control over $N_0(f) \overset{\text{def}}{=} N_0(D_f^2)$ and $N_\pm(f)$ with the techniques of 2.2.4, and there isn't (?) any h-principle in sight to improve the mere approximation claim of the lemma.

The space of skew-isometric injective homomorphisms $T_v^2(V) \to T_w(W)$, $w = f_0(v)$, is identical to the space R in (C) of 3.3.1, while the inequalities $N_0 + N_\pm < q_\pm$ distinguish the subspace $R(q_+ - 1, q_- - 1)$. Furthermore, the bound $N_0 < r_0 = q - [(n+1)(n+4)/2]$ defines the complement

$$R(r_0, q_+ - 1, q_- - 1) \overset{\text{def}}{=} R(q_+ - 1, q_- - 1) \setminus \left(\bigcup_{r \geq r_0} R_i \right).$$

Hence, this complement is $(n - 1)$-connected by (C′) in 3.3.1, provided

$$(*) \qquad q_\pm \geq n_\pm(v) + n_0(v) + n \qquad q \geq \frac{n(n+5)}{2} + n_0(v)$$

and

$$(**) \qquad \operatorname{codim} R_r \geq n + 1 \qquad \text{for } r \geq r_0.$$

Corollary. *If the inequalities* (∗) *and* (∗∗) *are satisfied, then every continuous map* $f_0\colon V \to W$ *can be* C^0*-approximated by free isometric* C^∞*-immersions.*

Remark. The bound $\operatorname{codim} R_r \geq n_0(v) + r$ in (C″) of 3.3.1 reduces (∗∗) to the inequality $q \geq [(n + 1)(n + 6)/2] + n_0(v)$, which already appears in (A″). But the estimate $\operatorname{codim} R_r \geq n_0(v) + n^{(1/2)-\varepsilon}$ for $n \to \infty$ (see (D″)) leads to more interesting results. See (B″) below for another approach.

Exercises. (a) Let $V_i = \{v \in V | [g | T_v(v)]_0 = i\}$ and let $\operatorname{codim} V_i \geq i$. [If the form g is *generic* then $\operatorname{codim} V_i \geq i(i + 1)/2$, see (B′) in 3.3.1.] Approximate f_0 by free isometric C^∞-immersions under the assumptions $q_\pm \geq n_\pm + 2n$ and $q \geq n(n + 7)/2$.

(b) Let V be a *parallelizable* manifold, let non-negative integers A_+ and A_- satisfy $A_+ + A_- = n(n + 1)/2$ and let $q_\pm \geq A_\pm + n_\pm + n_0 + n$. Show that an arbitrary continuous map $f_o\colon V \to W$ lifts to a fiberwise injective skew-isometric homomorphism $\varphi\colon T^2(V) \to T(W)$ for which $N_+ \geq A_+ + n_+$ and $N_- \geq A_- + n_-$. Then obtain the C^0-approximation, provided $q_\pm \geq n_\pm + n_0 + n$ and $q \geq [n(n + 7)/2] + n_0$.

Hint. Use a quadratic form $\tilde{h}$ on $T^2(V)$, such that

(i) $\ker \tilde{h} = T(V) \subset T^2(V)$
(ii) $\tilde{h}$ is 4-symmetric as a quadratic form on $(T(V))^2 = T^2(V)/T(V)$ (compare 3.1.5 and 3.3.3).
(iii) $[\tilde{h}]_+ = A_+$ and $[\tilde{h}]_- = A_-$.

(B″) *Non-free* 0-*Extensions.* The topological techniques in 2.2 apply to the (microflexible!) sheaf Φ of those isometric C^∞-immersions $f\colon (V \times \mathbb{R}, g \oplus 0) \to (W, h)$ at which the metric inducing operator $f \mapsto f^*(h)$ is infinitesimally invertible. The sheaf Φ strictly includes the free isometric immersions $V \times \mathbb{R} \to W$; moreover, Φ is sometimes non-empty for $q < (n + 1)(n + 4)/2$, $n = \dim V$, where no free immersion $V \times \mathbb{R} \to W$ is possible [see (E′) in 2.3.8]. Furthermore, a non-free immersion $(V, g) \to (W, h)$ obtained with Φ can be C^∞-perturbed for $q \geq [n(n + 5)/2] + n_0$ to a free isometric one (see the transversality theorem in 2.3.2). Thus one may obtain free isometric immersions $V \to W$ without (being able to prove) the h-principle for them. The sheaf Φ is expected to be non-empty (and thus useful) for

$q > [n(n+3)/2] + n_0 + 2$, but the complexity of the jet structure of Φ makes such a result hard to prove.

Exercise. Prove Φ to admit a global section for $q_\pm \geq n_\pm + n + n_0 + 1$, $q \geq [n(n+7)/2] + n_0 + 2$ and then extend (A″) to $q = [n(n+7)/2] + n_0 + 2$.

3.3.3 Immersions with a Prescribed Curvature and the C^1-Approximation

Let us generalize Riemannian constructions of 3.1 to pseudo-Riemannian manifolds.

(A) $(\delta \times \delta)$-*Extensions*. Fix $\delta = -1$ or 1 and let $f: (V, g) \to (W, h)$ be a free isometric C^∞-immersion. Let binormal mutually orthogonal fields X, X' and Y along V satisfy

(i) $\langle X, X \rangle = \langle Y, Y \rangle = \delta$;
(ii) the field Y is orthogonal to the covariant derivatives $\nabla_i X$, $i = 1, \ldots, n$, for all systems of local coordinates u_i in V;
(iii) the fields X, $\nabla_i X$, X' and Y are independent from the osculating bundle $T^2(V) \subset T(W)|V$.

(A′) **Lemma.** *There exists a C^∞-immersion $F: V \times \mathbb{R}^2 \to W$, such that* (a) $F|V = f$ *for* $V = V \times 0$ *and* $\nabla_t F|V = X$, $\nabla_\Theta F|V = Y$, *where t and Θ are the Euclidean coordinates in* $\mathbb{R}^2 = \mathbb{R} \times \mathbb{R}$.
(b) *The map F is infinitesimally isometric of infinite order along $V \subset V \times \mathbb{R}^2$ for the form $g + \delta(dt^2 + d\Theta^2)$ on $V \times \mathbb{R}^2$.*

Proof. First we extend f to a free infinitesimally isometric map $\tilde{f}$: $(V \times \mathbb{R}, g + \delta\, dt^2) \to W$, such that $\nabla_t \tilde{f}|V = X$ and such that the second derivative $\nabla_{tt} \tilde{f}|V$ is orthogonal to Y. We require, moreover, this derivative to be independent from the span of $T^2(V)$ and the vector X, $\nabla_i X$ snd Y. The existence of $\tilde{f}$ is immediate with the discussion in (E) of 3.3.1 which also insures an extension of $\tilde{f}$ to the required F (compare 3.1.7). Q.E.D.

Now, we claim the above tripples of fields X, X' and Y to satisfy the h-principle. Indeed, the only *differential* relation is imposed on X and it requires the independence of the derivatives $\nabla_i X$ from the span of $T^2(V)$ and the fields X, X' and Y. The pertinent relation $\mathscr{R}$ lies *over* (see 2.4.3) the 1-jets of fields X which are normal to $T^2(V)$ and independent from $T^2(V)$. Namely, the 1-jets in question are given at each point $v \in V$ by the vectors X and X_i, $i = 1, \ldots, n$, where $X \in T_w(W)$, $w = f(v)$, is a binormal vector independent from $T_v^2(V)$, and where the vectors X_i satisfy

$$
(*) \qquad \begin{aligned}
\langle X_i, X \rangle &= 0 \\
\langle X_i, \nabla_k f \rangle &= -\langle X, \nabla_{ik} f \rangle \\
\langle X_i, \nabla_{kl} f \rangle &= -\langle X, \nabla_{kli} f \rangle,
\end{aligned}
$$

[compare (5) in (E) of 3.3.1]. The relation $\mathscr{R}$ is given over V by the vectors X, X_i, X' and Y which satisfy (i)–(iii) with X_i in place $\nabla_i X$ and where X_i satisfy $(*)$. This

$\mathscr{R}$ goes to the 1-jets by the map $(X, X_i, X', Y) \to (X, X_i)$. A straightforward check up shows $\mathscr{R}$ to be ample over the first jet space (of vectors X) and the C^0-dense h-principle follows (see 2.4.3).

Next, we impose an additional condition on our fields over an open subset $U \subset V$ with local coordinates $u_1, \ldots, u_n$. This is done with the linear combinations $Z_i = Z_i(s, c) = s(u_i)\nabla_i X + c(u_i)\nabla_i Y$, $i = 2, \ldots, n$, by requiring the fields Z_i to be independent from the span of $T^2(V)$, X, X' and Y for all pairs of continuous functions $s(u_1)$ and $c(u_1)$, such that $s^2 + c^2 = 1$, (compare 3.1.7).

(A″) **Lemma.** *Triples (X, X', Y) which satisfy the above $\mathscr{R}$ over U as well as the additional condition abide by the C^0-dense h-principle*

Proof. The new condition, called $\mathscr{R}'$, is differential in X and in Y and it naturally projects to $\mathscr{R}$, say by the map $\rho'\colon \mathscr{R}' \to \mathscr{R}$. Thus $\mathscr{R}$ also lies over the 1-jets of fields X and it is easily seen to be ample over this space of 1-jets. Hence, every section $\varphi'\colon U \to \mathscr{R}'$ can be homotoped to a section $\varphi\colon U \to \mathscr{R}'$ whose projection to $\mathscr{R}$ is holonomic which means the holonomy of the X-component of (X, X', Y). Now we want the Y-component also to be holonomic. For every fixed X, the condition $\mathscr{R}'$ becomes an additional condition on Y, called $\mathscr{R}'_X$, over the 1-jets of fields Y. This one also is ample (over the jets of fields Y) and so there is the required homotopy of φ within $\mathscr{R}'_X$ to a holonomic [over the space of 1-jets of triples (X, X', Y)] section. Q.E.D.

Next, sections $U \to \mathscr{R}'$ are easily obtained with *binormal* fields X, X_i, X', Y and Y_i over U, $i = 1, \ldots, n$, which satisfy the following four conditions:

(1) *X, X' and Y are mutually orthogonal and* $\langle X, X\rangle = \langle Y, Y\rangle = \delta$;
(2) $\langle X_i, Y\rangle = 0$, $i = 1, \ldots, n$;
(3) *the fields X, X_i X', $i = 1, \ldots, n$, are independent from* $T^2(V)$;
(4) *the fields X, X', Y are $Z_i = sX_i + cY_i$, $i = 2, \ldots, n$ are independent* from $T^2(V)$ for all continuous functions $s = s(u_1)$ and $c = c(u_1)$, such that $s^2 + c^2 = 1$.

Such fields are easy to construct [compare (A) and (C′) in 3.3.1] for

$$(**) \qquad q \geq \frac{(n+2)(n+3)}{2} + N_0(f) + n,$$

where $N_0 = [h \mid T^2(V)]_0$, provided

$$(+) \qquad q_+ \geq N_+(f) + N_0(f) + 2 + n$$

in the case $\delta = +1$, and

$$(-) \qquad q_- \geq N_-(f) + N_0(f) + 2 + n$$

for $\delta = -1$, where $N_\pm = [h \mid T^2(V)]_\pm$. Therefore, the Lemmas (A′) and (A″) yield the following

(A‴) **Corollary.** *Let $f\colon (V, g) \to (W, h)$ be a free isometric C^∞-immersion which satisfies the inequalities (**) and (+) over $U \subset V$. Then $f \mid U$ extends to C^∞-map F: $U \times \mathbb{R}^2 \to W$ such that*

(a) *F is infinitesimally isometric of infinite order along $U = U \times 0 \subset U \times \mathbb{R}^2$ for the metric $g + \delta(dt^2 + d\Theta^2)$, where $\delta = +1$;*
(b) *the vectors $\nabla_i F, \nabla_{ij} F, \nabla_t F, \nabla_\Theta F$ and $Z_k = s(u_1)\nabla_{tk} F + c(u_1)\nabla_{\Theta k} F$, for $1 \leq i, j \leq n$ and $k = 2, \ldots, n$, are linearly independent. Furthermore, the same is true for $\delta = -1$ with $(-)$ instead of $(+)$.*

Exercises. (a) Assume $N_0(f) \equiv 0$ and prove the above corollary for $q \geq (n+2)(n+3)/2$ and $q_\pm \geq N_\pm + 2$.

(b) Generalize the Riemannian extension theorems in 3.1.6 to the pseudo-Riemannian (i.e. $[g]_0 \equiv 0$) case.

(B) *Adding* $\delta\varphi^2(du_1)^2$. Let $s = \sin \varepsilon u_1$, $c = \cos \varepsilon u_1$ and let $f_\varepsilon(u_1, \ldots, u_n) = F(u_1, \ldots, u_n, \varepsilon\varphi s, \varepsilon\varphi c)$, where F is a smooth map $U \times \mathbb{R}^2 \to W$ which satisfies (a) and (b) of the above Corollary and where $\varphi\colon U \to \mathbb{R}$ is a C^∞-function with a compact support. Then a straightforward computation gives us the following formulae (compare 3.1.5, 3.1.7):

$$\nabla_1 f_\varepsilon = \nabla_1 F + \varphi(c\nabla_t F - s\nabla_\Theta F) + \varepsilon B_1,$$

$$\nabla_i f_\varepsilon = \nabla_i F + \varepsilon B_i, \qquad i = 2, \ldots, n.$$

$$(+) \qquad \nabla_{11} f_\varepsilon = \nabla_{11} F - \varepsilon^{-1}\varphi(s\nabla_t F + c\nabla_\Theta F) + \frac{\partial\varphi}{\partial u_1}(c\nabla_t F - s\nabla_\Theta F) + \varphi B_{11},$$

$$\nabla_{1i} f_\varepsilon = \nabla_{1i} F + \varphi(cF_{ti} - sF_{\Theta i}) + \frac{\partial\varphi}{\partial u_i}(cF_t - sF_\Theta) + \varepsilon B_{1i}, \qquad i = 2, \ldots, n,$$

$$\nabla_{ij} f_\varepsilon = \nabla_{ij} F + \varepsilon B_{ij},$$

where the fields B_i and B_{ij} stay bounded for $\varepsilon \to 0$.

It follows that the maps f_ε are *uniformely free* on U for $\varepsilon \to 0$ (compare 3.1.7). Furthermore, the induced quadratic form is

$$f_\varepsilon^*(h) = g + \delta\varphi^2\, du_1^2 + \varepsilon^2\, d\varphi^2 + \mu(\varepsilon),$$

where $\varepsilon^{-k}\|\mu(\varepsilon)\|_{C^k} \to 0$ as $\varepsilon \to 0$ for all $k = 0, 1, 2, \ldots$. Now, we argue as in 3.1.5 and 3.1.7 and conclude to the following

(B') **Lemma.** *If $\varepsilon > 0$ is sufficiently small then there is a C^∞-perturbation f' of f_ε, such that $(\bar{f}')^*(h) = g + \delta\varphi^2\, du_1^2$ and such that f' equals f outside a compact subset in U.*

(C) *Adding $\delta\varphi^2(du_1)^4$ to the Curvature of f.* Consider two skew-isometric (see 3.3.2) homomorphisms ψ_1 and ψ_2 of $T^2(V)$ to $T(V)$ and let $\Delta = \Delta(\psi_1, \psi_2) = \psi_2^*(h) - \psi_1^*(h)$. This form Δ (obviously) vanishes on $T(V) \subset T^2(V)$ and so it is, in fact, defined on the symmetric square $(T(V))^2 = T^2(V)/T(V)$. Furthermore, the quadratic form Δ on $(T(V))^2$ is 4-symmetric when viewed as a 4-form on $T(V)$ (compare 3.1.5).

Now we use $\varepsilon^2\varphi$ instead of $\varepsilon\varphi$ in $(+)$ and we conclude by the argument in 3.1.5. to the following

(C′) **Lemma.** *There exists a C^∞-smooth family of free isometric C^∞-immersions $f_{\varepsilon,\tau}\colon V \to W$ for small $\varepsilon > 0$ and $0 \le \tau \le \varepsilon$, such that $f_{\varepsilon,0}$ equals the above f for all $\varepsilon > 0$ and $f_{\varepsilon,\tau} = f$ outside U for all ε and τ. Furthermore, the maps $f_{\varepsilon,\tau}$ C^1-converge to f for $\varepsilon \to 0$ and the form $\Delta(D_f^2, D_{f_{\varepsilon,\tau}}^2)$ C^0-converges to $\delta\varphi^2(du_1)^4$, while the C^2-distance between f and f_τ is bounded by* const $\varphi^2(du_1)^4$ C^0 *for all ε and τ.*

(D) *Admissible Deformations of the Curvature.* Consider a continuous homotopy of forms H_t on $T^2(V)$, $t \in [0, \infty]$, such that $\Delta = \Delta(v, t_1, t_2) = (H_{t_2} - H_{t_1}) T_v^2(V)$ for $0 \le t_1 \le t_2 < \infty$ has $\ker \Delta \supset T(V)$ and the associated form on $(T(V))^2$ is 4-symmetric. Denote by $\tilde{U}_\pm \subset V \times [0, \infty)$ the (open) subset of those pairs (v, t), where $[H_t | T_v^2(V)]_0 \le q - [(n+2)(n+3)/2] - n$ as well as $[H_t | T_v^2(V)]_\pm + [H_t | T_v^2(V)]_0 \le q_\pm - 2 - n$. Call the homotopy H_t *admissible* (*semi-admissible*) if the form $\Delta(v, t, t + t')$ satisfies for all small $t' = t'(v, t) \ge 0$,

$$\Delta(v, t, t + t') = \sigma \Delta_+,$$

where the form Δ_+ (which may depend on v, t and t') is positive definite (semi-definite) on $(T(V))^2 = T^2(V)/(T(V)$ and where the number $\sigma = \sigma(v, t)$ abides: $\sigma \ge 0$ for (v, t) outside $\tilde{U}_-$ and $\sigma \le 0$ outside $\tilde{U}_+$. In particular $\Delta = 0$ for all $(v, t) \in (V \times [0, \infty)) \setminus (\tilde{U}_+ \cup \tilde{U}_-)$ and all $t' \ge 0$.

The following fact generalizes the Riemannian curvature theorem [see (A) in 3.1.5].

(D′) **Proposition.** *Let $f_0\colon (V, g) \to (W, h)$ be a free isometric C^∞-immersion and let H_t, $0 \le t \le \infty$, be an admissible homotopy of forms on $T^2(V)$, such that $H_0 = (D_{f_0}^2)^*(h)$. Then there exists a C^∞-homotopy of free isometric C^∞-immersions $f_t\colon V \to W$, $0 \le t < \infty$, with the following three properties.*

(i) *The immersions f_t, $0 \le t < \infty$, lie in a given C^1-fine neighborhood of f_0 in the space of C^1-map $V \to W$.*
(ii) *The homotopy $(D_{f_t}^2)^*(h)$ lies in a given C^0-fine neighborhood of H_t where the homotopies of forms are viewed as C^0-section of the pertinent bundle over $V \times [0, \infty)$.*
(iii) *The immersions f_t C^2-converge for $t \to \infty$ to a free C^2-immersion f_∞. This f_∞ is necessarily isometric and $D_{f_\infty}^2(h) = H_\infty$ due to* (ii).

Proof. The forms $(du)^2$ for local coordinates $u\colon U \to \mathbb{R}$ on small open subsets $U \subset V$ span the bundle of symmetric 4-forms on $T(V)$ (compare 3.1.5). This allows one (by using a partition of unity argument) to C^0-approximate the homotopy H_t by a semi-admissible homotopy which linearly interpolates consecutive additions of the forms $\pm\varphi^2(du)^4$ for various coordinates u on small subsets $U \subset V$ and for smooth functions $\varphi\colon U \to \mathbb{R}$ with compact supports. Since the admissibility condition meets the dimension assumptions in (A‴), one can use (C′) to construct homotopies of maps $V \to W$ which approximately match linear homotopies of forms, like $(1-t)H_0 + t\varphi^2(du)^4$, $t \in [0, 1]$. In fact, since the form $\Delta(D_f^2, D_{f_{\varepsilon\varepsilon}}^2)$, $\varepsilon > 0$, in (C′) only *approximates* $\delta\varphi^2(du)^4$, one should keep compensating the error with appropriate choices of u and φ on the following stages. The positivity part of the admissibility condition leaves enough room for this. Finally, the controle over $\text{dist}_{C^2}(f, f_{\varepsilon,\varepsilon})$ in (C′) insures the convergence $f_t \to f_\infty$, while the uniform freedom of $f_{\varepsilon\varepsilon}$ yields the freedom of f_t.

Exercises. (a) Fill in the detail in this proof.

(b) Assume $[H_t]_0 \equiv 0$ and show the inequalities $q \geq (n+2)(n+3)/2$ and $[H_t] \pm \leq q_\pm - 2$ to suffice for the existence of the homotopy f_t.

(D″) **Corollary.** *Let $f_0\colon V \to W$ be a free isometric C^∞-immersion and let H_∞ be a continuous quadratic form on $T^2(V)$, such that the kernel of $\Delta = H_\infty - H_0$ for $H_0 = (D_{f_0}^2)^*(h)$ contains $T(V) \subset T^2(V)$ and the form Δ [now on $(T(V))^2 = T^2(V)/T(V)$] is* 4-symmetric. *If $[H_\infty]_0 + [H_\infty]_\pm \leq q_\pm - 2 - n$ and if $q \geq [(n+2)(n+3)/2] + n_0 + 2n + 1$, then the immersion f_0 admits a fine C^1-approximation by free isometric C^2-immersions $f\colon V \to W$ for which $(D_f^2)^*(h) = H_\infty$.*

Proof. We may assume $N_0(f_0) = [H_0]_0 \leq n_0 + n$ [see (D′) in 3.3.1] and also $[H_\infty]_0 \leq n_0 + n$, since q generic perturbation H'_∞ of H_∞ does satisfy $[H'_\infty]_0 \leq n_0 + n$ and such a H'_∞ can be admissibly homotoped to H_∞.

Next we semi-admissibly homotope H_0 to a form H_1 for which $[H_1]_0 \leq n_0 + n$ and $[H_1]_0 + [H_1]_\pm \leq q_\pm - 2$. This is done by a (locally finite sequence) of the following elementary deformations. Take a form H on $T^2(V)$, which may be assumed [for the genericity reason, see (C″) in 3.3.1] to have $[H_0] \leq n_0 + n$, and consider a point $v_- \in V$ (if there is any) where the inequality $[H]_0 + [H]_+ \leq q_+ - 2 - n$ is violated. Then $[H]_0 + [H]_- \leq q_- - 3 - n$ in a small neighborhood $U_- \subset V$ of v_- and so the form $H - l^4$ for an arbitrary linear form l on $T(V)$ satisfies

$$[H - l^4]_0 + [H - l^4]_- \leq q_- - 3 - n \quad \text{on } U_-.$$

Thus we obtain with a linear interpolation semi-definite negative deformations of H (to $H - l^4$) which will eventually force the inequality $[H]_0 + [H] + \leq q - 2 - n$ without destroying $[H]_0 + [H]_- \leq q_- - 2 - n$. In a similar way one achieves the inequality $[H]_0 + [H]_- \leq q_- - 2 - n$ at all $v \in V$, and then one makes the homotopy of H_0 to H_1 admissible by a generic perturbation

Finally, we observe with (C′) in 3.3.1 that the space of pertinent forms H on $T_v^2(V)$ for which $[H]_0 \leq n_0 + n + 1$ and $[H]_0 + [H]_\pm \leq q - 2 - n$ is n-connected. Hence, there is a homotopy H_t of H_1 to H_∞ which satisfies the above inequality at all points $v \in V$ and for all $t \in [1, \infty]$. This homotopy clearly is admissible and thus we get an admissible homotopy of H_0 to H_∞ to which (D′) applies.

(E) **C^1-Approximation.** *If $q_\pm \geq n_\pm + n_0 + 2 + n$ and $q \geq [(n+2)(n+3)/2] + n_0 + 2n + 1$, then every isometric C^1-immersion $f_0\colon V \to W$ admits a fine C^1-approximation by free isometric C^∞-immersions $f\colon V \to W$.*

Proof. If $q_- > q_+$ we reverse the signs of the forms g and h and thus we assume $q_+ \geq q_-$. Since $q_- > n_- + n_0$ and $q > n(n+5)/2$, we can C^1-approximate f by a free C^∞-immersion $f_1\colon V \to W$, such that the form $g - f_1^*(h)$ is positive definite on V. Then there is a locally finite decomposition $g - f_1^*(h) = \sum_\mu \varphi_\mu^2\, du_\mu^2$, $\mu = 1, 2, \ldots$.

where u_μ and φ_μ a C^∞-functions on V and the support of φ_μ is a compact subset in a small open set $U_\mu \subset V$, $\mu = 1, \ldots$.

Next, by applying (D′) and (D″) [or rather the first part of the *proof* of (D″)] we find another free C^∞-immersion f_1' which is arbitrarily C^1-close to f_1 and such that $(f_1')^*(h) = f_1^*(h)$, $N_0(f_1') \leq n_0 + n$ and $N_0(f_1', v) + N_+(f_1', v) \leq q_+ - n - 2$ for all $v \in V$. Hence, according to (A‴) and (B′) there exists a free C^∞-immersion $f_2\colon V \to W$ whose C^1-distance to f_1' is bounded by const $\|\varphi_1^2\, du_1^2\|_{C^0}$ and such that $f_2^*(h) = f_1^*(h) + \varphi_1^2\, du_1^2$.

The same operation applies to f_2 over U_2 then to f_3 over U_3 and we eventually arrive at a free C^∞-immersion $f\colon V \to W$ whose C^1-distance to f_0 is bounded by $\sum_\mu \|\varphi_\mu^2\, du_\mu^2\|_{C^0}$ and such that $f^*(h) = f_1^*(h) + \Sigma_\mu \varphi_\mu^2\, du^2 = g$. Q.E.D.

Questions. How does one improve the above argument in order to obtain the parametric C^1-dense h-principle for free isometric C^∞-immersions $V \to W$ for the above q and $q_\pm$? How does one reduce the bound on q to $q \geq [(n+2)(n+3)/2] + n_0$? (In fact, the h-principle may be true for $q_\pm \geq n_\pm + n_0 + 1$ and $q \geq [n(n+3)/2] + 1$, but this is hardly attainable by the present techniques.)

Exercises. (a) Study (by the convex integration) free C^∞-immersions $f\colon V \to W$ for which $N_0(f) + N_+(f) \leq A_+$ and $N_0(f) + N_-(f) \leq A_-$.

(b) Let $n_0 \equiv 0$ and call a free isometric immersion $f\colon V \to W$ *positive* if $N_+(f) = n_+ + [n(n+1)/2]$. Assume $q_+ \geq [(n+2)(n+3)/2] - n_-$ and generalize the h-principles in 3.1 to positive isometric C^∞-immersions $V \to W$.

3.3.4 Isotropic Maps and Non-unique Isometric Immersions

A C^1-map $f\colon V \to (W', h')$ is called *isotropic* if $f^*(h') \equiv 0$. An interesting case is that of $(W', h') = (W_1 \times W_2, h_1 \oplus h_2)$, where isotropic maps correspond to pairs of immersions, $f_1\colon V \to W_1$ and $f_2\colon V \to W_2$, such that $f_1^*(h_1) = f_2^*(-h_2)$. We specialize further to $(W \times W, h \oplus -h)$ and obtain (with the results in 3.3.2 and 3.3.3) the following

(A) **Theorem.** *If $q \geq [n(n+9)/4] + 2$, then arbitrary C^0-maps φ_1 and φ_2 of V into (W, h) admit C^0-approximations by C^∞-smooth (C^{an}-smooth if h is C^{an}) immersions f_1 and f_2 respectively, such that $f_1^*(h) = f_2^*(h)$. Furthermore, if φ_1 and φ_2 a C^1-immersions for which $\varphi_1^*(h) = \varphi_2^*(h)$, then the maps f_1 and f_2 can be chosen arbitrarily C^1-close to φ_1 and to φ_2 respectively.*

(B) **Remarks.** Denote by $\mathscr{X}$ the space of $C^\infty(C^{an})$-immersion $V \to W$, let G be the space of quadratic forms on V and let $D\colon \mathscr{X} \to G$ stand for the differential operator $D\colon x \mapsto x^*(h)$. The above theorem claims the C^0-density of the subset $\Sigma^2 \subset \mathscr{X} \times \mathscr{X}$ of the *double points* (for which $Dx = Dy$) of the operator D. This sharply contrasts with the following heuristic argument concerning the C^∞-density.

Quadratic forms g on V are locally given by s-tuples of functions for $s = n(n+1)/2$ while maps $x\colon V \to W$ are given by q-tuples. This may be expressed by writing "dim" $G = s$ and "dim" $\mathscr{X} = q$. Hence, the subset $\Sigma^2 \subset \mathscr{X} \times \mathscr{X}$ is likely to be nowhere C^∞-dense for $q < s$, and, moreover, to be empty for $2q < s$, provided h is a *generic* metric on W. [In fact, the discussion in (F) of 2.3.8 suggests a rigorous approach to these conjectures.]

(C) **Exercises.** (a) Assume $q \geq n(n+5)/2$ and find a free isotropic $C^\infty(C^{an})$-immersion in each homotopy class of maps $V \to W$, provided V is an open parallelizable manifold and the bundle $T(W)$ is h-trivial.

(b) Prove the C^0-density of the subset of triple points of D, called $\Sigma^3 \subset \mathscr{X} \times \mathscr{X} \times \mathscr{X}$, for $q \geq \frac{2}{3}s + 4n$, where $s = n(n+1)/2$. Generalize this to k-multiple points for all $k \geq 2$.

(c) Call a C^r-immersion $f\colon V \to (W, h)$ C^i-*unique* if every C^i-immersion $f_1\colon V \to W$ for which $f_1^*(h) = f^*(h)$ comes from f by a global isometry of (W, h). Show that an arbitrary C^1-immersion $V \to W$ can be C^1-approximated by non-$C^\infty(C^{an})$-unique $C^\infty(C^{an})$-immersions, provided $q \geq [n(n+9)/4] + 2$.

(d) C^1-approximate an arbitrary C^1-immersion $f\colon V \to \mathbb{R}^q$, $q \geq n+1$, by non-C^∞-unique C^∞-immersions.

(d′) Assume $q \geq n+2$ and then C^1-approximate f by those C^∞-immersions f', for which the space of isometric C^∞-immersions $(V, (f')^*(h) \to (\mathbb{R}^q, h)$, $h = \sum_{i=1}^q dx_i^2$, is infinite dimensional.

(d″) Find, for every $i = 1, 2, \ldots$, and $n = 1, 2, \ldots$, a closed C^{an}-hypersurface $V \subset \mathbb{R}^{n+1}$ which is non-C^i-unique. (This V is necessarily C^{an}-unique for $n \geq 3$.) Then find for a given $i = 1, 2, 3, \ldots$, a C^i-immersion $f\colon S^{n-1} \times S^1 \to \mathbb{R}^{n+1}$, $n \geq 3$, such that the induced metric $f^*(h)$ is C^{an} but no isometric C^{an}-immersion $V \to \mathbb{R}^{n+1}$ exists.

(D) *Perturbations of Isotropic Immersions.* Let V be a compact manifold and let $f_0\colon V \to (W, h)$ be a free isotropic C^∞-immersion. Then by Nash's implicit function theorem every C^α-small metric g on V for $\alpha > 2$ admits a free isometric immersion $f\colon (V, g) \to W$ which is C^2-close to f_0. This also is true for those non-free immersion for which the differential operator $D\colon f \mapsto f^*(h)$ is infinitesimally invertible at f_0, provided α is sufficiently large.

Example. Let g be an arbitrary C^α-form on V. Then we obtain with a free isotropic immersion $f_0\colon V \to W$ an *isometric* C^α-immersion $f\colon (V, \lambda^2 g) \to W$ for all small constants $\lambda > 0$. Moreover, if $W = \mathbb{R}^{q_+, q_-}$, then we have the isometric map $\lambda^{-1} f\colon (V, g) \to \mathbb{R}^{q_+, q_-}$.

Exercises. (a) Let (V, g) be a compact C^α-manifold for $\alpha > 2$. Prove the existence of a free isometric C^α-immersion $f\colon (V, g) \to \mathbb{R}^{q_+, q_-}$, provided $q_\pm \geq 2n+1$, $q \geq (n^2/2) + \frac{9}{2}n + 3$.

(b) Assume V to be a compact parallelizable manifold with a non-empty boundary and obtain the above C^α-immersion f for $q_\pm \geq n$ and $q \geq (n^2/2) + \frac{5}{2}n$. Show this bound on q to be the best possible for *free* isometric immersions $V \to \mathbb{R}^{q_+, q_-}$. Prove the existence of a (non-free!) isometric C^α-immersion $V \to \mathbb{R}^{q_+, q_-}$ for $q_\pm \geq n$ and $q \geq (n^2/2) + \frac{5}{2}n - 1$, provided $\alpha \geq \alpha_0(n)$.

(D′) *Expansion of Isotropic Immersions with Bounded Geometry.* Consider an immersed n-dimensional submanifold $V' \looparrowright \mathbb{R}^N$ and normally project a small neighborhood $U' \subset V'$ of a point $v \in V'$ into the tangent space $T_v \approx \mathbb{R}^n \subset \mathbb{R}^N$ to V'. Thus the (Euclidean) coordinates in T_v induce some local coordinates x_i', $i = 1, \dots, n$ near v' in V, called *standard* coordinates. Denote by $G_r(V', v')$ the maximum of the norms of the derivatives ∂^i of orders $i = 1, \dots, r$ of the immersion $V \looparrowright \mathbb{R}^N$ at v' in the standard coordinates.

Next, fix a proper C^∞-embedding $I\colon W \to \mathbb{R}^N$ and say a family $\mathscr{F}$ of immersions $f\colon V \to W$ to have (uniformely) *bounded geometry* of order r if there is a continuous function C on $W \times V$ such that $G_r(I \circ f(V), I \circ f(v)) \leq C(I \circ f(v), v)$ for all $v \in V$ and all $f \in \mathscr{F}$. Observe that this "boundness" of $\mathscr{F}$ does not depend on a choice of the embedding I (though the function C does). The geometry of $\mathscr{F}$ is called *bounded* if the above holds for each $r = 1, \dots,$ (with $C = C_r$).

A family $\mathscr{F}$ is called *expanding* of order r if for an arbitrary positive continuous function λ on V and for an arbitrary embedding $J\colon V \to \mathbb{R}^{2n}$ there exists an immersion $f \in \mathscr{F}$, such that the map $\bar{f} = J \circ (I \circ f)^{-1}\colon V' \to \mathbb{R}^{2n}$, for the immersed manifold $V' = (I \circ f)(V) \looparrowright \mathbb{R}^N$, has $\|\partial^i \bar{f}\|(v') \leq \lambda(v)$, $v = \bar{f}(v')$, where ∂^i, $i = 1, \dots, r$, denote the derivatives of $\bar{f}$ in the *standard* coordinates.

The property of being expanding is invariant (like the boundness) under the choice of the embedding I.

Examples. Let V be compact and let $f_0\colon V \to \mathbb{R}^q$ be a C^∞-immersion. Then the family $\{\lambda f_0\}$ for $\lambda \in [1, \infty)$ has bounded geometry and it is expanding for all $r = 1, 2, \dots$.

Exercise. Let $f\colon V \to W$ be an arbitrary C^1-immersion (embedding) and let $\dim W > \dim V$. Prove for all $r = 1, 2, \dots$, the existence of an expanding family of immersions (embeddings) $V \to W$ of bounded geometry which lie in a given fine C^0-neighborhood of f_0.

Finally, call a family $\mathscr{F}$ of immersions $f\colon V \to (W, h)$ *uniformly free* if there is a continuous positive function ε on $W \times V$ such that all $f \in \mathscr{F}$ have $\operatorname{Hess}_v(\nabla_i f, \nabla_{ij} f) \geq \varepsilon(f(v), v)$, where ∇_i and ∇_{ij}, $i, j = 1, \dots, n$, are the pseudo-Riemannian [for (W, h)] covariant derivatives in the standard coordinates (in $V' \approx V$) and where the Hessian is taken relative to the Riemannian metric in W induced by the embedding I.

Lemma. *Let $f_0\colon V \to W$ be a continuous map and assume the existence of a uniformly free expanding family of isotropic C^∞-immersions $f\colon V \to (W, h)$ of bounded geometry, such that* $\operatorname{dist}(f_0(v), f(v)) \leq \rho(v)$ *for some distance on W* and *for some continuous* function $\rho = \rho(v) > 0$. *If the implied order r satisfies $r \geq 5$, then an arbitrary C^α-form g on V for $2 < \alpha < r - 2$ admits a free isometric C^α-immersion $f'\colon (V, g) \to (W, h)$ such that* $\operatorname{dist}(f', f_0) \leq 2\rho$.

Proof. If a map $f\colon V \to W$ is sufficiently expanding, then the push-forward form $(I \circ f)_*(g)$ on $V' \approx V$ becomes C^r-small in the standard coordinates. Hence, the implicit function theorem applies (due to the "locality" of D^{-1} in 2.3.2).

Example. Let V be the n-torus T^n and let $f_0: T^n \to W$ be a free isotropic C^∞-map. Then every C^α-form g on T^n, $\alpha > 2$, admits an isometric C^α-immersion f': $(T^n, g) \to W$.

Proof. Consider the (expanding) map $E: T^n \to T^n$ for $E: t \mapsto 2t$ and apply the lemma to the family $\mathscr{F} = \{f_0 \circ E^i\}$, $i = 1, 2, \ldots$.

Exercises. (a) Generalize this example to the manifolds V which admit expanding endomorphisms [compare (E) in 2.4.4]. Generalize further to *branched* manifolds with expanding endomorphisms.

(b) Let $f_0: V \to W$ be a free isotropic C^∞-immersion and let $V_0 \subset V$ be a submanifold of *positive* codimension. Prove for all C^α-forms g_0 on V_0, $\alpha > 0$, the existence of a free isometric C^α-immersion $(V_0, g_0) \to W$ which is homotopic to $f_0 | V_0$.

(b′) Let g be a C^α-form on V, $\alpha > 2$. Prove the existence of a free isometric C^α-immersion $(V, g) \to W$ in each homotopy class of maps $V \to W$, provided $q_\pm \geq 2n + 1$, $q \geq (n^2/2) + \frac{9}{2}n + 3$.

(b″) Assume the forms g in V and h in W to be *positive* definite and C^0-approximate a given strictly short immersion $V \to W$ by free isometric C^α-immersions, provided $q \geq (p + 2)(p + 3)/2$ for $p = [(n + 2)(n + 3)/2] + [n(n + 5)/2]$ (compare 3.1.1). One does not know how to obtain an isometric immersion $V \to W$ for $2 < \alpha \leq 4$ with a more realistic bound on q, say for $q \geq (n + 2)(n + 3)/2$.

3.3.5 Isometric C^∞-Immersions $V^n \to W^q$ for $q \geq [n(n + 3)/2] + 2$

The basic construction of immersions $(V, g) \to W$ (see 3.3.2) which depends on the theory of invariant sheaves requires the existence of (germs of) *free* 0-cylindrical immersions $(V \times \mathbb{R}, g \oplus 0) \to W$. These do not exist for $q \leq (n + 1)(n + 4)/2$; however, we need, in fact, something less than the freedom of the cylinders. Namely, we only need the freedom on some hypersurfaces in $V \times \mathbb{R}$ which are used in building up our immersion $V \to W$. Non-free 0-cylinders may exist for $q_\pm \geq n_\pm + n_0 + 1$ and $q \geq [n(n + 3)/2] + n_0 + 2$. The h-principle for free isometric C^∞-immersions $V \to W$ becomes quite plausible in this range of dimensions. The examples studied below support this conjecture.

(A) *Isotropic Immersions $V^n \to W^q$ for $q \geq [n(n + 5)/2] + 2$.* We assume V to be a *parallelizable* manifold and we assume the tangent bundle $T(W)$ to be h-trivial, which is so, for example, for contractible manifolds W.

(A′) **Theorem.** *If $q_\pm \geq n + 1$, $q \geq [n(n + 5)/2] + 2$ and if the manifold (W, h) is real analytic, than an arbitrary continuous map $f_0: V \to W$ can be C^0-approximated by (possibly non-free) isotropic C^{an}-immersions $f: V \to W$.*

Corollary. *If $p \geq [n(n + 5)/4] + 1$, then arbitrary continuous maps φ_1 and φ_2 of V into $\mathbb{R}^p$ can be C^0-approximated by C^{an}-immersions f_1 and f_2 respectively, such that* *the induced metrics satisfy* $f_1^*(h') = f_2^*(h')$ *for the form* $h' = \sum_{i=1}^p dx_i^2$ *in* $\mathbb{R}^p$. (Compare 3.3.3.)

Question. Does this corollary hold true for $n = 2$ and $p = 3$?

Proof of (A'). The sheaf of free isotropic immersions $V \to W$ is microflexible and (Diff V)-invariant. Hence the h-principle holds for *folded* free isotropic immersions (see 2.2.7). Moreover, the h-principle still holds true (for the same reason) if we additionally require 0-fields [see (E) in 3.3.1] along our immersions. Thus, using the h-principle and the triviality of the bundles in question, we C^0-approximate f_0 by a C^∞-map $f_1: V \to W$ which is free isotropic outside a collection of disjoint closed hypersurfaces in V and near such a hypersurface, say $V' \subset V$, the map looks as follows. There is a tubular neighborhood $U' = V' \times [-1, 1] \subset V$ of $V' = V \times 0$ which admits a free isotropic C^∞-immersion $f': U' \to W$, such that the map $f_1 | U'$ is the composition of f' with the standard folding $\sigma: U' \to U'$, that is $\sigma: (v', t) \mapsto (v', t^2)$. Moreover, we may assume the map f' to be C^{an} and then we can obtain with a 0-field along U' a (non-free) isotropic C^{an}-immersion $F': U' \times \mathbb{R} \to W$ such that $F' | U' = f'$ for $U' = U' \times 0$ [see (E) in 3.3.1]. Finally, we unfold σ to an immersion $\sigma': U' \to U' \times \mathbb{R}$ that is $\sigma': (v', t) \mapsto (v', t^2, \beta(t))$, where $\beta(t)$ is a C^∞-function which is zero outside $[-\frac{1}{2}, \frac{1}{2}]$ and whose derivative does not vanish at $t = 0$. Thus we modify f_1 to $F' \circ \sigma'$ near each V' and get an isotropic C^∞-immersion $f_1': V \to W$. Since this map f_1' is free away from the neighborhood U' and is free on the hypersurfaces $V' \times t$ in U', $t \in [-1, 1]$, there is [see (E') in 2.3.8] a C^∞-approximation of f_1' by a *real analytic* isotropic map $f: V \to W$.

Exercises. (a) Assume the above (W, h) to be C^∞-smooth and approximate f_0 by (non-free) isotropic C^∞-immersions f on which the operator $D: f \mapsto f^*(h)$ is infinitesimally invertible.

(b) Construct an isometric C^∞-immersion $(T^n, g) \to W$ for an arbitrary C^∞-form g on the torus T^n.

(A'') *Control of the Freedom.* If for every hypersurface $H \subset U' \times \mathbb{R}$ which transversally meets $U' = U' \times 0$ across some submanifold $V' \times t$, $t \in [0, 1]$, the map $F | H$ is free on $\mathcal{O}_{\not p} V_t' \subset H$, then the map f_1' is free for a sufficiently small function β. This freedom can be insured by an appropriate (quasi)-regularity of the 0-field in question (compare 3.1.9, 3.1.7) and the (quasi)-regularity is not hard to achieve (an exercise for the reader) for $q \geq [n(n + 5)/2] + 4$. (Probably, the inequality $q \geq [n(n + 5)/2] + 3$ suffices.)

Exercise. Prove for $q \geq [n(n + 5)/2] + 4$ the existence of an expanding uniformely free family of isotropic C^∞-immersions $V \to W$ of bounded geometry and then approximate the map f_0 by free isometric C^∞-immersions $(V, g) \to W$ for an arbitrary C^α-form g on V, provided $\alpha > 2$.

(B) *Immersions of Pseudo-Riemannian Manifolds.* Let V be a (topological) product, $V = V_0 \times S$ where V_0 is compact and where S is a one-dimensional manifold, and let g be a quadratic C^∞-form on V for which $n_0 \equiv 0$. Moreover, assume the form g to be non-singular on the hypersurfaces $V_s = V_0 \times s \subset V$ for all $s \in S$ and require the bundle $T(V_s)$ to be g_s-trivial for $g_s = g | T(V_s)$. We study isometric immersions $(V, g) \to (W, h)$, where the bundle $T(W)$ is assumed h-trivial.

(B′) **Theorem.** *If the manifolds* (V, g) *and* (W, h) *are real analytic and if* $q_\pm \geq n_\pm + 1$ *and* $q \geq [n(n+3)/2] + 2$, *then an arbitrary continuous map* $f_0\colon V \to W$ *can be* C^0-*approximated by isometric* C^{an}-*immersions* $f\colon V \to W$

Proof. Let $\tilde{V} = (V \times \mathbb{R}, g \oplus 0)$, and consider the hypersurface $\tilde{V}_s = (V_s \times \mathbb{R}, g_s \oplus 0)$ in V for some $s \in S$ and take a small neighborhood $\tilde{U}_s \subset \tilde{V}$ of $\tilde{V}_s$, say $\tilde{U}_s = \tilde{V}_s \times (s - \varepsilon, s + \varepsilon)$. We start with a study of C^∞-immersions $\tilde{F}\colon \tilde{U}_s \to W$ which are free [compare (A″)] along $\tilde{V}_s$ (this means the freedom of $\tilde{F} | \tilde{V}_s$ as well as the independence of the field $V_s F$ from the osculating bundle $T^2(\tilde{V}_s) \subset T(W)$) and infinitesimally isometric of order three along $\tilde{V}_s$. The freedom insures infinitesimal isometric extension from $\tilde{V}_s$ to $\tilde{U}_s$. This, in addition to the microflexibility of free isometric immersions $(\tilde{V}_s, g_s \oplus 0) \to W$, implies the microflexibility of the maps $\tilde{F}\colon \tilde{U}_s \to W$. Hence, we have the h-principle which provides (with our assumptions on V and on W) a C^{an}-continuous family of C^{an}-maps $\tilde{F}_s\colon \tilde{U}_s \to W$, $s \in S$, such that

(a) the map $\tilde{F}_s$ is free along $\tilde{V}_s$ in the above sense and it is isometric on a (possibly) smaller neighborhood $\tilde{U}_s' \subset \tilde{U}_s$ of $\tilde{V}_s$, say on $\tilde{V}_s \times [s - \varepsilon', s + \varepsilon']$ for $0 < \varepsilon' = \varepsilon'(s) \leq \varepsilon$, and for all $s \in S$.
(b) The map $\bar{F}\colon V = V_0 \times S \to W$, for $\bar{F}(v_0, s) = \tilde{F}_s(v_0, s, 0)$, lies in a given C^0-fine neighborhood of f_0.

The map $\tilde{F}_s | U_s$ for $U_s = \tilde{U}_s \cap V \subset V$ is isometric for all $s \in S$ but these maps $U_s \to V$ do not agree on intersections of the neighborhoods $U_s = V_s \times (s - \varepsilon', s + \varepsilon') \subset V$. Let us make them agree on successive neighborhoods, U_{s_1} and U_{s_2}, where s_2 is sufficiently close to s_1, such that $s_1 + \varepsilon'(s_1) > s_2 - \varepsilon'(s_2)$. Consider two C^∞-functions β_1 on $(s_1 - \varepsilon'(s_1), s_1 + \varepsilon'(s_1))$ and β_2 on $(s_2 - \varepsilon'(s_2), s_2 + \varepsilon'(s_2))$ which range in $[0, 1] \subset \mathbb{R}$ and such that

(i) there is a small subinterval I'' of length ε'' in the intersection of (s_1, s_2) with $(s_2 - \varepsilon'(s_2), s_1 + \varepsilon'(s_1))$ on which $\beta_1 = \beta_2$ and $d\beta_1/ds = d\beta_2/ds = 1/2\varepsilon''$;
(ii) The functions β_1 and β_2 are zero away from (a slight enlargement of) the above subinterval I''.

Now we define (isometric) maps $f_{s_i}'\colon U_{s_i} \to W$, $i = 1, 2$, by $f_{s_i}'\colon (v, s) \mapsto \tilde{F}_{s_i}(v, s\beta_i(s))$. If $s_2 \to s_1$, then the maps f_{s_i}', $i = 1, 2$, on $V_0 \times I''$ keep bounded geometry and the uniform freedom, while the C^∞-distance between them goes to zero. Hence, there is a small isometric perturbation of the map f_{s_2}' on I'' which makes it equal to f_{s_1}' on I'' and thus we obtain an isometric C^∞-map on $U_{s_1} \cup U_{s_2} = V_0 \times (s_1 - \varepsilon', s_2 + \varepsilon')$. This "fitting operation" applies to all pairs of nearby points in S and then it delivers an isometric C^∞-immersion $f'\colon V \to W$, which can be C^∞-approximated by desired isometric C^{an}-immersions f.

(B″) **Exercises.** (a) Make the above map f free as well as isometric for $q \geq [n(n+3)/2] + 4$ and also generalize (B′) (for $q \geq [n(n+3)/2] + 4$) to C^∞-manifolds and maps.

(b) Drop the assumption of the non-degeneracy of g on $T(V_s)$ and obtain the isometric immersion f, provided $q_\pm \geq n_\pm + 2$ and $q \geq [n(n+3)/2] + 4$.

(b′) Generalize further by considering a Morse function $s\colon V \to S$ (instead of the projection $V_0 \times S \to S$) which may have isolated critical points, and by applying the above construction outside these points. Prove that *an arbitrary pseudo-Riemannian manifold* (V, g) *with a* g*-trivial bundle* $T(V)$ *admits an isometric* C^∞-(C^{an})*-immersion into* W *for* $q_\pm \geq n_\pm + 2$, $q \geq [n(n+3)/2] + 4$.

(c) Prove (B′) for extensions of isometric maps from $\mathcal{O}p\,\partial(V_0 \times S)$ to $V_0 \times S$ for compact manifolds V_0 with a boundary.

(c′) Assume a pseudo-Riemannian manifold (V, g) to admit a codimension one foliation such that g is non-degenerate on the leaves and such that the tangent bundle of this foliation, say $T' \subset T(V)$, is g'-trivial for $g' = g|T'$. Prove in the C^{an}-analytic case the existence of an isometric C^{an}-immersion $V \to W$ for $q_\pm \geq n_\pm + 1$, $q \geq [n(n+3)/2] + 2$.

(c″) Consider an arbitrary C^∞-form g on V and let $T' \subset T(V)$ be an arbitrary (non-integrable) codimension one C^∞-subbundle. Study the sheaf Φ' of the C^∞-maps $F\colon V \times \mathbb{R} \to W$ (where W is an arbitrary pseudo-Riemannian manifold), such that

(1) the maps F are infinitesimally isometric of order 3 along $V = V \times 0 \subset (V \times \mathbb{R}, h \oplus 0)$;
(2) the maps F enjoy the above freedom along T', that is for arbitrary independent vector fields X_i on $V \times \mathbb{R}$, $i = 1, \ldots, n+1$, where the fields $X_i|V \times 0$ lie in T' for $i = 1, \ldots, n-1$, the covariant derivatives $\nabla_{X_i} F$ for $i = 1, \ldots, n+1$ and $\nabla_{X_k}\nabla_{X_l} F$ for $k, l = 1, \ldots, n$ are linearly independent along $V = V \times 0 \subset V \times \mathbb{R}$.

Prove the restriction $\Phi'|V$ to be a flexible sheaf and obtain the h-principle for Φ' as a corollary. Then derive the following.

Theorem. *If* $q_\pm \geq n_\pm + n_0 + n + 1$ *and* $q \geq [(n+2)(n+3)/2] + n_0 + 2$, *then an arbitrary continuous map* $f_0\colon V \to W$ *can be* C^0*-approximated by free isometric* C^∞*-immersions* $f\colon V \to W$.

Remark. The techniques developed for pseudo-Riemannian immersions are likely to apply to more general differential relations, [e.g. to isometric C^∞-immersions of symmetric tensors of degree $d \geq 2$, compare (B) in 2.4.9 and Gromov (1972)] but no systematic study has been conducted so far.

3.4 Symplectic Isometric Immersion

Here we study isometric immersions $f\colon (V, g) \to (W, h)$ for *exterior* differential forms of a fixed degree $\deg g = \deg h$. The word "isometric" refers, as earlier, to the form inducing equation $f^*(h) = g$. Though, the convex integration does not apply to this equation (see in 2.4.9), the techniques of 2.2, 2.3 and of 3.3 work with minor modifications. In fact, isometric immersions for forms of degrees one and two (which arise in the contact and symplectic geometry) do not require Nash's implicit function theorem. The h-principle for these immersions only requires the theory of topological sheaves.

3.4.1 Immersions of Exterior Forms

Denote by $\bigwedge^k L$ the space of k-linear anti-symmetric forms ω on a given linear space L and denote by $x \cdot \omega$ for $x \in L$ the *interior product*, that is the $(k-1)$-form defined by $(x \cdot \omega)(x_1, \ldots, x_{k-1}) = \omega(x, x_1, \ldots, x_{k-1})$. The resulting linear map $x \mapsto x \cdot \omega$ is called I_ω: $L \to \bigwedge^{k-1} L$. A subspace $L' \subset L$ is called *ω-regular* if the composition of I_ω with the restriction homomorphism $\bigwedge^{k-1}(L) \to \bigwedge^{k-1}(L')$ sends L *onto* $\bigwedge^{k-1}(L')$. A form $\omega \in \bigwedge^2 L$ is called *non-singular or symplectic* if the homomorphism I_ω: $L \to \bigwedge^1(L) = L^*$ is an isomorphism. In general one defines rank ω = rank I_ω, and one observes that the rank of a 2-form ω equals twice the greatest integer r for which the *exterior* power ω^r is non-zero. Thus, the existence of a symplectic form on L makes the dimension of L even, say $\dim L = 2m$, and the non-singularity condition amounts to the non-vanishing of the top dimensional (volume) form ω^m.

The direct sum of an arbitrary linear space L_0 with its dual, $L = L_0 \oplus L_0^*$, carries a canonical symplectic form, that is $\sum_{i=1}^m x_i \wedge y_i$ for $m = \dim L_0$, where the vectors $y_i \in L_0$ constitute a basis of linear forms on L_0 and where $(x_1, \ldots, x_m)$ is the dual basis in L_0. If L is identified with the complexification $\mathbb{C}L_0 = L_0 \oplus \sqrt{-1}L_0$, then $\omega(a, b) = \langle a, \sqrt{-1}b\rangle$ for all a and b in L, where the scalar product $\langle\ ,\ \rangle$ is given by the symmetric form $\sum_{i=1}^m x_i^2 + y_i^2$ on L.

The group of linear transformations of L, $\dim L = 2m$, is transitive on the space of symplectic forms ω on L since any such ω (obviously) reduces to ω_0 for some splitting $L = L_0 \oplus L_0^*$. The isotropy subgroup, called the *symplectic* group $\mathcal{S}p\ell_m = \mathcal{S}p\ell(L, \omega_0)$, clearly has $\dim \mathcal{S}p\ell_m = 4m^2 - m(2m-1) = 2m^2 - 1$ and it *transitively* acts on $L \setminus \{0\}$. The maximal compact subgroup in $\mathcal{S}p\ell_m$ can be easily identified with the unitary group $U_m = U(\mathbb{C}L_0, \sum_{i=1}^m z_i \bar{z}_i)$ for $z_i = x_i + \sqrt{-1}y_i$, which is also the maximal compact subgroup in the full linear group $GL_m\mathbb{C}$. Thus the groups $\mathcal{S}p\ell_m$, U_m and $GL_m\mathbb{C}$ have the same homotopy type.

(A) *Regular Immersions.* Let h be a C^∞-smooth k-form on a manifold W, that is a C^∞-section h: $W \to \bigwedge^k(W) = \bigwedge^k T(W)$ and let $\mathscr{D} = \mathscr{D}_h$: $f \mapsto f^*(h)$ be the form inducing (non-linear differential) operator on smooth maps f: $V \to (W, h)$ for a given manifold V. Then the linearized operator $L = L_{f_0}\mathscr{D}$ at a given map f_0 sends vector fields ∂ in W along V (mapped to W by f_0) to k-forms on V according to the following formula

$$(*) \qquad L(\partial) = (d(\partial \cdot h) + \partial \cdot dh) | V,$$

where d is the exterior differential and $\partial\cdot$ is the interior product with ∂. This is an obvious corollary of the well-known expression for the Lie derivative of any form ω by a (globally defined) vector field ∂,

$$(**) \qquad \partial\omega = d(\partial \cdot \omega) + \partial \cdot d\omega$$

Definitions. An *immersion* f: $V \to (W, h)$ is called *h-regular* if the image of the tangent space $T_v(V)$ under the differential of f is h_w-regular in the receiving space $T_w(W)$, $w = f(v)$, for all $v \in V$. For example, if h is a nowhere singular 2-form on W, then every immersion is h-regular. We call f an *(h, dh)-regular immersion* if the linear map $T_w(W) \to \bigwedge_v^{k-1}(V) \oplus \bigwedge_v^k(V)$ defined by $\tau \mapsto ((\tau \cdot h) \oplus (\tau \cdot dh)) | T_v(V)$ is surjective for all

$v \in V$, where $\tau \in T_w(W)$ for $w = f(v)$ and $T_v(V)$ is embedded into $T_w(W)$ by the differential of f. Clearly, this regularity implies the h-regularity as well as the dh-regularity of f.

Example. If $(W, h) = (W_1 \times W_2, h_1 \oplus h_2)$ and if the projection of f to (W_1, h_1) is h_1-regular while the projection to W_2 is dh_2-regular, then f is (h, dh)-regular.

Lemma. *The operator $\mathscr{D}: f \mapsto f^*(h)$ is infinitesimally invertible on the space of (h, dh)-regular immersions $f: V \to (W, h)$.*

Proof. The regularity implies the solvability in ∂ of the linear system

$$\partial \cdot h | V \equiv 0, \qquad \partial \cdot dh | V = \tilde{g},$$

for all k-forms $\tilde{g}$ on V, which solves the equation $L(\partial) = \tilde{g}$ (compare 2.3.1).

Now the results in 2.2 and 2.3.2 yield (compare 3.3) the following

(A′) **Theorem.** *If $V = V_0 \times \mathbb{R}$ and if the form g on V is induced from a C^∞-form on V_0 by the projection $V = V_0 \times \mathbb{R} \to V_0$, then isometric (h, dh)-regular C^∞-immersions $f: (V, g) \to (W, h)$ satisfy the parametric h-principle. Furthermore, if V is an open manifold, then (h, dh)-regular isotropic (i.e. $f^*(h) \equiv 0$) C^∞-immersions $f: V \to (W, h)$ satisfy this h-principle.*

Exercise. Let $W = \mathbb{R}^q$ for $q = Mk + N(k+1)$ and $h = \sum_{i=1}^{M} dx_{i,1} \wedge \cdots \wedge dx_{i,k} + \sum_{j=1}^{N} x_{j,1}\, dx_{j,2} \wedge \cdots \wedge dx_{j,k+1}$. Analyse the above h-principle for $M \geq [2(n!)/k!(n-k)!] + 2n$ and $N \geq [2(n!)/(k+1)!(n-k-1)] + 2n$, where $n = \dim V$ and show for $k \geq 1$ that any continuous map $f: V_0 \to \mathbb{R}^q$ admits a fine C^0-approximation by isometric C^∞-immersions $(V_0, g_0) \to (\mathbb{R}^q, h)$ for an *arbitrary* C^∞-form g_0 on V. Prove this property of the form h on $\mathbb{R}^q$ to be stable under small C^1-perturbations.

The Cauchy Problem. Consider an *arbitrary* k-form g on $V = V_0 \times \mathbb{R}$ and let a C^∞-map $f: V \to (W, h)$ satisfy the following equations with the field ∂ in W along V (mapped to W by f) which is the image of the field $\partial_t = \partial/\partial t$ in $V = V_0 \times \mathbb{R}$ under the differential of f,

$$\begin{aligned} \partial \cdot h | V_t &= \partial_t \cdot g | V_t \\ \partial \cdot dh | V_t &= \partial_t \cdot dg | V_t \end{aligned} \tag{1}$$

for $V_t = V_0 \times t \subset V \times \mathbb{R}$, $t \in \mathbb{R}$. The system (1) implies with (∗) and (∗∗) that $\partial_t f^*(h) = \partial_t g$ and hence each solution of (1) with the initial condition $f^*(h) | V_0 = g | V_0$ is an *isometric* map $(V, g) \to (W, h)$. If the map $f_0 = f | V_0 : V_0 \to W$ is (h, dh)-regular as well as isometric for $g_0 = g | V_0$, then the Eqs. (1) can be resolved formally in ∂ which reduce them to the Cauchy-Kovalevskaya form. This yields the following

Proposition. *Let (V, g) and (W, h) be real analytic and let $f_0: (V_0, g_0) \to (W, h)$ be a (h, dh)-regular C^{an}-immersion. Then f_0 extends to a (possibly non-regular) isometric C^{an}-map $f: (U, g | U) \to (W, h)$ of some small neighborhood $U \subset V$ of $V_0 \subset V$.*

In order to control the regularity of f which is needed for an application of the microextension theorem (see 2.2.4), we introduce the following

Definition. A subspace $L \subset T_w(W)$ is called *(h, dh)-biregular* if there exists a vector $\tau' \in T_w(W)$ outside L for which $\tau' \cdot h_w|L = 0$ and such that the span $L' = \mathrm{Span}(L, \tau') \subset T_w(W)$ is (h, dh)-regular (i.e. the pertinent map $T_w(W) \to \bigwedge^{k-1} L' \oplus \bigwedge^k L'$ is onto). Then we obviously define the biregularity of immersions $f: V \to W$ in terms of the subspaces $T_v(V) \subsetneq T_w(W)$, $w = f(v)$ and we obtain with 2.2.4 and 2.3.2 (compare 3.3) the following

(A″) **Theorem.** *Let g_0 be an arbitrary C^∞-smooth k-form on V_0. Then isometric (h, dh)-biregular C^∞-immersions $(V_0, g_0) \to (W, h)$ satisfy the C^0-dense h-principle.*

Exercise. Give examples where (A″) yields isometric immersions but where (A′) does not apply.

(B) *Immersions of Closed Forms.* Let the form h on W be closed, that is $dh = 0$. Then no map into W is (dh)-regular and so the above considerations do not apply.

Exercise. Show for $\dim V \geq 1$ and $k \geq 2$ that the sheaf of h-regular isometric C^∞-immersions $(V, g) \to (W, h)$ contains no open non-empty microflexible subsheaf whenever $dh = 0$.

Now, if the forms g and h are exact, $g = dg_1$ and $h = dh_1$ for some C^∞-smooth $(k-1)$-forms g_1 on V and h_1 on W then the equation $f^*(h) = g$ can be rewritten as $f^*(h_1) + d\varphi = g_1$ with an auxiliary (unknown) $(k-2)$-form φ on V.

Lemma. *The (non-linear differential) operator $\mathscr{D}_1: (f, \varphi) \to f^*(h_1) + d\varphi$ is infinitesimally invertible at those pairs (f, φ) where f is an h-regular immersion $V \to (W, h)$.*

Proof. The linearized operator $L_1 = L_f \mathscr{D}_1$ applies to the pairs $(\partial, \tilde{\varphi})$, where ∂ is a field in W along V (mapped to W by f) and $\tilde{\varphi}$ is a $(k-2)$-form on V, by the formula

$$L_1(\partial, \tilde{\varphi}) = \partial \cdot h + d(\partial \cdot h_1) + d\tilde{\varphi}.$$

If f is h-regular, then the system

$$\begin{aligned} \partial \cdot h | V &= \tilde{g} \\ \partial \cdot h_1 + \tilde{\varphi} &= 0 \end{aligned} \tag{2}$$

is solvable for all $(k-1)$-forms $\tilde{g}$ on V and every solution $(\partial, \tilde{\varphi})$ of Eqs. (2) satisfies $L(\partial, \tilde{\varphi}) = \tilde{g}$.

Generalization. Let $[g] \in H^k(V; \mathbb{R})$ and $[h] \in H^k(W, \mathbb{R})$ denote the respective cohomology classes of the closed (but now not necessarily exact) forms g and h. Fix a continuous map $f_0: V \to W$ for which $f_0^*[h] = [g]$. Denote by $\Gamma_0 \subset X = V \times W$ the graph of f_0 and let $\bar{g}$ and $\bar{h}$ be the pull-backs of the forms g and h to X under the obvious projections. Take a small neighborhood $Y \subset X$ of Γ_0 which contracts to Γ_0 and observe that the form $\bar{h} - \bar{g}$ is exact on Y. Write $\bar{h} - \bar{g} | Y = d\hat{h}$ for some C^∞-smooth $(k-1)$-form $\hat{h}$ on Y and let $\bar{\mathscr{D}}(\bar{f}, \varphi) = \bar{f}^*(\hat{h}) + d\varphi$ for sections $\bar{f}: V \to Y$ and $(k-2)$-forms φ on V. Observe that the maps $f: V \to W$ underlying $\bar{f}$ satisfy $f^*(h) = g + d\hat{h} = g + d\bar{\mathscr{D}}(\bar{f}, \varphi)$. It easily follows that the space of sections $\bar{f}: V \to Y$ for which $f^*(h) = g + dg_1$ for a given $(k-1)$-form g_1 on V has the same homotopy type as the space of solutions to the equation $\bar{\mathscr{D}}(\bar{f}, \varphi) = \bar{g}_1$ for the pull-back $\bar{g}_1$ of g_1 to Y under the projection $Y \to V$. In particular, the equation $f^*(h) = g$ reduces to the equation $\bar{\mathscr{D}}(\bar{f}, \varphi) = 0$, in so far as the unknown map f is C^0-close to f_0

Lemma. *The operator $\bar{\mathscr{D}}$ is infinitesimally invertible at those pairs $(\bar{f}, \varphi)$ for which the underlying maps $f: V \to W$ are h-regular.*

Indeed, the proof of the previous lemma applies.

Exact Diffeotopies. A vector field ∂ on an open subset $U \subset V$ is called *g-isometric* if $\partial g = 0$, which is equivalent (g is closed!) to $d(\partial \cdot g) = 0$. We call ∂ *exact* if the $(k-1)$-form $\partial \cdot g$ on U is exact, that is $\partial \cdot g = d\alpha$ for some $(k-2)$-form α on U.

Consider a C^∞-diffeotopy of g-isometric diffeomorphisms $\delta_t: U \to V$, $t \in [0,1]$, $\delta_0 = \mathrm{Id}$, and call δ_t *exact* if the vector field $\delta_t' = (d/dt)\delta_t$ on $\delta_t(U) \subset V$ is exact for all $t \in [0,1]$. Then there is a smooth family of forms α_t on $\delta_t(U_t) \subset V$, such that $d\alpha_t = \delta_t' \cdot g$. Call δ_t *strictly exact* if one can choose these forms α_t equal to zero on the maximal open subset $U_0 \subset U$, where the diffeotopy δ_t is constant in t, that is $\delta_t(u) = \delta_0(u)$ for $u \in U_0$. Lift δ_t to $U \times W \subset X$ by $\bar{\delta}_t(u, w) = (\delta_t(u), w)$ and observe $\bar{\delta}_t$ to be $(\bar{h} - \bar{g})$-exact (strictly exact) for each g-exact (strictly exact) diffeotopy δ_t.

Take an open subset $Y' \subset U \times W$ such that $\bar{\delta}_t(Y') \subset Y$ for all $t \in [0,1]$ and let $\hat{h}_t'$ be the pull-back of the form $\hat{h}$ on Y under the diffeomorphism $\bar{\delta}_t: Y' \to Y$. Since $\bar{\delta}_t'\hat{h} = d(\bar{\delta}_t' \cdot \bar{h}) - \bar{\delta}_t' \cdot \bar{g}$, there exists, for a strictly exact diffeotopy δ_t, a smooth family of $(k-2)$-forms φ_t' on Y', $t \in [0,1]$, which vanishes near the subset where δ_t is constant in t and such that $\hat{h}_t' = \hat{h} + d\varphi_t'$. Moreover, if δ_t is constant for $t \geq t_0$ on all of U, then one can (and does) choose $\varphi_t = 0$ for $t \geq t_0$.

Let $\bar{\Phi}_0$ be the sheaf of sections $V \to Y$. Then there is a (partially defined) action of diffeotopies in V on $\bar{\Phi}_0$ (in 2.2.3). In particular, if $\bar{\delta}_t(Y') \subset Y$, $t \in [0,1]$, and if a section $\bar{f}_0 \in \bar{\Phi}_0(U)$ sends U to Y' then the underlying diffeotopy $\delta_t: U \to V$ does act on $\bar{f}_0$. Now, if the diffeotopy δ_t is strictly exact, then this action extends to the sheaf $\bar{\Phi}$ of the C^∞-solutions of $\bar{\mathscr{D}}(\bar{f}, \varphi) = 0$ by $\delta_t^*(\bar{f}, \varphi) = (\delta_t^*\bar{f}, \varphi - \bar{f}(\varphi_t'))$. In fact, this extension agrees with $\bar{\mathscr{D}}$, as $\bar{\mathscr{D}}(\delta_t^*\bar{f}, \varphi - \bar{f}^*(\varphi_t')) = \bar{f}^*(\hat{h}_t') + d(\varphi - \bar{f}^*(\varphi_t')) = \bar{f}^*(h) + d\varphi = \bar{\mathscr{D}}(\bar{f}, \varphi)$.

(B′) *The h-Principle for Regular Isometric Immersions.* Let V_0 be a C^∞-smooth submanifold in V which is sharply movable (see 2.2.3) by strictly *exact* diffeotopies in V.

Theorem: *The h-principle for regular isometric C^∞-immersions $f: (\mathcal{O}_{\!\mathscr{p}} V_0, g | \mathcal{O}_{\!\mathscr{p}} V_0) \to (W, h)$ (where $\mathcal{O}_{\!\mathscr{p}} V_0 \subset V$ is an arbitrarily small neighborhood of V_0) is dense in the space of those continuous maps $f_0: V_0 \to W$ for which $f_0^*[h] = [g] | V_0 \in H^r(V_0; \mathbb{R})$.*

Proof. Let $\bar{\Phi}_{\text{reg}}$ be the subsheaf in the above Φ for which the maps $f: V \to W$ underlying the sections $\bar{f}: V \to Y$ are h-regular. The infinitesimal invertibility of $\mathscr{D}$ implies the microflexibility of $\bar{\Phi}_{\text{reg}}$ and hence (see 2.2.3) the flexibility of Φ_{reg} on V_0. This gives us the h-principle for $\bar{\Phi}_{\text{reg}}|V_0$. The (obvious) homomorphism of Φ_{reg} to the sheaf Φ_{reg} of h-regular isometric immersions $f: V \to W$ for which f sends V to Y clearly is a weak homotopy equivalence. The resulting h-principle for Φ_{reg} for the small neighborhoods $Y \subset X = V \times W$ of the graphs of the maps $f_0: V \to W$ easily yields the required dense h-principle. Q.E.D.

Examples (*Sharply Movable Submanifolds*). (a) Take an arbitrary closed form g on V_0 and let $V = (V_0 \times \mathbb{R}, g_0 \oplus 0)$. Then, clearly, $V_0 = V_0 \times 0 \subset V$ is sharply movable by strictly exact diffeotopies in V.

(b) The total space of the bundle $\Lambda^{k-1}V_0 \to V_0$ (obviously) admits a unique $(k-1)$-form, λ^{k-1} on $\Lambda^{k-1}(V_0)$, such that every $(k-1)$-form μ on V_0 is induced from λ^{k-1} by the form μ itself viewed as a section $\mu: V_0 \to \Lambda^{k-1}V_0$, that is $\mu = \mu^*(\lambda^{k-1})$. Then every $(k-2)$-form υ on V_0 defines an exact diffeotopy δ_t on $(\Lambda^{k-1}V_0, d\lambda^{k-1})$ by the formula $\delta_t: (v, x) \mapsto (v, (1-t)x + t(d\upsilon)(v))$, for all $v \in V_0$ and $x \in (\Lambda^{k-1}V_0)_v \subset \Lambda^{k-1}V_0$, and where the $(k-1)$-form $d\upsilon$ is viewed as a section $d\upsilon: V_0 \to \Lambda^{k-1}(V_0)$. It is easily seen that the zero section $V_0 \subsetneq \Lambda^{k-1}V_0$ is sharply movable by these diffeotopies.

(c) Let ω be a symplectic (see 2.4.2) form on V and let $g = \omega^l$ for some $l \geq 1$. Then (see 2.4.2) every submanifold $V_0 \subset V$ of *positive* codimension is sharply movable by strictly exact diffeotopies.

(d) Let g be a non-vanishing n-form on V for $n = \dim V$. Then every submanifold $V_0 \subset V$ of positive codimension (obviously) is sharply movable.

Exercise. Let $(W, h) = (\mathbb{R}^{Mk}, h_0 + \varepsilon h_1)$, $\varepsilon \geq 0$, where $h_0 = \sum_{i=1}^{M} dx_{i,1} \wedge \cdots \wedge dx_{ik}$ and where h_1 is an arbitrary *exact* C^∞-smooth k-form. Let $k \geq 2$, $M \geq [2(n+1)!/k!(n+1-k)!] + 2n + 2$ and let ε be small, $0 \leq \varepsilon \leq \varepsilon_0(h_1) > 0$. Prove for an arbitrary C^∞-smooth exact k-form g on V, $\dim V = n$, the existence of a regular isometric C^∞-immersion $(V, g) \to (W, h)$.

(B″) *Biregular Immersions.* An immersion $f: V \to W$ is called *h-biregular* if there exists a vector $\tau' \in T_w$, for $w = f(v)$ and for all $v \in V$, which lies outside the image of the tangent space $T_v(V)$ in $T_w(W)$, for which $\tau' \cdot h_w | T_v(V) = 0$ and such that the span of $T_v(V)$ and τ' is an h_w-regular subspace in $T_w(W)$.

Theorem. *Let g be a closed C^∞-smooth k-form on V. Then the h-principle for h-biregular isometric C^∞-immersions $f: (V, g) \to (W, h)$ is C^0-dense near those continuous maps $f_0: V \to W$ for which $f_0^*[h] = [g]$.*

Proof. Combine (B′) with the proof of (A″).

Exercises. (a) State and prove the parametric h-principle for biregular isometric immersions. Study extensions of these immersions and prove the pertinent h-principle.

(b) Let h be a k-form, $k \geq 2$, on $\mathbb{R}^q$ with constant coefficients (which is equivalent to the invariance of h under parallel translations of $\mathbb{R}^q$). Let $f_0: V \to \mathbb{R}^q$ be an h-biregular C^∞-immersion and let g_0 be the induced form $f_0^*(h)$ on V. Show an arbitrary continuous map $V \to \mathbb{R}^q$ to admit a fine C^0-approximation by isometric C^∞-immersions $(V, g_0) \to (\mathbb{R}^q, h)$.

(b′) Study (bi)regular isotropic immersions $T^n \to \mathbb{R}^q$ [compare (D′) in 3.3.4].

3.4.2 Symplectic Immersions and Embeddings

Let h be a *symplectic* 2-form on W that is $dh = 0$ and h is nowhere singular on W. The h-principle [see (B′) in 3.4.1] for isometric immersions $(V, g) \to (W, h)$ is especially useful in this case due to an abundance of *symplectic* (i.e. h-isometric) diffeotopies of W. Indeed, since h is non-singular, the homomorphism $I_h: T(W) \to \Lambda^1 W = T^*(W)$ for $I_h(\partial) = \partial \cdot h$ is an isomorphism. Hence, every *closed* 1-form φ on W defines a *symplectic (i.e.* h-isometric) field by $\partial = I^{-1}(l)$, as $\partial h = d(\partial \cdot h) = dI_h(\partial) = dl = 0$, and the one-parameter subgroup generated by ∂ is a symplectic diffeotopy. In particular, every smooth function, called a *Hamiltonian*, $H: W \to \mathbb{R}$ defines an *exact* symplectic field $\partial = I_h^{-1}(dH)$, which gives us a one-to-one correspondence between exact fields on W and smooth functions modulo (additive) constants. This agrees with the following heuristic "dimension" count [compare (B) in 3.3.4]. The "dimension" of Diff W equals $q = \dim W$, while the space Ω of closed 2-forms [which are 1-forms/d(functions)] has "dim" $\Omega = q - 1$. Hence the isotropy subgroup $\mathrm{Sympl}_h \subset \mathrm{Diff}\, W$ should have "dimension" one.

Exercise. Study ω-isometric fields for closed $(q-1)$-forms ω on W, for $q = \dim W$.

Now, let (V, g) be another *symplectic* manifold (i.e. g is a symplectic form) and let $V_0 \subset V$ be a submanifold of *positive* codimension.

Lemma. *The strictly exact diffeotopies of* (V, g) *sharply move* V_0.

Proof. To move a closed hypersurface S lying in a small neighborhood $U_0 \subset V_0$ we start with a vector $\partial_0 \in T_{v_0}(V)$ transversal to V_0 at some point $v_0 \in U_0$. This ∂_0 (obviously) extends to an exact field $\partial = I_g^{-1}(dh)$ on V which is transversal to U_0, since the neighborhood $U_0 \subset V_0$ is chosen small. In order to make the corresponding exact isotopy δ_t for $d\delta_t/dt = \partial$ *sharply* (see 2.2.3) move S, we take the union $S_\varepsilon = \bigcup_t \delta_t(S) \subset V$ over $t \in [0, \varepsilon]$ and then multiply the Hamiltonian H by a C^∞-function a on V which vanishes outside an (arbitrarily) small neighborhood $\mathcal{O}\!p\, S_\varepsilon \subset V$ and which equals one in a smaller neighborhood of S_ε. This makes the diffeotopy corresponding to the field $I_g^{-1}(d(aH))$ as sharp as we want. Q.E.D.

Since every *immersion* into the (symplectic!) manifold (W, h) is h-regular, the lemma shows the h-principle (B′) of 3.14. to apply to isometric immersions $f: \mathcal{O}\!p\, V_0 \to W$ for all symplectic manifolds (V, g). Now, the implied sections of the pertinent jet bundle are fiber-wise injective isometric homomorphisms $F_0: (T_0, g | T_0) \to (T(W), h)$, for $T_0 = T(V)|V_0$. Hence, *such an* F_0 *is homotopic to the differential* $D_f|T_0$ *of the same isometric immersion* $\mathcal{O}_{\not{p}}V_0 \to W$ *if and only if the continuous map* $f_0\colon V_0 \to W$ *underlying* F_0 *sends the cohomology class* $[h] \in H^2(W, h)$ *to* $[g]|V_0 \in H^2(V_0; \mathbb{R})$. Furthermore, the h-principle (B″) of 3.4.1 is refined in the symplectic case by the following

(A) **Theorem.** *Let g be an arbitrary (possibly singular) closed C^∞-smooth 2-form on V and let $F_0\colon (T(V), g) \to (T(W), h)$ be a fiberwise injective isometric homomorphism for which $f_0^*[h] = [g]$. If $\dim V < \dim W$, then the map f_0 admits a fine C^0-approximation by isometric C^∞-immersions $f\colon (V, g) \to (W, h)$ whose differentials $D_f\colon T(V) \to T(W)$ are homotopic to F_0 in the space of fiberwise injective isometric homomorphisms.*

Proof. Assume without loss of generality the homomorphism F_0 to be C^∞-smooth, consider the quotient bundle $X = f_0^*(T(W))/T(V) \to V$ and embed V into X by the zero section $V \hookrightarrow X$. Then the homomorphism F_0 extends (non-uniquely) to a fiberwise isomorphic homomorphism $\tilde{F}_0$ of the bundle $T(X)|V \supset T(V)$ into $T(W)$. Since F_0 is isometric, the induced form $\tilde{F}_0^*(h)$ on $T(X)|V$ satisfies $\tilde{F}_0^*(h)|T(V) = g$. Therefore, there exists a *closed* C^∞-form $\tilde{h}$ on X such that $\tilde{h}|V = g$ and which, moreover, equals $\tilde{F}_0^*(h)$ on $T(X)|V$. Indeed, start with the pull-back $p^*(g)$ on X for the projection $p\colon X \to V$ and let $\tilde{h} = p^*(g) + dl$ for some 1-form l on X whose differential on $T(X)|V$ equals $\tilde{F}_0^*(h) - p^*(g)$. This $\tilde{h}$ is clearly symplectic (i.e. non-singular) near $V \subset X$ and hence, the above h-principle applies to isometric immersions $\mathcal{O}_{\not{p}}V \to W$. Furthermore, the microextension theorem of 2.2.4 which works here as well as in (B″) of 3.4.1, "descends" the h-principle from $\mathcal{O}_{\not{p}}V$ to V. Q.E.D.

Corollaries. (a) *Let the form g on V be exact and let $\operatorname{rank}(g|T_v)) \geq r$ for some $0 \leq r \leq n = \dim V$ and for all $v \in V$. Then there exists an isometric immersion*

$$f\colon (V, g) \to \left(\mathbb{R}^{2m}, \sum_{i=1}^{m} dx_i \wedge dy_i\right),$$

provided $2m \geq 3n - r$.

Proof. The space of injective isometric homomorphisms $T_v(V) \to \mathbb{R}^{2m}$ is (easily seen to be) k-connected for $k = 2m - 2n + r - 1$ and (A) in 3.3.1 applies

(b) *An n-dimensional manifold V admits an isotropic (i.e. $f^*(h) \equiv 0$) immersion f into $(\mathbb{R}^{2n}, h = \sum_{i=1}^{n} dx_i \wedge dy_i)$ if and only if the bundle $T(V) \oplus T(V) \to V$ is trivial.* [Compare Lees (1976).]

Proof. Identify $\mathbb{R}^{2n}$ with $\mathbb{C}^n = \mathbb{R}^n \oplus \sqrt{-1}\mathbb{R}^n$ and observe that the isotropy condition to the orthogonality of the bundle $\sqrt{-1}T(V) \subset T(\mathbb{C}^n)|V$ to the tangent bundle $T(V)$. This makes the "only if" part obvious, and the "if" claim follows from (A).

Remark. Isotropic immersions $V \to (W, h)$ for $\dim W = 2 \dim V$ are called *Lagrange* immersions. The self-intersection points of these play an important role in the symplectic geometry (see 3.4.4).

(c) *Let h be the standard symplectic form on $\mathbb{C}P^m$ (which is uniquely characterized by the invariance under the action of the unitary group $U(m+1)$ on $\mathbb{C}P^m$ and by the normalization $\langle [h], [\mathbb{C}P^1] \rangle = 1 \rangle$) and let a closed form g on V be integral, which means the integrality of $\int_Z h$ for all 2-cycles Z in V. If $\operatorname{rank}_v g \geq 3n - 2m$ for all $v \in V$, then there exists an isometric immersion $f: (V, g) \to (\mathbb{C}P^m, h)$.*

Proof. The integrality (obviously) is equivalent to the existence of a continuous map $f_0: V \to \mathbb{C}P^m$ for which $f_0^*[h] = [g]$ and then the proof of the above (a) applies

Exercises. (1) [Compare Tischler (1977)]. Show every symplectic manifold (V, g) with the integral form g to admit an isometric *embedding* into $(\mathbb{C}P^m, h)$ for $m = 2n + 1$.

Hint. Use a generic symplectic perturbation of the immersion f in the above (c).

(2) Prove the parametric and the extension versions of the h-principle claimed by (A).

(2′) Consider a (non-closed) C^∞-smooth 2-form on V which is closed on the leaves of a given C^∞-foliation. Generalize (A) to C^∞-maps $V \to (W, h)$ which are isometric immersions on the leaves.

(3) Let V be an open manifold and let $2m = \dim V = \dim W$. Homotope the map $f_0: V \to W$, which satisfies the assumptions of (A), to an isometric C^∞-immersion $f: (V, g) \to (W, h)$ which is no longer required to be C^0-close to f_0. In particular, construct an isometric immersion of V into a small ball in $(\mathbb{R}^{2m}, \sum_{i=1}^m dx_i \wedge dy_i)$, provided V is topologically contractible.

Hint. Study isometric immersions of a symplectic manifold with a boundary into itself.

(3′) Prove the h-principle for those microflexible sheaves over a symplectic manifold (V, g) which are invariant under g-isometric diffeomorphisms (compare 2.2.2). Apply this to submersions $f: (V, g) \to X$ for which the submanifold $f^{-1}(x) \subset V$ is symplectic [i.e. $g | f^{-1}(x)$ is non-singular for all $x \in X$], where X is an arbitrary manifold of dimension $q \leq \dim V$.

(3″) Since the linear symplectic group is homotopy equivalent to $GL_m\mathbb{C}$, each symplectic structure g on V defines a (unique up to a homotopy) classifying map $C_g: V \to Gr_m\mathbb{C}^N$ for $2m = n = \dim V$ and for all $N > 2n$. Let (V, g) be an open symplectic manifold and let the map C_g be $(m+1)$-contractible. Construct [compare (C) in 2.2.7] an (integrable!) complex structure $J: T(V) \circlearrowleft$ such that J is a *g-isometric* automorphism and such that the quadratic form $g(x, Jy)$ is positive definite.

(B) *Isometric Embeddings.* Let (V, g) be an n-dimensional symplectic manifold and let $f_0: V \to W$ be a C^∞-embedding such that $f_0[h] = [g]$. Let $\varphi_t: T(V) \to T(W)$, $t \in [0, 1]$, be a homotopy of fiberwise injective homomorphisms whose underlying maps $V \to W$ equals f_0 for all $t \in [0, 1]$ and such that φ_0 equals the differential D_{f_0}, while the homomorphism φ_1 is isometric for the forms g on $T(V)$ and h on $T(W)$.

Theorem. (1) *If $q = \dim W \geq n + 2$ and if the manifold V is open, then there exists a homotopy of C^∞-embeddings $f_t: V \to W$, such that the embedding f_1 is isometric and*

the differential D_{f_1} *can be joined with* φ_1 *by a homotopy of isometric homomorphisms* $(T(V), g) \to (T(W), h)$.
(2) *Let* $q \geq n + 4$. *Then the above homotopy* f_t *exists for all (possibly closed) manifolds* V. *Moreover, one can choose the embeddings* f_t, $t \in [0, 1]$ *as* C^0*-close to* f_0 *as one wishes.*

Corollary. *If* V *is a contractible manifold and if* $n \leq q - 4$, *then every (nonisometric!)* C^1*-embedding* $V \to W$ *admits a fine* C^0*-approximation by isometric* C^∞*-embeddings.*

Proof. The construction of f_t proceeds in two steps. First f_0 is isotoped to a *symplectic* embedding $\bar{f}$ for which the induced form $\bar{f}^*(h)$ is nowhere singular (symplectic) on V. Then $\bar{f}$ is isotoped to the desired f_1 by a homotopy of symplectic embeddings.

Since the symplecticity of $\bar{f}$ (i.e. non-singularity of $\bar{f}^*(h)$) is an *open* differential relation, the existence of $\bar{f}$ in the case (1) follows from the results in (C′) of 2.4.6. Moreover, those results insure the existence of such an $\bar{f}$ for which the induced form $\bar{g}_0 = \bar{f}^*(h)$ admits a homotopy of non-singular (possibly nonclosed) forms $\bar{g}_t$, $t \in [0, 1]$ with $\bar{g}_1 = g$. The existence of $\bar{g}_t$ implies [since V is open, see (B‴) 2.2.3] the existence of an *exact* homotopy of *symplectic* form g_t between $\bar{g} = g_0$ and $g = g_1$, where "exact" means the existence of a homotopy of 1-forms, say α_t on V, such that $g_t = g_0 + d\alpha_t$ for all $t \in [0, 1]$. Now, an arbitrary homotopy α_t can be reduced by a small perturbation to a piecewise linear homotopy which interpolates a locally finite sequence of forms, say $\alpha_i = 0, 1, \ldots$, such that $\alpha_{i+1} = \alpha_i + x_i\, dy_i$, where x_i and y_i are C^∞-functions on V with support in a small open subset $U_i \subset V$. Indeed, the forms dy over small open subsets $U \subset V$ span the cotangent bundle $\Lambda^1(V)$ and the partition of unity applies. Thus the second step in the proof of (1) is reduced to the following.

(B′) **Lemma.** *Let* $g_t = g_0 + t\, dx \wedge dy$, $t \in [0, 1]$ *be symplectic forms on* V, *where the supports of the functions* x *and* y *lie in a contractible open subset* $U \subset V$, *and let* $f_0 \colon (V, g_0) \to (W, h)$ *be an isometric embedding of positive codimension. Then there exists a homotopy of isometric embeddings* $(V, g_t) \to (W, h)$ *which equal* f_0 *outside* U *and which lie in a given* C^0*-neighborhood of* f_0 *for all* $t \in [0, 1]$.

Proof. Let $da \wedge db$ be the standard area form on $\mathbb{R}^2$ with the coordinates a and b and let τ_ε^* be a C^∞-immersion of $\mathbb{R}^2$ into the disk $D_\varepsilon = \{a^2 + b^2 \leq \varepsilon\} \subset \mathbb{R}^2$, such that $\tau_\varepsilon^*(da \wedge db) = da \wedge db$. (The existence of τ_ε is obvious for all $\varepsilon > 0$.) Compose the map $(x, y) \colon U \to \mathbb{R}^2$ with τ_ε and thus obtain a C^∞-map, say $z_\varepsilon \colon U \to D_\varepsilon$ for which $z_\varepsilon^*(da \wedge db) = dx \wedge dy$. Denote by $\Gamma_t \colon U \to U \times D_\varepsilon$ the graph of the map tz_ε, $t \in [0, 1]$, and observe that $\Gamma_t^*(\tilde{g} = g_0 + da \wedge db) = g_0 + t^2\, dx \wedge dy$. Thus, the lemma follows (compare 3.1.2) from the existence of an isometric extension $\tilde{f}_\varepsilon \colon (U \times D_\varepsilon, \tilde{g}) \to W$ of the map f_0. Since the pertinent operator $\mathscr{D}$ is infinitesimally invertible [see (B) in 3.4.1], the local extension problem satisfies the h-principle. This reduces the existence of $\tilde{f}_\varepsilon$ (for small $\varepsilon > 0$) to that of an isometric homomorphism $\tilde{F}$: $T(U \times D_\varepsilon) \to T(W)$, such that $\tilde{F} | T(U)$ equals the differential of f_0. for $U = U \times 0 \subset U \times D_\varepsilon$. Since U is contractible, $\tilde{F}$ obviously exists and the lemma follows.

Proof of (2). The above argument provides the required homotopy in a small neighborhood of the $(n-1)$-skeleton of a given triangulation of V (for this $q \geq n-2$ suffices). This reduces (2) for $n \geq 4$ to the following

Relative *h*-Principle. *Let V be homeomorphic to the n-ball, let the map f_0 be isometric near the boundary $\partial V \approx S^{n-1}$ and let the homotopy $\varphi_t\colon T(V) \to T(W)$ be constant near ∂V. If $n \geq 4$ and $q \geq n+4$, then there exists a homotopy f_t as required in* (2) *which, moreover, is constant near ∂V.*

Proof. Let $\tilde{V}_\varepsilon = (V \times D_\varepsilon, g + da \wedge db) \supset V = V \times 0$. Then the map f_0 on $\mathcal{O}\!\mathit{p}\,\partial V \supset V$ extends, for small $\varepsilon > 0$, to an isometric embedding $\tilde{f}_0\colon \mathcal{O}\!\mathit{p}\,\partial V \times D_\varepsilon \to W$. Next, by applying (C′) of 2.4.6 as previously we extend this $\tilde{f}_0$ to a symplectic embedding $\tilde{f}$ of a small neighborhood $\mathcal{O}\!\mathit{p}\,V \to \tilde{V}_\varepsilon$ into W. Since $n > 2$ the relative cohomology $H^2(V, \partial V)$ vanishes which implies [see (C‴) in 2.2.3] the existence of an exact symplectic homotopy of forms g_t on $\mathcal{O}\!\mathit{p}\,V$, such that the (properly chosen) 1-forms α_t vanish on $\mathcal{O}\!\mathit{p}\,V \subset \tilde{V}_\varepsilon$. Then the earlier argument provides an isometric embedding $\mathcal{O}\!\mathit{p}\,V \to W$, whose restriction to $V \subset \mathcal{O}\!\mathit{p}\,V$ is the desired f_1.

Exercise. Fill in the detail in this argument.

The Case $n = 2$. If $q \geq 6$, then generic isometric immersions obviously are embeddings and the above (A) applies.

Exercise. Prove the *parametric* h-principle for isometric embeddings with a special consideration for the case $n = 2$.

Remark. The embedding h-principles (1) and (2), probably hold true for singular closed forms g on V if $\operatorname{rank} g \geq 2n - q + 2$ in the case (1) and $\operatorname{rank} g \geq 2n - q + 4$ for (2). However, codimension 2 symplectic embeddings of closed manifolds are related to some problems in the symplectic geometry which go beyond the h-principle.

Exercises. (a) Let g be a closed form of *constant* rank on V. Prove the h-principle (1) for $\operatorname{rank} g \geq 2n - q + 2$ and prove (2) for $\operatorname{rank} g \geq 2n - q + 4$.

(b) Let V be diffeomorphic to $\mathbb{R}^n$ and let U be an open subset in $(\mathbb{R}^{2p}, h = \sum_{i=1}^{p} dx_i \wedge dy_i)$ with a smooth (possibly empty) boundary. Prove the existence of a *proper* isometric embedding $(V, g) \to (U, h)$ for an arbitrary symplectic form g on V, provided $2p \geq n + 2$.

(C) *Stability of Symplectic Forms.* The Nash implicit function theorem can be replaced in the symplectic geometry by the following more elementary.

Stability Theorem (Darbaux-Moser-Weinstein, see Weinstein 1977). *Let g_t, $t \in [0,1]$, be a C^∞-homotopy of symplectic forms on V such that $g_t = g_0 + d\alpha_t$ where the 1-forms α_t vanish outside a compact subset $U \subset V$. Then there exists a C^∞-diffeotopy $\delta_t\colon V \to V$, for $t \in [0,1]$, $\delta_0 = \mathrm{Id}$, which is constant in t outside U and such that $\delta_t^*(g_t) = g_0$ for*

$t \in [0,1]$. *Furthermore, if the forms* α_t *vanish on a given submanifold* $V_0 \subset V$, *while* $d\alpha_t$ *vanish on the bundle* $T(V)|V_0$, *then one may choose* δ_t *constant on* V_0 *as well.*

Proof. Perturb the forms α_t, if necessary, in order to make them zero on $T(V)|V_0$. Then the vector field $\alpha_t^* = -I_{g_t}^{-1}(d\alpha_t/dt)$ (for which $\alpha_t^* \cdot g_t = d\alpha_t/dt$) vanishes on V_0 as well as outside U and $\alpha_t^* g_t = dg_t/dt$. Since the field α_t^* has compact support, it integrates to a unique diffeotopy δ_t such that $d\delta_t/dt = \alpha_t^*$ and $\delta_0 = \mathrm{Id}$. Then $(d/dt)(\delta^*(g_t)) = \delta_t^*(\alpha_t^* g_t + (dg_t/dt)) = 0$. Q.E.D.

Exercises. (a) Prove the results (A) and (B) using the stability theorem instead of Nash's theorem.

(b) Construct local coordinates $x_1, \dots, x_m, y_1, \dots, y_m$ in a small neighborhood of a given point in a $2m$-dimensional symplectic manifold (V, g), such that $g = \sum_{i=1}^m dx_i \wedge dy_i$.

(c) Let g_t, $t \in [0,1]$, be a homotopy of symplectic forms on a *closed* manifold V, such that $[g_0] = [g_1] \in H^2(V; \mathbb{R})$. Assume $H^2(V; \mathbb{R}) \approx \mathbb{R}$ and construct an isometry $(V, g_0) \to (V, g_1)$.

(d) Let g_0 and g_1 be symplectic C^{an}-forms on V. Show that every C^α-isometry $(V, g_0) \to (V, g_1)$ admits a fine C^∞-approximation by C^{an}-isometries.

(e) (Moser 1965; Green-Shiohama 1979). Let ω_0 and ω_1 be non-vanishing $C^\infty(C^{\mathrm{an}})$-smooth n-form on a connected oriented n-dimensional (possibly non-compact with or without boundary) manifold V, such that $\int_V \omega_0 = \int_V \omega_1 > 0$. Furthermore, if $\int_V \omega_0 = \infty$, assume

$$\int_U \omega_0 = \infty \Leftrightarrow \int_U \omega_1 = \infty,$$

for all open subsets $U \subset V$ with *compact* boundaries $\partial U = (\mathrm{Cl}U) \backslash U \subset V$. Construct a $C^\infty(C^{\mathrm{an}})$-diffeomorphism $f\colon V \to V$, such that $f^*(\omega_1) = \omega_0$.

State and prove a similar result for smooth *measures* (which are locally given by non-vanishing n-forms) on non-orientable manifolds V.

3.4.3 Contact Manifolds and Their Immersions

Let L be a C^∞-subbundle of the tangent bundle $T(W)$, let $L' = T(W)/L$ and let $\lambda\colon T(W) \to L'$ be the quotient homomorphism. There (obviously) exists a unique anti-symmetric bilinear map of L to L', called $d'\lambda\colon \Lambda^2 L \to L'$ such that every linear form α on L' satisfies $d(\alpha \circ \lambda) = \alpha \circ (d'\lambda)$. For example, if $\operatorname{codim} L = 1$, $k = 1$ and if the bundle L' is identified with the trivial line bundle, then λ reduces to a 1-form on W and $d'\lambda = d\lambda|L$. [Compare Gray (1959).]

Define for every vector $l \in L_w \subset T_w$, $w \in W$, the homomorphism $I(l) = I_{d'\lambda}(l)$: $L_w \to L'_w$ by $I(l)(l') = (l \cdot d'\lambda)(l') = d'\lambda(l, l')$ for all $l' \in L_w$. A subspace $K_0 \subset L_w$ is called $(d'\lambda)$-*regular* if the homomorphism $L_w \to \mathrm{Hom}(K_0, L')$ defined by $l \to I(l)|K_0$ is surjective.

Next, consider a C^∞-subbundle $K \subset T(V)$ and call a smooth map $f\colon (V, K) \to (W, L)$ *contact* if the differential D_f sends K to L. Such an f is called *regular* if D_f is injective on K_v and the D_f-image of K_v in L_w, $w = f(v)$ is $(d'\lambda)$-regular for all $v \in V$.

(A) **Lemma.** *Regular contact C^∞-maps $f: V \to W$ form a microflexible sheaf.*

Proof. Fix a local basis of 1-forms $\alpha_i: L' \to \mathbb{R}$, $i = 1, \ldots, r = \operatorname{codim} L$ and define a differential operator $\mathscr{D}$: (C^∞-maps $V \to W$) $\to$ (r-tuples of 1-forms on K) by $\mathscr{D}$: $f^* \mapsto \{f(\alpha_i \circ \lambda)|K\}$. The linearization of $\mathscr{D}$ acts on the tangent fields ∂ in $T(W)|V$ by [see (A) in 3.4.1].

$$\partial \mapsto \{d\partial \cdot (\alpha_i \circ \lambda) + \partial \cdot d(\alpha_i \circ \lambda)\}|K,$$

which reduces for fields ∂ in $L|V \subset T(W)|V$ to $\partial \mapsto \{\alpha_i \circ I(\partial)\}|K$. Therefore [compare (A) in 3.4.1], the operator $\mathscr{D}$ is infinitesimally invertible at regular contact maps f. Since these are solutions of $\mathscr{D}f = 0$, the lemma follows.

Exercise. Define *biregular* contact maps [compare (A′) in 3.4.1] and prove the h-principle for them.

(B) Let, $L = \operatorname{Ker} \lambda \subset T(W)$ for a 1-form λ on W. Then the subbundle L (as well as the form λ) is called *contact* if the 2-form $d'\lambda = d\lambda|L$ is nowhere singular. Then $\operatorname{Ker} d\lambda$ is a one-dimensional subbundle in $T(W)$ transversal to L, called $L_\lambda^\perp \subset T(W)$.

A vector field ∂ on W is called *contact* if it preserves the subbundle L. This is equivalent to

$$\partial\lambda|L = (\partial \cdot d\lambda + d(\partial \cdot \lambda))|L = 0.$$

Let ∂_0 and $\partial_\perp$ be the respective L and $L_\lambda^\perp$-components of ∂. Then $\partial\lambda|L = I(\partial_0) + d(\lambda(\partial_\perp))|L$. Hence, every function H on W defines a unique contact field ∂ (depending on the choice of λ) for which $\lambda(\partial_\perp) = H$, and $\partial_0 = -I^{-1}(dH)$. It follows, (compare 3.4.2) that diffeotopies of contact diffeomorphisms $(W, L) \supset$ sharply move all submanifolds in W. Thus, we conclude [compare (A) in 3.4.2] to the following

(B′) **Theorem.** *Let $L \subset T(W)$ be a contact subbundle of codimension one. Then for every subbundle $K \subset T(V)$ contact C^∞-immersions $(V, K) \to (W, L)$ satisfy the C^0-dense parametric h-principle, provided* $\dim V < \dim W$.

Corollaries and Exercises. (a) If $K = T(V)$, then contact immersions $V \to W$ are everywhere tangent to L. For $\dim W = 2n + 1$, $n = \dim V$, these are called *Legendre immersions*. According to the C^0-dense h-principle, one obtains (Duchamp 1983) the following

Approximation Theorem. *Let F_0: $T(V) \to K$ be a fiberwise injective homomorphism, such that $F_0^*(d'\lambda) \equiv 0$. Then the underlying continuous map f_0: $V \to W$ admits a fine C^0-approximation by Legendre immersions $f: V \to W$.*

(b) Let $W = (\mathbb{R}^{2p+1}, L = \operatorname{Ker} \lambda)$ for $\lambda = dz + \sum_{i=1}^p x_i\, dy_i$ and let K be a co-orientable [i.e. $K' = T(V)/K$ is orientable] contact subbundle in $T(V)$. *If $p \le m + n$ for* $\dim V = n = 2m + 1$, *then an arbitrary continuous map f_0: $V \to W$ admits a fine C^0-approximation by contact immersions.* Indeed, the inequality $p \ge m + n$ insures the existence of a homomorphism F_0: $(T(V), K) \to (T(W), L)$ which pulls back the form $d'\lambda$ on L to (a non-zero multiple of) the respective form $d'\kappa$ on K (where $\operatorname{Ker}\kappa = K$).

(b′) Prove (b) by the techniques of (E) in 2.1.3.

(c) Extend (C) of 3.4.2 to contact manifolds.

3.4.4 Basic Problems in the Symplectic Geometry

We have seen that the h-principle applies to symplectic isometric embeddings for codim ≥ 4. Furthermore, the construction of symplectic forms on *open* manifolds V is possible with the h-principle [see (B‴) in 2.2.3]. But the further study of symplectic manifolds brings forth new phenomena.

(A) *The Existence Problem.* Let a closed manifold V of dimension $n = 2m$ satisfy

(i) there is a cohomology class $\alpha \in H^2(V, \mathbb{R})$ such that $\alpha^m \neq 0$;

(ii) there exists a (possibly non-closed) nowhere singular 2-form g_0 on V. [This is equivalent to a reduction of the structure group of the tangent bundle $T(V)$ to the complex linear group $GL_m\mathbb{C}$, that is an *almost complex* structure on V.]

Does there exists a homotopy of non-singular forms g_t, $t \in [0, 1]$, for which g_1 is symplectic? (The h-principle predicts such a homotopy but one may expect further obstructions to the existence of g_t.) Does V support any symplectic form at all?

(A′) A similar problem arises for the contact structure. Namely, let V admit a codimension one subbundle $K \subset T(V)$ with a nowhere singular 2-form g_0: $K \to T(V)/K$ (if K is co-orientable one speaks of ordinary $\mathbb{R}$-valued 2-forms). Does K admit a homotopy to a contact subbundle? [The answer is "yes" for $\dim V = 3$, see Lutz (1971), Martinet (1971).]

(B) *Standard Examples of Symplectic and Contact Manifolds.* Let h be the standard symplectic form on $\mathbb{C}P^q$ [which is $U(q+1)$-invariant] and $\langle h, [\mathbb{C}P^1]\rangle = 1$. Then every *complex* submanifold $V \subset \mathbb{C}P^q$ is symplectic, since the induced form $h|V$ is nowhere singular on V. Thus every *complex algebraic manifold* admits a symplectic structure.

(B′) Let a Lie group L admit a left invariant symplectic form $\tilde{g}$ and let $\Gamma \subset L$ be a discrete subgroup. Then, by passing to the quotient, we get a symplectic form on $V = L/\Gamma$. This is especially interesting if V is compact.

Exercise. Let L be a nilpotent Lie group and let some $\alpha \in H^2(V; \mathbb{R})$ have $\alpha^m \neq 0$ for $2m = \dim L$. Prove the existence of an invariant symplectic form on L and thus obtain a symplectic form on $V = V/\Gamma$ for all discrete subgroup $\Gamma \subset L$. [See Cordero-Fernandez-Gray (1985) for further results of this type.]

(B″) Let V be a strictly pseudo-convex hypersurface in a complex manifold W. Then the subbundle $T(V) \cap \sqrt{-1}\,T(V) \subset T(V) \subset T(W)$ is (easily seen to be) contact. For example, the unit sphere $S^{2m+1} \subset \mathbb{C}^{m+1}$ carries a $U(m+1)$-invariant contact structure which, moreover, is invariant under the group $U(m+1, 1)$ (of automorphisms of the form $z_1\bar{z}_1 - \sum_{i=2}^{n+2} z_i\bar{z}_i$ in $\mathbb{C}^{m+2}$) which acts on S^{2m+1} by linear fractional transformations. This structure agrees (in an obvious way) with the symplectic form on $\mathbb{C}P^m$ for the Hopf map $S^{2m+1} \to \mathbb{C}P^m$ and the pull-back of every symplectic submanifold $V_0 \subset \mathbb{C}P^m$ is a contact submanifold in S^{2m+1}. Furthermore, if X is a complex analytic submanifold in $\mathbb{C}^{m+1}$ which transversally meets S^{2m+1} (or any pseudo-convex hypersurface for this purpose) then the intersection $X \cap S^{2m+1}$ is a contact submanifold in S^{2m+1}. In particular, every (germ of a) subvariety $X \subset \mathbb{C}^m$ with an isolated singularity at the origin meets every small sphere $S_\varepsilon^{2m+1} \subset \mathbb{C}^{m+1}$ over a contact submanifold. [In fact, Eliashberg (1983) proved every contact structure on a 3-dimensional manifold V to be the above $T(V) \cap \sqrt{-1}\, T(V)$ for some complex structure on $V \times \mathbb{R} \supset V$.]

Exercises. (a) Recall [see (C) of 3.2.2] the canonical contact structure on the oriented Grassmann bundle $Q = Gr_{q-1}P$ for a given q-dimensional C^{an}-manifold P and obtain this structure by a pseudo-convex embedding of P into a complexification $\mathbb{C}P \supset P$.

(b) Find local coordinates z, x_i, y_i, $i = 1, \ldots, m$, on S^{2m+1} minus a point, such that $\mathrm{Ker}(dz + \sum_{i=1}^m x_i\, dy_i)$ equals the subbundle $T(S^{2m+1}) \cap \sqrt{-1}\, T(S^{2m+1})$.

(c) Construct a contact structure on $S^{2m} \times S^1$, $m = 1, 2, \ldots$, which equals $\mathrm{Ker}(dz + \sum_{i=1}^m x_i\, dy_i)$ on the universal covering $\widetilde{S^{2m} \times S^1} = \mathbb{R}^{2m+1} \setminus \{0\}$.

(C) *Symplectic Fibrations.* Let (V, g) be a symplectic manifold, let $p\colon X \to V$ be a smooth fibration and let h_v be a symplectic form on the fiber $X_v = p^{-1}(v) \subset X$ which is smooth in v for all $v \in V$.

(C′) (Thurston 1973). *If there is a cohomology class $\beta \in H^2(X, \mathbb{R})$, such that $\beta | X_v = [h_v]$, then X admits a symplectic structure.*

Proof. There obviously exists a smooth closed 2-form h on X, such that $h | X_v = h_v$ for all $v \in V$. Then, assuming X is compact, (the open case is covered by the h-principle) the form $p^*(g) + \varepsilon h$ is non-singular on X for all small $\varepsilon > 0$. Q.E.D.

Example. (a) Let $Y \to V$ be a k-dimensional complex vector bundle and let $PY \to V$ be the associated projective bundle with the fiber $\mathbb{C}P^{k-1}$. Then one has, with a Hermitian metric in Y, the canonical symplectic forms on the fibers $X_v = \mathbb{C}P^{k-1}$. Furthermore, the Chern class c_1 of the canonical complex line bundle $X \to PY$ satisfies $c_1 | PY_v = [h_v]$ and, hence, (C′) applies.

(b) (Thurston 1973). Let X be a bundle of closed oriented surfaces over a symplectic manifold V, such that the homology class $[X_v] \in H_2(X)$ has an infinite order. Then the above h_v and β obviously exist and yield a symplectic structure on X.

Exercise (Thurston). Find a closed symplectic manifold X of dimension 4 which admits no Kähler structure.

(D) *Blow up*. Take the canonical line bundle $p\colon X_0 \to \mathbb{C}P^m$ and consider the obvious holomorphic map $\sigma_0^*\colon X_0 \to \mathbb{C}^{m+1}$. This σ_0 is biholomorphic outside the origin $0 \in \mathbb{C}^{m+1}$ and $\sigma_0^{-1}(0)$ equals $\mathbb{C}P^m$ (embedded into X_0 by the zero section). Then the pull-back $\sigma_0^*(g_0)$ on X of the standard form $g_0 = \sum_{i=1}^{m+1} dx_i \wedge dy_i$ on $\mathbb{C}^{m+1}$ can be perturbed to a *symplectic* form on X_0 as follows. Take the standard symplectic form g_1 on $\mathbb{C}P^m$ and observe that the form $p^*(g_1)$ on X_0 is cohomologous to zero outside $\mathbb{C}P^m \subset X_0$. Hence, there exists a closed form $\tilde{g}_1$ on X_0 which vanishes away from $\mathbb{C}P^m$ and which equals $p^*(g_1)$ near $\mathbb{C}P^m \subset X_0$. Then, obviously, the form $\sigma_0^*(g_0) + \varepsilon\tilde{g}_1$ is nowhere singular on X_0 for all small $\varepsilon > 0$.

Next, take a point v_0 in an arbitrary symplectic manifold (V, g) of dimension $2m + 2$. Then there is a small neighborhood $U_0 \subset V$ of v_0 which is isometric to the 2ε-ball B_{2_ε} in $(\mathbb{C}^{m+1}, \sum_{i=1}^{m+1} dx_i \wedge dy_i)$ around the origin. Delete the neighborhood $U \subset U_0$ of v_0 which corresponds to the ε-ball in $\mathbb{C}^{m+1}$ and then attach $\varepsilon_0^{-1}(B_{2_\varepsilon})$ to V by the map σ_0 over $U_0 \backslash U = B_{2_\varepsilon} \backslash B_\varepsilon$. The resulting *blow up* manifold $\tilde{V}$ (which is diffeomorphic to the connected sum $V \# \mathbb{C}P^{m+1}$) comes with a natural map $\tilde{\sigma}_0\colon \tilde{V} \to V$ which is a diffeomorphism of $\tilde{\sigma}_0^{-1}(V \backslash \{v_0\})$ onto $V \backslash \{v_0\}$ and for which $\tilde{\sigma}_0^{-1}(v_0) \approx \mathbb{C}P^m$. The above local construction provides a symplectic form on $\tilde{V}$ which equals $\tilde{\sigma}^*(g)$ outside any given neighborhood of $\tilde{\sigma}^{-1}(v_0) \subset \tilde{V}$.

Let us apply the "blow up" $\sigma_0\colon X_0 \to \mathbb{C}^{m+1}$ to the fibers of a complex vector bundle $Y \to V_0$. That is, we take the associated projective bundle $PY \to V_0$ with the canonical line bundle $X \to PY$ and then take the obvious map $\sigma\colon X \to Y$. This σ is diffeomorphic outside $V_0 \subsetneq Y$ (embedded by the zero section) while $\sigma^{-1}(V_0) = PY$. If V_0 is a symplectic manifold, then we equip PY with a symplectic form according to the above Example (a) and we extend this form to X by a closed form $\tilde{g}$ vanishing away from $PY \subset X$. Then we lift the symplectic form g_0 on V_0 to X and observe (an exercise to the reader) that the form $\sigma^*(g_0) + \varepsilon\tilde{g}$ is nowhere singular on X for small $\varepsilon > 0$.

Finally, we take a *symplectic* submanifold $V_0 \subset (V, g)$ (i.e. $g_0 = g | V_0$ is symplectic) and let $Y \to V_0$ be the g-normal bundle of V_0 in V. We equip Y with a *complex* bundle structure for which the multiplication by $\sqrt{-1}$ is isometric for $g | Y$ and we blow up this Y to the above $X \to Y$. Then, as earlier, we delete a small neighborhood $U \subset V$ of V_0 and attach some neighborhood of the submanifold $PY \subset X$. Thus we *blow up V along V_0* and we construct (the detail is up to the reader) a symplectic form on the blown up manifold.

Exercises. (a) (Mc Duff 1984). Construct, by blowing up $\mathbb{C}P^5$ along some 4-dimensional symplectic submanifold $V_0 \subset \mathbb{C}P^5$, a closed simply connected symplectic manifold which admits no Kähler structure.

(b) An immersed symplectic submanifold $V_0 \to V$, by definition, has a *symplectic crossing* if the self-intersection is transversal and if the k-intersection, say $V_k \to V$ *symplectically* immerses into V. Assume V_k to be empty for $k \geq k_0 + 1$ and blow up V along the (embedded!) symplectic submanifold $V_{k_0} \subset V$. Show that V_0 lifts to an immersed symplectic submanifold $\tilde{V}_0$ in the blown up manifold $\tilde{V}$, such that the self-intersection $\tilde{V}_k$ of $\tilde{V}_0$ is symplectic for $k < k_0$ and V_k is empty for $k \geq k_0$. Then blow up $\tilde{V}$ along $\tilde{V}_{k-1}$ and keep blowing up until the self intersection is completely eliminated.

(b′) Let V_1 be a symplectic submanifold in (V_0, g_0), such that $\operatorname{codim} V_1 = \dim V - \dim V_0 \geq 4$ and let the group $\mathbb{Z}_2 = \mathbb{Z}/2\mathbb{Z}$ freely and isometrically act on V_1. State and prove the embedding type h-principle for isometric immersions $(V_0, g_0) \to (V, g)$ which symplectically self intersect along V_1 according to the given $\mathbb{Z}_2$-action and have no self intersection apart from V_1.

Question. Can one define singular symplectic (sub) varieties? Can one resolve their singularities?

(c) Let (V, K) be a contact manifold and let $V_0 \subset V$ be a contact submanifold, i.e. the subbundle $K_0 = K \cap T(V_0) \subset T(V)$ is contact. Then the normal bundle Y of V_0 in V equals the $(d'\kappa)$-orthogonal complement of K_0 in $K | V_0$. Equip the bundle Y with a complex structure compatible with $d'\kappa | Y$ [i.e. the multiplication by $\sqrt{-1}$ is $(d'\kappa)$-conformal] and define the blow up of V along V_0 as earlier. Assume the existence of a free contact S^1-action on (V_0, K_0) which lifts to a fiberwise complex linear action on Y and construct a contact structure on the blow up manifold V. In particular, prove the existence of a contact structure on the blow up V, in case (V_0, K_0) is isomorphic to S^{2k+1} with the standard [$U(k+1)$-invariant] contact structure.

(E) *Ramified Coverings.* Let V_0 be a codimension two symplectic submanifold in a symplectic manifold (V, g) and let $p\colon \tilde{V} \to V$ be a ramified covering which ramifies along V_0. ($\tilde{V}$ is obtained from a finite covering of the complement $V \setminus V_0$ by completing with respect to a metric induced from some Riemannian metric in V.) The induced form $p^*(g)$ on $\tilde{V}$ can be easily perturbed to a *symplectic* form $\tilde{g}$ on $\tilde{V}$ which equals $p^*(g)$ away from $p^{-1}(V_0) \subset \tilde{V}$. Unfortunately, this construction gives few new examples, as symplectic *embeddings* $V_0 \to V$ violate the h-principle ($\operatorname{codim} V_0 = 2$!) and so the submanifolds $V_0 \subset V$ are hard to come by.

Exercise. Study symplectic ramified coverings with a *singular* ramification locus $V_0 \subset V$.

(E′) Consider two disjoint codimension 2 symplectic embeddings of V_0 into V and let Y_1 and Y_2 be their respective normal bundles. Take small tubular neighborhoods T_1 and T_2 of these embeddings, let $\delta\colon \partial T_1 \to \partial T_2$ be a diffeomorphism induced by an *orientation reversing* isomorphism $Y_1 \to Y_2$ (the forms $g | Y_1$ and $g | Y_2$ endow the bundles with natural orientations), and let $\tilde{V}$ be obtained from V by deleting $T_1 \cup T_2$ and gluing the boundary of $V \setminus (T_1 \cup T_2)$ by δ. Then the form g outside $T_1 \cup T_2$ easily extends to a *symplectic* form on the manifold $\tilde{V}$.

Exercises. (a) Generalize the above to *immersed* submanifolds $V_0 \subset V$ with g-normal crossings.

(b) Generalize (E) and (E′) to codimension 2 contact submanifolds in (V, K).

(c) (Meckert 1982). Construct a contact structure on the connected sum of two contact $(2m+1)$-dimensional manifold.

Hint. Use the contact action of $U(m+1, 1)$ on S^{2m+1}.

(F) The results indicated in (I) below provide many pairs of open subsets U_1 and U_2 in $(\mathbb{R}^{2m}, h_0 = \sum_{i=1}^m dx_i \wedge dy_i)$, $m \geq 2$, which admit a volume preserving diffeomorphism $U_1 \to U_2$ but not a symplectic (i.e. h_0-isometric) one. For example, the open bicylinders

$$U_i = \{x_1^2 + y_1^2 < a_i^2, x_2^2 + y_2^2 < b_i^2\} \subset \mathbb{R}^4, \qquad i = 1, 2, \text{ and } 0 < a_i < b_i,$$

are symplectically diffeomorphic if and only if $a_1 = a_2$ and $b_1 = b_2$.

A similar fact is known for contact 3-manifolds.

(F′) (Bennequin 1983). *There exists a contact subbundle K on $\mathbb{R}^3$ which admits no contact embedding into* $(\mathbb{R}^3, \mathrm{Ker}(dz + x\,dy))$.

(G) *Double Points of Lagrange Submanifolds.* The h-principle provides Lagrange immersions $f: V \to (W, h)$, $\dim W = 2 \dim V = 2n$, which have, after a generic Lagrange perturbation, a discrete subset of transversal double points. The lower bound on the possible number of these points is an interesting invariant of the symplectic form h (and of the topology of V) which is not reducible to the h-principle.

If the form h is exact, $h = d\alpha$, then the form $f^*(\alpha)$ is closed on V [since $f^*(h) = 0$] and the cohomology class $[f^*(\alpha)] \in H^1(V, \mathbb{R})$ depends on f and h but not on α. Then one distinguishes *exact* Lagrange immersions for which $[f^*(\alpha)] = 0$.

Example. The obvious Lagrange embedding of the torus T^n into $(\mathbb{R}^{2m} \sum_{i=1}^n dx_i \wedge dy_i)$ is not exact. In fact, no closed manifold admits an *exact* Lagrange *embedding* into $(\mathbb{R}^{2n}, \sum_{i=1}^n dx_i \wedge dy_i)$ (see Gromov 1985).

Exercises. (a) Show that the circle S^1 admits no Lagrange embedding into $\mathbb{R}^2$.

(b) Let V be a parallelizable manifold. Prove for $7 \neq n = \dim V \geq 2$ (compare totally real embeddings in 2.4.5) the existence of a symplectic form h on $\mathbb{R}^{2n}$ which receives an exact Lagrange embedding $f: V \to (\mathbb{R}^{2n}, h)$.

(b′) Show for the above V the existence of a (non-exact!) Lagrange embedding $V \times S^1 \to (\mathbb{R}^{2n+2}, \sum_{i=1}^{n+1} dx_i \wedge dy_i)$.

Hint. Start with an *isotropic* embedding $V \to (\mathbb{R}^{2n+2}, \sum_{i=1}^{n+1} dx_i \wedge dy_i)$.

(b″) Let V admit a Morse function with k critical points. Then construct an *exact* Lagrange immersion $V \times S^1 \to (\mathbb{R}^{2n+2}, \sum_{i=1}^{n+1} dx_i \wedge dy_i)$ with precisely k transversal double points.

(c) Construct for an arbitrary closed manifold V a closed symplectic manifold (W, h), $\dim W = 2 \dim V$ which receives a Lagrange embedding $f: V \to (W, h)$.

Hint. Represent V by the $\mathbb{R}$-points of a complex algebraic manifold W defined over $\mathbb{R}$.

(G′) *Fixed Points of Exact Diffeomorphisms.* An isometric diffeomorphism of a symplectic manifold, $f: (V, g) \circlearrowleft$ is called *exact* if there exists an exact diffeotopy (see 3.4.1) $f_t: (V, g) \circlearrowleft$, such that $f_0 = \mathrm{Id}$ and $f_1 = f$. For example, if ∂ is an exact field, $\partial = I_g^{-1}(H)$ for a C^∞-function H on V, then the one parameter subgroup of diffeomorphisms f_t: $V \circlearrowleft$ is clearly exact. Then every critical point of H is fixed under the diffeomorphisms f_t and so the number of the fixed points of f_1 can be estimated from below by Morse theory, assuming V is a closed manifold. Observe that the isometry condition $dI_g(\partial) = 0$ is equivalent to (the graph of) the 1-form $I_g(\partial)$: $V \to \bigwedge^1 V$ to be a Lagrange embedding for the canonical symplectic form h_0 on $\bigwedge^1 V$. In fact, Laudenbach and Sicorav (1985) extended the Morse theory to Lagrange submanifolds in $\bigwedge^1 V$ obtained from the zero section by an exact diffeomorphism.

Furthermore, fixed points of an arbitrary g-isometric diffeomorphism $f: V \to V$ are exactly the intersection of the following two Lagrange submanifolds in $(V \times V, g \oplus -g)$: the first is the diagonal and the second is the graph of f.

Exercise (Arnold 1974; Weinstein 1976). Let an exact diffeomorphism f of a closed symplectic manifold V be sufficiently C^1-close to the identity. Then there exists a C^∞-function $H = H_f$: $V \to \mathbb{R}$ (classically known as a *generating function* for f) whose critical points are exactly the fixed points of f.

The above C^1-closeness condition was relaxed by Weinstein (1983) to the C^0-*closeness* of the pertinent exact diffeotopy to the identity. The crucial new ingradient is due to Conley and Zehnder (1983) who solved the following

Arnold's Conjecture. *Let (T^{2m}, g) be a symplecting torus obtained from $(\mathbb{R}^{2m}, \sum_{i=1}^m dx_i \wedge dy_i)$ by dividing by a lattice $\mathbb{Z}^{2m} \subset \mathbb{R}^{2m}$ and let f be an exact g-isometric C^∞-diffeomorphism of T^{2m}. Then there exists a C^∞-function $H = H_f$ on V whose all critical points are fixed under f. Moreover, if the fixed points of f are non-degenerate, then the critical points of H also are nondegenerate. (In the latter case the number of the fixed points of f is bounded from below by 2^{2m} and in the former by $2m + 1$.)*

The idea of the proof (due to Conley and Zehnder) is indicated in the following

Exercise. (a) Consider an N-dimensional vector bundle over a closed manifold, say $Z \to V$ and let A: $Z \to \mathbb{R}$ be a C^∞-function whose restriction to each fiber $Z_v \approx \mathbb{R}^N$, $v \in V$, is a non-singular quadratic form (i.e. $\sum_{i=1}^k z_i^2 - \sum_{i=k+1}^N z_i^2$ in some linear basis in z_v). Let h be an arbitrary *bounded* C^∞-function on X. Show that the number of critical points of the function $A + h$ on Z is bounded from below by the Morse theory for functions on V. In particular, $A + h$ has at least two critical points (and at least $2m + 1$ if $V \approx T^{2m}$).

(a′) Generalize (a) by allowing fibrations X with infinite dimensional Hilbert space fibers $z_v \approx \mathbb{R}^\infty$.

(b) Let f_t, $t \in [0, 1]$ be an exact diffeotopy of a symplectic manifold (V, g), such that $df_t/dt = I_g^{-1}(dH)$, where $H = H(v, t)$ is a C^∞-function in $v \in V$ and $t \in \mathbb{R}$, which is periodic with period $= 1$ in t, and where the differential dH applies to the v-variable. Establish a one-to-one correspondence between the fixed points of f_1 and those C^∞-maps z: $S^1 \to V$, for $S^1 = \mathbb{R}/\mathbb{Z}$, which satisfy

$$(*) \qquad \frac{dz(t)}{dt} = I_g^{-1}(dH(z(t), t).$$

(b′) (Hamilton principle). Assume $V = T^{2m} = \mathbb{R}^{2m}/\mathbb{Z}^{2m}$ and define for contractible maps z, which lift to maps $\tilde{z}\colon S^1 \to (\mathbb{R}^{2m}, \sum_{i=1}^m dx_i \wedge dy_i)$, the action $A(y) = \int_{S^1} \tilde{z}^*(\alpha)$ for the 1-form $\alpha = \sum_{i=1}^m x_i\,dy_i$ on $\mathbb{R}^{2m}$. Show $(*)$ to be the Euler-Lagrange equation for the functional $A(z) + h(z)$ for $h(z) = \int_{S^1} H(z(t), t)\,dt$.

(c) Prove, by suitably adopting (a), that the above functional has as many critical points z, as Morse theory predicts for functions on T^{2m}. Prove Arnold's conjecture by approximating the space $Z_v \approx \mathbb{R}^\infty$ of maps $\tilde{z}\colon S^1 \to \mathbb{R}^{2m}$ with $\int_{S^1} \tilde{z}(t)\,dt = v \in \mathbb{R}^{2m}$ by the (finite dimensional!) spaces of truncated Fourier expansions of $\tilde{z}$. Prove the above-mentioned result by Weinstein for C^0-diffeotopies [see Chaperon (1983, 1984) for further results in this direction].

Remark. Arnold's conjecture for all closed *surfaces* was proven by Eliashberg (1978) by a more direct (and more complicated) method. [Compare Floer (1984), Sikorav (1984), Fortune-Weinstein (1984), Laudenbach-Sikorav (1985).]

(G″) *Contact Embeddings.* One expects the codimension two contact *embeddings* of closed manifolds to display certain "rigidity" (similar to that of Lagrange embeddings) which is not accountable for by the h-principle. The following result by Benniquin (1983) [compare Eliashberg (1981)] confirms this belief.

Theorem. *Let $S^1 \subset (\mathbb{R}^3, K = \operatorname{Ker}(dz + x\,dy))$ be a closed unknotted Legendre curve. Then S^1 has a strictly negative linking number with its small parallel translate in the direction of the z-coordinate.*

Exercise. Derive (F′) from this theorem.

(H) *The Group* Diff_g *of Sympletic Diffeomorphisms of* (V, g). Eliashberg (1981) announced (among many deep facts) that $\operatorname{Diff}_g^\infty$ is a C^0-*closed* subgroup in the group $\operatorname{Diff}_{g^m}$ of diffeomorphisms preserving the volume form g^m where $2m = \dim V$. One can recapture Eliashberg's result by combining the techniques indicated in (I) with the following.

Maximality Theorem. *Let (V, g) be a closed connected symplectic manifold. Let a (possibly non-closed) subgroup $\mathscr{G} \subset \operatorname{Diff}_{g^m}^\infty$ contain all g-exact diffeomorphisms and let some element $\psi \in \mathscr{G}$ be neither g-isometric nor anti-isometric, that is $\psi^*(g) \not\equiv \pm g$. (If m is odd, then no $\psi \in \operatorname{Diff}_{g^m}^\infty$ is anti-isometric.) If $H^1(V; \mathbb{R}) = 0$, then the subgroup $\mathscr{G}$ contains the connected component of the identity element $\operatorname{Id} \in \operatorname{Diff}_{g^m}^\infty$. Furthermore, if $H^1(V; \mathbb{R}) \neq 0$, then $\mathscr{G}$ contains the subgroup of g^m-exact diffeomorphisms which are defined in the course of the proof.*

Proof. A vector field ∂ on V preserves g^m iff $\partial g^m = d(\partial \cdot g^m) = m\partial \cdot g \wedge g^{m-1} = 0$. Such a ∂ is called g^m-*exact* if the cohomology class $[\partial \cdot g^m] \in H^{2m-1}(V, \mathbb{R})$ is zero. Thus every g^m-exact field ∂ equals $I_{g^m}^{-1}(d\varphi)$ for some $(2m-2)$-form φ on V.

Next consider N-tuples of symplectic forms $\{g_v\}$ on V such that $g_1^m = g_2^m = \cdots = g_N^m = \omega$. Call an N-tuple $\{g_v\}$ *large* [compare (D) in 2.3.8] if the forms g_v^{m-1} span the bundle $\Lambda^{2m-2} V$.

(H′) **Lemma.** *Let $\{g_\nu\}$ be a large N-tuple. Then every ω-exact field ∂ on V is a sum, $\partial = \sum_{\nu=1}^N \partial_\nu$ where ∂_ν is a g_ν-exact g_ν-isometric field for all $\nu = 1, \ldots, N$.*

Proof. Let $\partial \cdot \omega = d\varphi = I_\omega(\partial)$. Then, by the largeness, there are C^∞-functions H_ν on V such that $\varphi = m\sum_{\nu=1}^N H_\nu g_\nu^{m-1}$. Put $\partial_\nu = I_{g_\nu}^{-1}(dH_\nu)$ and compute,

$$\partial = I_\omega^{-1}(d\varphi) = mI_\omega^{-1}\left(\sum_{\nu=1}^N dH_\nu \wedge g^{m-1}\right) = mI_\omega^{-1}\left(\sum_{\nu=1}^N \partial_\nu \cdot g_\nu \wedge g_\nu^{m-1}\right)$$

$$= mI_\omega^{-1}\left(m^{-1}\sum_{\nu=1}^N \partial_\nu \cdot g_\nu^m\right) = I_\omega^{-1}\left(\sum_{\nu=1}^N I_\omega(\partial_\nu)\right) = \sum_{\nu=1}^N \partial_\nu. \quad \text{Q.E.D.}$$

(H″) **Remark.** One obtains a *canonical* decomposition $\partial = \sum_\nu \partial_\nu$ with a canonical choice of φ for $\partial \cdot \omega$. This is possible, for example, with the Hodge-De Rham theory.

Lemma. *There exist diffeomorphisms $f_\nu \in \mathscr{G}$, $\nu = 1, \ldots, N$, for some N, such that the N-tuple of the induced forms $g_\nu = f_\nu^*(g)$ is large.*

Proof. Consider diffeomorphisms in $\mathscr{G}$ which keep a given point $v_0 \in V$ fixed and let G_0 consist of the linear transformations of the tangent space $T_{v_0}(V) \approx \mathbb{R}^{2m}$ which are the differentials of all these diffeomorphisms at v_0. Since g-exact diffeomorphisms *transitively* act on V, one reduces the Lemma with a standard partition of unity argument, to showing that the G_0-orbit of the form $g_0^{m-2} = g^{m-2} | T_{v_0}(V)$ linearly spans the space $\Lambda^{2m-2} T_{v_0}(V)$. Since $\mathscr{G}$ does not preserve $\pm g$, the G_0-orbit of g_0 contains a 2-form g' which is *not* a scalar multiple of g_0. Then (an exercise in linear algebra) the span of the orbit of g' under linear symplectic transformations of $(T_{v_0}(V), g_0)$ equals $\Lambda^2 T_{v_0}(V)$. Hence, the span of the orbit $G_0(g_0)$ also equals $\Lambda^2 T_{v_0}(V)$ which implies (another exercise) the equality Span $G_0(g_0^{m-2}) = \Lambda^{2m-2} T_{v_0}^2(V)$. Q.E.D.

Let $\mathscr{X}$ be the Cartesian product of N copies of Diff_g^∞ and let $\mathscr{D}\colon \mathscr{X} \to \mathrm{Diff}_{g^m}^\infty$ be defined by $\mathscr{D}\colon (x_1, x_2, \ldots, x_N) \to \varphi_1 \cdot x_1 \cdot \varphi_2 \cdot x_2 \cdot \ldots \cdot \varphi_N \cdot x_N$ for some fixed $\varphi_\nu \in \mathscr{G}$ and for the group product in $\mathrm{Diff}_{g^m}^\infty$. If we identify the tangent space $T_f(\mathrm{Diff}_{g^m}^\infty)$ for $f = \mathscr{D}(\mathrm{Id}) = f = \varphi_1 \cdot \varphi_2 \cdot \ldots \cdot \varphi_N$ with $T_{\mathrm{Id}}(\mathrm{Diff}_{g^m}^\infty)$ by the right translation $\delta \mapsto \delta f^{-1}$ in $\mathscr{D}\mathrm{iff}_{g^m}^\infty$, then the linearization (differential) L of $\mathscr{D}$ at $\mathrm{Id} \in \mathscr{X}$ will send N-tuples of g-isometric fields to g^m-isometric fields by the formula

$$L\colon \{\partial_\nu\} \mapsto \sum_{\nu=1}^N D_{f_\nu}^{-1} \partial_\nu,$$

for $f_1 = \varphi_1, f_2 = \varphi_1 \cdot \varphi_2, \ldots, f_N = f = \varphi_1 \cdot \varphi_2 \cdot \ldots \cdot \varphi_N$.

We arrange the diffeomorphisms f_ν to make the N-tuple $\{f_\nu^*(g)\}$ large and then we solve the equation $\partial = L\{\partial_\nu\}$ for all g^m-exact fields ∂ on V by using (H′) and by observing that the differential D_{f_ν} bijectively maps g_ν-isometric fields, $g_\nu = f_\nu^*(g)$, onto g-isometric ones. Hence, the linearization L is surjective at $\mathrm{Id} \in \mathscr{X}$ and the above argument also shows this L to be surjective near $\mathrm{Id} \in \mathscr{X}$. Furthermore, (H″) provides a right inverse to L, to which the implicit function theorem in 2.3.2 applies. Strictly speaking, the results in 2.3.2 do not apply to our $\mathscr{D}$ which *is not* a differential operator. However, this operator (obviously) satisfies the estimates needed for the construction of (non-local) inversion $\mathscr{D}^{-1}$. Thus we come to the following conclusion.

Let $\delta_t \in \mathscr{D}\mathrm{iff}^\infty_{g^m}$, $0 \le t \le 1$, be a *g^m-exact diffeotopy, which means the exactness of the field $d\delta_t/dt$ for all $t \in [0,1]$, such that* $\delta_0 = \mathrm{Id}$. *Then δ_t for $t \in [0,\varepsilon]$, $\varepsilon > 0$, lies in the image of $\mathscr{D}$.* If $H^1(V,\mathbb{R}) \approx H^{2m-1}(V,\mathbb{R}) = 0$, then these diffeotopies (obviously) generate the connected component of $\mathrm{Id} \in \mathscr{D}\mathrm{iff}^\infty_{g^m}$. In general, they generate a subgroup of codimension $\le \dim H^1(V,\mathbb{R})$ in $\mathscr{D}\mathrm{iff}_{g^m}$ which *by definition* consists of all g^m-exact diffeomorphisms. Q.E.D.

Exercises. (a) Fill in the details in this proof. Then extend the maximality theorem to the subgroups $\mathscr{D}\mathrm{iff}^\infty_\omega \subset \mathscr{D}\mathrm{iff}^\infty$, for an arbitrary volume form ω on V, and to $\mathscr{D}\mathrm{iff}^\infty_K \subset \mathscr{D}\mathrm{iff}^\infty$ for contact manifolds (V, K).

(b) Let a subgroup $\mathscr{G} \subset \mathscr{D}\mathrm{iff}^\infty_{g^m}$, which is not assumed to contain $\mathscr{D}\mathrm{iff}^\infty_g$, be normalized by $\mathscr{D}\mathrm{iff}^\infty_g$, that is $f\mathscr{G}f^{-1} = \mathscr{G}$ for all $f \in \mathscr{D}\mathrm{iff}_g$, and let $\mathscr{G}$ contain a diffeomorphism ψ which is neither g-isometric nor anti-isometric. Show all g^m-exact diffeomorphism to lie in $\mathscr{G}$.

Hint. Show $\mathscr{G}$ to be normalized by *all* exact diffeomorphisms and apply Thurston's theorem (1974) on normal subgroups in $\mathscr{D}\mathrm{iff}_{g^m}$.

Remark. If one tries a direct proof of (b) one arrives, after the linearization, to a linear system of *difference* equations to which the formalism of 2.3.8 does not apply. It would be interesting to extend the technique of 2.3.8 to difference equations (e.g. by an approximation of differences by differentials), in order to prove (b) without (in fact, together with) Thurston's theorem. See Mc Duff (1984) and Banyaga (1978) for a further study of symplectic diffeomorphisms.

(I) *Pseudo-Holomorphic Curves in Symplectic Manifolds.* A 2-dimensional subset $S \subset V$ is called a *J-curve* for a given almost complex structure $J\colon T(V) \to T(V)$ [See (C) in 2.2.7] if there exists a (connected or not) smooth surface $\tilde{S}$ with an almost complex structure $\tilde{J}\colon T(\tilde{S}) \to T(\tilde{S})$ and a C^∞-map $f\colon \tilde{S} \to V$, such that

(i) f sends $\tilde{S}$ onto S and the set of double points, $\{(x,y) \in \tilde{S} \times \tilde{S} \mid f(x) = f(y)\}$, is discrete,

(ii) f is *pseudo-holomorphic*, which means $J \circ D_f(\tau) = D_f \circ \tilde{J}(\tau)$ for all tangent vectors $\tau \in T(\tilde{S})$.

Such an $(\tilde{S}, \tilde{J})$, if it exists, clearly is unique up to an isomorphism; the *parametrizing* map f is unique up to an automorphism of $(\tilde{S}, \tilde{J})$.

If the structure J is C^∞-smooth, then the space Σ of *closed* J-curves in V is (easily seen to be) locally finite dimensional. This is also true for *compact* J-curves S with the boundary condition $\partial S \subset V'$, where V' is *totally real* submanifold in V, that is $T(V') \cap JT(V') = 0$. Furthermore, if V is a closed symplectic manifold whose 2-form g is *J-positive*, which means $g(\tau, J\tau) > 0$ for all non-zero vectors $\tau \in T(V)$, then the area of closed J-curves $S \subset V$ (relative to a fixed Riemannian metric in V) obviously abides

$$(*) \qquad \operatorname{Area} S \le \mathrm{const} \, \|[S]\|,$$

where $[S] \in H_2(V)$ is the fundamental homology class of S and $\| \ \|$ denotes a fixed norm on the homology of V. (A similar inequality holds true for compact J-curves S whose boundary ∂S lies in a fixed *Lagrange* submanifold $V' \subset V$.) The inequality $(*)$ insures a weak compactness (in the current topology) of the space Σ' of closed J-curves $S \subset V$ with $\|[S]\| \leq$ const' $< \infty$, which leads [compare (B″) in 3.2.4] to certain existence theorems of closed J-curves in V (and of compact J-curves whose boundaries lie in a prescribed Lagrange submanifold $V' \subset V$). Here is a simple example [see Gromov (1985) and Mc Duff (1985) for further applications of J-curves to symplectic geometry].

(I′) *Let the symplectic (J-positive!) form g split,*

$$(V, g) = (S^2 \times V_1, g_0 + g_1),$$

where (V_1, g_1) is a closed symplectic manifold and where g_0 is an area form on the sphere S^2. Then, for each point $v \in V$, there exists a connected J-curve $S = S_v \subset V$ which contains v and such that $[S] = [S^2 \times v_1] \in H_2(V)$.

Exercise. Let a round ball $B(r)$ of radius r in the standard symplectic space $(\mathbb{R}^n, \omega_0)$, $n = \dim V$, admit a symplectic *embedding* $(B(r), \omega_0) \to (V, g)$. Show that $\pi r^2 \leq \int_{S^2} g_0$.

(I″) **Remark.** The above J-curve S_v may be singular (it may even have $\tilde{S}_v$ disconnected) and, in general, it is not unique. However, *if* $\dim V = 4$, *if the surface V_1 is connected and if $\int_{V_1} g_1 = \int_{S^2} g_0$, then S_v is unique. Moreover, this $S_v \subset V$ is a smoothly embedded sphere which is also smooth as a function in the variable J (as long as the inequality $g(\tau, J\tau) > 0$ holds).*

Exercise. Show the group of symplectic diffeomorphisms of $(S^2 \times S^2, g_0 + g_1)$ to be weakly homotopy equivalent to the subgroup of orientation preserving isometries of $S^2 \times S^2$ (which is a $\mathbb{Z}/2\mathbb{Z}$-extension of $SO(3) \times SO(3)$), provided $\int_{S^2} g_1 = \int_{S^2} g_0$. (If $\int_{S^2} g_1 \neq \int_{S^2} g_2$ the two groups are not w.h. equivalent.)

References

Adachi, M. (1979) Construction of complex structures on open manifolds, Proc. Jap. Acad. **55**, Ser. A, pp. 222–224.

Ahlfors, L.V. (1935) Zur Theorie der Überlagerungsflächen, Acta Math. **65**, pp. 157–194.

Alexander, J. (1920) Note on Riemann spaces, Bull. Am. Math. Soc. **26**, pp. 370–372.

Alexandrov, A. (1944) Intrinsic metrics of convex surfaces in spaces of constant curvature, Dokl. Akad. Nauk S.S.S.R., 45:1, pp. 3–6.

Allendoerfer, C.B. (1937) The embedding of Riemann spaces in the large, Duke Math. J. **3**, pp. 317–333.

Aminov, J.A. (1975) On estimates of the volume and diameter of a submanifold in Euclidean space, (Russian), Ukr. Geom. Sbornik **18**, pp. 3–15, 1982

Aminov, J.A. (1982) Embedding problems: Geometric and topological aspects, Itogi Nauki Tekh. Ser. Probl. Geom. (in Russian).

Arnold, V. (1957) On functions of three variables, Dokl. Akad. Nauk U.S.S.R., **114**, #4, pp. 679–681.

Arnold, V. (1974) Mathematical methods in classical mechanics, Moscow.

Arnold, V. (1980) Lagrange and Legendre cobordisms, J. Funct. Analysis and Applications, **14**, pp. 1–14.

Asimov, D. (1975) Round handles and non-singular Morse-Smale flows, Ann. Math. (1) **102**, pp. 41–54.

Asimov, D. (1976) Homotopy to divergence free vector fields, Topology **15**, pp. 349–352.

Atiyah, M. (1976) Elliptic operators, discrete groups and Von Neumann algebras, Asterisque 32–33, Soc. Math. de France, pp. 43–72.

Atiyah, M., Patodi, V. and Singer I. (1975) Spectral asymmetry and Riemannian geometry, Math. Proc. Camb. Phil. Soc. **77**, pp. 43–69.

Audin, M. (1984) Quelques calculs en cobordisme Lagrangien, Prépublication, Orsay.

Banyaga, A. (1978) On fixed points of symplectic maps, Preprint.

Baldin, Y. and Mercuri, F. (1980) Isometric immersions in codimension two with non-negative curvature, Math. Zeit. **173**, pp. 111–117.

Barnette, D.A. (1973) Proof of the lower bound conjecture for convex polytopes, Pac. J. Math. **46**:2, pp .349–354.

Barth, W. (1975) Larsen's theorem on the homotopy groups of projective manifolds of small embedding codimension, Proc. Symp. in Pure Math., XXIX, A.M.S., pp. 307–315.

Bennequin, D. (1983) Entrelacements et équations de Pfaff, Soc. Math. de France, Astérisque 107–108, pp. 87–161.

Berger, E. (1981) The Gauss map and isometric embedding, Theses, Harvard.

Berger, E., Bryant, E. and Griffiths, P. (1983) The Gauss equation and rigidity of isometric embeddings, Duke Math. Journ. **50**:3, pp. 803–892.

Bierstone, E. (1973) An equivariant version of Gromov's theorem, BAMS **79**, pp. 924–929.

Blank. S. (1967) Thesis, Brandeis.

Blanusa, D. (1955) Über die Einbettung hyperbolischer Raüme in euclidische Räume, Mon. Math. **59**, #3, pp. 217–229.

Boardman, J. (1967) Singularities of differentiable maps, Publ. Math. IHES n°33, pp. 21–57.

Borisenko (1977) Complete l-dimensional surfaces of non-positive extrinsic curvature in a Riemannian space, Math. **56**. (N.S.) 104 (146):4, pp. 559–576.

Borisov, Ju. (1965) $C^{1,\alpha}$-isometric immersions of Riemannian spaces, Dokl. Akad. Nauk, S.S.S.R., **163**, #, pp. 11–13.

Bott, R. (1969) Lectures on $K(X)$, W.A. Benjamin.

Bott, R. (1970) On topological obstructions to integrability, Proc. Symp. Pure Math. **16**, A.M.S., pp. 127–131.

Briant, L., Griffiths, P. and Yang, D. (1983) Characteristics and existence of isometric embeddings, Duke Math. Journ., **50**:4, pp. 893–995.

Brown, E. (1962) Cohomology theories, Ann. Math. **75**, pp. 467–484.

Burago, Yu. (1968) Inequalities of isoperimetric type in the theory of surfaces of bounded exterior curvature. Proc. Leningrad, Steklov Inst., v. 10.

Burago, Yu. and Zalgaller, V. (1960) Polyhedral imbedding of a net-vest., Leningrad University, **15**, #7, pp. 66–80.

Burago, Yu. and Zalgaller, V. (1980) Geometric Inequalities, Leningrad (The English translation by Springer-Verlag is in preparation).

Burlet, O. (1976) Propos au sujet des applications différentiables, Singularités d'applications différentiables, Springer Lect. Notes Math., **535**, pp. 187–204.

Chaperon, M. (1983) Quelques questions de géométrie symplectique, Sém. Bourbaki, Astérisque 105–106, pp. 231–249.

Chaperon, M. (1984) Questions de géométry symplectique, Sem. Sud-Rhodanien IV, Balarue, to appear in Travaux en cours, Hermann, Paris.

Cartan, H. (1953) Variétés analytiques complexes et cohomologie, 2° Coll. de géométrie algébrique. (Liège) Centre Belge de Rech. Math. pp. 41–55.

Cartan, H. (1957) Variétés analytiques réelles et variétés analytiques complexes, Bull. Soc. Math. Fr. **85**, pp. 77–99.

Cartan, H. (1958) Espaces fibrés analytiques, Symp. Int. de Top. Alg., Univ. Nac, de Mexico, pp. 97–121.

Cheeger, J. and Gromoll, D. (1971) The splitting theorem for manifolds of non-negative Ricci curvature, J. Diff. Geom. **6**, pp. 119–128.

Chen, K.T. (1971) Differential forms and homotopy groups, J. Diff. Geom., **6**, #2., pp. 231–246.

Chen, K.T. (1973) Iterated integrals of differential forms and loop space homology, Ann. Math. (2) **97**, pp. 217–246.

Cerf, J. (1984) Suppression des singularités de codimension plus grande que 2 dans les familles de functions differéntiables réelles, Sém. Bourbaki, Juin, Paris.

Chern, S.S. and Lashosf, R. (1957) On the total curvature of immersed manifolds, Am. J. Math. **79**, pp. 306–318.

Chern, S.S. and Kuiper N.H. (1952) Some theorems on the isometric imbeddings of compact Riemannian manifolds in Euclidean space, Ann. Math. **56**, #, pp. 422–430.

Clarke, C. (1970) On the global isometric embeddings of pseudo-Riemannian manifolds, Proc. Royal Soc. Lond. A-314, pp. 417–428.

Cohen, R.L. (1984) The homotopy theory of immersions, Proc. ICM in Warszawa, p. 627–639, North Holland.

Cohn-Vossen, S. (1933) Sur la courbure totale des surfaces ouvertes, C.R. Ac. Sc. (Paris) **197**, pp. 1165–1167.

Conley, C. and Zehnder, E. (1983) The Birkhoff-Lewis fixed point theorem and a conjecture of V.I. Arnold, Invent. Math. **73**, pp. 33–49.

Connelly, R. (1977) A counterexample to the rigidity conjecture for polyhedra, Publications Mathématiques IHES *n*° **47**, pp. 333–338.

Cordero L., Fernandez M. and Gray A. (1985) Symplectic manifolds with no Kähler structure, Preprint.

Cowen, M. and Griffiths, P. (1976) Holomorphic curves and metrics of negative curvature, J. An. Math. **29**, pp. 93–153.

D'Ambra (1985) Construction of connection inducing maps between principal bundles I, Preprint M.S.R.I., Berkeley.

Dajczer, M. and Gromoll, D. (1984) On spherical submanifolds with nullity, Preprint, SUNY at Stony Brook.

Deligne, P. and Sullivan, D. (1975) Fibrés vectoriels complexes à groupe structural discret, C.R. Ac. Sc. (Paris) **281**, Ser. A-1081.

Duchamp, T. (1984) The classification of Legendre immersions, Preprint.

Duplessis, A. (1975) Maps without certain singularities, Comm. Math. Helv. **50**, pp. 363–382.

Duplessis, A. (1976) Contact invariant regularity conditions, Singularités d'Applications Différentiables, Springer Lect. Notes Math. **535**, pp. 205–236.

Efimov, N. (1949) Qualitative problems in the local deformation theory of surfaces, Proc. Steklov Inst. XXX.
Efimov, N. (1964) Generation of singularities on surfaces of negative curvature, Math. Sb. (N.S.) **64**, pp. 286–320.
Eliashberg, J. (1970) On maps of the folding type, Izv. Akad. Nauk S.S.S.R. **34**, pp. 1111–1127.
Eliashberg, J. (1972) Surgery of singularities of smooth maps, Izv. Akad. Nauk S.S.S.R. **36**, pp. 1321–1347.
Eliashberg, J. (1978) Estimates of the number of fixed points of area preserving maps, preprint, Syktyvkar (in Russian).
Eliashberg, J. (1981) Rigidity of symplectic and contact structures, Preprint.
Eliashberg, J. (1984) Cobordisme des solutions de relations différentielles, Séminaire Sud-Rhodanien de géométrie symplectique I, Hermann, Paris.
Farrell, F.T. and Jones, L.E. (1981) Expanding immersions on branched manifolds, Am. Journ. of Math. **103**, pp. 41–101.
O'Farrell, A.G. (1986) C^∞-maps may increase C^∞-dimension, Preprint IHES, France.
Feit, S. (1969) k-mersions of manifolds, Acta Math. **122**, #3–4, pp. 173–195.
Feldman, E. (1968) Deformations of closed space curves, J. Diff. Geom. **2**, pp. 67–75.
Feldman, E. (1971) Nondegenerate curves on a Riemannian manifold, J. Diff. Geom. **5**, pp. 187–210.
Ferus, D. (1975) Isometric immersions of constant curvature manifolds, Math. Ann. **217**:2, pp. 155–156.
Ferus, D., Karcher, H. and Munzner, H. (1981) Cliffordalgebren und neue isoparametrische Hyperflächen, Math. Z. **177**, pp. 479–502.
Fiala, F. (1941) Le problème des isomérimétries sur les surfaces ouvertes à courbure positive, Comm. Math. Helv. **13**, pp. 293–346.
Floer, A. (1984) Proof of the Arnold conjecture for surfaces and generalizations for certain Kähler manifolds, Preprint, Bochum.
Forster, O. (1970) Plongements des variétés de Stein, Comm. Math. Helv. **45**, pp. 170–184.
Forster, O. (1971) Topologische Methoden in der Theorie Steinscher Räume, ICM-1970, Nice, pp. 613–618.
Fortune, B. and Weinstein, A. (1984) A symplectic fixed point theorem for complex projective spaces. To appear.
Friedman, A. (1961) Local isometric imbeddings of Riemannian manifolds, with indefinite metrics, J. Math. Mech. **10**, #4, pp. 625–649.
Friedman, A. (1965) Isometric imbeddings into Euclidean spaces, Rev. Mod. Phys. **37**, #1, pp. 201–203.
Fuchs, D. (1977) Non-trivialité des classes caractéristiques de g-structures, C.R. Ac. Sc. (Paris) **284**, pp. A-1105–1109.
Fuchs, D. (1981) Foliations, Itogi Nauki Tekh., Ser. Probl. Geom., **18**, pp. 191–213.
Garsia, A. (1961) An embedding of closed Riemann surfaces in Euclidean space, Comm. Math. Helv. **35**, #2, pp. 93–110.
Gluck, H. (1975) Almost all simply connected closed surfaces are rigid, Geom. Topology (Proc. Conf. Park-City, Utah), Springer Lect. Notes Math. **438**, pp. 225–239.
Godbillon, C. and Vey, J. (1971) Un invariant des feuilletages de codimension 1, C.R. Ac. Sci. (Paris) Ser. A–B 273, pp. A 92–A 95.
Godement, R. (1958) Topologie algébrique et théorie des faisceaux, Hermann, Paris.
Golubitsky, M. and Guillemin, V. (1973) Stable mappings and their singularities, graduate Text in Math. Springer-Verlag.
Grauert, H. (1957) Holomorphe Funktionen mit Werten in komplexen Lieschen Gruppen, Math. Ann. **133**, pp. 450–472.
Grauert, H. (1958) On Levi's problem and the embedding of real analytic manifolds, Ann. Math. (2) **68**, pp. 460–472.
Gray, A. (1969) Isometric immersions in symmetric spaces, J. Diff. Geom. **3**, pp. 237–244.
Gray, J.W. (1959) Some global properties of contact structures, Ann. Math. **69**, pp. 512–540.
Greene, R. (1970) Isometric embeddings of Riemannian and pseudo-Riemannian manifolds, Mem. Am. Math. Soc. #97.
Greene, R. and Jacobowitz, H. (1971) Analytic isometric embeddings, Ann. Math **93**, pp. 189–203.
Greene, R. and Wu, H. (1975) Whitney's imbedding theorem by solutions of elliptic equations and geometric consequences, Proc. Symp. Pure Math. XXVII, #2, A.M.S., pp. 287–297.

Greene, R. and Shiohama, K. (1979) Diffemorphisms and volume preserving embeddings of non-compact manifolds, Trans. Am. Math. Soc. **255**, pp. 403–414.
Griffiths, P. (1974) Entire holomorphic mappings in one and several complex variable, Ann. of Math. Studies, **85**, Princeton.
Gromoll, D., Klingenberg, W. and Meyer, W. (1968) Riemannsche Geometrie im Grossen, Springer Lect. Notes Math. **55**.
Gromoll, D. and Meyer, W. (1969) On complete open manifold of positive curvature, Ann. Math. **90**, pp. 75–90.
Gromov, M. (1969) Stable maps of foliations into manifolds, Izv. Akad. Nauk, S.S.S.R. **33**, #4, pp. 707–734.
Gromov. M. (1971) A topological technique for the construction of solutions of differential equations and inequalities, ICM. 1970, Nice, vol. 2, pp. 221–225.
Gromov, M. (1972) Smoothing and inverting differential operators, Mat. Sbornik, pp. 383–441.
Gromov, M. (1973) Convex integration of differential relations, Izv. Akad. Nauk S.S.S.R. **37**, #2, pp. 329–343.
Gromov, M. (1981) Curvature, diameter and Betti numbers, Comm. Math. Helv. **56**, pp. 179–195.
Gromov, M. (1983) Filling Riemannian manifolds, J. Diff. Geom. **18**, pp. 1–147.
Gromov, M. (1985) Pseudo-holomorphic curves in symplectic manifolds, 82, pp. 307–347 Inv. Math.
Gromov, M. and Eliashberg, J.(1971), Construction of nonsingular isoperimetric films, Trudy Steklov Inst. **116**, pp. 18–33.
Gromov, M. and Eliashberg, J.(1971), Removal of singularities of smooth maps, Izv. Akad. Nauk, S.S.S.R. **35**, #5, pp. 600–627.
Gromov, M. and Eliashberg, J.(1971) Nonsingular maps of Stein manifolds, Func. Anal. and Applications **5**, #2, pp. 82–83.
Gromov, M. and Eliashberg, J.(1973) Construction of a smooth map with prescribed Jacobian, Func. Anal. and Applications **7**, #1, pp. 27–32.
Gromov. M. and Lawson, B. (1980) The classification of simply connected manifolds of positive scalar curvature, Ann. of Math. **111**, pp. 423–434.
Gromov, M. and Lawson, B. (1983) Positive scalar curvature and the Dirac operator on complete Riemannian manifolds, Publications Mathématiques IHES $n°$ **58**, pp. 295–408.
Gromov, M. and Rochlin, V. (1970) Embeddings and immersions in Riemannian geometry, Uspechi, XXV, #5, pp. 3–62.
Gunning, R. and Narasimhan, R. (1967) Immersion of open Riemann surfaces, Math. Ann. **174**, pp. 103–108.
Gunning, R. and Rossi, H. (1965) Analytic functions of several complex variables, Prentice-Hall Inc. New Jersey.
Haefliger, A. (1962) Plongements différentiables dans le domaine stable, Comm. Math. Helv. **37**, pp. 155–176.
Haefliger, A. (1962) Variétés, feuilletés, Ann. Sc. Norm. Sup. Pisa (3), **16**, pp. 367–397.
Haefliger, A. (1970) Homotopy and integrability, Springer Lect. Notes Math., **197**, pp. 133–164.
Haefliger, A. and Hirsch, M. (1962) Immersions in the stable range, Ann. of Math. (2) **75**, pp. 231–241.
Halpern, B. and Weaver, C. (1977) Inverting a cylinder through isometric immersions and isometric imbeddings, Trans. Ann. Math. Soc. **230**, pp. 41–70.
Hamenstädt U. (1985) Non-degenerate closed curves in S^n. To appear in Composito Mathem.
Hamilton, R. (1982) The inverse function theorem of Nash and Moser, BAMS **7**:1, pp. 65–222.
Hartman, P. (1971) On the isometric immersions in Euclidean space of manifolds with non-negative sectional curvatures II, Trans. Am. Math. Soc. **147**:2, pp. 529–540.
Hayman, W. (1964) Meromorphic functions, Oxford Math. Monographs, Oxford.
Hironaka, H. (1973) Subanalytic sets, Numb. Theory, Alg. Geom. and Comm. Alg., in honour of Y. Akizuki, Tokyo, pp. 453–493.
Hirsch, M. (1959) Immersions of manifolds, Trans. Am. Math. Soc. **93**, pp. 242–276.
Hirsch, M. (1961) On imbedding differential manifolds into Euclidean space, Ann. Math. **73**, pp. 566–571.
Hitchin, N. (1974) Harmonic spinors, Adv. Math. **14**, pp. 1–55.
Hörmander, L. (1963) Linear partial differential operators, Springer, Berlin Heidelberg New York.

Hörmander, L. (1976) The boundary problems of physical geodesy, Arch. Rat. Mech. Anal. **62**, #1, pp. 1–52.
Hopf, H. (1925) Über die Curvatura Integra geschlossener Hyperflächen, Math. Ann. **95**, pp. 340–367.
Huber, A. (1957) On subharmonic functions and differential geometry in the large, Comm. Math. Helv. **32**, #1, pp. 13–72.
Ionin, V. (1969) Isoperimetric and certain other inequalities for manifolds of bounded curvature, Sib. Math. J. **10**, pp. 329–342.
Jacobowitz, H. (1972) Implicit function theorems and isometric embeddings, Ann. Math. **95**, pp. 191–225.
Jacobowitz, H. (1973) Isometric embedding of a compact Riemannian manifold into Euclidean space, Proc. Ann. Math. Soc. **40**: 1, pp. 245–246.
Jacobowitz, H. (1974) Extending isometric imbeddings, J. Diff. Geom. **9**, pp. 291–307.
Jacobowitz, H. (1976) Equivariant embeddings of flat tori, Preprint.
Jacobowitz, H. (1982) Local isometric embeddings, Sem. on Diff. Geom., pp. 381–393, Ann. of Math. Studies 102, Princeton.
Janet, M. (1926) Sur la possibilité de plonger un espace riemannien donné dans un espace euclidien, Ann. Soc. Pol. Math **5**, pp. 38–43.
Jost, J. and Karcher, H. (1982) Geometrische Methoden zur Gewinnung von A-priori-schranken für harmonische Abbildungen, Manuscripta Math. **40**, pp. 27–77.
Jorge, L.P. and Xavier, F.V. (1981) An inequality between the exterior diameter and the mean curvature of bounded immersions, Math. Z. **178**, pp. 77–82.
Kalai, G. (1986) Rigidity and the lower bound theorem I, preprint Hebrew University, Jerusalem.
Källen, A. (1978) Isometric embedding of a smooth compact manifold with a metric of low regularity, Arch. Mat. **16**, #1, pp. 29–50.
Kazdan, J. and Warner, F. (1975) Prescribing curvatures, Proc. Symp. Pure Math. XXVII, #2, A.M.S., pp. 309–321.
Kervaire, M. and Milnor, J. (1960) Bernoulli numbers, homotopy groups and a theorem of Rohlin, Proc. ICM 1958, pp. 454–458, Cambridge Univ. Press New York.
Kirby, R. and Siebenmann, L. (1977) Foundational essays on topological manifolds, smoothing and triangulations, Ann. Math. St. Princeton.
Kobayashi, S. and Nomizu, K. (1963, 1969) Foundations of differential geometry, v. I and II, Wiley-Inc. New York.
Kodaira, K. (1964) On the structure of compact complex analytic surfaces, I, Am. J. of Math. **86**, pp. 751–798.
Kolmogorov, A. (1956) On representation of continuous functions of several variables by superpositions of continuous functions of fewer variables, Dokl. Akad. Nauk, S.S.S.R. **108**, #2, pp. 179–182.
Kuiper, N.H. (1955) On C^1-isometric imbeddings, I. Proc. Koninkl. Nederl. Ak. Wet., A-58, pp. 545–556.
Kuiper, N.H. (1958) Immersions with minimal total absolute curvature, Coll. Géom. Diff. Glob., Centre Belge de Rech. Math., pp. 75–88.
Labouri, P. (1984) Thesis, Paris.
Landweber, P. (1974) Complex structures on open manifolds, Topology **13**, pp. 69–75.
Lannes, J. (1982) La conjecture des immersions, Sém. Bourbaki, Astérisque 92–93, Soc. Math. de France, pp. 331–347.
Laudenbach, F. and Sikorav J-C. (185) Persistance d'intersection avec la section nulle au cours d'une isotopie hamiltonienne dans un fibre cotangent, Préprint, Orsay.
Lawson, B. (1974) Foliations, Bull. Ann. Math. Soc. **80**, #3, pp. 417.
Lees, J. (1976) On the classification of Lagrange immersions, Duke Math. J. **43**, pp. 217–224.
Levin, H. (1965) Elimination of cusps, Topology **3**, pp. 263–296.
Levin, H. (1971) Blowing up singularities, Springer Lect. Notes Math. **192**, pp. 90–104.
Levin, H. (1971) Singularities of differentiable mappings, Springer Lect. Notes Math. **192**, pp. 1–90.
Little, J. (1970) Nondegenerate homotopies of curves on the unit 2-sphere, J. Diff. Geom. **4**, pp. 339–348.
Little, J. (1971) Third order nondegenerate homotopies of space curves, J. Diff. Geom. **5**, pp. 503–515.
Lojasievicz, S. (1965) Ensembles semi-analytiques, Lecture Notes IHES, Bures-sur-Yvette.
Lutz, R. (1971) Sur quelques propriétés des formes différentielles en dimension 3, Thèse Strasbourg.
Martinet, J. (1971) Formes de contact sur les variétés de codimension 3, Springer Lect. Notes Math. **209**, pp. 142–163.

Mather, J. (1971) On Haefliger classifying space, Bull. Ann. Math. Soc. **77**, pp. 1111–1115.
Mather, J. (1973) Solutions of generic linear equations, Dynamical System, ed. Peixoto. Acad. Press.
Mattila, P. and Rickman, S. (1979) Averages of the counting function of a quasiregular mapping, Acta Math. **143**, pp. 273–305.
Mc Duff, D. (1978) On the group of volume preserving diffeomorphisms of $\mathbb{R}^n$.
Mc Duff, D. (1979) Foliations and monoids of embeddings, Geom. Top. (ed. Contrell) Ac. Press.
Mc Duff, D. (1981) On groups of volume preserving diffeomorphisms and foliations with transverse volume form, Proc. London Math. Soc. **43**, pp. 295–320.
Mc Duff, D. (1984*) Application of Convex integration to symplectic and contact geometry, Preprint, SUNY at Stony Brook.
Mc Duff, D. (1984**) Examples of simply connected symplectic manifolds which are not Kähler, Preprint, SUNY at Stony Brook.
Mc Duff, D. (1985) Examples of symplectic structures, Preprint SUNY.
Meckert, C. (1982) Forme de contact sur la somme connexe de deux variétés de contact de dimension impaire, Ann. Inst. Four. **32**:3, pp. 251–260.
Milnor, J. (1963) Morse theory, Ann. Math. St. **51**, Princeton.
Milnor, T. (1972) Efimov's theorem about complete immersed surfaces of negative curvature, Adv. Math. **8**, pp. 474–543.
Mishachev, N. (1979) Constructions of flags of foliations, Uspechy **34**, #1 (205), pp. 237–238.
Mishachev, N. and Eliashberg, J. (1977) Surgery of singularities of foliations, Func. Anal. and Applications **11**, #3, pp. 43–53.
Montogomery, D. and Zippin, L. (1955) Topological transformation groups, Interscience Publ., Wiley and Sons N.Y.
Moore, J.D. (1972) Isometric immersions of space forms in space forms, Pac. J. Math. **40**, pp. 157–166.
Moore, J.D. (1978) Codimension two submanifolds of positive curvature Proc. A.M.S. **70**, p. 72–74.
Moore, J.D. and Schlafly, R. (1980) On equivariant isometric embeddings, Math. Z. **173**, pp. 119–133.
Morin, B. (1965) Formes canoniques des singularités d'une application différentiable, C.R. Ac. Sci. (Paris) **260**, pp. 5662–5665.
Morse, A.P. (1939) The behavior of a function on its critical set, Ann. Math. **40**, pp. 62–70.
Moser, J. (1961) A new technique for the construction of solutions of nonlinear differential equations, Proc. Nat. Ac. Sci. U.S.A., **47**, #11, pp. 1824–1831.
Moser, J. (1965) On the volume elements on a manifold, Trans. Am. Math. Soc. **120**, pp. 286–294.
Narasimhan, R. (1967) On the homology groups of Stein spaces, Inv. Math. **2**, pp. 377–385.
Narasimhan, M. and Ramanan, S. (1961) Existence of universal connections, Am. J. Math. **83**, pp. 563–572.
Nash, J. (1954) C^1-isometric imbeddings, Ann. Math. **60**, #3, pp. 383–396.
Nash, J. (1956) The imbedding problem for Riemannian manifolds, Ann. Math. **63**, #1, pp. 20–63.
Nash, J. (1966) Analyticity of the solutions of implicit function problems with analytic data, Ann. Math. **84**, #2, pp. 345–355.
O'Neil, B. (1962) Isometric immersions of flat Riemannian manifolds in Euclidean space, Michig. Math. J. **9**, pp. 199–205.
Novikov, S.P. (1965) The homotopy and topological invariance of certain rational Pontryagin classes, Dokl. Akad. Nauk, S.S.S.R., **162**, pp. 1248–1251.
Otsuki, T. (1953) On the existence of solutions of a system of quadratic equations and its geometric application, Proc. Jap. Ac. **29**, pp. 99–100.
Pasternak, J. (1975) Classifying spaces for Riemannian foliations, Proc. Symp. Pure Math., A.M.S., v. 27, part 1, pp. 303–310.
Phillips, A. (1967) Submersions of open manifolds, Topology **6**, #2, pp. 170–206.
Phillips, A. (1969) Foliations on open manifolds, Comm. Math. Helv. **44**, pp. 367–370.
Phillips, A. (1974) Maps of constant rank, Bull. Ann. Math. Soc. **80**, #3, pp. 513–517.
Poenaru, V. (1966) Regular homotopy in codimension 1, Ann. Math. (2) **83**, pp. 257–265.
Poenaru, V. (1968) Extensions des immersios en codimension 1 (d'apres S. Blanck) Sem. Bourbaki 342-01–342-33.
Poenaru, V. (1970) Homotopy theory and differential singularities, Springer Lect. Notes Math. **197**, pp. 106–133.

Pogorelov, A. (1969) Exterior geometry of convex surfaces, Nauka, (Russian).
Pogorelov, A. (1971) An example of a 2-dimensional metric admitting no local realization in E_3, Dokl. Akad. Nauk S.S.S.R. 198, 1, pp. 42–43.
Poznjak, E. (1973) Isometrical immersions of 2-dimensional metrics into Euclidean spaces, Uspechy **28**, #4, pp. 47–76.
Poznjak, E. and Sokolov, D. (1977), Isometrical immersions from a Riemannian space into Euclidean spaces, (Itogi Nauki), Tekh. Ser. Probl. Geom. pp. 173–211.
Ramspott, K.J. (1962) Über die Homotopieklassen holomorpher Abbildungen in homogene komplexe Manningfaltgkeiten, Bayer. Ak. Wiss. Math. Natur. Kl. S.-B. Abt II, pp. 57–62.
Reinhart, B. (1983) Foliation, Erg. Math. Springer-Verlag.
Ritt, J. (1950) Differential algebra, Ann. Math. Soc. Publications, V. XXXIII.
Rolfsen, D. (1976) Knots and links, Math. Lect. Series, Publish or Perish, Inc.
Rosendorn, E. (1960) Realization of the metric $ds^2 = du^2 + f^2(u)\,dv^2$ in the five-dimensional Euclidean space, Dokl. Akad. Nauk Arm. S.S.S.R.
Rosendorn, E. (1961) Construction of a bounded complete surface of non-positive curvature, Uspechy **16**, #2, pp. 149–156.
Rosendorn, E. (1966) Weakly non-regular surfaces of negative curvature, (Russian), U.M.N. 5 (131), (Russian Math. Surveys), pp. 59–116.
Sard, A. (1942) The measure of the critical values of differentiable maps, Bull. Ann. Math. Soc. **48**, pp. 8883–890.
Schaft, U. (1984) Einbettungen Steinscher Mannigfaltigkeiten, Manuscripta Math. **47**, pp. 175–186.
Schilt, H. (1937) Über die isolierten Nullstellen der Flächenkrümmung Comp. Math. **5**, #2, pp. 105–119.
Schoen, R. (1984) Minimal surfaces and positive scalar curvature. ICM 1983, Warszawa, p. 575–579, North-Holland.
Schoen, R, and Yau, S.T. (1979) Existence of incompressible minimal surfaces and the topology of three dimensional manifolds of positive scalar curvature, Ann. Math. **110**, pp. 127–142.
Schoen, R. and Yau, S.T. (1979) On the structure of manifolds with positive scalar curvature, Manuscripta Math. **28**, pp. 159–183.
Schur, F. (1886) Über die Deformation der Räume konstanter Riemannschen Krümmungsmass, Math. Ann. **27**, pp. 170.
Schwartz, J. (1960) On Nash's implicit function theorem, Comm. Pure and Appl. Math. **13**, pp. 509–530.
Seeley, R.T. (1964) Extensions of C^∞-functions defined in a half space, Proc. Ann. Math. Soc. **15**, pp. 625–626.
Segal, G. (1978) The classifying space for foliations, Topology **17**, #4, pp. 367–383.
Serre, J.-P. (1953) Groupes d'homotopie et classes de groupes abéliens, Ann. Math. **58**, pp. 258–294.
Sergeraert, F. (1972) Un théorème de fonctions implicites sur certains espaces de Fréchet et quelques applications, Ann. Ec. Norm. Sup., 4ème sér., v. 5, pp. 599–660.
Siebenmann, L. (1973) Approximating cellular maps by homeomorphisms, Topology **11**, pp. 271–294.
Singer, I.M. (1960) Infinitesimally homogeneous spaces, Comm. Pure Appl. Math. **13**, pp. 685–697.
Smale, S. (1958) A classification of immersions of the two-sphere, Trans. Ann. Math. Soc. **90**, pp. 281–290.
Smale, S. (1959) The classification of immersions of spheres in Euclidean spaces, Ann. Math., pp. 327–344.
Shub, M. (1969) Endomorphisms of compact differential manifolds, Am. J. Math. XCI. #1, pp. 175–199.
Shulman, H. (1972) Characteristic classes and foliations, thesis, U.C. Berkeley.
Sikorav, J.C. (1984) Points fixes d'un symplectomorphisme homologue à l'identité, C.R. Ac. Sc., to appear.
Spanier, E.H. and Whitehead, J.H.C. (1957) Theory of carriers and S-theory, Alg. Geom. and Top. (A symp. in Honour of S. Lefschets), Princeton Univ. Press, Princeton N.J.
Spivak, M. (1979) A comprehensive introduction to differential geometry, Publish or Perish Inc., Berkeley.
Spring, D. (1983) Convex integration of non-linear systems of partial differential equations, Ann. Inst. Fourier, **33**:3, pp. 121–177.
Spring, D. (1984) C^∞-solutions to non-linear undetermined systems of partial differential equations, Preprint York University, Canada.
Stiel, E. (1965) Isometric immersions of manifolds of non-negative constant sectional curvature, Pac. J. Math. **15**, pp. 1415–1419.
Sullivan, D. (1973) Differential forms and topology of manifolds, Proc. Conf. on Manifolds, Tokyo.

Szuecs, A. (1982) The Gromov-Eliashberg proof of Haefliger's theorem, Studia Sci. Math. Hung. **17**, pp. 303–3/8.

Szuecs, A. (1983) On detached immersions, Ibid, to appear.

Szuecs, A. (1984) Cobordism groups of immersions with restricted self-intersections, Osaka Journ. of Math. **21**, pp. 71–80.

Szuecs, A. (1976) Cartan-De Rham homotopy theory, Astérisque 32–33, pp. 227–253.

Thom, R. (1955, 1956) Les singularités des applications différentiables, Ann. Inst. Fourier **6**, pp. 43–87.

Thom, R. (1959) Remarques sur les problèmes comportant des inéquations différentielles globales, Bull. Soc. Math. Fr. **87**, pp. 455–461.

Thurston, W. (1973) Some simple examples of symplectic manifolds, Preprint.

Thurston, W. (1974) Foliations and groups of diffeomorphisms, Bull. Ann. Math. Soc. **80**, #2, pp. 304–307.

Thurston, W. (1974) The theory of foliations of codimension greater than one, Comm. Math. Helv. **49**, pp. 214–231.

Thurston, W. (1976) Existence of codimension-one foliations, Ann. Math. **104**, pp. 249–268.

Thurston, W. (1978) Geometry and topology of 3-manifolds, Lecture Notes, Princeton.

Tischler, D. (1977) Closed 2-forms and an embedding theorem for symplectic manifolds, J. Diff. Geom. **12**, pp. 229–235.

Toda, H. (1962) Composition methods in homotopy groups of spheres, Princeton Univ. Press.

Tomkins, C, (1939) Isometric imbedding of flat manifolds in Euclidean space, Duke Math. J. **5**, pp. 38–61.

Verner, A. (1970) Non-boundedness of a hyperbolic horn in Euclidean space, (Russian) Sib. Math. Journ. **9**: 1, pp. 20–29.

Wall, C.T.C. (1971) Stratified sets, Springer Lect. Notes Math. **192**, pp. 133–141.

Weinstein, A. (1970) Positive curved n-manifolds in $\mathbb{R}^{n+2}$, Journ. Diff. Geom. **4**, pp. 1–4.

Weinstein, A. (1973) Lagrangian manifolds and Hamiltonian systems, Ann. Math. **98**: 3, pp. 377–410.

Weinstein, A. (1977) Lectures on symplectic manifolds, North Carolina, Regional Conf. Ser. in Math., N 29, A.M.S., Providence.

Weinstein, A. (1984) C^0-perturbation theorems for symplectic fixed points and Lagrangian intersections, Géometrie Symplectique et Contact, Edited by P. Dazord and N. Desolneux-Moulis, Hermann, Paris, pp. 140–145.

Wells, R. (1966) Cobordisms of immersions, Topology **5**, #3, pp. 281–294.

Whitney, H. (1934) Analytic extensions of differentiable functions defined in closed sets, Trans. Ann. Math. Soc. **36**, pp. 63–89.

Whitney, H. (1936) Differentiable manifolds, Ann. Math. **37**, pp. 645–680.

Whitney, H. (1937) On regular closed curves in the plane, Comp. Math. **4**, pp. 276–284.

Whitney, H. (1944) The singularities of a smooth n-manifold in $(2n - 1)$-space, Ann. Math. **45**, pp. 247–293.

Whitney, H. (1955) On singularities of mappings of Euclidean spaces I. Mappings of the plane into the plane, Ann. Math. **62**, pp. 374–410.

Whitney, H. (1957) Elementary structure of real algebraic varieties, Ann. Math. **66**, pp. 545–556.

Wintgen, D. (1978) Über von höherer Ordnung reguläre Immersionen, Math. Nachr. **85**, pp. 177–184.

Yoshifumi, A. (1982) Elimination of certain Thom-Boardman singularities of order two, Journ. Math. Soc. Jap. **34**: 2. pp. 241–267.

Yau, S.T. (1976) Parallelizable manifolds without complex structure, Topology **15**, pp. 51–53.

Zalgaller, V. (1958) Isometric imbedding of polyhedra, Dokl. Akad. Nauk. S.S.S.R. **123**, pp. 599–601.

Zeghib, A. (1984) Feuilletages géodésiques de variétés localement symétriques et applications, Thèse de 3e cycle, Dijon.

Author Index

Subject Index

Ergebnisse der Mathematik und ihrer Grenzgebiete, 3. Folge

A Series of Modern Surveys in Mathematics

1. A. Fröhlich: **Galois Module Structure of Algebraic Integers.** ISBN 3-540-11920-5
2. W. Fulton: **Intersection Theory.** ISBN 3-540-12176-5
3. J. C. Jantzen: **Einhüllende Algebren halbeinfacher Lie-Algebren.** ISBN 3-540-12178-1
4. W. Barth, C. Peters, A. Van de Ven: **Compact Complex Surfaces.** ISBN 3-540-12172-2
5. K. Strebel: **Quadratic Differentials.** ISBN 3-540-13035-7
6. M. Beeson: **Foundations of Constructive Mathematics.** Metamathematical Studies. ISBN 3-540-12173-0
7. A. Pinkus: **n-Widths in Approximation Theory.** ISBN 3-540-13638-X
8. R. Mañe: **Ergodic Theory.** ISBN 3-540-15278-4
9. M. Gromov: **Partial Differential Relations.** ISBN 3-540-12177-3
10. A. L. Besse: **Einstein Manifolds**

Springer-Verlag Berlin Heidelberg New York London Paris Tokyo

Ergebnisse der Mathematik und ihrer Grenzgebiete, 3. Folge

A Series of Modern Surveys in Mathematics

Forthcoming titles:

11. M. D. Fried, M. Jarden: **Field Arithmetic**
ISBN 3-540-16640-8

12. J. Bochnak, M. Coste: **Géométric Algébrique Réelle**
ISBN 3-540-16951-2

K. Diedrich, J. E. Fornaess, R. P. Pflug: **Convexity in Complex Analysis**
ISBN 3-540-12174-9

E. Freitag, R. Kiehl: **Etale Kohomologietheorie und Weil-Vermutung**
ISBN 3-540-12175-7

G. A. Margulis: **Discrete Subgroups of Lie Groups**
ISBN 3-540-12179-X

Springer-Verlag
Berlin Heidelberg New York
London Paris Tokyo